Methods in Enzymology

Volume 133
BIOLUMINESCENCE AND CHEMILUMINESCENCE
Part B

METHODS IN ENZYMOLOGY

EDITORS-IN-CHIEF

Sidney P. Colowick   Nathan O. Kaplan

Methods in Enzymology

Volume 133

Bioluminescence and Chemiluminescence

Part B

EDITED BY

Marlene A. DeLuca

DEPARTMENT OF CHEMISTRY
UNIVERSITY OF CALIFORNIA AT SAN DIEGO
LA JOLLA, CALIFORNIA

William D. McElroy

DEPARTMENT OF BIOLOGY
UNIVERSITY OF CALIFORNIA AT SAN DIEGO
LA JOLLA, CALIFORNIA

1986

ACADEMIC PRESS, INC.
Harcourt Brace Jovanovich, Publishers
Orlando   San Diego   New York   Austin
Boston   London   Sydney   Tokyo   Toronto

ACADEMIC PRESS, INC.
Orlando, Florida 32887

United Kingdom Edition published by
ACADEMIC PRESS INC. (LONDON) LTD.
24–28 Oval Road, London NW1 7DX

LIBRARY OF CONGRESS CATALOG CARD NUMBER: 54-9110

ISBN 0–12–182033–5

PRINTED IN THE UNITED STATES OF AMERICA

86 87 88 89   9 8 7 6 5 4 3 2 1

Table of Contents

Section I. Bioluminescence

Section II. Chemiluminescence

Section III. Instrumentation

Contributors to Volume 133

Article numbers are in parentheses following the names of contributors.
Affiliations listed are current.

MUSHTAQ AHMAD (10), *Bioluminescence Laboratory, Department of Biochemistry, University of Georgia, Athens, Georgia 30602*

ROBERT C. ALLEN (36), *Laboratory Division, U.S. Army Institute of Surgical Research, Fort Sam Houston, Texas 78234*

Y. AMIR-ZALTSMAN (25), *Department of Hormone Research, The Weizmann Institute of Science, Rehovot 76100, Israel*

THOMAS O. BALDWIN (9, 22), *Department of Biochemistry and Biophysics, Texas A&M University, College Station, Texas 77843*

G. BARNARD (25), *Department of Obstetrics and Gynaecology, King's College Medical School, London SE5 8RX, England*

E. A. BAYER (25), *Department of Biophysics, The Weizmann Institute of Science, Rehovot 76100, Israel*

ANNE MARIE BOUSSIOUX (18), *Institut National de la Sante et de la Recherche Médicale U.58, 34100 Montpellier, France*

MICHAEL BOYLAN (7), *Department of Biochemistry, McGill University, Montreal, Quebec, Canada H3G 1Y6*

JÖRG BRAUN (30), *Klinik für Innere Medizin, Medizinische Universität zu Lübeck, D-2400 Lübeck 1, Federal Republic of Germany*

RICHARD C. BROWN (31), *Department of Medical Biochemistry, University of Wales College of Medicine, Cardiff CF4 4XN, Wales*

DAVID BYERS (15), *Clinical Research Centre, Atlantic Research Centre for Mental Retardation, Halifax, Nova Scotia, Canada B3H 6H7*

A. L. CALDINI (40), *Department of Clinical Physiopathology, Andrology Unit, University of Florence, Florence 50134, Italy*

LUC CAREY (15), *Department of Biochemistry, McGill University, Montreal, Quebec, Canada H3G 1Y6*

GIACOMO CARREA (21), *Istituto di Chimica degli Ormoni, C.N.R., 20131 Milano, Italy*

MILTON J. CORMIER (26), *Department of Biochemistry, University of Georgia, Athens, Georgia 30602*

ANDRÉ CRASTES DE PAULET (18), *Institut National de la Sante et de la Recherche Médicale U.58, 34100 Montpellier, France*

PATRICK DE BAETSELIER (38), *Institute of Molecular Biology, Free University of Brussels, B-1640 St. Genesius Rode, Belgium*

J. DE BOEVER (32), *Akademisch Ziekenhuis, Vrouwenkliniek/Poli III, B-9000 Gent, Belgium*

M. DELUCA (1, 17, 19), *Department of Chemistry, University of California at San Diego, La Jolla, California 92093*

JAY C. DUNLAP (28), *Department of Biochemistry, Dartmouth Medical School, Hanover, New Hampshire 03755*

JOANNE ENGEBRECHT (8), *Microbial Genetics Section, The Agouron Institute, La Jolla, California 92037*

SANDRO GHISLA (12), *Fakultät Biologie, University of Konstanz, D-7780 Konstanz, Federal Republic of Germany*

STEFANO GIROTTI (21), *Istituto di Scienze Chimiche, Università di Bologna, 15-40127 Bologna, Italy*

ANGUS GRAHAM (7), *Department of Biochemistry, McGill University, Montreal, Quebec, Canada H3G 1Y6*

GARY GRANT (16), *Canadian Forestry Service, Forest Pest Management Institute, Sault Ste-Marie, Ontario, Canada P6A 5M7*

BRUCE A. HANNA (3), *Department of Pathology, New York University School of Medicine, Bellevue Hospital, New York, New York 10016*

UWE HANTKE (30), *Henning Berlin GmbH., Komturstrasse 58-62, 1000 Berlin 42, Federal Republic of Germany*

MICHAEL HASENSON (4), *Department of Medical Nutrition, Karolinska Institute, Huddinge University Hospital, S-141 86 Huddinge, Sweden*

J. WOODLAND HASTINGS (12, 28), *Department of Cellular and Developmental Biology, Harvard University, Cambridge, Massachusetts 02138*

DONALD R. HELINSKI (1), *Department of Biology, University of California at San Diego, La Jolla, California 92093*

RITA B. HOLZMAN (9, 22), *Fermentation Bioprocess Research and Development, The Upjohn Company, Kalamazoo, Michigan 49001*

THOMAS F. HOLZMAN (9, 22), *Biotechnology Research and Development, Control Division, The Upjohn Company, Kalamazoo, Michigan 49001*

JAN C. HUMMELEN (39), *Department of Organic Chemistry, University of Groningen, 9747 AG Groningen, The Netherlands*

KAZUHIRO IMAI (35), *Branch Hospital Pharmacy, University of Tokyo, Bunkyo-ku, Tokyo 112, Japan*

J. B. KIM (32), *Department of Animal Products Science, College of Animal Husbandry, Kon-kuk University, Seoul 133, Korea*

F. KOHEN (25, 32), *Department of Hormone Research, The Weizmann Institute of Science, Rehovot 76100, Israel*

LARRY J. KRICKA (20, 29, 33), *Department of Clinical Chemistry, Wolfson Research Laboratories, University of Birmingham, Birmingham B15 2TH, England*

MANFRED KURFÜRST (12), *BASF Company, 6700 Ludwigshaven, Federal Republic of Germany*

FRANKLIN R. LEACH (6), *Department of Biochemistry, Oklahoma State University, Stillwater, Oklahoma 74078*

JOHN LEE (10, 13), *Bioluminescence Laboratory, Department of Biochemistry, University of Georgia, Athens, Georgia 30602*

THEO M. LUIDER (39), *Department of Organic Chemistry, University of Groningen, 9747 AG Groningen, The Netherlands*

ARNE LUNDIN (4), *Research Centre and Department of Medicine, Karolinska Institute, Huddinge University Hospital, S-141 86 Huddinge, Sweden*

IAIN B. C. MATHESON (10), *Bioluminescence Laboratory, Department of Biochemistry, University of Georgia, Athens, Georgia 30602*

RICHARD O. MCCANN (26), *Department of Biochemistry, University of Georgia, Athens, Georgia 30602*

EDWARD MEIGHEN (7, 14, 15, 16), *Department of Biochemistry, McGill University, Montreal, Quebec, Canada H3G 1Y6*

G. MESSERI (40), *Clinical Chemistry Laboratory, USL 10/D, Florence 50134, Italy*

CAROL MIYAMOTO (7), *Department of Biochemistry, McGill University, Montreal, Quebec, Canada H3G 1Y6*

DAVID MORSE (16), *The Biological Laboratories, Harvard University, Cambridge, Massachusetts 02138*

JEAN-CLAUDE NICOLAS (18), *Institut National de la Sante et de la Recherche Médicale U.58, 34100 Montpellier, France*

MASATO NOGUCHI (27), *Research Laboratory for Genetic Information, Kyushu University, Fukuoka 812, Japan*

DENNIS J. O'KANE (10, 13), *Bioluminescence Laboratory, Department of Biochemistry, University of Georgia, Athens, Georgia 30602*

C. ORLANDO (40), *Andrology Unit, Department of Clinical Physiopathology, University of Florence, Florence 50134, Italy*

MARIO PAZZAGLI (34), *Endocrinology Unit, Department of Clinical Physiopathology,*

University of Florence, 50135 Florence, Italy

JÖRGEN PERSSON (4), Research Centre and Department of Medicine, Karolinska Institute, Huddinge University Hospital, S-141 86 Huddinge, Sweden

GARY H. POSNER (41), Department of Chemistry, The Johns Hopkins University, Baltimore, Maryland 21218

ÅKE POUSETTE (4), Department of Clinical Chemistry I, Karolinska Institute, Huddinge University Hospital, S-141 86 Huddinge, Sweden

DOUGLAS C. PRASHER (26), Department of Biochemistry, University of Georgia, Athens, Georgia 30602

VICKI A. RIDDLE (9), Department of Biochemistry and Biophysics, Texas A&M University, College Station, Texas 77843

DENIS RIENDEAU (14), Merck Frosst Inc., 16701 Transcanada Highway, Pointe Claire, Quebec, Canada H9H 3L1

A. RODA (19, 21), Istituto di Scienze Chimiche, Università di Bologna, 15-40127 Bologna, Italy

ANGEL RODRIGUEZ (14, 15), Department of Medicine, Royal Victoria Hospital/McGill University, Montreal, Quebec, Canada H3A 1A1

YOSHIYUKI SAKAKI (27), Research Laboratory for Genetic Information, Kyushu University, Fukuoka 812, Japan

JAMES M. SCHAEFFER (5), Department of Reproductive Medicine, School of Medicine, University of California at San Diego, La Jolla, California 92093

J. SCHOELMERICH (19), Medizinische Universitätsklinik, University of Freiburg, D-7800 Freiburg, Federal Republic of Germany

ERIC SCHRAM (38), Institute of Molecular Biology, Free University of Brussels, B-1640 St. Genesius Rode, Belgium

H. R. SCHROEDER (40), Ames Research and Development Laboratory, Division of Miles Laboratories, Elkhart, Indiana 46515

HOWARD H. SELIGER (41), Department of Biology, The Johns Hopkins University, Baltimore, Maryland 21218

MARIO SERIO (34), Department of Clinical Physiopathology, University of Florence, 50135 Florence, Italy

MICHAEL SILVERMAN (8), Microbial Genetics Section, The Agouron Institute, La Jolla, California 92037

PHILIP E. STANLEY (2, 42), 48 Glisson Road, Cambridge CB1 2HF, England

MARIA STURGESS (31), Department of Medical Biochemistry, University of Wales College of Medicine, Cardiff CF4 4XN, Wales

ROSE SZITTNER (16), Department of Biochemistry, McGill University, Montreal, Quebec, Canada H3G 1Y6

AMBLER THOMPSON (41), Department of Biology, The Johns Hopkins University, Baltimore, Maryland 21218

GARY H. G. THORPE (29, 33), Department of Clinical Chemistry, Wolfson Research Laboratories, Queen Elizabeth Medical Centre, Birmingham B15 2TH, England

FREDERICK I. TSUJI (27), Scripps Institution of Oceanography, University of California at San Diego, La Jolla, California 92093, and Veterans Administration Medical Center Brentwood, Los Angeles, California 90073

SHIAO-CHUN TU (11), Department of Biochemical and Biophysical Sciences, University of Houston-University Park, Houston, Texas 77004

S. ULITZUR (23, 24), Department of Food Engineering and Biotechnology, Technion, Haifa 32000, Israel

CHRIS VAN DYKE (37), Department of Pharmacology/Toxicology, West Virginia University Medical Center, Morgantown, West Virginia 26506

KNOX VAN DYKE (37), Department of Pharmacology/Toxicology, West Virginia University Medical Center, Morgantown, West Virginia 26506

DANIEL C. VELLOM (20), *Department of Chemistry, University of California at San Diego, La Jolla, California 92093*

LEE WALL (14), *Laboratory of Gene Structure and Expression, National Institute for Medical Research, Mill Hill, London NW7 1AA, England*

JoANN J. WEBSTER (6), *Department of Biochemistry, Oklahoma State University, Stillwater, Oklahoma 74078*

IAN WEEKS (31), *Department of Medical Biochemistry, University of Wales College of Medicine, Cardiff CF4 4XN, Wales*

JEFFREY R. DE WET (1), *Department of Biological Sciences, Stanford University, Stanford, California 94305*

G. WIENHAUSEN (17), *Department of Chemistry, University of California at San Diego, La Jolla, California 92093*

M. WILCHEK (25), *Department of Biophysics, The Weizmann Institute of Science, Rehovot 76100, Israel*

KEITH V. WOOD (1), *Department of Chemistry, University of California at San Diego, La Jolla, California 92093*

W. GRAHAM WOOD (30), *Department of Internal Medicine, Medical University of Lübeck, D-2400 Lübeck 1, Federal Republic of Germany*

J. STUART WOODHEAD (31), *Department of Medical Biochemistry, University of Wales College of Medicine, Cardiff CF4 4XN, Wales*

HANS WYNBERG (39), *Department of Organic Chemistry, University of Groningen, 9747 AG Groningen, The Netherlands*

Preface

Since the publication of Volume LVII of *Methods in Enzymology* on Bioluminescence and Chemiluminescence in 1978 (edited by Marlene A. DeLuca), many new and exciting developments have occurred in the field. For example, the genes encoding firefly luciferase, bacterial luciferase, and aequorin have recently been cloned and expressed. Novel new approaches to chemiluminescent assays have been reported during the past six years. Several international symposia have been devoted to luminescent assays. Judged by the response from our colleagues, the time seemed appropriate for an updating of the various techniques that have been used in this research.

In Section I we attempt to cover the major developments in bioluminescence. This includes basic science and some applications of the firefly, bacterial, *Renilla, Aequorea,* and dinoflagellate luciferases. Special techniques on the cloning of luciferases and the immobilization of the light-emitting systems leading to automated continuous flow assays have been included. Space prevents a comprehensive coverage of all of the applications of these techniques; however, we have tried to include representative types of all assays. A more complete coverage can be found in publications resulting from recent symposia. Section II is devoted to the major recent advances in chemiluminescence. Included in these papers is excellent coverage of various acridinium esters and their use in immunoassays, haptens and protein interaction, horseradish peroxidase-enhanced chemiluminescence, phagocytic leukocyte oxygenation as revealed by luminescence, and chemiluminescent probes for singlet oxygen in biological reactions.

Section III contains a chapter on the characteristics of commercial luminometers.

We would like to take this opportunity to thank our colleagues for their many suggestions and, in particular, recognize the cooperation of the authors for their timely reviews. As usual the staff of Academic Press was very helpful in organizing the material for publication.

MARLENE A. DeLUCA
WILLIAM D. McELROY

METHODS IN ENZYMOLOGY

EDITED BY

Sidney P. Colowick and Nathan O. Kaplan

VANDERBILT UNIVERSITY
SCHOOL OF MEDICINE
NASHVILLE, TENNESSEE

DEPARTMENT OF CHEMISTRY
UNIVERSITY OF CALIFORNIA
AT SAN DIEGO
LA JOLLA, CALIFORNIA

METHODS IN ENZYMOLOGY

EDITORS-IN-CHIEF

Sidney P. Colowick and Nathan O. Kaplan

xvii

VOLUME XVIII. Vitamins and Coenzymes (Parts A, B, and C)
Edited by DONALD B. MCCORMICK AND LEMUEL D. WRIGHT

VOLUME XIX. Proteolytic Enzymes
Edited by GERTRUDE E. PERLMANN AND LASZLO LORAND

VOLUME XX. Nucleic Acids and Protein Synthesis (Part C)
Edited by KIVIE MOLDAVE AND LAWRENCE GROSSMAN

VOLUME XXI. Nucleic Acids (Part D)
Edited by LAWRENCE GROSSMAN AND KIVIE MOLDAVE

VOLUME XXII. Enzyme Purification and Related Techniques
Edited by WILLIAM B. JAKOBY

VOLUME XXIII. Photosynthesis (Part A)
Edited by ANTHONY SAN PIETRO

VOLUME XXIV. Photosynthesis and Nitrogen Fixation (Part B)
Edited by ANTHONY SAN PIETRO

VOLUME XXV. Enzyme Structure (Part B)
Edited by C. H. W. HIRS AND SERGE N. TIMASHEFF

VOLUME XXVI. Enzyme Structure (Part C)
Edited by C. H. W. HIRS AND SERGE N. TIMASHEFF

VOLUME XXVII. Enzyme Structure (Part D)
Edited by C. H. W. HIRS AND SERGE N. TIMASHEFF

VOLUME XXVIII. Complex Carbohydrates (Part B)
Edited by VICTOR GINSBURG

VOLUME XXIX. Nucleic Acids and Protein Synthesis (Part E)
Edited by LAWRENCE GROSSMAN AND KIVIE MOLDAVE

VOLUME XXX. Nucleic Acids and Protein Synthesis (Part F)
Edited by KIVIE MOLDAVE AND LAWRENCE GROSSMAN

VOLUME XXXI. Biomembranes (Part A)
Edited by SIDNEY FLEISCHER AND LESTER PACKER

VOLUME XXXII. Biomembranes (Part B)
Edited by SIDNEY FLEISCHER AND LESTER PACKER

VOLUME XXXIII. Cumulative Subject Index Volumes I–XXX
Edited by MARTHA G. DENNIS AND EDWARD A. DENNIS

VOLUME XXXIV. Affinity Techniques (Enzyme Purification: Part B)
Edited by WILLIAM B. JAKOBY AND MEIR WILCHEK

VOLUME XXXV. Lipids (Part B)
Edited by JOHN M. LOWENSTEIN

VOLUME XXXVI. Hormone Action (Part A: Steroid Hormones)
Edited by BERT W. O'MALLEY AND JOEL G. HARDMAN

VOLUME XXXVII. Hormone Action (Part B: Peptide Hormones)
Edited by BERT W. O'MALLEY AND JOEL G. HARDMAN

VOLUME XXXVIII. Hormone Action (Part C: Cyclic Nucleotides)
Edited by JOEL G. HARDMAN AND BERT W. O'MALLEY

VOLUME XXXIX. Hormone Action (Part D: Isolated Cells, Tissues, and Organ Systems)
Edited by JOEL G. HARDMAN AND BERT W. O'MALLEY

VOLUME XL. Hormone Action (Part E: Nuclear Structure and Function)
Edited by BERT W. O'MALLEY AND JOEL G. HARDMAN

VOLUME XLI. Carbohydrate Metabolism (Part B)
Edited by W. A. WOOD

VOLUME XLII. Carbohydrate Metabolism (Part C)
Edited by W. A. WOOD

VOLUME XLIII. Antibiotics
Edited by JOHN H. HASH

VOLUME XLIV. Immobilized Enzymes
Edited by KLAUS MOSBACH

VOLUME XLV. Proteolytic Enzymes (Part B)
Edited by LASZLO LORAND

VOLUME LX. Nucleic Acids and Protein Synthesis (Part H)
Edited by KIVIE MOLDAVE AND LAWRENCE GROSSMAN

VOLUME 61. Enzyme Structure (Part H)
Edited by C. H. W. HIRS AND SERGE N. TIMASHEFF

VOLUME 62. Vitamins and Coenzymes (Part D)
Edited by DONALD B. MCCORMICK AND LEMUEL D. WRIGHT

VOLUME 63. Enzyme Kinetics and Mechanism (Part A: Initial Rate and Inhibitor Methods)
Edited by DANIEL L. PURICH

VOLUME 64. Enzyme Kinetics and Mechanism (Part B: Isotopic Probes and Complex Enzyme Systems)
Edited by DANIEL L. PURICH

VOLUME 65. Nucleic Acids (Part I)
Edited by LAWRENCE GROSSMAN AND KIVIE MOLDAVE

VOLUME 66. Vitamins and Coenzymes (Part E)
Edited by DONALD B. MCCORMICK AND LEMUEL D. WRIGHT

VOLUME 67. Vitamins and Coenzymes (Part F)
Edited by DONALD B. MCCORMICK AND LEMUEL D. WRIGHT

VOLUME 68. Recombinant DNA
Edited by RAY WU

VOLUME 69. Photosynthesis and Nitrogen Fixation (Part C)
Edited by ANTHONY SAN PIETRO

VOLUME 70. Immunochemical Techniques (Part A)
Edited by HELEN VAN VUNAKIS AND JOHN J. LANGONE

VOLUME 71. Lipids (Part C)
Edited by JOHN M. LOWENSTEIN

VOLUME 72. Lipids (Part D)
Edited by JOHN M. LOWENSTEIN

Section I

Bioluminescence

A. Firefly Luciferase
Articles 1 through 6

B. Bacterial Luciferase
Articles 7 through 25

C. *Renilla* and *Aequorea* Luciferases
Articles 26 and 27

D. Dinoflagellate Luciferase
Article 28

[1] Cloning Firefly Luciferase

By JEFFREY R. DE WET, KEITH V. WOOD, DONALD R. HELINSKI, and
MARLENE DeLUCA

Introduction

Firefly luciferase catalyzes the ATP-dependent oxidative decarboxylation of luciferin (LH_2) resulting in the production of light as shown in reaction (1) where P denotes the product oxyluciferin.

$$LH_2 + ATP + O_2 \xrightarrow{Mg^{2+}} P + AMP + PP_i + CO_2 + h\nu \tag{1}$$

The most well-characterized firefly luciferase is that isolated from the common North American firefly *Photinus pyralis*. The reaction catalyzed by this enzyme has a quantum yield of 0.88 with respect to LH_2 making it the most efficient bioluminescent reaction known.[1] At neutral pH the light emission of *P. pyralis* luciferase is yellow-green with a peak at 562 nm. A variety of factors including heat, low pH, and divalent metal ions have been shown to shift the color of the emitted light to red (610 nm).[2] Furthermore, it has been shown that other species of fireflies emit different colors of light ranging from 552 to 582 nm. Since the luciferin which is utilized by all of these enzymes is the same, the different colors of light emitted must be due to variation in the structure of the enzymes. It would be of considerable interest to be able to compare the structures and active sites of luciferases from different species of insects since this might give some insight as to the structural features of the protein that affect the color of the light emitted.

In addition to being of interest in studies of the relationships between structure and function in bioluminescent enzymes, firefly luciferase is useful in a variety of applications. Because of its specificity for ATP, firefly luciferase can be used to measure the amount of ATP present in biological samples with no interference from other nucleotide triphosphates. This has served as a basis for quantitating bacteria in urine, milk, wine, and polluted waters. The enzyme has also been used in a bioluminescent immunoassay where it is just as sensitive as the comparable radioimmunoassay.[3]

[1] H. H. Seliger and W. D. McElroy, *Arch. Biochem. Biophys.* **88,** 136 (1960).
[2] H. H. Seliger and W. D. McElroy, *Proc. Natl. Acad. Sci. U.S.A.* **52,** 75 (1964).
[3] J. Wannlund, J. Azari, L. Levine, and M. DeLuca, *Biochem. Biophys. Res. Commun.* **96,** 440 (1980).

METHODS IN ENZYMOLOGY, VOL. 133

Isolating luciferase cDNA clones will allow the determination of the nucleotide sequence of this gene from which the amino acid sequence of the enzyme can be deduced. When the luciferase genes of other species of fireflies are cloned, comparison of their sequences can help pinpoint regions of the enzyme that are essential to its function and increase our understanding of bioluminescent reactions. The cloning of luciferase cDNA and the expression of the enzyme in an active form in *Escherichia coli* would provide an unlimited source of this enzyme. In addition, cloned luciferase cDNA could be very useful for the monitoring of gene expression. The simple and sensitive assay for luciferase may allow the *in vivo* detection of this enzyme obviating the need to destroy the cells being monitored. Since this is a single gene system, it is likely to be particularly useful in eukaryotic systems as an indicator of promoter activity in plant, animal, and lower eukaryote cells using both *in vitro* and *in vivo* tests.

We have isolated the poly(A)+ RNA from the lanterns of *P. pyralis* and constructed a cDNA library in the *E. coli* expression vector λgt11. The luciferase cDNA from one of the clones that was isolated was inserted into an expression plasmid, and this plasmid was able to direct the synthesis of active luciferase in *E. coli*.

Strains and Growth Conditions

Escherichia coli strains Y1088 [*supE supF metB trpR hsdR⁻ hsdM⁺ tonA21 ΔlacU169 proC::Tn5* (pMC9)], Y1090 [*ΔlacU169 proA⁺ Δlon araD139 strA supF trpC22::Tn10* (pMC9)], and λgt11 were obtained from Rick Young.[4,5] The plasmid cloning vector pUC13 has been described.[6] The expression vector pKJB824.17 consisting of pBR322 carrying λcI_{857} and λP_R was obtained from Ken Buckley.[7] *E. coli* strain TB1 [*ara Δ (lac proA,B) strA Φ80dlacZΔM15 hsdR⁻ hsdM⁺*] was provided by Thomas Baldwin (Texas A & M University).

The following media were used for propagating *E. coli* and bacteriophage λ: LB (per liter: 10 g tryptone, 5 g yeast extract, 5 g NaCl, pH 7.4), NZYM (per liter: 10 g NZ-amine, 5 g yeast extract, 5 g NaCl, 2 g MgSO₄ · 7H₂O, pH 7.4), bottom agar (LB plus 15 g/liter agar for bacterial colonies or LB plus 12 g/liter agar and 2 g/liter MgSO₄ · 7H₂O for phage plaques), top agar (NZYM plus 7 g/liter agarose), and SM (50 mM Tris, pH 7.5, 100 mM NaCl, 10 mM MgSO₄, and 0.1% gelatin).

E. coli strains to be used as hosts for bacteriophage λ were grown to

[4] R. A. Young and R. W. Davis, *Proc. Natl. Acad. Sci. U.S.A.* **80,** 1194 (1983).
[5] R. A. Young and R. W. Davis, *Science* **222,** 778 (1983).
[6] J. Messing, this series, Vol. 101, p. 20.
[7] K. Buckley, Ph.D. Thesis, University of California, San Diego (1985).

stationary phase in NZYM broth plus 0.2% maltose at 37°. Phage (in ≤200 μl of SM) was mixed with 200 μl of fresh, stationary cells and incubated at 37° for 15 min without shaking. Liquid top agar (3 ml) at 50° was added to the cells, mixed and poured onto a bottom agar plate that had been pre-warmed to 37°. After the top agar had solidified the plates were inverted and incubated (usually at 37°).

Firefly Lantern cDNA Library Construction

RNA Isolation

Live fireflies (*P. pyralis*) were obtained from William Biggley, Johns Hopkins University. The fireflies were frozen in liquid nitrogen and the lanterns were removed. The lanterns were ground to a powder under liquid nitrogen with a mortar and pestle. The powdered lanterns were homogenized in 4 M guanidinium thiocyanate[8] and particulate material was removed from the homogenate by centrifugation at 10,000 g for 10 min. Total RNA was isolated by sedimenting the RNA through cesium chloride.[9] Poly(A)$^+$ RNA was selected by chromatography on oligo(dT)-cellulose.[10] From 200 lanterns we obtained 24 mg of total RNA, and oligo(dT) chromatography of 8 mg of total RNA yielded 60 μg of poly(A)$^+$ RNA.

cDNA Synthesis

A modification of the method of Gubler and Hoffman[11] was used to synthesize cDNA using *P. pyralis* lantern poly(A)$^+$ RNA as template. This method avoids the use of S1 nuclease by using *E. coli* RNase H and DNA polymerase I to synthesize the second strand of the cDNA. Ten micrograms of lantern poly(A)$^+$ RNA in 20 μl of H_2O was heated at 70° for 10 min to disrupt secondary structure and cooled on ice. The first strand of the cDNA was synthesized from this template in a total volume of 200 μl under the following conditions: 50 mM Tris, pH 8.3, 100 mM KCl, 25 mM 2-mercaptoethanol, 6 mM MgCl$_2$, 4 mM sodium pyrophosphate, 200 μM each dATP, dCTP, dGTP, and dTTP, 500 μCi/ml [α-^{32}P]dCTP (3000 Ci/mmol), 500 U/ml RNasin, 1000 U/ml AMV reverse transcriptase (Boehringer Mannheim Biochemicals), 50 μg/ml oligo(dT)$_{18}$, and 50 μg/ml

[8] T. Maniatis, E. F. Fritsch, and J. Sambrook, "Molecular Cloning: A Laboratory Manual," p. 189. Cold Spring Harbor Lab., Cold Spring Harbor, New York, 1982.
[9] T. Maniatis, E. F. Fritsch, and J. Sambrook, "Molecular Cloning: A Laboratory Manual," p. 196. Cold Spring Harbor Lab., Cold Spring Harbor, New York, 1982.
[10] H. Aviv and P. Leder, *Proc. Natl. Acad. Sci. U.S.A.* **69**, 1408 (1972).
[11] U. Gubler and B. J. Hoffman, *Gene* **25**, 263 (1983).

poly(A)$^+$ RNA. The reaction was incubated at 42° for 60 min and was terminated by extraction with an equal volume of phenol : chloroform (1 : 1, neutralized and equilibrated with 10 mM Tris, pH 8.0, and 1 mM EDTA). The tube was centrifuged in a microcentrifuge for 3 min to separate the phases, and the aqueous phase was transferred to a clean tube (after any organic extraction, the aqueous phase should be transferred to a clean tube). The cDNA was extracted with an equal volume of chloroform, centrifuged for 1 min, and the aqueous phase was transferred to a clean tube. Occasionally a precipitate [probably pyrophosphate which is included as an inhibitor of RNase H[12]] forms during the synthesis of the first strand of the cDNA. This does not seem to affect the reaction or the quality of the product, but care should be taken during the organic extraction steps to exclude the precipitate from the cDNA containing aqueous phase. The pyrophosphate is included as an inhibitor of RNase H.[12] The cDNA was precipitated by the addition of $\frac{1}{2}$ volume of 7.5 M ammonium acetate and 2 volumes of ethanol (two times the volume of the cDNA plus ammonium acetate). The tube was placed on dry ice for 15 min and then allowed to thaw. The cDNA was pelleted by centrifugation in a microfuge for 10 min. The supernatant was carefully removed and the pellet was dissolved in 100 µl of water. The cDNA was precipitated and pelleted a second time as described above. The pellet was washed once with 0.5 ml cold 75% ethanol, dried under vacuum, and dissolved in 50 µl of water. The yield of the first strand reaction was ≈1 µg of cDNA.

The conditions for the synthesis of the second strand of the cDNA were as follows: 20 mM Tris, pH 7.5, 100 mM KCl, 10 mM $(NH_4)_2SO_4$, 5 mM $MgCl_2$, 40 µM dNTPs, 50 µg/ml BSA, 8 U/ml RNase H, 230 U/ml DNA polymerase I, and 5 µg/ml cDNA; total reaction volume was 200 µl. The reaction was incubated at 12° for 60 min followed by 60 min at 22°. The reaction was terminated by the addition of EDTA (pH 8.0) to a final concentration of 10 mM followed by extractions with equal volumes of phenol : chloroform (1 : 1) and chloroform. The cDNA was precipitated twice in the presence of $\frac{1}{2}$ volume of ammonium acetate and 2 volumes of ethanol, the pellet was washed once with 0.5 ml of cold 75% ethanol, and dried under vacuum. The cDNA was treated with EcoRI methylase before the addition of the linkers in order to prevent cleavage of EcoRI restriction sites that may be present. The double-stranded cDNA was incubated with 20 units of EcoRI methylase (New England BioLabs) in 20 mM Tris, pH 8.0, 10 mM EDTA, and 80 µM S-adenosylmethionine at 37° for 30 min in a volume of 50 µl. The methylated cDNA was extracted with an equal volume of phenol : chloroform followed by an extraction with an

[12] M. G. Muttay, L. M. Hoffman, and N. P. Jarvis, *Plant Mol. Biol.* **2**, 75 (1983).

equal volume of chloroform. The cDNA was precipitated by the addition of $\frac{1}{10}$ volume of 3 M sodium acetate (pH adjusted to 5.2 with glacial acetic acid) and 2.5 volumes of ethanol. The tube was placed on dry ice for 15 min, thawed, and centrifuged for 5 min in a microfuge. The supernatant was removed, the pellet was dissolved in 50 μl of water, and the precipitation was repeated. The pellet was washed once with 200 μl of cold 75% ethanol and dried under vacuum. The methylated cDNA was dissolved in water and ligated to phosphorylated EcoRI linkers (pGGAATTCC, New England Biolabs) at 4° for 12 hr under the following conditions: 33 μg/ml double-stranded cDNA (total conversion of single-stranded to double-stranded cDNA was assumed), 50 μg/ml EcoRI linkers, 66 mM Tris, pH 7.5, 6.6 mM MgCl$_2$, 10 mM dithiothreitol, 1 mM ATP, and 160 Wiess units/ml T$_4$ DNA ligase; total volume of the reaction was 60 μl. The ligase was inactivated by heating at 65° for 15 min, and the buffer was adjusted to 100 mM Tris, pH 7.5, 50 mM NaCl, and 6 mM MgCl$_2$ in a total volume of 100 μl. Fifty units of EcoRI was added for each microgram of linkers used, and the reaction was incubated at 37° for 4 hr.

The reaction was terminated by extraction with an equal volume of phenol : chloroform (1 : 1) followed by extraction with an equal volume of chloroform. The cDNA was precipitated by the addition of $\frac{1}{10}$ volume of 3 M sodium acetate (pH 5.2) and 2.5 volumes of ethanol. This was placed on dry ice for 15 min, thawed, and the cDNA was pelleted by centrifugation for 5 min in a microfuge. The supernatant was removed, and the pellet was dried under vacuum. The pellet was dissolved in 40 μl of 20 mM Tris, pH 8.0, 1 mM EDTA after which 10 μl of 7.5 M LiCl was added. The cDNA was separated from the cleaved linker fragments by chromatography on a 1 ml column of Sepharose CL-4B that had been equilibrated with the running buffer (1.5 M LiCl, 20 mM Tris, pH 8.0, 1 mM EDTA, and 0.1% SDS). Three drop fractions were collected as long as radioactivity could be detected in them with a Geiger counter. A 1 μl sample of each fraction was analyzed by autoradiography following electrophoresis on an alkaline/1.1% agarose gel.[13] Fractions containing cDNA greater than 500 bases in length were pooled and precipitated by the addition of 2.5 volumes of ethanol. The tube was placed on dry ice for 15 min, thawed, and centrifuged for 5 min. The supernatant was removed and the pellet was washed once with 1 ml of cold 75% ethanol. The pellet was dried under vacuum and was resuspended in 50 μl of 10 mM Tris, pH 8.0, and 0.1 mM EDTA. Beginning with 10 μg of poly(A)$^+$ RNA, this procedure yielded approximately 1 μg of double-stranded, EcoRI linkered cDNA.

[13] T. Maniatis, E. F. Fritsch, and J. Sambrook, "Molecular Cloning: A Laboratory Manual," p. 171. Cold Spring Harbor Lab., Cold Spring Harbor, New York, 1982.

In order to reduce the background of phage lacking inserts, the λgt11 DNA was treated with calf intestinal alkaline phosphatase. The cohesive ends of the λgt11 DNA were first ligated, the DNA was digested with EcoRI, and then was treated with phosphatase as described.[14] EcoRI linkered cDNA (0.3 μg) was mixed with 7 μg of EcoRI cut, phosphatase-treated λgt11 DNA. The DNAs were precipitated by the addition of sodium acetate and ethanol as described above and pelleted by centrifugation. The pellet was washed with cold 75% ethanol, dried, and dissolved in 10 μl of water. The cDNA was ligated to the vector with 3 Weiss units of T4 DNA ligase at 4° for 12 hr in a total volume of 15 μl. The buffer conditions were the same as those used to ligate the linkers to the cDNA. One microgram of the ligated λgt11–cDNA was packaged using λ in vitro packaging extracts according to the supplier's (Bethesda Research Laboratories) instructions. The packaged library was titered on Y1088 cells in the presence of 1 mM isopropyl-β-D-galactopyranoside (IPTG) and 40 μg/ ml 5-bromo 4-chloro 3-indolyl-β-D-galactopyranoside (X-Gal). Under these conditions, λgt11 phage containing cDNA inserts produces colorless plaques while the parent vector which has an intact β-galactosidase gene produces blue plaques. The library contained a total of 3×10^5 phage of which 1×10^5 contained cDNA inserts. The library was amplified by plating 3×10^4 phage per 100-mm dish on a lawn of Y1088 cells and incubating the plates at 37° for 8 hr. The phage was then eluted from the top agar by adding 5 ml of SM to each plate. After 8 hr at 4°, the SM was removed from the plates and pooled. The amplified library was titered, and it was found that the percentage of recombinant phage had decreased during the amplification step to ≈10%.

Screening the Library

The λgt11 firefly lantern cDNA library was initially screened for clones producing antigens recognized by rabbit anti-P. pyralis antibodies using a previously described chromogenic immunodetection technique.[15] In the initial round of screening up to 50,000 phage were plated on a single 100-mm dish on a lawn of Y1090 cells. Proteins produced by the phage were bound to nitrocellulose circles, the filters were treated with rabbit anti-P. pyralis antibody (2 μg/ml), and the bound antibody was detected with horseradish peroxidase conjugated goat anti-rabbit IgG. Sixteen plaques reacting with anti-luciferase antibody were detected among the

---

[14] T. Maniatis, E. F. Fritsch, and J. Sambrook, "Molecular Cloning: A Laboratory Manual," p. 133. Cold Spring Harbor Lab., Cold Spring Harbor, New York, 1982.

[15] J. R. de Wet, H. Fukushima, N. N. Dewji, E. Wilcox, J. S. O'Brien, and D. R. Helinski, DNA 3, 437 (1984).

150,000 phage (15,000 recombinant phage) screened. Eight of these plaques were purified to homogeneity through repeated rounds of screening. DNA isolated from plate lysates of each of these clones was analyzed by digestion with *Eco*RI followed by electrophoresis on a 1.0% agarose gel. One of these clones, λLuc1 contained two *Eco*RI inserts. The larger (Luc1A) fragment was 1.2 kilobases (kb) in length and the smaller (Luc1B) fragment was 270 bases in length. The Luc1A *Eco*RI cDNA fragment hybridized to the other seven cDNA clones, and both Luc1A and Luc1B were found to hybridize to a 1.9–2.0 kb firefly lantern poly(A)$^+$ RNA.[16] We also demonstrated that hybridization of total lantern poly(A)$^+$ RNA to the Luc1A and Luc1B cDNAs resulted in the selection of an RNA that directed the *in vitro* synthesis of active luciferase. SDS–polyacrylamide gel electrophoresis of the products of the *in vitro* translation showed that a single polypeptide that comigrated with *P. pyralis* luciferase had been synthesized.

The total length of the cDNA insert in λLuc1 was only 1.47 kb in length while the luciferase mRNA detected on Northern blots measured 1.9–2.0 kb. In order to locate longer luciferase cDNA clones the λgt11 lantern cDNA library was screened using the Luc1B cDNA fragment as probe. The Luc1B *Eco*RI fragment was subcloned into the *Eco*RI site of pUC13 to produce the plasmid pLuc1B. The Luc1B *Eco*RI fragment was purified from *Eco*RI-digested pLuc1B DNA by electrophoresis on an agarose gel. The cDNA band was then isolated from the gel by electrophoresis onto DE81 paper.[17] Purified Luc1B cDNA was labeled with [α-^{32}P]dCTP by nick translation[18] and was used to screen the library using the plaque hybridization protocol of Benton and Davis.[19] Approximately 2 × 10^4 phage were plated on each 100-mm dish using strain Y1088 as the host. A total of 1 × 10^5 phage (1 × 10^4 recombinant phage) were screened, and 24 Luc1B homologous plaques were detected. Ten of these clones were analyzed after they had been purified to homogeneity through repeated rounds of screening. These clones contained Luc1B homologous *Eco*RI fragments that ranged from 200 to 600 bases in length. All but one of these clones contained a 1.2–1.3 kb *Eco*RI fragment that was homologous to Luc1A cDNA. The simplest explanation for these clones having "A" fragments with a constant size and "B" fragments that vary in size is that the constant fragments are complementary to the 3' end of the luciferase

[16] J. R. de Wet, K. V. Wood, D. R. Helinski, and M. Deluca, *Proc. Natl. Acad. Sci. U.S.A.* **82**, 7870 (1985).

[17] G. Dretzen, M. Bellard, P. Sassone-Corsi, and P. Chambon, *Anal. Biochem.* **112**, 295 (1981).

[18] P. W. J. Rigby, M. Dieckman, D. Rhodes, and P. Berg. *J. Mol. Biol.* **113**, 237 (1977).

[19] W. D. Benton and R. W. Davis, *Science* **196**, 180 (1977).

mRNA. The largest of these clones, λLuc23, contained a total of 1.8 kb of cDNA composed of two *Eco*RI fragments: a 600 base (Luc23B) fragment and a 1.2 kb (Luc23A) fragment.

Expression of Firefly Luciferase in *E. coli*

Foreign sequences inserted into the *Eco*RI site in the *lacZ* gene carried on the λgt11 chromosome are expressed as fusion proteins whose N-terminal portions are composed of 114 kDa of β-galactosidase sequences.[4] This system works very well for producing fusion proteins that can be detected by immunochemical means but is not well suited for expressing foreign proteins in an active form in *E. coli*. We chose to use the plasmid vector pKJB824.17 to express the luciferase cDNA because it can direct the synthesis of fusion proteins with a minimal number of added N-terminal amino acids. This vector consists of the plasmid pBR322 carrying the temperature-sensitive repressor of λ (λcI$_{857}$) and the λ promoter P$_R$ plus the beginning of the λ *cro* gene. The expression of genes placed downstream from λP$_R$ can be induced by raising the temperature of the culture which inactivates the λ repressor and allows transcription to initiate at λP$_R$. pKJB824.17 has a single *Eco*RI site located downstream from the initiation codon of the *cro* gene. Foreign genes inserted into this site are expressible as fusion proteins whose first six amino acids are determined by λ *cro* gene sequences.

The nucleotide sequence of the ends of the Luc23 cDNA was determined by the dideoxy sequencing method as described by Messing.[6] Both of the *Eco*RI cDNA fragments (Luc23A and Luc23B) were cloned in both orientations into the *Eco*RI site of M13mp19, and the nucleotide sequence was determined. A poly(A) tail was located on the end of the 1.2 kb Luc23A *Eco*RI fragment confirming that this portion of the clone was homologous to the 3′ end of the message. Only one open reading frame could be found in the Luc23B fragment, and ligation of the *Eco*RI site at the end of the Luc23 cDNA to the *Eco*RI site in pKJB824.17 would fuse this open reading frame to the reading frame of the *cro* gene in the plasmid. λLuc23 was partially digested with *Eco*RI, and the 1.8 kb fragment (Luc23A plus Luc23B) was isolated from an agarose gel. The plasmid pKW101 was constructed by ligating the isolated Luc23 cDNA to pKJB824.17 DNA that had been cut with *Eco*RI (Fig. 1). Heat induction of cells containing pKW101 should result in the synthesis of a fusion protein consisting of the amino acids encoded by the Luc23 cDNA preceded by 8 amino acids encoded by the vector and the *Eco*RI synthetic linker. Since the N-terminus of *P. pyralis* luciferase has not been sequenced we cannot determine what portion of luciferase is lacking in

```
met asp gly ser pro gly ile pro ILE LYS LYS GLY
ATG GAC GGA TCC CCG GGA ATT CCC ATA AAG AAA GGC
```
                              Eco RI

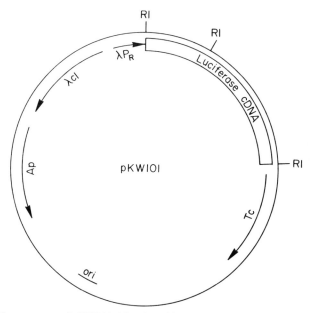

FIG. 1. The structure of pKW101. The plasmid pKW101 was constructed by inserting the cDNA from λLuc23 into the *Eco*RI site of pKJB824.17 as described in the text. This places the luciferase cDNA downstream from the λ promoter P_R and places the open reading frame in the cDNA in frame with the start of the *cro* gene of λ. Transcription from P_R is under the control of the temperature-sensitive λ repressor cI_{857} designated λcI in the figure. Heat induction inactivates the repressor allowing transcription to initiate at P_R resulting in the expression of the luciferase cDNA sequence. Shown at the top of the figure is the sequence at the junction of the *cro* gene with the luciferase cDNA. The sequence begins with the start codon of the *cro* protein. The amino acids shown in lower case are determined by the sequence of the vector and the *Eco*RI linker that is present at the 5′ end of the cDNA. The initial amino acids determined by the luciferase cDNA are shown in upper case letters. Labeled on the diagram are the *Eco*RI sites (RI), the ampicillin-resistance gene (Ap), the tetracycline-resistance gene (Tc), and the pBR322 origin of replication (*ori*).

the fusion protein produced by the expression of pKW101 in *E. coli* cells.

  E. coli cells harboring pKW101 synthesize active luciferase as assayed by the production of light in the presence of ATP and luciferin as shown in Figs. 2 and 3. TB1(pKW101) and TB1(pKJB824.17) cells were grown overnight at 30° in LB broth plus 250 μg/ml penicillin G. One-tenth vol-

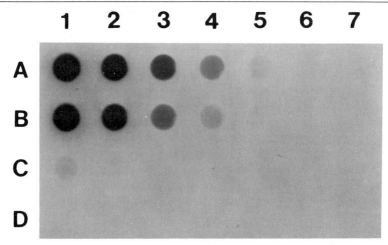

FIG. 2. Detection of light produced by active luciferase synthesized in *E. coli*. Firefly (*P. pyralis*) luciferase and extracts of *E. coli* TB1 cells were assayed as described in the text. Light production was detected with Kodak OG-1 X-ray film that had been flashed before exposure to increase its sensitivity. The spots in row A were produced by the light emitted by the following amounts of purified *P. pyralis* luciferase: 1, 10 ng; 2, 3.2 ng; 3, 1.0 ng; 4, 0.32 ng; 5, 0.10 ng; 6, 0.032 ng; 7, 0.010 ng. Rows B, C, and D show the light produced by the following amounts of 5-fold concentrated bacterial extract; 1, 100 μl; 2, 32 μl; 3, 10 μl; 4, 3.2 μl; 5, 1.0 μl; 6, 0.32 μl; 7, 0.1 μl. The extract in row B was obtained from TB1(pKW101) cells heat induced at 45°, the extract in row C was from TB1(pKW101) cells maintained at 30°, and the extract in row D was from TB1(pKJB824.17) cells that had been heat induced at 45°.

umes of these cultures were used to inoculate fresh LB broth and the cultures were incubated at 30° with shaking until they reached an OD_{660} of 0.8. The cultures were then transferred to a 45° shaker bath and incubated for 30 min. A portion of the TB1(pKW101) culture was maintained at 30° without heat induction. The cultures were chilled on ice, and 1 volume of each of the cultures was centrifuged at 5000 g for 10 min at 4°. The supernatant was removed, and the cell pellets were resuspended in $\frac{1}{5}$ volume of cold lysis buffer (100 mM KPO_4, pH 7.8, 2 mM EDTA, 1 mM dithiothreitol, lysozyme at 1 mg/ml, and protamine sulfate at 0.2 mg/ml) and were transferred to a 1.5-ml microfuge tube. The cells were incubated on ice for 15 min and were then frozen on dry ice. The lysates were allowed to thaw at room temperature and were cleared by centrifugation in a microfuge for 1 min.

The luciferase activity in the *E. coli* extracts was assayed by two methods: the light emitted by the luciferase reaction was detected by film, and the time course of the reaction was monitored with an LKB luminometer equipped with a chart recorder. A modified microtiter plate was

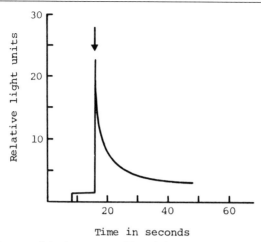

Time in seconds

FIG. 3. Luciferase activity in extracts of heat-induced TB1(pKW101) cells. *E. coli* cell extract was prepared and assayed in the presence of excess ATP and luciferin as described in the text. The time course of light emission was monitored in an LKB luminometer equipped with a chart recorder. The arrow indicates the time at which ATP was injected into the tube containing extract and luciferin.

constructed from a plate of black acrylic plastic with holes drilled through it, and this plate was sealed to a thin sheet of clear plastic with a film of petroleum jelly. The extracts and purified *P. pyralis* luciferase were serially diluted in 25 mM glycylglycine containing bovine serum albumin (BSA) at 1 mg/ml. Sample (100 μl) was placed in each well, and 100 μl of substrate mix (25 mM glycylgylcine, pH 7.8, 10 mM MgCl$_2$, 8 mM ATP, and 0.1 mM luciferin) was added. The sample plate was placed over flashed Kodak OG-1 X-ray film (flashing the film increased its sensitivity >10-fold) which was exposed to the light produced by the luciferase reactions for 3 hr (Fig. 2). From the relative intensities of the spots on the film produced by the luciferase reactions we have estimated that 1 ml of heat-induced TB1(pKW101) cells ($\approx$5 × 10^8 cells) contained approximately 10 ng of active recombinant luciferase. A comparison of the amount light emitted by extracts of heat induced cells and uninduced cells showed that the heat induction resulted in about a 50-fold increase in the activity of firefly luciferase present in the *E. coli* extracts. No light production could be detected in extracts of heat induced cultures of TB1 cells carrying the parent vector pKJB824.17. Figure 3 shows the results of assaying 20 μl of extract from the heat-induced TB1(pKW101) cells in a luminometer. The extract was added to 300 μl of 25 mM glycylglycine, pH 7.8, 5 mM MgCl$_2$, and 0.1 mM luciferin. The test tube with the sample was placed in the luminometer, 100 μl of 20 mM ATP, pH 7.0, was injected, and the time course of light emission was recorded. Using a luminometer it was also possible to detect the light emitted by intact bacteria in the presence of luciferin although the intensity of the light was greatly reduced in comparison to that emitted by the luciferase in cell extracts (data not shown). Presumably, this is due to a low permeability of the *E. coli* membrane to luciferin.

The smallest amount of firefly luciferase that we could detect on OG-1 X-ray film was 30 pg or 5×10^{-16} mol. This range of sensitivity could very likely be extended by using a high-speed film such as the ASA 20,000 Polaroid Land Type 612 instant film.[20] Much greater levels of sensitivity can be attained using a photomultiplier tube to detect the light emitted by the luciferase. One light unit is produced by 8.5 pg (1.4×10^{-16} mol) of luciferase when assayed in an Analytical Luminescence Laboratories Monolight 401 luminometer, and assay mixes lacking luciferase produced a background signal of 0.005 light units in the same instrument. A signal 10 times greater than the background was produced by 0.43 pg (7×10^{-18} mol) of firefly luciferase. The ease of performing assays for luciferase coupled with the great sensitivity of light detection systems make the firefly luciferase gene potentially very useful as a means of monitoring promoter activity in cells.

[20] G. H. G. Thorpe, T. P. Whitehead, R. Penn, and L. J. Kricka, *Clin. Chem.* **30**, 807 (1984).

[2] Extraction of Adenosine Triphosphate from Microbial and Somatic Cells

By PHILIP E. STANLEY

There are many reasons for extracting and measuring adenosine triphosphate (ATP) from cells, but they may be placed into one of two main categories.

1. The level of endogenous adenosine triphosphate (ATP) in a cell may be used as an index of energy status. It is therefore useful in metabolic and physiological studies.[1]

2. Estimates of cell numbers in microbial and tissue cultures may be obtained after assuming that the ATP per cell remains at a fairly constant (within a factor of five) and known value under defined conditions. Thus

[1] M. J. Harber, *in* "Clinical and Biochemical Luminescence" (L. J. Kricka and T. J. N. Carter, eds.), p. 189. Dekker, New York, 1982.

by measuring total ATP in a sample of culture, cell numbers may be rapidly obtained. This is the basis of rapid microbiology using the ATP-firefly luminescence technique.[1–5]

Extraction of ATP from cells and its subsequent measurement using the firefly luciferase procedure is often used. However there have been few studies to critically test the effectiveness of the adopted protocols for a wide range of cells.[6,7]

Properties of the Ideal Extractant

1. It should penetrate the cell wall and membrane more or less instantaneously.

2. It should extract ATP more or less instantaneously.

3. It should extract the target intracellular ATP pool completely.

4. It should instantaneously and irreversibly inactivate all enzymes that use ATP as a substrate or produce ATP from other substrates.

5. It should not cause breakdown of ATP (e.g., hydrolysis) either in short term (at the extraction time) or long term (during storage).

6. It should not have an inhibitory (quenching) effect on firefly luciferase during ATP assay.

7. It should not have an effect on the kinetics of firefly reaction.[8] Such an effect will cause problems of signal distortion and consequently internal standardization.

8. It should not extract undue quantities of extraneous materials which in themselves affect the firefly assay (by quenching or inhibition) and/or the result, e.g., colored agents, turbidity.[9]

In mixed cell populations, where, for example, bacteria and somatic cells exist, there may also be a need to selectively extract ATP from

[2] D. M. Karl, *Microbiol. Rev.* **44**, 739 (1980).

[3] A. Lundin, *in* "Clinical and Biochemical Luminescence" (L. J. Kricka and T. J. N. Carter, eds.), p. 43. Dekker, New York, 1982.

[4] H. Van de Werf and W. Verstraete, *in* "Analytical Applications of Bioluminescence and Chemiluminescence" (L. J. Kricka, P. E. Stanley, G. H. G. Thorpe, and T. P. Whitehead, eds.), p. 33. Academic Press, London, 1984.

[5] D. Slawinska and J. Slawinski, *in* "Chemi- and Bioluminescence" (J. G. Burr, ed.), p. 533. Dekker, New York, 1985.

[6] A. Lundin and A. Thore, *Appl. Microbiol.* **30**, 713 (1975).

[7] A. Lundin, *in* "Analytical Applications of Bioluminescence and Chemiluminescence" (L. J. Kricka, P. E. Stanley, G. H. G. Thorpe, and T. P. Whitehead, eds.), p. 491. Academic Press, London, 1984.

[8] L. J. Kricka and M. DeLuca, *Arch. Biochem. Biophys.* **217**, 674 (1982).

[9] W. W. Nichols, G. D. W. Curtis, and H. H. Johnston, *Anal. Biochem.* **114**, 433 (1981).

microbial or somatic cells. This imposes further constraints on the properties of the extractant.

When the sample itself is complex and perhaps variable, e.g., soil[10,11] or rumen contents,[12] this will put additional requirements on the properties of the extractant and extraction procedure.

It is worthwhile noting that properties 1 to 4 are associated with the extraction process alone whereas 5–8 have an affect on the ATP-firefly assay.

Note that properties 4 and 6 are apparently contradictory. However, the use of boiling buffer or cold acid (the latter requires immediate neutralization following extraction) does fulfill these needs but unfortunately also introduces sample dilution (and thus loss in sensitivity) or extra manipulation.

Treatment of Cells Which Affect ATP Level

Before extraction is performed it is usually necessary to take a sample; if this is not done properly ATP levels may change during this process. In addition a wide range of other factors have been shown to effect ATP levels[13] and the worker should be aware of these: (1) change of growth rate,[14] (2) change of nutient(s) or their concentration(s),[14–19] (3) change of gaseous environment, e.g., oxygen tension,[20,21] (4) change of temperature,[22] (5) change of pH,[23] (6) change of pressure,[24] and (7) change of light flux (for photosynthetic organisms).[17,25]

Laboratory techniques which one might employ to harvest or sample cells may well involve one or more of the above. I am thinking of centrifuging and filtration.

[10] J. M. Oades and D. S. Jenkinson, *Soil Biol. Biochem.* **11,** 201 (1979).
[11] J. J. Webster, G. J. Hampton, and F. R. Leach, *Soil Biol. Biochem.* **16,** 335 (1984).
[12] D. E. Nuzback, E. E. Bartley, S. M. Dennis, T. G. Nagaraja, S. J. Galitzer, and A. D. Dayton, *Appl. Environ. Microbiol.* **46** 533 (1983).
[13] P. C. T. Jones, *J. Theor. Biol.* **34,** 1 (1972).
[14] J. S. Franzen and S. B. Binkley, *J. Biol. Chem.* **236,** 515 (1961).
[15] L. Gustafsson, *Arch. Microbiol.* **120,** 15 (1979).
[16] M. Ohmori and A. Hattori, *Arch. Microbiol.* **117,** 17 (1978).
[17] A. Lewenstein and R. Bachofen, *Arch. Microbiol.* **116,** 169 (1978).
[18] C. M. M. Franco, J. E. Smith, and D. R. Berry, *J. Gen. Microbiol.* **130,** 2465 (1984).
[19] M. B. Nair and R. R. Eady, *J. Gen. Microbiol.* **130,** 3063 (1984).
[20] R. E. Strange, H. E. Wade, and F. A. Dark, *Nature (London)* **199,** 55 (1963).
[21] R. R. Mathis and O. R. Brown, *Biochem. Biophys. Acta* **440,** 723 (1976).
[22] Y. N. Lee and M. J. Colston, *J. Gen. Microbiol.* **131,** 3331 (1985).
[23] K. D. Beaman and J. D. Pollack, *J. Gen. Microbiol.* **129,** 3103 (1983).
[24] J. V. Landau, *Exp. Cell Res.* **23,** 539 (1961).
[25] P. C. T. Jones, *Cytobios* **6,** 89 (1970).

Therefore the experimental design should encompass a phase to ascertain whether or not the harvesting or sampling technique itself affects the ATP level.

Other Factors Which Affect ATP Level

These factors include the (1) age of cells or stage of growth,[23,26,27] (2) stage of cell division,[28–29] (3) density of cells,[30] (4) phage and virus[31–33] and microbial infections,[34] (5) action of agents which change cell type, e.g., tumor-promoting agents, and other agents such as (6) antibiotics[35,36] and certain drugs,[37] (7) metabolic inhibitors, cogeners of metabolites and toxins, (8) disinfectants,[38] pesticides,[39] and herbicides,[40] etc., (9) heavy metals, and (10) radiation, e.g., ultraviolet, microwave, X- and gamma rays.

General Considerations

A wide range of extractants have been used but few have been extensively investigated as far as effectivity. As well as those mentioned earlier[6,7] the group at NASA have published a series of reports[41–43] and at an

[26] K. W. Hutchison and R. S. Hansen, *J. Bacteriol.* **119**, 70 (1974).

[27] W. W. Forrest, *J. Bacteriol.* **90**, 1013 (1965).

[28] B. Chin and I. A. Bernstein, *J. Bacteriol.* **96**, 330 (1968).

[29] C. Edwards, M. Statham, and D. Lloyd, *J. Gen. Microbiol.* **88**, 141 (1975).

[30] L. Huzyk and D. J. Clark, *J. Bacteriol.* **108**, 74 (1971).

[31] I. Vlodavsky, M. Inbar, and L. Sachs, *Proc. Natl. Acad. Sci. U.S.A.* **70**, 1780 (1973).

[32] R. Wahl and L. M. Kozloff, *J. Biol. Chem.* **237**, 1953 (1962).

[33] E. Egberts, P. B. Hackett, and P. Traub, *J. Virol.* **22**, 591 (1977).

[34] Y. Tsuchiya and H. Sugai, *Biochem. Med.* **28**, 256 (1982).

[35] P. McWalter, *J. Appl. Bacteriol.* **56**, 145 (1984).

[36] R. Guerrero, M. Llagostera, A. Villaverde, and J. Barbé, *J. Gen. Microbiol.* **130**, 2247 (1984).

[37] G. E. Thomas, S. Levitskyy, and H. Feinberg, *J. Mol. Cell Cardiol.* **15**, 621 (1983).

[38] J. E. Cairns, S. G. Nutt, and B. K. Afghan, *in* "International Symposium on Analytical Applications of Bioluminescence and Chemiluminescence" (E. Schram and P. Stanley, eds.), p. 303. State Printing and Publishing, Westlake Village, California 91361, 1979.

[39] R. D. Gruenhagen and D. E. Moreland, *Weed Science* **119**, 319 (1971).

[40] T. J. Clegg and J. L. Koevenig, *Bot. Gaz.* **135**, 368 (1974).

[41] E. A. Knust, E. W. Chappelle, and G. L. Picciolo, *in* "Analytical Applications of Bioluminescence and Chemiluminescence," p. 27 (and other references). NASA Document SP-388, 1975.

[42] G. L. Picciolo *et al.*, *in* "Applications of Luminescence Systems to Infectious Disease Methodology." Goddard Space Flight Center, Greenbelt Maryland, Document X-726-76-212, 1976.

[43] E. W. Chappelle, G. L. Picciolo, and J. W. Deming, this series, Vol. 57, p. 65.

early date highlighted the fact that bacterial ATP turns over rather quickly.[44] A number of comparative studies have been made for environmental samples (marine waters, etc.).[45–47]

In general, many bacteria have ATP levels around 1 fg (10^{-15} g) per cell. Yeasts have around 100 times more and many animal cells contain around a picogram ATP (10^{-12} g). Some actual values are given in refs. 1–3.

I will now consider, albeit briefly, the main extractants and give a few references wherein details can be obtained.

1. Boiling buffer, usually Tris–Cl, with EDTA; marine microbial samples,[2] freshwater microbial samples,[48] nematodes,[49] yeast,[50] rumen contents,[12] mycoplasmas,[23] mycobacteria.[22]

2. Various dilute acids including nitric acid (mainly clinical samples,[41–43] sulfuric acid [seawater samples,[2,45,47] clinical samples (ref. 47, p. 189)], perchloric acid (microbial samples,[14,27,51] erythrocytes,[34] tumor,[33] plants[52]), trichloroacetic acid (microbial samples,[6,7] somatic cells[3,7]), and formic acid (microbial samples[44]).

3. Organic compounds including dimethyl sulfoxide (bacteria),[26] ethanol (alga),[53] acetone (yeast),[54] chloroform (mycobacteria),[55,56] and butanol (bacteria).[57]

4. Surfactants including Triton X-100 (somatic cells)[3,6,7] and benzalkonium chloride (yeast).[58]

In addition various mixtures have been described which include some of those extractants mentioned above and which have been used for com-

[44] W. Klofat, G. Picciolo, E. W. Chappelle, and E. Freese, *J. Biol. Chem.* **244,** 3270 (1969).
[45] O. Holm-Hansen and D. M. Karl, this series, Vol. 57, p. 73.
[46] "ATP Methodology Seminar" (G. A. Borun, ed.). SAI Technology Co., San Diego, California, 1975.
[47] "2nd Bi-Annual ATP Methodology Symposium" (G. A. Borun, ed.). SAI Technology Co., San Diego, California, 1977.
[48] B. R. Taylor and J. C. Roff, *Freshwater Biol.* **14,** 195 (1984).
[49] H. J. Atkinson and A. J. Ballantyne, *Ann. Appl. Biol.* **87,** 167 (1977).
[50] A. Cockayne and F. C. Odds, *J. Gen. Microbiol.* **130,** 465 (1984).
[51] M. Statham and D. Langton, *Process Biochem.* **10,** Oct. 25 (1975).
[52] P. E. Stanley, *in* "Liquid Scintillation Counting" (M. A. Crook and P. Johnson, eds.), Vol. 3, p. 253. Heyden and Son, London, 1974.
[53] J. B. St John, *Anal. Biochem.* **37,** 409 (1970).
[54] L. F. Miller, M. S. Mabee, H. S. Gress, and N. O. Jangaard, *J. Am. Soc. Brew. Chem.* **36,** 59 (1978).
[55] A. M. Dhople and E. E. Storrs, *Int. J. Leprosy* **50,** 83 (1982).
[56] R. P. Prioli, A. Tanna, and I. N. Brown, *Tubercle* **66,** 99 (1985).
[57] E. C. Tifft, Jr., and S. J. Spiegel, *Environ. Sci. Technol.* **10,** 1268 (1976).
[58] M.-R. Siro, H. Romar, and T. Lovgren, *Eur. J. Appl. Microbiol. Biotechnol.* **15,** 258 (1982).

plex samples such as soil.[10,11] Further a number of commercial extractants of undisclosed content are available but little has been published about their efficiency in comparison with those listed above.[4,56]

In all cases it is important not to overload the extractant with too much sample. Usually a few milligrams dry weight of sample per milliliter of extractant is satisfactory. It is also important to check the sampling/ extraction procedure to ascertain that the process itself does not change the ATP level or influence the measured result. An example of poor sampling might be in drawing a blood sample from an animal and allowing the sample to hemolyze, thereby releasing ATP into the plasma. Another might be to allow a microbial sample to become nutrient deficient during sampling.

As mentioned previously there have been few critical studies in this area. One of the first[6] involved the use of 5 microbial species with 10 different extractants. From this study, the workers concluded that only trichloroacetic acid (TCA) was entirely satisfactory for all species. In a follow up,[7] 5 different extractant systems were tested on 7 species of microorganisms and 9 different types of somatic cells. From both studies it was concluded that trichloroacetic acid should be the standard against which other extractants should be tested. In the second of the studies the authors suggest 10, 5, 2.5, and 1.25% TCA (final concentration: equal volumes of TCA and sample) should be employed initially as a standard before searching for better extractants of ATP. In coming to a decision, the total luminescence as well as the lowest degree of quenching of luciferase by the extractant need to be taken into account when choosing operational parameters. It is not possible to cover all types of extractants and their use in a wide range of cells. I will therefore indicate some guidelines to follow when designing extraction protocols.

Somatic Cells

Cells which are separate from one another (e.g., blood cells, certain tissue cultures) do not generally need special treatment prior to extraction and can be mixed directly with the extractant. An exception would be a case in which extracellular (free) ATP was present in amounts that could not be neglected and here cells would need washing or the ATP removed enzymically. However, cells that are formed together in a more or less solid tissue from living organisms (e.g., muscle, kidney, plant leaves) should be frozen to stop the action of ATPases and then homogenized or thinly sliced (<0.2 mm). Freezing is best done in liquid nitrogen and homogenizing with a blender or a simple mortar and pestle. Slicing can be done with a razor blade or on a cryomicrotome.

If small amounts of ATP are expected it is important that equipment be kept scupulously clean and free from microbial contamination. Following homogenization or slicing, an aliquot of the sample can then be allowed to thaw in the presence of the extractant so long as they are well mixed and no clumps of tissue are formed. It is essential that extractant be quickly and intimately associated with all cells. If the cells clump together the extractant may take a long time to penetrate to the inner cells by which time considerable changes in ATP content may have taken place. Alternatively the final result may be too low because little or no ATP has been extracted from those cells. The problem of clumping may be solved in some cases by subjecting the extracting sample to ultrasound.

Alternatively tissue samples may be homogenized at ice water temperature in a glass–glass homogenizer, blender, etc. Another approach would be to homogenize the sample in boiling buffer (e.g., nematodes[49]). For large samples of tissue, 10 g tissue in 100 ml buffer may be homogenized as described in ref. 59 at room temperature or in a cold room.

Microbial Cells

Bacterial cultures can often be added directly to the extractant. For acidic extractants 1 ml culture to 1 ml acid is generally satisfactory and if boiling buffer is used then 1 ml culture into 10 ml boiling buffer should be employed.

Fungal cultures with mycelia and algal cultures with multicellular filaments may require homogenizing in a blender in the presence of the extractant.[60]

Because yeast cells and microbial spores have thicker walls than bacteria they consequently often require more drastic extractants or a longer extraction time.

When yeast and bacteria are both present in the sample (e.g., urine) one method of measuring both is to use a differential filtration procedure.[61] I am not aware of any extraction procedures which permit differentiation of two or more microbial species in the same sample without complex manipulation, e.g., cell separation or growth in selective media.

When microbial numbers are small a concentration step may be required and filtration or centrifuging may be convenient. Both methods tend to lead to a decrease in ATP content because of oxygen and nutrient depletion. The problem can be usually solved by adding a small volume of a suitable culture medium to the filter or pellet and allowing the microbial cells to recover for 5–10 min before extracting them.

[59] C. J. Stannard and P. A. Gibbs, *J. Biolumin. Chemilumin.* **1,** 1 (1986).
[60] N. A. Hendy and P. P. Gray, *Biotechnol. Bioeng.* **21,** 153 (1979).
[61] T. S. Tsai and L. J. Everett, *in* "Analytical Applications of Bioluminescence and Chemiluminescence" (L. J. Kricka, P. E. Stanley, G. H. G. Thorpe, and T. P. Whitehead, eds.), p. 75. Academic Press, London, 1984.

Somatic and Microbial Cells Mixed Together with Nonliving Material

This is the most complex system. Examples in this section include soil,[4] nematodes in soil,[49] meat,[59] milk,[62] rumen contents,[12] fruit juices,[63] and clinical samples.[1,3,42,43]

Suppose you wish to determine microbial ATP in one of the above samples. In most situations the somatic ATP will be present in large excess (perhaps 10^6-fold) and this can pose considerable technical problems, especially if a number of samples are to be processed. Two approaches have been used. In the first, somatic cells are selectively extracted (e.g., using Triton X-100) and the somatic ATP together with any free ATP can then be hydrolyzed with an ATPase.[43] The second method involves physical separation of somatic and microbial cells by differential centrifuging or filtration or addition of resin.[59] In both cases the remaining microbes can then be extracted.

As far as differential extraction is concerned use is generally made of the substantial difference in cell wall/membrane in microbial and somatic cells. Care however must be exercised since so-called somatic extractants (e.g., Triton X-100) have been shown to extract ATP from some bacteria (e.g., *Pseudomonas aeruginosa*.)[7] Microbial extractants will in most cases also extract ATP from somatic cells.

There are two other sources of ATP in many samples. First, free ATP, that is ATP which is in true solution. This will generally be measured and not distinguished from cellular ATP unless it is removed enzymically or by washing. Second, there is the more problematic "bound" ATP which is sequestered on protein and other macromolecules and surfaces in a fairly firm fashion. Bringing the pH of the sample below 4.5 using malic acid has been used to release this ATP so that intereference by it can be removed.[43]

Some Other Points Concerning Extractants

The use of boiling buffer is popular because of simplicity and is quite satisfactory if very high sensitivity is not required. If sensitivity is a requirement then the dilution involved (usually 10-fold) is a disadvantage. Be sure to have the buffer boiling before addition of sample (at least 1 : 5, or better 1 : 10 sample : buffer) so that enzyme denaturation is immediate. If it is not ATPases and other enzymes will change the ATP content even if they are active for say 10 sec.

It is best to avoid as much as possible any postextraction processing as this inevitably leads to losses of ATP during manipulation and to prob-

[62] R. Bossuyt, *Milchwissenschaft* **36**, 257 (1981).
[63] J. G. H. M. Vossen and H. D. K. J. Vanstaen, *Forum Mikrobiol.* **4**, 280 (1981).

lems of contamination because of the ubiquitous nature of ATP (e.g., in sweat, contaminating microbes, etc.) which may be a problem if you are working at high sensitivity.

The use of perchloric acid has led to two reports which indicate other types of problems. Instead of adding the perchloric acid extract directly to the firefly assay system as is usually the case it is possible to reduce the quenching caused by the acid if one removes the acid by precipitating it as the insoluble potassium salt and finishing with a sample at neutral pH. However, one report indicates that some ATP coprecipitates with potassium perchlorate so that final ATP measurements are too low.[64] Another set of workers have reported that on neutralizing perchloric acid extracts from *Bacillus brevis* a phosphatase is reactivated which then proceeds to hydrolyze extracted ATP![65]

If organic solvents are used, some of them may be readily evaporated and this provides an easy way to increase sensitivity should that be necessary.[55]

[64] S. Wiener, R. Wiener, M. Urivetzky, and E. Meilman, *Anal. Biochem.* **59,** 489 (1974).
[65] J. A. Davison and G. H. Fynn, *Anal. Biochem.* **58,** 632 (1974).

[3] Detection of Bacteriurea by Bioluminescence

By BRUCE A. HANNA

Urinary Tract Infection

Microbial colonization of the urogenital tract, resulting in bacteriurea, is an increasingly common event in modern medical management. While normally sterile, the urinary tract which includes the kidneys, ureters, urinary bladder, and urethra may readily become infected with a wide variety of microbes. Frequently, such infections are preceded by instrumentation and manipulation of the urinary tract, often by insertion of a urinary catheter. As a result, normal flora microbes, particularly skin and enteric bacteria, may gain entry to urinary tract tissues. The severity of such infections may range from those which are asymptomatic except for the presence of bacteriurea, to overt clinical infections of the bladder (cystitis) or kidney (pyelonephritis) which may be accompanied by severe systemic symptoms. Evaluation of the urine to determine the presence and concentration of microorganisms in the urine is an important adjunct in the diagnosis and treatment of urinary tract infections.

Significant Bacteriurea

The definition of what represents clinically significant bacteriurea has been the subject of much discussion. In actuality, the criteria for defining significant bacteriurea is dependent on the patient from whom the sample is derived.[1] The conventional criterion invoked to detect urinary tract infections in asymptomatic patients where there is a low prevalence of disease in the population, is $10^4–10^5$ colony-forming units (CFU)/ml in a freshly voided, first morning specimen.[2] In this select patient population counts of $>10^5$ are almost certainly significant, while counts of $<10^4$ have a high probability of representing urethral contamination. In symptomatic patients, on the other hand, where the prevalance of urinary tract infection in the population is high, colony counts of 10^3 or even 10^2 CFU/ml may be considered significant.[3] Frequently in such patients the urine will contain numerous polymorphonuclear leukocytes and other blood cell components as well as bacteria. In addition to bacteria, urinary tract infections may on occasion be caused by fungi, especially *Candida* species, *Chlamydia trachomatis, Ureaplasma urealyticum, Mycoplasma hominis, Mycobacteria* and viruses.

Detection of Bacteriurea

The techniques available to determine the presence and quantity of microorganisms in a urine specimen can be divided into those that are growth dependent and those that are non-growth dependent. Growth-dependent methods, by definition, require dilution, inoculation onto a suitable medium, and an incubation period of 18–24 hr. This will result in an enumeration as well as a prelude to the identification of the microbes present. Non-growth-dependent methods, in contrast, do not require cultivation of the organism, but rather provide a direct enumeration of the bacterial population present. Since the noncultivation methods provide the user with a quantitation but not an identification. they are termed screening tests.

In a typical clinical microbiology laboratory as many as 70% or more of urine specimens may not contain significant populations of bacteria. In such a setting, bacteriurea screening tests are very useful in eliminating these samples from further analysis. Intrinsic to these methods is their ability to rapidly identify such samples on the same day as they are collected from the patient. Conversely, as the majority of urinary tract

[1] R. C. Bartlett and R. C. Galen, *Am. J. Clin. Pathol.* **79,** 756 (1983).
[2] R. Plott, *Am. J. Med.* **75**(1B), 44 (1983).
[3] W. E. Stamm, G. W. Counts, K. R. Running, S. Fihn, M. Turck, and K. K. Holmes, *Am. J. Med.* **307,** 463 (1982).

pathogens can be successfully treated with an empiric antibiotic regimen, urine samples determined to have significant bacteriurea by a screening test can be a rapid signal to institute therapy against the probable pathogen.

Bioluminescence Screening for Bacteriurea

The development of the firefly luciferin–luciferase system for the assay of adenosine triphosphate (ATP) has had a marked impact on microbiology. Since the amount of ATP within bacterial cells was readily able to be calculated, and the luciferase assay was shown to be sufficiently sensitive to detect the ATP content of as few as 10 CFU/ml, the utility of this method of bacterial quantitation was quickly recognized.[4] Its applicability to clinical material, however, was severely limited by the inability of this method to discriminate between the ATP of microbial origin, and that of host cell origin.

It is perhaps prophetic, that the first paper ever published in the *Journal of Clinical Microbiology* described a new method for the detection of bacteriurea by the luciferase assay of ATP. In 1975 Thore *et al.*[5] reported that by preincubating urine in the presence of the detergent Triton X-100 and apyrase a loss of ATP from a variety of human cells but not bacterial cells resulted. Subsequent extraction of the remaining ATP from the urine correlated well with the concentrations of ATP attributable to only the microbial component of the sample. These authors calculated the average ATP content of a single bacterial cell recovered from the urine of patients studied to be 3.6×10^{-18} M, with, however, a variation of about one order of magnitude between various species at different concentrations. Using these data they suggested that at a concentration of 10^5 CFU/ml, indicative of significant bacteriurea, an ATP level of approximately 3.6×10^{-10} M would be expected. Since the sensitivity of the luciferase assay has been shown to be on the order of 10^{-14} M ATP, a level of detection of 10 CFU/ml might very well be expected.[4]

In practical experience, however, ATP analysis at such high levels of sensitivity have been found to display poor specificity. When compared to the numbers of cultivatable microorganisms recovered from urine samples by conventional growth-dependent techniques a high incidence of luciferase-positive, culture-negative samples have been reported.[6] Many of these so called false-positive specimens have been shown to be attributable to fastidious organisms, not normally recovered by conventional methods, i.e., anaerobes, or organisms which were in fact present but

[4] E. W. Chappelle and G. V. Levin, *Biochem. Med.* **2**, 41 (1968).

[5] A. Thore, S. Ansehn, A. Lundin, and S. Bergman, *J. Clin. Microbiol.* **1**, 1 (1975).

[6] A. Thore, A. Lundin, and S. Ansehn, *J. Clin. Microbiol.* **17**, 218 (1983).

likely were inhibited by antimicrobial agents present in the urine.[7] The early study by Thore *et al.*[5] in fact reported 4% of samples tested were falsely negative, while 30% were falsely positive. Similarly, another comparative study by Alexander *et al.,*[8] using an ATP concentration limit of 6 × 10^{-14} *M* ATP as representative of 10^5 CFU/ml, found 7% false-negative samples and 27% false-positive samples. Both studies revealed that in attempting to correlate ATP levels with sensitivity and specificity at 10^5 CFU/ml, a decrease in the number of false-positive values was accompanied by an increase in the number of false-negative values. During the ensuing years comparitive studies have since confirmed these early observations.[9]

In a similar fashion, 140 midstream, clean catch urine specimens submitted to the clinical microbiology laboratory of Bellevue Hospital were evaluated for the presence of significant bacteriurea. In this study the Microscreen Bacteriurea Screening Kit and Monolight 500 luminometer (Analytical Luminescence Laboratory, San Diego, CA) were used to screen the samples. Briefly, this technique is based upon the classic firefly luciferin–luciferase assay for ATP available as an *in vitro* diagnostic kit. To perform the test, 25 μl of urine was added to an assay cuvette containing Somalight, a detergent-like compound which selectively releases somatic cell ATP, and apyrase to convert the liberated ATP to ADP. Following a 10-min incubation period the cuvette was placed into the measuring chamber of the Monolight 500 which automatically injects Extralight, another detergent-like compound which releases the ATP from the bacteria in the urine, followed by a Firelight reagent, which consists of the firefly luciferin–luciferase mixture. The resulting bioluminescence was read in the Monolight 500 over a 10 sec span with the result displayed in relative light units (RLU). A threshold of 1.5 RLU on the Monolight 500 is indicative of significant bacteriurea. Simultaneously, each urine specimen studied was inoculated to the surface of a trypticase soy agar plate containing 5% sheep blood (BAP) using a 0.001 ml calibrated loop. After overnight incubation at 35° the number of colonies on the BAP was counted to determine the number of CFU/ml in the original specimen.

The results of the paired assays are listed in Tables I and II. As shown, of the 140 urine samples compared, depending on the colony count breakpoint used to define significant bacteriurea in comparison to 1.5 RLU, 66–73% of the samples were culture negative (true negative + false positive). Of these culture negative samples, the bioluminescence assay detected better than 75% of the samples correctly at all breakpoints (specificity). Of the samples determined to be positive for significant bacteriurea (true

[7] B. A. Hanna, L. Larrier, C. F. Fam, and A. C. Coxon, *Pathologist* (in press).
[8] D. N. Alexander, G. M. Ederer, and J. M. Matsen, *J. Clin. Microbiol.* **3,** 42 (1976).
[9] G. Szilagyi, V. Aning, and A. Karman, *J. Clin. Lab. Autom.* **3,** 117 (1983).

TABLE I

COMPARISON OF MONOSCREEN RESULT WITH COLONY
COUNT RESULT[a]

Interpretation	Number of positive urines (%)		
	10^4	5×10^4	10^5
False positive[b]	22(15.7)	24(17.1)	25(17.9)
False negative[c]	9(6.4)	6(4.3)	3(2.1)
True negative[d]	70(50)	75(53.1)	77(55)
True positive[e]	39(27.9)	35(25)	35(25)

[a] At three different breakpoints for significant bacteriurea ($n = 140$).
[b] False positive, luciferase positive, colony count negative.
[c] False negative, luciferase negative, colony count positive.
[d] True negative, luciferase negative, colony count negative.
[e] True positive, luciferase positive, colony count positive.

positive + false negative) between 81 and 92% were correctly identified (sensitivity) by bioluminescence. The predictive value of a positive result ranged from approximately 59 to 64%, while the predictive value of a negative result was 87 to 96%, It is important to realize in evaluating these data that a significant number of so-called false-positive urines do in fact

TABLE II

COMPARISON OF SENSITIVITY, SPECIFICITY,
AND PREDICTIVE VALUES

Interpretation	Monoscreen versus calibrated loop (%)		
	10^4	5×10^4	10^5
Sensitivity[a]	81.3	85.4	92.1
Specificity[b]	76.1	75.8	75.5
Predictive value positive[c]	63.9	59.3	58.3
Predictive value negative[d]	88.6	92.6	96.3

[a] Sensitivity, true positive/true positive + false negative × 100.
[b] Specificity, true negative/true negative + false positive × 100.
[c] Predictive value positive, true positive/true positive + true negative × 100.
[d] Predictive value negative, true negative/true negative + false negative × 100.

contain substantial numbers of microorganisms. As we have previously reported, the clinical significance of the organisms actually present in the luciferase-positive, culture-negative samples is a provocative but as yet unanswered question. The organisms which were recovered in significant numbers from the urine samples screened as luciferase negative may represent uneven sample distribution or low levels of bacterial ATP present during periods of low metabolic activity. The presence of inhibitors in the urine, or of antibody-coated bacteria have not been found to be responsible for this phenomenon.[7] The standard against which bioluminescence assays are usually compared, the calibrated loop method, has been demonstrated to be fraught with potential technique dependent variables which may introduce an error rate of $\pm 50\%$.[10] It is likely, therefore, that bioluminescence screening is not only a more rapid, but may also be a more reliable indicator of the absence of significant bacteriurea.

Acknowledgment

Supported in part by The Foundation For Scientific Research In The Public Interest.

[10] A. C. Albers and R. D. Fletcher, *J. Clin. Microbiol.* **18,** 40 (1983).

[4] Estimation of Biomass in Growing Cell Lines by Adenosine Triphosphate Assay

By ARNE LUNDIN, MICHAEL HASENSON, JÖRGEN PERSSON, and ÅKE POUSETTE

Introduction

During the last decade the interest in cell cultures has been continuously increasing. This is partly due to simplified cell culturing techniques making large numbers of assays possible. The use of cell lines as models for different human diseases has gained special interest, e.g., the use of cell lines as models for human prostatic carcinomas.[1] The use of a cell cultures makes it possible to compare effects of a large number of different treatments or combinations of treatments. For convenience and economic reasons this is often done by culturing the cells in tissue culture plates.

The techniques usually used for estimation of growth are microscopic cell counting and protein and DNA determinations. These techniques are,

[1] J. S. Horoszewicz, S. S. Leong, E. Kawinski, J. P. Karr, H. Rosenthal, T. M. Chu, E. A. Mirand, and G. P. Murphy, *Cancer Res.* **43,** 1809 (1983).

however, not as sensitive, specific, and easy to handle as may be required. In our studies[2] on the effect of different steroidal and pituitary hormones of various concentrations and combinations on the growth of a human prostatic carcinoma cell line, LNCaP-r, a sensitive assay capable of handling large numbers of samples was needed. We therefore investigated the possibility of using ATP for estimating growth. This had been successfully done for microbial cells.

The purpose of the present chapter is to describe the steps involved in estimation of biomass in growing cell lines by luminometric ATP assay. Methods for studying the reliability of the results are described. In particular estimation of energy charge (EC) by a new method for assay of ATP, ADP, and AMP was found to be valuable.

Principles

The ATP concentration in the protoplasm is similar in all living cells. Data on the amount of intracellular ATP in cell suspensions are reasonably well correlated to biomass parameters such as cellular carbon or dry or fresh weight.[3-6] Several applications of ATP assays for biomass estimation mainly in microbiology have been described previously.[7,8]

The activity of several adenosine nucleotide converting enzymes as a function of the energy charge, i.e. (ATP + 0.5 ADP)/(ATP + ADP + AMP), indicates that the energy charge in intact metabolizing cells should be stabilized in the interval 0.8–0.9.[9] Actual measurements confirm that growth can occur only at energy charge values above 0.8, that viability is maintained at values above 0.5, and that cells die at values below 0.5.[10] By measuring all three adenine nucleotides, one obtains a measure of the physiological status of the cells that provides information on whether ATP can be used to estimate biomass or if the energy metabolism is disturbed. By the method described in the present chapter all three ade-

[2] M. Hasenson, B. Hartley-Asp, C. Kihlfors, A. Lundin, J.-Å. Gustafsson, and Å. Pousette, *Prostate* **7**, 183 (1985).

[3] O. Holm-Hansen and C. R. Booth, *Limnol. Oceanogr.* **11**, 510 (1966).

[4] O. Holm-Hansen, *Plant Cell Physiol.* **11**, 689 (1970).

[5] O. Holm-Hansen, *in* "Estuarine Microbial Ecology" (L. H. Stevenson and R. R. Colwell, eds.), p. 73. Univ. of South Carolina Press, Columbia, 1973.

[6] O. Holm-Hansen and D. M. Karl, this series, Vol. 57, p. 73.

[7] A. Lundin, *in* "Clinical and Biochemical Luminescence" (L. J. Kricka and T. J. N. Carter, eds.), p. 43. Dekker, New York, 1982.

[8] M. J. Harber, *in* "Clinical and Biochemical Luminescence" (L. J. Kricka and T. J. N. Carter, eds.), p. 189. Dekker, New York, 1982.

[9] D. E. Atkinson, *Annu. Rev. Microbiol.* **23**, 47 (1969).

[10] A. G. Chapman, L. Fall, and D. E. Atkinson, *J. Bacteriol.* **108**, 1072 (1971).

nine nucleotides are determined in a single aliquot from the cell extract. This and the fact that energy charge is calculated as a quotient makes the energy charge value insensitive to experimental errors in, e.g., pipetting. Relating ATP to other biomass parameters, e.g., cell counts, would involve larger experimental errors. Measurement of energy charge gives a reliable estimation of the accuracy of using ATP as a biomass parameter.

Growth monitoring by assay of adenine nucleotides includes the following steps: (1) cultivation of cells, (2) sampling and pretreatment of samples, (3) extraction of adenine nucleotides from cells, (4) luminometric assay of adenine nucleotides, and (5) correlation to other biomass estimations. For a reliable result all five steps have to be correctly performed. In the following, methodological aspects on the individual steps will be discussed with special reference to pitfalls that may occur.

Cultivation of Cells

From an analytical point of view the only potential problem that may appear in this step would be culture medium or culture vessels containing adenine nucleotides. We have not encountered this problem in our experiments, but it is recommended that a "blank culture," i.e., a culture without inoculation of cells, be run before starting with a new culture system. Adenine nucleotides released from the cells during growth can be estimated by removing the cells from the medium (cf. below) and assaying the nucleotides in the cell-free medium.

In the present study a human prostatic carcinoma cell line, LNCaP-r, was used. This cell line was kindly donated by Dr. Horoszewicz, Roswell Park Memorial Institute, Buffalo, NY. The cells were cultivated in a constant environment (37.3°, 7.0% CO_2) and RPMI 1640 medium (Flow Lab, Scotland) supplemented with 10% (v/v) inactivated calf serum (Gibco, Scotland), 50 μg/ml penicillin, 50 μg/ml streptomycin, and 2.0 μmol/ml L-glutamine. Culture vessels (A/S Nunc, Roskilde, Denmark) used were culture flasks (25, 80, and 175 cm^2) and tissue culture plates (24 and 96 wells). Cells were enumerated with a Bürker chamber in a 0.3 μl volume (corresponding to 48 B-squares).

Sampling and Pretreatment of Samples

If there are no adenine nucleotides in medium or released from the cells during growth the most reliable procedure is to avoid sampling and pretreatment of samples. In our routine procedure 1 volume of 5% trichloroacetic acid (TCA) is added to 2 volumes of culture directly in the culture vessel. At this TCA concentration (1.67% TCA final concentra-

TABLE I
EFFECT OF CHANGE OF CULTURE MEDIUM

Treatment of culture	Nucleotide concentration (μM)				
	ATP	ADP	AMP	ΣAXP	EC[a]
No change of medium	5.74	0.44	0.04	6.22	0.96
Change to warm medium (37°)	5.08	0.38	0.02	5.48	0.96
Change to cold medium (6–8°)	5.10	0.34	0.10	5.54	0.95

[a] EC, Energy charge.

tion) a complete extraction of adenine nucleotides is obtained (cf. below). Since temperature as well as other conditions are changed when the culture vessel is taken from the incubator it may be critical to add the extractant within a short period of time. However, in our culture system there was no measurable change of the ATP level within the time interval 0–60 min. This finding should be confirmed in each new culture system.

If adenine nucleotides are present in the medium or if other assays not compatible with the extractant should be performed on the culture medium, this medium may be decanted provided that this does not affect the nucleotide levels of the cells. This may be confirmed by measuring adenine nucleotides in the whole culture, in the medium, and in the cells after removing the medium. Changing to fresh medium may possibly affect nucleotide levels. Table I shows that 1 min after changing to warm (37°) or cold (6–8°) medium the energy charge was not affected in our culture system. The reduction of the sum of all three adenine nucleotides (ΣAXP) in cultures with fresh medium is due to part of the cells being removed with the old medium (cf. below). The possibility that changing the medium may affect energy charge or remove part of the cells should be considered when making schedules for sampling.

In some situations it is necessary to perform other assays on cells as well as medium. In the present study microscopic cell counts should be compared with ATP results in the same culture flasks. For this purpose the medium was decanted and cells dispersed with either 0.5% trypsin–0.2% EDTA (Gibco, Scotland) or 85 mM sodium citrate (one-tenth of the culturing volume). One aliquot of this suspension was used for extraction and assay of adenine nucleotides and one aliquot for obtaining microscopic cell counts. Assays were also performed on the decanted medium directly or after sterile filtration. In Table II results from such an experiment is shown. Table II also includes data on the effects of collecting the cells by centrifugation, resuspending the cells in new medium for 5 min,

TABLE II
EFFECT OF DISPERSION AND CENTRIFUGATION ON ADENINE NUCLEOTIDE
LEVELS AND CELL NUMBERS[a]

Sample	Nucleotide conc. (μM)					Cells per μl	ATP per cell (fmol)
	ATP	ADP	AMP	ΣAXP	EC		
1. Whole culture	1.91	0.03	0.03	1.97	0.98	70[b]	27
2. Medium	0.17	0.01	0.07	0.25	0.72	7	26
3. Filtrated medium	0.003	0.002	0.05	0.06	0.07	0	—
4. Trypsinated culture	1.49	0.02	0.03	1.54	0.98	80	19
5. Na-citrated-treated culture	1.30	0.05	0.04	1.38	0.96	45	29
6. Supernatant from trypsinated culture	0.31	0.01	0.04	0.36	0.88	0	—
7. Supernatant from Na-citrate-treated culture	0.28	0.01	0.03	0.32	0.88	0	—
8. Trypsinated cells after centrifugation	1.10	0.03	0.05	1.18	0.94	93	12
9. Na-citrate-treated cells after centrifugation	0.92	0.02	0.04	0.98	0.95	54	17
10. Intact cells[c]	1.91	0.03	(−0.02)	1.94	0.99	70	27
11. Trypsinated cells[d]	1.18	0.04	0.00	1.22	0.98	80	15
12. Na-citrate-treated cells[e]	1.02	0.03	0.00	1.05	0.99	45	23

[a] Two culture flasks were extracted without pretreatment. In four culture flasks medium was decanted and filtered (0.2 μm). Cells from these flasks were dispersed with one-tenth of the culture volume of trypsin-EDTA or Na-citrate (two flasks each). Cells were then collected by centrifugation and resuspended in fresh medium for 5 min. In each step aliquots were taken for determination of adenine nucleotides and cell numbers. Values for samples obtained after dispersion of cells have been divided by 10 to correspond to the original culture volume. Calculations of quotients were based on values before rounding to estimated accuracy.

[b] Average from values in rows 4 and 5 plus cells in medium (row 2).

[c] Calculated from rows 1 and 3. Negative AMP value due to underestimation in row 1 or overestimation in row 3 (negative value estimated to 0.00 in calculations).

[d] Calculated from rows 4 and 6.

[e] Calculated from rows 5 and 7.

and then determining the adenine nucleotides in an extract of the resuspended cells. Results show the following:

1. Some cells are lost when decanting the medium (rows 1–2).
2. The filtrated medium contains very little ATP (row 3).

3. During treatment of cells with trypsin or Na-citrate some ATP is released into the medium, since ATP but no cells can be found in the supernatant (rows 6–7).

4. Collection of cells by centrifugation further decreases the ATP level (rows 8–9).

5. Comparing intact, trypsinated, and Na-citrate-treated cells (rows 10–12) shows that although part of the cells are lost during dispersion, the remaining cells have an unchanged energy status (EC 0.98–0.99).

6. In contrast to energy charge the values on ATP/cell are rather scattered as might be expected from the poor accuracy of microscopic cell counts.

It is concluded that physiological changes during sampling and pre-treatment of samples may cause cell membrane damage or other effects on the energy metabolism resulting in changes of intracellular adenine nucleotide levels. Furthermore cells may be lost or lysed during pretreatment. As shown above such effects are easily detected by assay of ATP, ADP, and AMP and should not cause problems to anyone who is aware of the effects. Collecting bacterial cells by centrifugation without resuspending them before extraction has been shown to decrease energy charge as well as the sum of all three adenine nucleotides considerably.[11] The experiment in Table II shows that at least with LNCaP-r cells resuspended in medium before extraction, collection of cells by centrifugation may be acceptable in the estimation of biomass by ATP assay. However, the need for concentrating the cells is questionable since the amount of ATP in a single cell (~20 fmol) can be detected by the firefly assay.

Extraction of Adenine Nucleotides

A reliable method for extraction of adenine nucleotides from cells should fulfill three requirements:

1. Complete release of intracellular adenine nucleotides from the cells.

2. Complete and irreversible inactivation of all adenine nucleotide-converting enzymes that may act on ATP, ADP, or AMP during extraction of samples, storage of extracts, or enzymatic assay.

3. No inhibition or inactivation of enzymes used in the assay of adenine nucleotides resulting in analytical interference.

An extractant fulfilling the second requirement can evidently not fulfill the third requirement unless extracts are highly diluted before the assay.

---

[11] A. Lundin and A. Thore, *Appl. Microbiol.* **30**, 713 (1975).

A large number of extractants have been suggested in the literature and several releasing reagents are commercially available that claim not to cause interference with the luciferase assay even in high concentrations. Since extraction is one of the most critical steps in the entire analytical procedure, it is recommended that several extractants of varying concentrations be compared before deciding on an extraction method. Such experiments have been performed in our laboratory with more than 16 different types of cells[12] using 5–7 different concentrations of TCA (trichloroacetic acid), PCA (perchloric acid), DTAB (dodecyltrimethylammonium bromide, Sigma Chemical Co., St. Louis, Mo), Triton X-100 (Scintillation Grade, Eastman Kodak Co., Rochester, NY), and saponin (Sigma Chemical Co.). For immediate inhibition of nucleotide-converting enzymes the last three extractants were supplied with EDTA in excess of divalent metal ions present in the culture. In essentially all these experiments the highest yields of ATP were obtained with 1.25–10% TCA (the best concentration dependent on type of cell). Since formation of ATP during extraction is unlikely, the high yield with TCA strongly indicates that this is the most reliable extractant. A high yield of ADP and AMP or even the sum of all three adenine nucleotides may be obtained with poor extractants, since ADP and AMP may be formed by degradation of other cell constituents.

Figure 1 shows results from an experiment of the type described above performed with LNCaP-r cells dispersed with Na-citrate and suspended in fresh medium. In the experiment (performed at room temperature) a 100-μl cell suspension was added to 900 μl of extractant. After approximately 30 min extracts were transferred to an ice bath until the assay was performed. Extracts were prepared in duplicate, the first one being assayed within a couple of hours and the second one within 24 hr. In some extracts nucleotide levels were somewhat different in the second extract indicating degradation caused by cellular enzymes not completely inactivated by the extractant or acid hydrolysis. Thus data given in Fig. 1 are based on the first extract. Results show that 2.3% TCA (final concentration) gives the highest yield of ATP as well as the sum of all three adenine nucleotides (ΣAXP). The somewhat lower yields obtained with the highest TCA concentrations are most likely due to coprecipitation of adenine nucleotides with protein (assays were performed on aliquots of the extracts).

In Table III results on the extractant concentration giving the highest yield of ΣAXP in the assay of the first extract have been collected for all

[12] A. Lundin, in "Analytical Applications of Bioluminescence and Chemiluminescence" (L. J. Kricka, P. E. Stanley, G. H. G. Thorpe, and T. P. Whitehead, eds.), p. 491. Academic Press, New York, 1984.

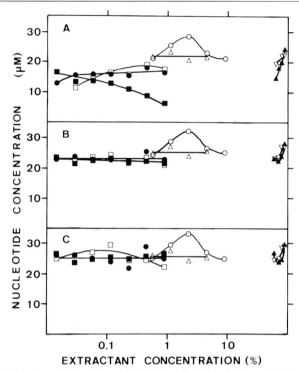

FIG. 1. Yields of ATP (A), ATP + ADP (B), and ATP + ADP + AMP (C) as obtained with different extractant concentrations. Extractants used were TCA (○), PCA (△), DTAB (□), Triton X-100 (●), saponin (■), dimethyl sulfoxide (▲), and ethanol (▽). A suspension in culture medium of LNCaP-r cells in logarithmic growth was used.

extractants. Extracts were not completely stable since energy charge as well as ΣAXP were lower in the second extract assayed the next day. This is in contrast to previous findings with TCA extracts of bacterial cultures kept in an ice bath showing degradations of ATP and ΣAXP less than 1% per day.[11] Results from Fig. 1 and Table III (first extract) can be summarized as follows:

1. The highest yields of ATP and ΣAXP were obtained with 2.3% TCA also giving the highest energy charge value.

2. Other extractants gave somewhat lower ΣAXP and with some extractants also a considerably lower energy charge.

3. Even low concentrations of mild extractants such as Triton X-100 released a major part of the adenine nucleotides as can be expected since LNCaP-r cells have fragile cell membranes.

TABLE III
EXTRACTION OF ADENINE NUCLEOTIDES WITH THE OPTIMAL CONCENTRATION OF
EACH EXTRACTANT[a]

Extractant	Extractant concentration (%)	First extraction			Second extraction		
		ATP (μM)	ΣAXP (μM)	EC	ATP (μM)	ΣAXP (μM)	EC
TCA	2.3	28.7	33.2	0.92	17.4	21.9	0.86
PCA	1.1	23.3	27.3	0.92	18.9	24.5	0.84
DTAB	0.11	16.5	29.5	0.69	8.9	26.7	0.52
Triton X-100	0.45	18.4	29.0	0.76	11.6	24.2	0.68
Saponin	0.9	6.6	26.6	0.53	2.7	23.0	0.33
DMSO[b]	90	24.2	29.7	0.88	16.9	21.8	0.84
Ethanol	90	22.5	29.1	0.86	20.8	27.5	0.84

[a] For each extractant the concentration resulting in the highest yield of AXP in the first extract (assayed within a couple of hours) was selected. Values for the second extract (stored in an ice bath and assayed the next day) are shown to illustrate the degradation of adenine nucleotides during storage.

[b] Dimethyl sulfoxide.

Previous experiments on extraction of bacterial cultures with various concentrations of TCA have shown that buffering capacity, salt concentration, and protein concentration may affect the extraction efficiency so that higher TCA concentrations were sometimes needed (A. Lundin, unpublished). Variations in sample composition with respect to these parameters could decrease the ATP yield dramatically at somewhat too low concentrations of the extractant. Therefore it was confirmed that our routine method for extraction (1 volume of 5% TCA to 2 volumes of culture) gave identical results as the method found to be best in the experiment described above (1 volume of culture in 9 volumes of 2.5% TCA).

In view of the data in Fig. 1 and Table III and the data previously reported,[11,12] it is recommended that any new extracting or releasing reagent be compared with 10, 5, 2.5, and 1.25% TCA (final concentrations in extract) selecting the method giving the highest yield of ATP and the highest energy charge. The selected method should also be studied with respect to time needed for complete extraction and stability of extracts. With *Escherichia coli* 5% TCA gave complete extraction of ATP within 5 min while 0.5% required more than 60 min. The 5% TCA extract was stable for 6 hr at room temperature and for 24 hr if kept in an ice bath (A. Lundin, unpublished).

Assay of Adenine Nucleotides

Principle. The ATP level in a sample can be continuously monitored by adding a suitable firefly luciferase reagent. Such a reagent should degrade only a minute percentage of the ATP per minute. Problems with product inhibition of luciferase can be taken care of by addition of pyrophosphate and a luciferin analog.[13] The commercially available firefly reagent used in the present study is of this type. When using this reagent for ATP assays the light emission is first measured after addition of extract and then after the subsequent addition of a known amount of ATP standard. From these two measurements the ATP concentration in the sample can be calculated by multiplying the ratio between first and second measurement by the concentration of the ATP standard and by the dilution factor.

The assay of ATP, ADP, and AMP with the above mentioned reagent can be performed in a single cuvette. This is done by end point conversion of ADP and AMP to ATP by the pyruvate and adenylate kinase reactions. At low concentrations of ATP and AMP the adenylate kinase reaction becomes extremely time consuming. This problem can be circumvented by adding an excess concentration of CTP to speed up the reaction. This can be done due to the poor specificity of the ATP site of adenylate kinase and the high specificity of firefly luciferase. The additions of the PK–PEP (pyruvate kinase–phosphoenolpyruvate) reagent and the AK–CTP (adenylate kinase–cytidine 5'-triphosphate) reagent may slightly affect the luciferase activity. To compensate for this effect the assay of a few blanks is performed with addition of ATP standard in each step. The ratios for luciferase activity obtained in this way are used in calculations of ATP, ADP, and AMP based on assays of samples in which the ATP standard has been added only after reaching the end point in the AK reaction.

$$AMP + CTP \rightarrow ADP + CDP \tag{1}$$
$$ADP + PEP \rightarrow ATP + pyruvate \tag{2}$$
$$ATP + \text{D-luciferin} + O_2 \rightarrow AMP + pyrophosphate + oxyluciferin + CO_2 + photon \tag{3}$$

Instrumentation. All instrumentation except computers were from LKB-Wallac (Turku, Finland). Manual assays were performed at room temperature using Luminometer 1250 supplied with Potentiometric Recorder 2210-032. Automatic assays were performed at 25° using Luminometer 1251 supplied with Potentiometric Recorder 2210-032 and four Dispensers 1291-001. A special device was needed for connecting more than

---

[13] A. Lundin, *in* "Luminescent Assays: Perspectives in Endocrinology and Clinical Chemistry" (M. Serio and M. Pazzagli, eds.), p. 29. Raven Press, New York, 1982.

three dispensers to the luminometer (for further information on how to obtain this unit as well as computer programs for running the luminometer automatically and for performing calculations contact the author, A.L.). Several personal computers can be connected on-line to the luminometer (for advice contact LKB-Wallac). In the present study a Galaxy 2 (Gemini Microcomputers Ltd., Bucks, England) and an OKI if800 model 20 (BMC America, Inc.) were used interchangeably.

Reagents. The lyophilized firefly reagent ATP Monitoring Reagent (LKB-Wallac, Turku, Finland) was reconstituted with 4 ml Tris-acetate buffer (cf. below) and 1 ml 1M potassium acetate and was kept at room temperature during the assays. The reagent could be kept for several days in a refrigerator. Tris-acetate buffer was prepared from 0.1 M tris(hydroxymethyl)aminomethane (E. Merck, Darmstadt, FRG) containing 2 mM EDTA (E. Merck) and adjusted to pH 7.75 with acetic acid. ATP Standard (LKB-Wallac) was reconstituted with 10 ml distilled water to give a final concentration of 10 μM ATP. The PK–PEP reagent was prepared by mixing equal volumes of 0.2 M phosphoenolpyruvate (Sigma Chemical Co.) and pyruvate kinase (10 mg/ml) in 50% glycerol (Boehringer Mannheim, FRG). The tri(cyclohexylammonium) salt of PEP was dissolved in distilled water and adjusted to pH 6.0 with acetic acid before being mixed with pyruvate kinase. Adenylate kinase (5 mg/ml) from rabbit muscle (Boehringer Mannheim) is prepared as a suspension in ammonium sulfate solution. Since ammonium sulfate is strongly inhibitory in the assay it was removed by collecting the enzyme by centrifugation, dissolving the pellet in an equal volume of 10-fold diluted Tris-acetate buffer, and dialyzing this solution overnight against the same buffer. Nine volumes of ammonium sulfate-free adenylate kinase preparation was mixed with 1 volume of 250 mM CTP (Sigma Chemical Co.) to obtain the AK–CTP reagent. Synthetic CTP gave the lowest blanks and was dissolved in 0.5 M tris(hydroxymethyl)aminomethane to obtain pH 6.8 before being diluted to 250 mM with distilled water.

ATP Assay Procedure. Manual and automatic assays of ATP were performed by essentially the same procedure and only the latter is described. After adding 20 μl extract to 900 μl Tris-acetate buffer, up to 25 cuvettes could be loaded into the temperature-controlled sample carousel. After a 10-min temperature equilibration time the following steps were performed automatically with each of the cuvettes: (1) addition of 100 μl of ATP Monitoring Reagent with mixing; measurement of light emission after 1 min; and (2) addition of 10 μl ATP Standard with mixing; measurement of light emission after 0.5 min. When all the data from the light measurements had been collected, ATP concentrations in the samples (with corrections for blanks) were calculated by a Basic Program on the personal computer. Blank values were less than 0.02 nM ATP in the cuvette.

ATP, ADP, and AMP Assay Procedure. This assay was always performed automatically. After addition of 20 μl extract to 900 μl Tris-acetate buffer, cuvettes were temperature equilibrated for 10 min in the sample carousel and the following steps were performed automatically:

1. Addition of 100 μl ATP Monitoring Reagent with mixing. Measurement of light emission after 1 min (corresponds to ATP level).
2. Addition of 10 μl PK–PEP reagent with mixing. Measurement of light emission after 1 min (corresponds to ATP + ADP concentration).
3. Addition of 10 μl AK–CTP reagent with mixing. Measurement of light emission after 5 min (corresponds to ATP + ADP + AMP concentration).
4. Addition of 10 μl ATP Standard with mixing. Measurement of light emission after 1 min.

In each series of assays a few blanks are run by a special assay procedure with addition of ATP Standard also after steps 1 and 2. Based on the relative increases of the light emission obtained at the addition of ATP Standard in these blanks and the ATP Standard measured in step 4 of all samples the personal computer calculated values of ATP, ADP, and AMP with subtraction of blanks.

In each series of assays it was confirmed that end points of the PK and AK reactions were attained. This was done by assaying a mixture of ATP, ADP, and AMP (0.05 μM of each in the cuvette). Since PEP and CTP are present in high concentrations and ADP and AMP in concentrations well below K_m values the times to reach the end points are constant and not influenced by the adenine nucleotide concentrations. This is in contrast to previously described assays based on conversion of ADP and AMP to ATP by the PK and AK reactions.[11,14] The linearity of the new assay extends from the detection limit up to 1 μM, i.e., the upper limit of the linear range of the luciferase reaction.

The sensitivity of the assay is limited by adenine nucleotide contaminations in the reagents. In particular the contamination of CTP with adenine nucleotides is a problem since the different batches of commercially available CTP contain different levels of adenine nucleotides. Typical blank values in the assay were 0.02 nM ATP, 0.3 nM ADP, and 10 nM AMP in the cuvette. We are presently trying to find a convenient method for removing the ~0.004% adenine nucleotide contamination of commercial CTP preparations.

[14] A. Lundin, A. Rickardsson, and A. Thore, *Anal. Biochem.* **75**, 611 (1976).

Comparison with Other Methods

It has been reported that the cellular ATP content in Ehrlich ascites tumor cells increases from 12 fmol at early G_1 phase to 28 fmol at $G_2 + M$ phases.[15] Thus the cellular ATP content may vary during growth depending on differences in either energy metabolism or intracellular volume.

In the experiment shown in Fig. 2 the cellular ATP content in LNCaP-r cells was ~15 fmol in the lag and stationary phases and slightly above 20 fmol in the logarithmic phase. The relatively low variation in cellular ATP content was also reflected as a good correlation between ATP and cell numbers obtained in several experiments with various additions to the medium as shown in Fig. 3.

Using a suspension of LNCaP-r cells and assaying ATP and cell numbers the coefficients of variation were determined to 5.5 and 11.9%, respectively.[2] In other similar experiments the variation in cell counts was even higher. LNCaP-r cells tend to aggregate under some conditions making it very difficult to estimate cell numbers microscopically. The ATP determination is not affected by this problem.

The linear range of the ATP assay is 0.02–1000 nM ATP (final concentration in 1 ml in cuvette) corresponding to 1–50,000 cells (20 fmol ATP/cell). With our routine procedure the culture is diluted 1.5 times in the extraction and 50 times in the assay. Thus the linear range with this procedure is 75–3,750,000 cells/ml in the cell suspension. If instead the medium is decanted, the cells extracted with a minimum volume of TCA (e.g., one-tenth of the culture medium) and 50 μl extract used in the assay the linear range would be 2–100,000 cells/ml in the original culture volume. The linear range of the determination of cell numbers in a 0.3 μl volume corresponding to 48 B-squares in the Bürker chamber is ~1–1000 cells or 3,333–3,333,333 cells/ml. Thus the detection limit in terms of number of cells is similar for both methods while the ATP method is several orders of magnitude more sensitive in terms of cell concentration (cells/ml), which is a more relevant measure. Furthermore the linear range of the ATP method is wider.

In a comparison between estimation of biomass by measurement of ATP and by microscopic cell counts the following statements can be made:

1. The ATP assay is more sensitive and has a wider linear range.
2. The ATP assay is more reproducible and in contrast to the microscopic method does not involve personal judgments.
3. The ATP assay is more convenient and can be automated. Fur-

[15] S. Skog, B. Tribukait, and G. Sundius, *Exp. Cell Res.* **141,** 23 (1982).

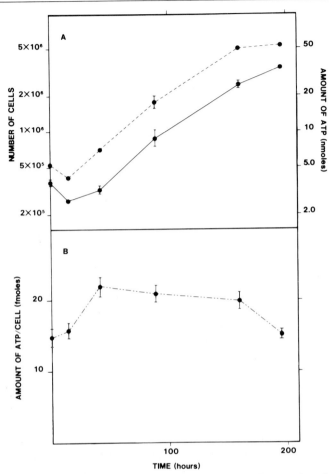

FIG. 2. Amount of ATP (●–––●) and number of cells (●——●) per culture flask (A) and amount of ATP per cell (B) during growth of the LNCaP-r cell line. Standard errors of means (SEM) from five determinations are indicated by vertical bars (SEM values less than 5% of the mean are not shown). From Hasenson et al.[2]

thermore cells do not have to be brought to a single cell suspension for the ATP assay, since cells may be extracted directly in the culture vessel.

4. Extracts for adenine nucleotide assays can in general be stored in an ice bath for a few hours or in the freezer for longer periods. Microscopic cell counts should be done as soon as possible, since dispersed cells will lyse.

5. The ATP method may be disturbed by changes in the physiological conditions and by extracellular ATP. The quantitative importance of these effects is easily estimated by assay of ATP, ADP, and AMP in cells

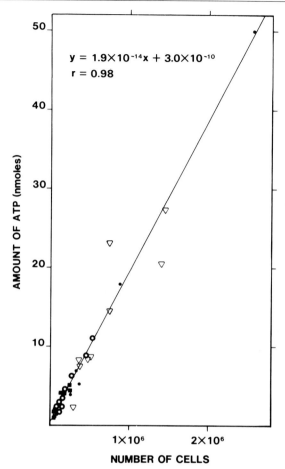

FIG. 3. Correlation between amount of ATP and number of cells. Results from four experiments (denoted by different symbols) using various steroid concentrations are shown. From Hasenson *et al.*[2]

and medium and calculations of energy charge. Microscopic estimation of cell numbers may be disturbed by cell aggregation. Cell aggregation is easily detected in the microscope but the quantitative importance may be difficult to estimate.

6. The intracellular ATP content depends on growth phase due to variations in cell volume or energy metabolism.

7. ATP is rapidly degraded when cells die while it is difficult to distinguish between dead and live cells in the microscope.

8. The ATP method is more expensive in terms of reagents and instrumentation while the microscopic method is more expensive in terms of labor cost.

In summary it may be said that ATP is a measure of the actively metabolizing biomass, while cell counts enumerate the number of structural units making up this biomass. Which of the two methods is preferred in a particular situation depends on the relative importance of the advantages and disadvantages given above. If a large number of growth conditions should be studied, it may be highly advantageous to use tissue culture plates for culturing the cells and automatic ATP assays for monitoring the growth. In this situation it is also worth while to study methods for sampling and pretreatment of samples, methods for extraction of adenine nucleotides, and to set up the assay of all three adenine nucleotides. If one has only a limited number of samples and is satisfied with the sensitivity and accuracy provided by microscopic cell counts there is no reason for changing.

Other methods for recording growth include measurement of DNA or protein either as increase of total concentration or as incorporation of radiolabeled precursors. For measurement of total concentration of DNA or protein ~10,000 cells are needed. The use of radioactive compounds is always limited by certain practical and legal restrictions. Measurement of specific marker proteins by enzymological or immunological methods can provide valuable information in certain situations. In general, however, these methods do not give accurate biomass estimations. The detection limit of radioimmunoassays is usually around a few micrograms of protein per liter of medium, which means that more than 1000 cells are needed. During the last decade experience with flow cytometry has shown that although this method is sensitive and specific at least 10,000 cells are needed. All the methods mentioned above are either more time consuming, require more pretreatment of samples, or are more expensive in terms of reagents or equipment as compared to the ATP method.

Conclusion

Estimation of biomass in growing cell lines by ATP assay is well suited for routine applications with large numbers of samples or when maximum sensitivity is required. Certain pitfalls that may appear, particularly in pretreatment of samples and in extraction of adenine nucleotides, are easily detected by determining the energy charge value using the described assay of ATP, ADP, and AMP.

Acknowledgment

This work was supported by a grant from the Swedish Medical Research Council.

[5] Sensitive Bioluminescent Assay for α-Bungarotoxin Binding Sites

By JAMES M. SCHAEFFER

α-Bungarotoxin (α-BTX) labeled with radioactive isotopes[1-6] or fluorescent dyes[7,8] has been used to study the localization and binding properties of nicotinic acetylcholine (nACh) receptors. Recently, our laboratory described the use of firefly luciferin-derivatized α-BTX as a nonradioactive probe for the study of nACh receptors in the cerebral cortex.[9] Firefly luciferase catalyzes a reaction with luciferin and ATP resulting in the emission of photons of light. In the presence of saturating concentrations of luciferase and ATP, the amount of light produced is proportional to the luciferin concentration. We have synthesized luciferin-derivatized α-BTX which retains its ability to specifically interact with the nicotinic binding site and also emit light in the presence of ATP and luciferase. α-BTX–luciferin provides a more sensitive probe for the study of nACh receptors than conventional ^{125}I-labeled α-BTX without the use of radiolabeled compounds. In this chapter I will describe the preparation, chromatographic purification, and characterization of α-BTX–luciferin for use in sensitive bioluminescent ligand receptor assays.

Conjugation of Luciferin to α-BTX

Luciferin was conjugated to α-BTX via a carbodiimide initiated reaction. α-Bungarotoxin (400 μg) was added to 25 mg of diisopropylcarbodiimide (Aldridge Chemical Co., Milwaukee, WI) in 250 μl of Krebs bicarbonate buffer (124 mM NaCl, 5 mM KCl, 26 mM NaHCO$_3$, 1.2 mM KH$_2$PO$_4$, and 1.3 mM MgSO$_4$), pH 7.4, containing 5% ethanol. After 15 min at 22°, 250 μg of luciferin (purchased from Sigma Chemical Co., St.

[1] C. Y. Lee and L. F. Tseng, *Toxicon* **3**, 281 (1966).

[2] J. C. Fertuck and M. Salpeter, *Proc. Natl. Acad. Sci. U.S.A.* **71**, 1376 (1974).

[3] A. J. Sytkowski, Z. Vogel, and M. W. Nirenberg, *Proc. Natl. Acad. Sci. U.S.A.* **70**, 270 (1973).

[4] R. J. Lukasiewicz, M. R. Hanley, and E. L. Bennett, *Biochemistry* **17**, 2308 (1978).

[5] S. L. Blanchard, U. Quast, K. Reed, T. Lee, M. I. Schimerlik, R. Valden, T. Claudio, C. D. Strader, H.-P. H. Moore, and M. A. Raftery, *Biochemistry* **18**, 1875 (1979).

[6] G.-K. Wang and J. Schmidt, *J. Biol. Chem.* **255**, 11156 (1980).

[7] J. B. Suszkin and M. Ichiki, *Anal. Biochem.* **73**, 109 (1976).

[8] P. Raudin and D. Axelrod, *Anal Biochem.* **80**, 585 (1977).

[9] J. M. Schaeffer and A. J. W. Hsueh, *J. Biol. Chem.* **259**, 2055 (1984).

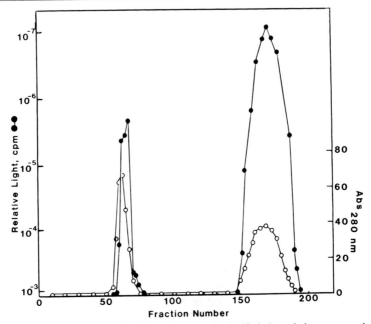

FIG. 1. Separation of luciferin and α-bungarotoxin–luciferin by gel chromatography. The α-bungarotoxin, luciferin, carbodiimide reaction mixture was placed onto a 1 × 120 cm Sephadex G-75 column and eluted with Krebs bicarbonate buffer. The eluate was collected in 1.2 ml fractions and the absorbance was measured at 280 nm (○). The amount of luciferin in each fraction (●) was estimated by measuring the release of light in the presence of excess luciferase and ATP as described in the text.

Louis, MO) dissolved in 250 µl distilled water was added to the α-BTX (Calbiochem, La Jolla, CA) and the incubation continued for 18 hr at 22°. The mixture was then placed on a Sephadex G-75 column (1 × 120 cm) and eluted with Krebs bicarbonate buffer. Optical densities and absorbance spectra were determined on a Gilford spectrophotometer. Two peaks were observed by measuring the absorbance at 280 nm (Fig. 1). The first peak (peak I) has an estimated molecular weight of 8000 and comigrates with α-BTX. The second peak (peak II) is greatly retarded by the Sephadex G-75 and comigrates with synthetic firefly luciferin. The luciferin activity of each fraction was quantitated by measuring the amount of light produced in the presence of excess luciferase and ATP (see below). Luciferin activity was present in both peaks I and II. Greater than 90% of the luciferin activity was recovered from the column with 6% present in peak I and the remainder in peak II. These results suggest that peak I contains α-BTX–luciferin and peak II is luciferin. Purified α-BTX–luciferin was aliquoted into Eppendorf tubes, frozen in dry ice, and stored at

−70°. The compound remains stable under these conditions for greater than 6 months.

Quantification of Luciferin Activity

Luciferin activity in individual samples was determined by measuring the amount of light produced in response to the addition of a saturating concentration of luciferase (200 μg/ml) and ATP (2 mM). Luciferin samples (50 μl) were placed into a conical Eppendorf tube and 10 μl of the luciferase–ATP solution was added. The Eppendorf tube was immediately placed into a glass scintillation vial and the luminesence quantitated with a Searle Scintillation spectrometer (0–1000 discriminator divisions for 0.5 min). Because of the rapid decay of light emission, it is critical that after the luciferase and ATP are added the vial is immediately placed into the scintillation counter (elapsed time 4 + 0.5 sec). The amount of light produced is then converted to moles of luciferin by use of a standard curve measuring light production as a function of luciferin concentration. Standard curves are determined by measuring the amount of light produced when a set concentration of luciferase is added to increasing concentrations of α-BTX–luciferin in a constant final volume.[9] The amount of light is plotted as a function of the α-BTX–luciferin concentration, and yields a straight line over at least three orders of magnitude.

Effect of Luciferase Concentration on Light Production

As shown in Fig. 2, in the presence of a fixed concentration of luciferase and ATP, light emission increases linearly with increasing concentrations of α-BTX–luciferin when plotted on a log–log scale. Conversely, if the concentrations of luciferin, ATP, and magnesium are maintained constant, increasing the luciferase concentration augments the light production until the concentration of luciferase exceeds 100 μg/ml, at which point no further increase in light production is observed (data not shown). Since the light production varies so greatly with small changes in the luciferase concentration at subsaturating concentrations of luciferase, we routinely use 200 μg/ml of luciferase for the quantitation of α-BTX–luciferin.

Fluorescent Properties of α-BTX–Luciferin

The fluorescence excitation and emission spectra for the material collected from peaks I and II were examined. The excitation spectra measured at an emission wavelength of 530 nm show maximal excitation at

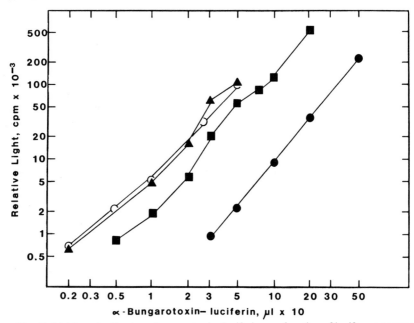

FIG. 2. Light production by α-bungarotoxin–luciferin as a function of luciferase concentration. Various concentrations of α-bungarotoxin–luciferin were added to a fixed concentration of luciferase and ATP and the amount of light production was quantitated. The concentrations of ATP and magnesium were 2 and 4 mM, respectively. The concentration of luciferase was 50 (●), 100 (■), 150 (▲), and 200 (○) μg/ml.

390 nm for both samples (Fig. 3A). Synthetic luciferin has the same pattern of excitation (data not shown). However, the emission spectra (Fig. 3B) is different for the material present in peaks I and II. At an excitation wavelength of 330 nm, the emission spectra of the material present in peak II has one maxima at 545 nm whereas the material in peak I has maxima at 440 and 545 nm. α-Bungarotoxin has no fluorescent properties. These results demonstrate that the conjugation of luciferin to α-BTX alters the fluorescent properties of luciferin.

Time Dependency of Light Production

As previously reported,[10–13] light emission from the luciferin–luciferase reaction is time dependent with a rapid decline in light within seconds of the initiation of the reaction. We measured the rate of decline

[10] D. M. Karl and O. Holm-Hansen, *Anal. Biochem.* **75**, 100 (1976).
[11] A. Lundin and A. Thore, *Anal. Biochem.* **66**, 47 (1975).
[12] W. D. McElroy, H. H. Seliger, and E. H. White, *Photochem. Photobiol.* **10**, 153 (1969).
[13] M. DeLuca and W. D. McElroy, *Biochemistry* **13**, 921 (1974).

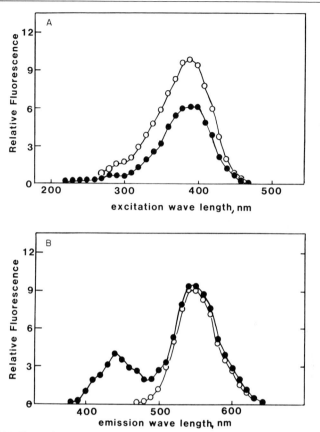

FIG. 3. (A) The excitation spectra of material from peak I (●) and peak II (○) measured at an emission wavelength of 530 nm. (B) Emission spectra of peak I (●) and peak II (○) measured at an excitation wavelength of 330 nm.

of light production of both luciferin and α-BTX–luciferin as a function of time (Fig. 4). Two concentrations of luciferin and α-BTX–luciferin were examined and their time course of ATP–luciferase-catalyzed light emission remained constant with a half-life of 54 sec. Our results suggest that conjugation of luciferin to α-BTX does not affect the time course of luciferin interaction with luciferase to produce light.

Specific α-BTX–Luciferin Binding to Retina and Skeletal Muscle Membranes

Rats were decapitated and the retinas immediately removed and placed in ice-cold Krebs bicarbonate buffer. The tissue was homogenized in 200 volumes of buffer with a glass–glass homogenizer. Muscle tissue

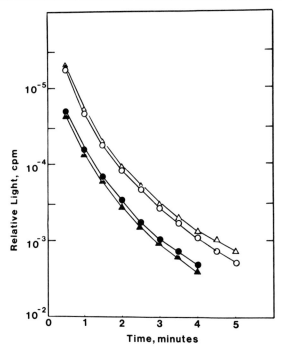

FIG. 4. Light emission by luciferin and α-BTX–luciferin as a function of time. α-Bungarotoxin–luciferin or luciferin was added to a saturating concentration of ATP, magnesium, and luciferase and the amount of light emitted was determined at various times after the initiation of the reaction. Two concentrations of α-bungarotoxin–luciferin [10^{-14} M (▲) and 10^{-13} M (△)] and luciferin [10^{-14} M (●) and 10^{-13} M (○)] were examined.

was minced thoroughly with a scissors before homogenization. Specific α-BTX–luciferin binding was determined by incubating the homogenate with α-BTX–luciferin in the presence (nonspecific binding) or absence (total binding) of a 500-fold molar excess of unlabeled α-BTX in a final volume of 400 μl for 45 min at 37°. The incubation was terminated and the membranes isolated by the addition of 1.5 ml of Krebs bicarbonate buffer and centrifugation for 3 min at 15,000 g. The membranes were washed a second time with 1.5 ml of buffer and the pellet was resuspended in 50 μl of Krebs buffer using a Kontes cell disrupter (Vineland, NJ). Luciferase, ATP, and magnesium were added to the resuspended membranes and the amount of α-BTX–luciferin was estimated as described above.

Specific α-BTX–luciferin binding to membranes prepared from rat retina and gastrocnemius muscle was saturable with increasing concentrations of α-BTX–luciferin (Fig. 5). The Scatchard analysis[14] of these data

[14] G. Scatchard, *Ann. N.Y. Acad. Sci.* **51**, 660 (1949).

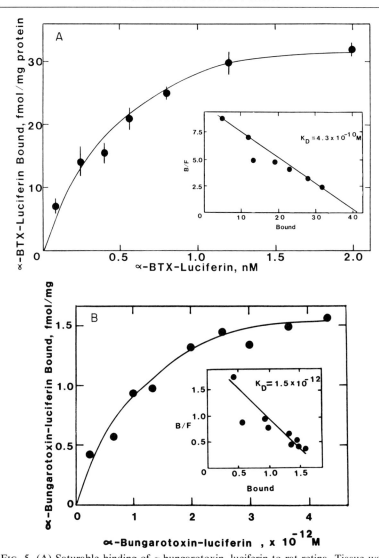

FIG. 5. (A) Saturable binding of α-bungarotoxin–luciferin to rat retina. Tissue was homogenized in Krebs bicarbonate buffer and aliquots were incubated with increasing concentrations of α-bungarotoxin–luciferin in the presence (nonspecific binding) or absence (total binding) of a 500-fold molar excess of α-bungarotoxin. Nonspecific binding was determined by subtracting nonspecific from total binding. Nonspecific binding is consistently less than 25% of the total binding. The amount of α-bungarotoxin–luciferin binding was quantitated using a standard curve as described in the text. Each point is the average of at least four determinations; replicate experiments gave similar results. A Scatchard analysis (inset) of the saturation curve data is shown. (B) Similar experiments were performed using membranes obtained from the rat gastrocnemeus muscle.

yielded a straight line in both sets of experiments, consistent with the existence of a single class of α-BTX–luciferin binding sites. In the muscle tissue, the binding site has an apparent dissociation constant of 1.5 $\times$ 10^{-12} M, whereas the apparent dissociation constant for binding to retinal tissue is 0.4 $\times$ 10^{-9} M. In addition, the number α-BTX–luciferin binding sites in the retina is much higher than in the muscle tissue (145 vs 2 fmol bound/mg protein, respectively). These data are consistent with reports of α-BTX binding sites using ^{125}I-labeled α-BTX as the radioligand.[15–18]

Comments

Firefly luciferin has been covalently bound to α-BTX for use in measurement of nicotinic acetylcholine receptors. α-BTX–luciferin contains approximately one molecule of active luciferin per α-BTX.[9] Consequently, the sensitivity of detection for α-BTX–luciferin is similar to that for luciferin (3 $\times$ 10^{-17} mol). ^{125}I-labeled α-BTX has a specific activity of 100–1000 Ci/mmol, providing a sensitivity of approximately 10^{-16} mol, therefore, the luciferin-derivatized probe is at least 30-fold greater than the standard ^{125}I-labeled α-BTX.

Luciferin has two reactive sites: a hydroxyl group at the 6′-position and a carboxyl group at the 4-position. Modification of the carboxyl group inactivates luciferin,[19] consequently, the active luciferin is presumed to be bound to α-BTX via the 6′-hydroxyl group. The most likely reaction is a condensation reaction of the 6′-hydroxyl group of luciferin with carboxyl groups of α-BTX, such as the γ-carboxyl group of glutamate and/or the β-carboxyl group of aspartate (there are 6 such functional groups in α-BTX).[6]

Due to the sensitivity of detection of α-BTX–luciferin, nicotinic receptors may be characterized in extremely small tissue samples. This is of particular importance in the study of nicotinic receptors in muscle tissue or discrete brain regions where the number of specific binding sites is too low to be easily measured using ^{125}I-labeled α-bungarotoxin.

Use of the luciferin–luciferase reaction for quantitative assays must consider several factors. Luciferase-catalyzed light emission is a time-dependent process, consequently the quantitation of light release after the initiation of the reaction (addition of luciferase) must be done precisely. In our protocol, the light is measured for 30 sec. The use of specialized

[15] G.-K. Wang and J. Schmidt, *Brain Res.* **114**, 524 (1976).
[16] A. Seto, Y. Arimatsu, and T. Amano, *J. Neurochem.* **37**, 210 (1981).
[17] D. K. Berg, R. B. Kelly, P. B. Sargent, P. Williamson, and Z. W. Hall, *Proc. Natl. Acad. Sci. U.S.A.* **69**, 147 (1972).
[18] B. J. Morley, G. E. Kemp, and P. Salvaterra, *Life Sci.* **24**, 859 (1979).
[19] E. H. White, H. Worther, and W. D. McElroy, *J. Org. Chem.* **30**, 2344 (1965).

equipment (such as commercially available luminometers) enables light to be quantitated at shorter time periods, resulting in greater sensitivity. Since the luciferin concentration is being estimated using an enzyme assay, it is imperative that the other factors (ATP and luciferase) be added in excess. This maximizes sensitivity and decreases the variability due to minute changes in the amount of luciferase or ATP added. This is demonstrated in Fig. 4, where at low luciferase concentrations the light emission is very sensitive to small incremental changes in luciferase concentration. Whereas at high luciferase concentrations (greater than 100 $\mu g/ml$), small changes in luciferase concentration have no effect on the amount of light production.

In addition to increased sensitivity, the use of luciferin-derivatized ligands has several other advantages over commonly used radiolabeling techniques. These include the elimination of hazardous radiolabeled compounds, increased speed of the assay, and decreased cost/assay. Because of the obvious advantages of the luciferin-derivatized compounds for use in ligand receptor assays (and luminescent immunoassays) much research will be done in this field over the next several years.

[6] Commercially Available Firefly Luciferase Reagents

By Franklin R. Leach and JoAnn J. Webster

Development of Commercial Reagents

Three factors have led to a greater utilization of bioluminescence methods for analysis in the 1980s. They are (1) the availability of commercially prepared reagents, (2) the commercial manufacture of suitable measuring instruments, and (3) the holding of conferences and the publishing of reports and monographs that extol the advantages of bioluminescence techniques. This article focuses on the commercial reagents; Stanley's article[1] in this volume examines the commercially available instruments. Three international symposia on Analytical Applications of Bioluminescence and Chemiluminescence,[2-4] two ATP methodology symposia,[5,6]

[1] P. E. Stanley, this volume [2].

[2] E. Schram and P. Stanley, eds., "Analytical Applications of Bioluminescence and Chemiluminescence." State Publishing and Printing, Westlake Village, California, 1979.

[3] M. A. DeLuca and W. D. McElroy, eds., "Bioluminescence and Chemiluminescence." Academic Press, New York, 1981.

METHODS IN ENZYMOLOGY, VOL. 133

this and the previous volume of *Methods in Enzymology* devoted to Bioluminescence and Chemiluminescence,[7] and a monograph[8] and a symposium[9] on clinical applications document many applications of bioluminescence techniques. In addition there are two recent reviews.[10,11]

Lowry and Passonneau[12] described a system of biochemical analysis for various metabolites based on spectrophotometric and fluorimetric determinations of pyridine nucleotides produced by various coupled reactions. A similar but much more sensitive system of analysis for ATP, pyridine nucleotides, and flavin nucleotides could be based on bioluminescence determinations using either firefly or bacterial luciferase. The increase in sensitivity would be on the order of 10^5 and 10^4 over the spectrophotometric and fluorimetric methods, respectively.[13] The collection, evaluation, and development of appropriate protocols for such a bioluminescence-based system of analysis would be a significant contribution to analytical biochemistry.

Some of the suppliers (DuPont, Schwartz, and Worthington—firefly only) of bioluminescent reagents during the 1960s no longer provide these reagents, but they along with Calbiochem and Sigma contributed to the acceptance of bioluminescence techniques. Many users during the 1960s purified their own luciferases. Most of the commercial preparations available were either firefly tails or lantern extracts.

Because of the presence of enzymes that converted other nucleoside triphosphates and both mono- and diphosphates to ATP in the crude

[4] L. J. Kricka, P. E. Stanley, G. H. G. Thorpe, and T. P. Whitehead, eds., "Analytical Applications of Bioluminescence and Chemiluminescence." Academic Press, New York, 1984.
[5] G. A. Borun, ed., "Proceedings of ATP Methodology Seminar." SAI Technology, San Diego, California, 1975.
[6] G. A. Borun, ed., "Proceedings of the 2nd Bi-annual ATP Methodology Symposium." SAI Technology, San Diego, California, 1977.
[7] M. A. DeLuca, ed., this series, Vol. 57.
[8] L. J. Kricka and T. J. N. Carter, eds., "Clinical and Biochemical Luminescence." Dekker, New York, 1982.
[9] M. Serio and M. Pazzagli, eds., "Luminescent Assays—Perspectives in Endocrinology and Clinical Chemistry." Raven Press, New York, 1982.
[10] F. R. Leach, *J. Appl. Biochem.* **3,** 473 (1981).
[11] W. D. McElroy and M. A. DeLuca, *J. Appl. Biochem.* **5,** 197 (1983).
[12] O. H. Lowry and J. V. Passonneau, "A Flexible System of Enzymatic Analysis." Academic Press, New York, 1972.
[13] F. R. Leach, *in* "Groundwater Pollution Microbiology" (G. Bitton and C. P. Gerba, eds.), p. 308. Wiley, New York, 1984.

lantern extracts, a rapid assay (peak height determination) was required.[14] Nielsen and Rasmussen[15] developed an abbreviated purification scheme that removed the endogenous ATP (and thus reduced the background) and interfering enzymes (adenylate kinase and nucleoside diphosphate kinase). Among the early (1977 or before) commercial suppliers of partially purified firefly luciferases were Calbiochem, DuPont, and Sigma. DeLuca articulated the need for commercial production of partially purified firefly luciferase.[16]

Suppliers of Commercial Bioluminescence-Based Reagents

Table I lists the suppliers of reagents for use with bioluminescence-based determinations. This article is based partially on replies to questionnaires sent to suppliers listed in *Chem Sources-U.S.A.*,[17] other known suppliers, and exhibitors at the 1985 FASEB meeting. There was also telephone follow-up. Our laboratory has compared many of the reagents. The availability of trouble-shooting or other technical information is noted, and appropriate telephone numbers are listed. The addresses are limited to those in the United States except for the major suppliers whose home offices are in another country.

Evaluation of Commercial Firefly Luciferase Reagents

In the middle to late 1970s several manufacturers started supplying partially purified and crystalline firefly luciferase preparations. This laboratory was beginning to use firefly luciferase to determine ATP and we wanted to select the best commercial preparation for our purpose. Since there was no comparative information available, we characterized several of these preparations.[18,19] Newer and modified ones have also been tested.[13] In the competitive market in supplying bioluminescence-related reagents, several manufacturers feel that their preparations are superior

[14] M. A. DeLuca, this series, Vol. 57, p. 3.

[15] R. Nielsen and H. N. Rasmussen, *Acta Chem. Scand.* **22,** 1757 (1968).

[16] M. A. DeLuca, *in* "Proceedings of ATP Methodology Seminar" (G. A. Borun, ed.), p. 1. SAI Technology, San Diego, California, 1975.

[17] "Chem Sources-U.S.A." Directories Publishing Company Inc., Ormond Beach, Florida, 1983.

[18] J. J. Webster, J. C. Chang, J. L. Howard, and F. R. Leach, *J. Appl. Biochem.* **1,** 471 (1979).

[19] J. J. Webster, J. C. Chang, J. L. Howard, and F. R. Leach, *in* "Bioluminescence and Chemiluminescence" (M. A. DeLuca and W. D. McElroy, eds.), p. 755. Academic Press, New York, 1981.

TABLE I
COMMERCIAL SUPPLIERS OF REAGENTS FOR BIOLUMINESCENCE-BASED ASSAYS

Abbreviation	Company, US address, phone number	Home address, phone number
AD	Aldrich Chemical 940 West St. Paul Ave. Milwaukee WI 53233 414 273-3850	
ALL	Analytical Luminescence Laboratory 11760 Sorrento Valley Rd. #E San Diego CA 92121 800 854-7050	
BM	Boehringer Mannheim Biochemicals 7941 Castleway Drive P.O. Box 50816 Indianapolis IN 46250 800 428-5433	6800 Mannheim West Germany 0621-7592876
CB	Calbiochem Biochemicals Behring Diagnostics P.O. Box 12087 San Diego CA 92112-4180 800 854-3417	
CD	Chemical Dynamics P.O. Box 395, Hadley Rd. South Plainfield NJ 07080 201 753-5000	
LKB	LKB Instruments 9319 Gaither Road Gaithersburg MD 20877 301 963-3200	LKB/Wallac P.O. Box 10 20101 Turku 10 Finland 921-678 111
L/3M	Lumac/3M Medical Products Div/3M 225-5S, 3M Center St. Paul MN 55101 800 328-1671	Lumac/3M bv Postbus 31101 6370 AC Schaesberg The Netherlands 31-45-318335
NBC	ICN Biochemicals P.O. Box 28050 Cleveland OH 44128 800 321-6842	
PICO	Packard Instrument Co. United Technologies 2200 Warrenville Rd. Downers Grove IL 60515 312 969-6000	
PP-L	Pharmacia P-L Biochemicals 800 Centennial Ave. Piscataway NJ 08854 800 558-0005	

TABLE I (*continued*)

Abbreviation	Company, US address, phone number	Home address, phone number
RE	Regis Chemical 8210 North Austin Ave. Morton Grove IL 60053 312 967-6000	
RP	Research Plus Labs P.O. Box 571 Denville NJ 07834 201 823-3592	
SC	Sigma Chemical P.O. Box 14508 St. Louis MO 63178 800 325-8070	
TD	Turner Designs 2247 Old Middlefield Way Mountain View CA 94043 415 965-9800	
USB	United States Biochemical P.O. Box 22400 Cleveland OH 44122 800 321-9322	

because of proprietary secrets used in their formulations. There are specific differences in composition and in the philosophy of reagent preparation. These differences are significant for certain experimental systems. We have ascertained some by direct experimentation. However, because of the intricacies involved, selection of a specific reagent should be made in one's own laboratory. General parameters for evaluating commercial preparations are presented below.

Components

Because of the properties of firefly luciferase, additives are required to stabilize the preparation and make it usable over a period of time. Other additives are required to optimize the reaction conditions.

Salts. Since firefly luciferase is a euglobulin[20] (insoluble in water), some salt must be present for the enzyme to be soluble. However, too much salt inhibits firefly luciferase activity.[21,22] Certain ions can change

[20] A. A. Green and W. D. McElroy, *Biochim. Biophys. Acta* **20,** 170 (1956).
[21] H. Holmsen, I. Holmsen, and A. Bernhardsen, *Anal. Biochem.* **17,** 456 (1966).
[22] J. L. Denberg and W. D. McElroy, *Arch. Biochem. Biophys.* **141,** 668 (1970).

the wavelength of the emitted light[23] so care must be taken to exclude them (primarily in the sample).

Arsenate. Strehler and Totter[24] used an arsenate buffer to obtain a fairly constant output of light. They[25] found that two factors were required for constant light production: (1) high magnesium concentrations and (2) high phosphate buffer concentrations. Since these reacted and magnesium phosphate precipitated, they used arsenate instead of phosphate. Both sulfate and arsenate diminish light production,[26] but yield systems with a more constant production of light with higher concentrations of ATP than reaction mixtures without a high salt concentration. Some reagents, particularly crude firefly lantern extracts, still contain arsenate that only reduces the sensitivity of the ATP determination. With modern instrumentation and faster recorders the slower light production by an arsenate-inhibited system is not needed. Therefore, the complicating factor of arsenate should be completely eliminated from all preparations. Then the *very poisonous* label could be removed from certain of the preparations and greater sensitivity would result. Since firefly luciferase has two ATP sites with different K_m values, the kinetics differ depending on the ATP concentration. High ATP gives a flash and decay, but low levels of ATP produce a constant and continuous low light emission, even without arsenate.[27]

Buffer. The pH optimum of firefly luciferase is 7.8. The buffers most often used are Tris,[27a] HEPES, glycylglycine, MOPS, Tricine, phosphate, and arsenate (in spite of its inhibiting effect). The enzyme is more stable in phosphate buffer than in Tricine buffer but is less active.

[23] H. H. Seliger and W. D. McElroy, *Proc. Natl. Acad. Sci. U.S.A.* **52,** 75 (1964).

[24] B. Strehler and J. Totter, *Arch. Biochem. Biophys.* **40,** 28 (1952).

[25] J. Totter, *in* "Bioluminescence and Chemiluminescence" (M. A. DeLuca and W. D. McElroy, eds.), p. 218. Academic Press, New York, 1981.

[26] M. DeLuca, J. Wannlund, and W. D. McElroy, *Anal. Biochem.* **95,** 194 (1979).

[27] M. DeLuca and W. D. McElroy, *Biochem. Biophys. Res. Commun.* **123,** 764 (1984).

[27a] Abbreviations: BSA, bovine serum albumin; DIPSO, 2,3-[N-bis(2-hydroxyethyl)amino]-2-hydroxypropanesulfonic acid; DTT, dithiothreitol; EDTA, ethylenediaminetetraacetic acid; Gly, glycine; Glgly, glycylglycine; G1P, glucose 1-phosphate; G6P, glucose 6-phosphate; G1,6-P_2, glucose 1,6-diphosphate; G6PD, glucose 6-phosphate dehydrogenase; HEPES, N-(2-hydroxyethyl)piperazine-N'-2-ethanesulfonic acid; HEPPSO, N-(2-hydroxyethyl)piperazine-N'-2-hydroxypropanesulfonic acid; LH_2, firefly D-luciferin; Lu, firefly luciferase; MOPS, 3-(morpholino)propanesulfonic acid; MOPSO, 3-(morpholino)-2-hydroxypropanesulfonic acid; PGM, phosphoglucomutase; POPSO, piperazine-N,N'-bis(2-hydroxy)-1-propanesulfonic acid; PP$_i$, inorganic pyrophosphate; TAPS, 3-[tris(hydroxymethyl)methylamino]-1-propanesulfonic acid; TAPSO, 3-[N-(Trishydroxymethyl)methylamino]-2-hydroxypropanesulfonic acid; Tricine, N-[tris(hydroxymethyl)methyl]glycine; Tris, tris(hydroxymethyl)aminomethane; UDPG, uridine 5'-diphosphoglucose; UPDGpp, uridine 5'-diphosphoglucose pyrophosphorylase; 6PG, 6-phosphogluconate.

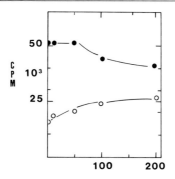

FIG. 1. Effect of bovine serum albumin concentration on firefly luciferase activity. Du-Pont firefly luciferase, ○, which already contained BSA was supplemented with additional BSA. The reagent was aged overnight after preparation. The light production was measured using the JRB Model 3000 photometer; a 1-ml reaction volume with 10 ng of ATP and Tricine buffer at pH 7.4 was used. Quadruplicate samples were assayed. For the Sigma crystalline firefly luciferase, ●, 50 μg of luciferin was added to the reaction mixture. The indicated amounts of BSA were used. Tricine buffer, pH of 7.6, and 1 ng of ATP were used. Assays were done in triplicate.

Bovine Serum Albumin. Firefly luciferase in dilute solutions is unstable because it is adsorbed to surfaces[28] and inactivated. Firefly luciferase is stabilized by bovine serum albumin (see Lundin[29]). Most partially purified commercial preparations contain considerable added bovine serum albumin. Figure 1 shows the effect of bovine serum albumin on both a crystalline firefly luciferase and a partially purified preparation that already contains bovine serum albumin. Too large an excess of bovine serum albumin inhibits firefly luciferase.

Metal Ions. Mg^{2+} is required for the firefly luciferase reaction since it combines with ATP to form the actual substrate MgATP. The metal ions are normally supplied with the luciferin–luciferase preparation. EDTA[20] is added to most commercial preparations and is used in most laboratory preparative procedures. EDTA prevents inhibition by metal ions that shift the wavelength of emitted light.

Luciferin. The crude lantern extracts contain D-luciferin, but light production can be increased with additional luciferin.[30] We have found that most partially purified firefly luciferase preparations can be improved in performance by addition of luciferin.[18] The major exception to this observation is the LKB preparation, probably due to the presence of L-

[28] C. H. Suelter and M. DeLuca, *Anal. Biochem.* **135,** 112 (1983).
[29] A. Lundin, this series, Vol. 57, p. 56.
[30] D. M. Karl and O. Holm-Hansen, *Anal. Biochem.* **75,** 100 (1976).

luciferin in LKB's preparation.[31] Additional luciferin seems appropriate for most preparations if low levels of ATP are being measured. Since the commercial preparations contain both firefly luciferase and luciferin, any dilution of the preparation to conserve reagents will decrease both the enzyme and one substrate's concentration; therefore, the light production will be linearly related to the square of the volume of the preparation used.[32] The luciferin content of a preparation can be estimated by measuring the absorbance at 325 nm.[33]

Sulfhydryl Compounds. Firefly luciferase contains two essential —SH groups.[34] Most commercial preparations contain an —SH-protective compound in concentration sufficient to prevent most complications. The mercapto group content of a preparation can be measured by the *p*-chloromercuribenzoate procedure of Boyer.[35]

Methods Used to Evaluate Firefly Luciferase Preparations

The methods used for the determination of —SH content and luciferin are cited above.

Protein Content. The protein content of the preparation may be important with respect to stability. Protein can be estimated by absorbance measurement (an absorbance of 0.75 at 280 nm in a 1-cm cell represents 1 mg/ml protein[14]). Of course other methods may be used to estimate the protein content. Because of the proprietary nature of many of the additives and their possible effect on protein estimation procedures, several different protein estimations may be required. We have employed the Lowry *et al.*[36] procedure, fluorescence measurements using *o*-phthalaldehyde,[37] and dye-binding using either bromosulfalein[38] or Coomassie Blue.[39] Usually the amount of bovine serum albumin added to the partially purified preparations far exceeds the content of firefly luciferase.

Activity. Two methods are used to determine firefly luciferase activity: (1) direct determination by measuring light production from a known amount of ATP under standard controlled conditions, and (2) measurement of the production of pyrophosphate via coupled enzyme reactions.

[31] A. Lundin, B. Jaderlund, and T. Lovgren, *Clin. Chem. (Winston-Salem, N.C.)* **28,** 609 (1982).
[32] J. J. Webster and F. R. Leach, *J. Appl. Biochem.* **2,** 469 (1980).
[33] R. A. Morton, T. A. Hopkins, and H. H. Seliger, *Biochemistry* **8,** 1598 (1969).
[34] M. DeLuca, G. W. Wirtz, and W. D. McElroy, *Biochemistry* **3,** 935 (1964).
[35] P. D. Boyer, *J. Am. Chem. Soc.* **76,** 4331 (1954).
[36] O. H. Lowry, N. J. Rosenbrough, A. L. Farr, and R. J. Randall, *J. Biol. Chem.* **193,** 265 (1951).
[37] K. Kutchai and L. M. Gaddis, *Anal. Biochem.* **77,** 315 (1977).
[38] J. McGuire, P. Taylor, and L. A. Greene, *Anal. Biochem.* **83,** 75 (1977).
[39] G. S. McKnight, *Anal. Biochem.* **78,** 86 (1977).

TABLE II
COMPONENTS OF THE BASIC REACTION MIX

Sigma catalog number	Compound	Concentration	Parts in cocktail
G58855	G6PD	11,000 U/ml	1 : 500
N0505	NADP	11 mg/ml	1 : 25
G5750	G1,6-P_2	1 mg/ml	1 : 100
P3397	PGM	500 U/ml	1 : 600
A5394	ATP	15 mg/ml	1 : 20
L9504	Luciferin	0.5 mg/ml	1 : 10
U4625	UDPG	9.6 mg/ml	1 : 20
P1754	Tween 80	10% solution	1 : 100
P9625	Phenazine methosulfate	0.5 mg in 1.25 ml	1 : 100
T4000	Tetranitroblue tetrazolium	0.45 mg/ml	1 : 10
A6756	Sodium arsenate	0.25 M	1 : 100
P8010	Sodium pyrophosphate	Add 3–10 nmol	
U8501	UDPG pyrophosphorylase	100 U/ml	
Luciferase	4 mg/ml	Use 10 μg/assay	

| | Cuvette | |
Component	Blank	Sample
Cocktail	0.5 ml	0.5 ml
		Read A_{550}
Water	3 μl	None
UDPG pyrophosphorylase	None	3 μl
PP$_i$	3–10 nmol	3–10 nmol
		Read A_{550}
Luciferase	3–15 μl	3–15 μl
		Read A_{550}

The system that this laboratory[31] uses for firefly luciferase activity determination by light production is 1 ml containing 1 ng ATP, 25 mM Tricine, pH 7.8, 5 mM MgSO$_4$, 0.5 mM EDTA, 0.5 mM dithiothreitol, 50 μg luciferin, and 100 μg bovine serum albumin.

An alternative assay used by several commercial suppliers (Boehringer Mannheim, Calbiochem, and Sigma) is based on determining the pyrophosphate produced. The following description is based on a personal communication from Sigma Chemical and describes the procedure that they have developed.

The basic reaction mix contains 0.05 M Tris, pH 7.3, 0.01 M MgSO$_4$, 0.001 M EDTA, and the components designated by the Sigma catalog numbers (the reagents from the other companies could be used) shown in Table II.

The molar absorption coefficient of reduced tetranitroblue tetrazolium is 33,000. The reaction sequence is as follows:

$$ATP + LH_2 \xrightarrow{Lu} AMP + PP_i + OL + light \tag{1}$$

$$PP_i + UDPG \xrightarrow{UDPGpp} G1P + UTP \tag{2}$$

$$G1P \xrightarrow[G1,6P_2]{PGM} G6P \tag{3}$$

$$G6P + NADP^+ \xrightarrow{G6PD} NADPH + H^+ + 6PG \tag{4}$$

$$NADPH + H^+ + tetranitroblue\ tetrazolium \rightarrow reduced\ form + NADP^+ \tag{5}$$

Because of the variations of the detector positions, individual spectral response of various photomultiplier tubes, and geometry of the individual photometers, there is no standard method of measuring light production. A secondary standard such as luminol can be used to calibrate the individual laboratory instruments. This is not worthwhile for most biochemical applications of bioluminescence. A standard curve is sufficient, with corrections being made for sample-contributed effects by using an internal standard.

Standard ATP Response Curve. Endogenous ATP in the luciferin–luciferase preparation or other reagents can produce light without the addition of ATP. This background or inherent light production will influence the sensitivity of the determination. By incubation of the reconstituted firefly luciferase preparation prior to use, the background production of light can be reduced. Most preparations yield a usable range of three to four orders of magnitude of ATP concentration (this is instrument dependent). The standard curves are plotted on log–log paper and are reproducible to $\pm8\%$. The relative positions of the curves of individual preparations and the position of a single luciferase–luciferin preparation can be changed by changing the reaction conditions—particularly the amounts of luciferin and luciferase (see Fig. 3).

Other materials present in the solution or sample can reduce the measured light production by either adsorption of a portion of the light or by scattering of the light away from the detector. Eliminate any quenching materials or use an internal standard for correction.

An indication of the specificity of the firefly luciferase preparation can be obtained by incubation with 4.9 μM ADP for adenylate kinase and 1.72 μM GTP and 4.4 μM ADP for nucleoside diphosphate kinase. Correct for ATP in other nucleotides.

Stability of Reagents. We find that several of the commercial luciferase preparations can be stored either refrigerated or frozen and used for at least a week with minimal loss of activity after reconstitution. When they are stored frozen, activity remains longer than in the refrigerator. For example, samples stored frozen are usable for at least a month. The most important factors in longevity of luciferase are the presence of the stabilizers mentioned above, the use of pure water, and sterile conditions. The sterile condition is extremely important. The water used in this laboratory is purified by reverse osmosis treatment, use of two ion-exchange columns, glass distillation into a sterile container, filter sterilization by membrane filtration (0.45-μm pore size filter), and finally autoclaving. The sterile material should not contain any ATP. We find the product expiration dates provided by the suppliers conservative. Luciferin is stable for months if stored refrigerated or frozen, in the dark, and under nitrogen. ATP is stable for years if stored in a sterile solution, preferably containing a buffer (some labs add EDTA), at a concentration of 10 μg/ml or greater.

Content of Other Contaminating Enzymes. The activity of contaminating enzymes can be determined by the appropriate assay (adenylate kinase,[17] nucleoside diphosphate kinase,[17] and pyrophosphatase[40]).

Desirable Properties for Commercial Luciferases

Lundin[41,42] has listed several requirements for a commercial firefly luciferase preparation (reagent).

1. There should be no degradation or production of ATP by other enzymes.

2. There should be no accumulation of inhibitors or appreciable concentration of the enzyme–product complex during the time of measurement of ATP concentration.

3. The reaction conditions should be controlled to those which are optimum: pH 7.8, 25°, ionic environment, and proper concentrations of the various components. Some of these are often provided in the formulation of the reagent.

4. A reagent should contain all the components of the assay except ATP. This should include the various stabilizers such as EDTA, bovine serum albumin, and an —SH-protective compound. The reagent should easily dissolve yielding a clear solution, have a low inherent light production, have a saturating luciferin concentration, and contain components stable in solution. The dried material or reagent should have a long shelf life (years) and after being dissolved should be usable for a month with proper storage. Bacteria and free ATP should have been eliminated and the quality of water used should be the best possible. Azide is often added to prevent bacterial growth.

5. Light production should be proportional to ATP concentration over at least four orders of magnitude. The light production should be constant

[40] O. H. Lowry, this series, Vol. 4, p. 366.
[41] A. Lundin, *in* "Clinical and Biochemical Luminescence" (L. J. Kricka and T. J. N. Carter, eds.), p. 43. Dekker, New York, 1982.
[42] A. Lundin, *in* "Luminescent Assays" (M. Serio and M. Pazzagli, eds.), p. 29. Raven Press, New York, 1982.

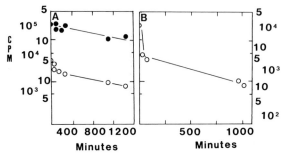

FIG. 2. Aging of firefly luciferase preparations after reconstitution. (A) Sigma FLE-50 firefly luciferase was supplemented with 250 μg of firefly luciferin per ml and incubated at room temperature and 20-μl samples were taken at the indicated times. The inherent light production, $\bigcirc$, was determined with no ATP addition and the response to 1 ng of ATP, $\bullet$, was also determined with the JRB Model 3000 photometer. (B) Boehringer Mannheim firefly luciferase was dissolved in Tricine buffer supplemented with BSA to give a final concentration of 80 μg/ml. This firefly luciferase preparation and luciferin were combined and incubated at 4°. The inherent light production was determined at the indicated times in the Pico-Lite Model 6100 photometer with a 1-sec delay and 30-sec assay time.

with time (at least with lower ATP concentrations) and the mode of mixing should not be critical. The consumption of ATP should be minimal during the measurement.

6. The reagent should be produced under strict quality control and the quality analysis data should be supplied to the purchaser of the reagent. Information concerning applications and trouble-shooting help should be available.

Using Commercial Firefly Luciferase Reagents

Aging of Preparations

Any ATP present in the reagents or on the glassware used produces inherent light (background). Another background source is light activation of either the glass containers or the firefly luciferin. Figure 2 shows the decrease in the inherent light production with aging. Figure 2A shows data with a Sigma FLE-50 preparation supplemented with additional luciferin to increase sensitivity. About 50% of the inherent light production is gone after 3 hr of incubation. Figure 2B shows results with reconstituted Boehringer Mannheim crystalline firefly luciferase and luciferin. Again a period of 3 hr gives an acceptable reduction of the inherent light production. For convenience we often age the preparations overnight at 4°.

Ratio of Firefly Luciferase and Luciferin

The production of light from a given amount of ATP can be changed by changing the ratio of firefly luciferase to luciferin concentrations. Fig-

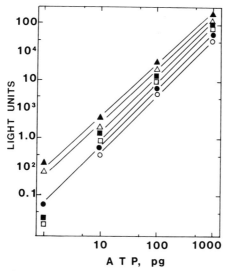

FIG. 3. Variation of light production with concentration of firefly luciferase and firefly luciferin. Crystalline Sigma brand luciferase (L 5256) and D-luciferin (L 6882) were used. The total reaction volume was 300 μl and Tricine buffer, pH 7.8, was used with the complete reaction mixture described in the text. Light production was determined on a Lumac/3M Biocounter M2010A operated in the manual mode with a 10-sec assay. (○) 1 μg firefly luciferase and 10 μg luciferin, (●) 1 μg firefly luciferase and 25 μg luciferin, (□) 2 μg firefly luciferase and 10 μg luciferin, (■) 2 μg firefly luciferase and 25 μg luciferin, (△) 4 μg firefly luciferase and 10 μg luciferin, (▲) 4 μg firefly luciferase and 25 μg luciferin.

ure 3 illustrates this point. An increase in either of the components results in an increased production of light. The relative positions of the lines on the graph can be changed by changing either the reaction volume or the amount of reagent used.

Effect of Reaction Conditions

The Reaction Vessel and Reaction Volume. Because of the different positions of the phototubes and arrangements of the light-gathering systems in the several instruments, there are dissimilar effects of variations in reaction volumes. Plastic containers are preferable because they are less susceptible to light activation. A smaller volume also allows conservation of reagents, but salt inhibition is reduced by dilution. These factors must be balanced for individual experimental conditions.

Optimization. Given the ranges of concentrations and conditions used for assays involving firefly luciferase and ATP, each laboratory may need to optimize the reaction conditions for its own experimental conditions (specific samples). We[31] have reported a general optimization for our

TABLE III
EFFECT OF BUFFERS ON FIREFLY LUCIFERASE[a]

Buffer	pK_a	Activity relative to Tricine
MOPSO	6.95	0.69
HEPES	7.55	0.92
DIPSO	7.60	0.85
TAPSO	7.70	0.92
POPSO	7.85	0.48
HEPPSO	7.90	0.85
Tricine	8.15	1.00
Glygly	8.20	0.82
TAPS	8.40	0.44

[a] Sigma crystalline firefly luciferase (catalog number L-5256) was dissolved in ammonium sulfate and diluted in the indicated buffer (0.05 M, pH 7.8) containing $MgSO_4$, DTT, BSA, and EDTA. The final buffer concentration (assay) was 0.037 mM for each buffer. $MgSO_4$ concentration was 6.67 mM, and the DTT and EDTA concentrations were both 1.33 mM. Firefly luciferase was used at a concentration of 2 μg per reaction mixture of 200 μl and that of luciferin was 50 μg. The assays were done on a Lumac/3M Biocounter Model M2010A with a 10-sec count in the manual operation mode. Values for three different ATP concentrations each assayed in triplicate on 2 days were averaged and expressed relative to the values obtained with Tricine.

laboratory that represents a starting point for new sets of experimental conditions.

Buffer. We[43] observed that different buffers yield different conformations of firefly luciferase with different activities. In phosphate buffer there was 12% β-pleated sheet; in Tricine buffer there was 43% β-pleated sheet conformation. Since several new zwitterionic buffers[44] with pK_a values close to the pH optimum for firefly luciferase are now commercially available, we determined their effect on the activity of this luciferase. Table III shows that none of these buffers is better than the commonly used HEPES, glycylglycine, and Tricine.

[43] J. J. Webster, J. C. Chang, E. R. Manley, H. O. Spivey, and F. R. Leach, *Anal. Biochem.* **106,** 7 (1980).
[44] W. J. Ferguson, K. I. Braunschweiger, W. R. Braunschweiger, J. R. Smith, J. J. McCormick, C. C. Wasmann, N. P. Jarvis, D. H. Bell, and N. E. Good, *Anal. Biochem.* **104,** 300 (1980).

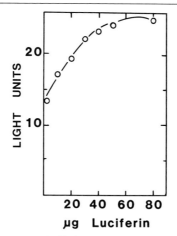

FIG. 4. Effect of supplementary luciferin on a commercial firefly luciferase preparation. A Lumit PM firefly luciferase preparation was used in a total reaction volume of 300 μl. The luciferin supplementation was with Calbiochem (438477) luciferin. The ATP concentration was 0.5 ng per reaction mix. Light production was measured on a Lumac/3M Biocounter M2010A operated in the automatic mode with a 10-sec assay.

Additional Luciferin. As has been mentioned, the sensitivity of most firefly luciferase preparations can be increased by the addition of more luciferin. Most preparations are not saturated with luciferin—in part because of its cost and also because additional luciferin is not needed for the accurate determination of most concentrations of ATP. Figure 4 shows the response of a typical commercial firefly luciferase preparation to addition of luciferin.

Commercial Firefly Luciferase Reagents

Firefly Luciferase Preparations

Table IV summarizes the information provided by the manufacturers. The firefly luciferase preparations are divided into four classes: (1) firefly tails or lanterns, which can be used as starting material for making laboratory preparations, (2) firefly lantern extracts, which represent crude extracts commercially prepared from tails, (3) purified firefly luciferase reagents, which are made in at least two different ways (partial purification of firefly luciferase and compounding of a reagent with other proteins and materials using crystalline firefly luciferase), and (4) crystalline firefly luciferase.

Commercial Sources of D-Luciferin (Firefly)

The commercial sources of firefly luciferin are listed in Table V. As already noted, supplementation with additional luciferin can often yield

TABLE IV
COMMERCIALLY AVAILABLE FIREFLY LUCIFERASE PREPARATIONS[a]

Company	Reagent name	Catalog number	Sizes supplied	Buffer content
Firefly tails (lanterns)				
NBC	Firefly lanterns, dried	100655	1 g	
LKB	Firefly lanterns	1243-228	0.5 g	
SC	Fireflies, desiccated tails	FFT	250, 500, 1000 mg	
Firefly lantern extracts (crude extracts)				
CB	Firefly lantern extract	341862	50-mg vials	5 ml each, 20 mM MgSO$_4$, 50 mM K arsenate
CD	Firefly lantern extract	44-5340-00	50-mg vials	5 ml each, 20 mM MgSO$_4$, 50 mM K arsenate, pH 7.4
LKB	Crude firefly luciferase	1243-229	From 250-mg lanterns	In ammonium sulfate
NBC	Firefly lantern extract	100654	50-mg vials	5 ml each, 20 mM MgSO$_4$, 50 mM K arsenate, pH 7.4
PP-L	Firefly lantern extract	27-0185-xx	50-mg vials	5 ml each, 20 mM MgSO$_4$, 50 mM K arsenate, pH 7.4
RP	Firefly lantern extract		50-mg vials	
SC	Firefly lantern extract	FLE-50	50, 250, or 1000 mg	Variable 20 mM MgSO$_4$, 50 mM K arsenate, pH 7.4
SC	Firefly lantern extract	F 4130	5, 15, 50 mg	Gly buffer with added luciferin
SC	Firefly lantern extract	F 4380		MOPS buffer
SC	Firefly lantern extract	F 5630	5 mg	Gly buffer with luciferin and mannitol
SC	Firefly lantern extract	F 5505	5 mg	Gly buffer with luciferin and human albumin
SC	Firefly lantern extract	F 4005	2, 5, 15, 50 mg	Gly buffer, no luciferin
SC	Firefly lantern extract	F 4255	5, 15, 50 mg	MOPS, no luciferin
USB	Firefly lantern extract	15775	50-mg vials	5 ml each, 20 mM MgSO$_4$, 50 mM K arsenate, pH 7.4

Purified firefly luciferases and/or prepared reagents (crystalline luciferase used with other proteins)

				Purpose
ALL	Firelight	2005, 2010	50 or 100 tests	ATP bioluminescence
BM	ATP bioluminescence CLS	567736	150 tests	Constant light signal
BM	ATP bioluminescence HS	567701	150 tests	Use with single photon counters
CB	ATP assay kit	119108	200 tests	ATP measurement
LKB	ATP monitoring reagent	1243-200	250 tests	Measure ATP
LKB	ATP assay kit	1243-107	250 tests	10^{-11} to 10^{-6} M ATP
LKB	ATP monitoring kit	1243-102	1250 tests	10^{-6} to 10^{-11} M ATP
L/3M	Lumit-PM	9225	70 tests	Use with standard photometers
L/3M	Lumit-HS	9255	50 tests	Use with photon counters
PICO	Picozyme F	6016710	50 tests	10^{-12} M ATP
SC	Luciferase–luciferin	L 0633	50 tests	ATP determination; different buffers available
TD	Luciferin–luciferase	20-2102	200 tests	ATP measurement

Crystalline firefly luciferase

				Notes
BM	Luciferase (firefly)	411523, 634409	1 and 5 mg, respectively	1 mg gives 5000 determinations
CB	Luciferase, *Photinus*	438449	1 mg	
LKB	Luciferase, *Photinus*	1233-214	1 mg	
SC	Luciferase (firefly)	L 5256	1, 2, 10, 25, mg	Crystallized and lyophilized
SC	Luciferase (firefly)	L 1759	1 mg	Lyophilized with buffer (Tris-aspartate), soluble
SC	Luciferase (firefly)	L 9009	1, 5 mg	Lyophilized with Tris-succinate, soluble

[a] Prepared from information supplied by the manufacturers.

TABLE V
COMMERCIAL SOURCES OF D-LUCIFERIN (FIREFLY)[a]

Company	Catalog number	Sizes supplied	Notes
BM	411 400 and 739 413	10 and 50 mg, respectively	
CB	438477	10 mg	
CD	56-1300-00	10 mg	
K&K	4134	100 mg	
LKB	1243-215	10 mg	
RE		5 mg	
SC	L 9504	1, 5, 10, 50, 100 mg	Crystalline
SC	L 6882	0.2, 2, 5, 10, 50, 100 mg	Na salt, readily water soluble

[a] Based on information supplied by the manufacturers.

greater sensitivity and of course is required when the crystalline preparations are used.

Firefly Luciferase-Containing Kits

Many of the manufacturers package the firefly luciferase preparations in kits to make application of bioluminescence techniques simpler. Table VI lists those kits and states the purpose(s) for which they were designed. Such packaging may be convenient for an investigator with a specific application.

Discussion of Reagents for Commercial Bioluminescence-Based Assays

In the case of bioluminescent reagents we do not recommend any one reagent as being markedly superior. The differences that exist may make one preparation better for a particular experimentally constrained system than another preparation. But just as surely, the opposite may be true for other circumstances. We have recorded our observations on comparison and use, and have tabulated information supplied by the manufacturers that will provide investigators with sufficient information to select several of the preparations for testing in their own experimental systems. For most experimental problems the commercial reagents can be applied without modification. The manufacturers provide high-quality reagents and technical advice to enhance application of these reagents. The bacterial luciferase reagents are discussed in this volume by Wienhausen and DeLuca.[45]

[45] G. Wienhausen and M. DeLuca, this volume [17].

TABLE VI
FIREFLY LUCIFERASE-CONTAINING KITS[a]

Company	Catalog number	Name	Function
ALL	2000-10	ATP analysis kit	ATP analysis
	501	ATP in platelets kit	Platelet ATP
	502	Red blood cell viability kit	RBC ATP
	8001	Microscreen	Bacteriurea screening
BM	567736	ATP bioluminescence CLS	Constant light—2.5×10^{-8} to 2.5×10^{-11} M ATP
	567701	ATP bioluminescence HS	High sensitivity—10^{-8} to 5×10^{-13} M ATP
CB	119108	ATP assay kit	With releasing agent—10^{-5} to 10^{-13} M ATP
LKB	1243-107	ATP assay kit	With extracting agent—10^{-11} to 10^{-6} M ATP
	1243-102	ATP monitoring kit	10^{-6} to 10^{-11} M ATP
	1243-100	CK kit	Creatine kinase in serum
	1243-101	CK-B kit	Total and B subunit of creatine kinase
L/3M	4641	Aquatox test kit	Viability of activated sludge
	4642	A.E.C. kit	Adenylate energy charge
	4645	LBE—100 kit	Enrichment of bacteria before analysis
	4651	Milk bacteria kit	Bacteria in milk
	9273	Bacteriurea screening kit	Bacteria in urine
	9283	IMC kit	Microorganisms in industrial samples
	9289	Fruit juice test kit	Bacteria in fruit juices
PICO	6016713	Picozyme yeast ATP kit	Yeast in carbonated beverages
	6016911	Picozyme kit I	ATP without extraction
	6016912	Picozyme kit II	ATP in bacterial cells
	6016913	Picozyme kit III	ATP in somatic cells
	6016914	Picozyme kit IV	ATP in both bacterial and somatic cells
	6016924	Picozyme ATP kit	Automated instruments
SC	FF-3, 10, 15	Bioluminescence demonstration	To demonstrate firefly bioluminescence
TD	20-3100	Bacteriurea screening	Bacteria in urine
	20-2192	ATP kit, normal	ATP analysis with extractant
	20-2193	ATP kit, high phosphatase	Contains a phosphatase inhibitor
	20-2194	ATP starter kit	Components of the two above

[a] Based on information supplied by the manufacturers.

Acknowledgments

Drs. Otis C. Dermer, Arne Lundin, Ulrich K. Melcher, E. C. Nelson, and William A. Weppner read the manuscript and made helpful suggestions. Some of the data used were provided by Ginger Hampton, Linda Staten, and Marliese Hall. The drawings were by Marliese Hall and Tom VanDoren. Additional proofreading was by Anna Belle Leach. The persons supplying manufacturers' information were ALL, Mr. Peter Ulrich and Dr. Jon Wannlund; BM, Dr. Karl Wulff, Gawehn, and Dr. Len Mascaro; CB, Dr. John Snow; CD, Ms Vicki Nodes; LKB, Mr. Dick Reid and Dr. Kalevi Kurkijarvi; L/3M, Dr. Henk Vanstaen; NBC, Ms. Ann Igo; PICO, Ms. Cindy Sanville; PP-L, Mr. Alan Gerstein; RE, Ms. Mary Kave; SC, Dr. Steve Magee and Dr. Ted Gimenez; TD, Dr. Bob Phillips; and USB, D. S. Richardson. The Oklahoma Agricultural Experiment Station supported this work through Project No. 1806. This is manuscript number J-4841 of the Oklahoma Agricultural Experiment Station.

[7] Cloning and Expression of the Genes from the Bioluminescent System of Marine Bacteria

By CAROL MIYAMOTO, MICHAEL BOYLAN, ANGUS GRAHAM, and EDWARD MEIGHEN

Genes encoding at least five polypeptides are required to produce the enzymes necessary for bioluminescence in marine bacteria. These include genes coding for the two subunits of luciferase (α and β) and the three polypeptides of a fatty acid reductase complex which supplies aldehyde for the bioluminescent reaction.[1-4] In addition, two regulatory polypeptides have been identified in the *Vibrio fischeri* luminescence system.[1,2] Transfer of the genes from luminescent bacteria into *Escherichia coli* and the study of their regulation can now readily be accomplished since very rapid and highly sensitive techniques are available for detection of the cloned genes and the analyses of the mRNAs and gene products.

Gene Transfer from Bioluminescent Bacteria into *E. coli*

Published protocols for the transfer and recombination of DNA[5] were closely followed. Chromosomal DNA is released from marine bacteria by

[1] J. Engebrecht, K. Nealson, and M. Silverman, *Cell* **32,** 773 (1983).
[2] J. Engebrecht and M. Silverman, *Proc. Natl. Acad. Sci. U.S.A.* **81,** 4145 (1984).
[3] A. Rodriguez, L. Wall, D. Riendeau, and E. A. Meighen, this volume [14].
[4] D. M. Byers, A. Rodriguez, L. Carey, and E. A. Meighen, this volume [15].
[5] T. Maniatis, E. F. Fritsch, and J. Sambrook, "Molecular Cloning. A Laboratory Manual." Cold Spring Harbor Lab., Cold Spring Harbor, New York, 1982.

using a standard procedure for the isolation of *E. coli* DNA.[6] A 20 ml culture of marine bacteria at $OD_{660\ nm} = 0.7$ yields 180 μg of DNA. The DNA is restricted with an appropriate endonuclease so that it can be recombined with plasmids such as pBR322. The plasmid, after a similar restriction, is dephosphorylated with calf intestinal phosphatase for 60 min at 37°. The restricted DNAs are extracted with phenol–chloroform (1 : 1) and then mixed together before or after precipitation with 60–70% ethanol at −20°. The DNA precipitates are washed in 75–90% ethanol, dried, and dissolved in a volume of distilled water such that the plasmid concentration is 20–30 μg/ml and that of the added DNA approximately 4-fold higher on a molar basis. The DNAs are incubated at 65° for 10 min prior to ligation under published conditions[5] with T_4 polynucleotide ligase at 14° for 2–16 hr. The recombinant plasmids can be stored at −20° indefinitely and used to transform $CaCl_2$ treated *E. coli*[5] whenever required. The transformed bacteria are grown on LB agar plates containing the appropriate antibiotic for plasmid selection.

If restriction fragments of chromosomal DNA ligated with plasmid are longer than ~2.5 kbp, the length of DNA that encodes luciferase, the transformed bacteria can be rapidly screened for light emission. Recombinant plasmids as large as 16 kbp can be maintained with no apparent difficulty. Approximately 90% of the plasmids constructed by the protocol described above contained recombinant DNA, and on the order of 1 in 10^3 to 10^5 contained both luciferase genes depending on the restriction enzyme used.

Screening for Clones Containing Genes from the Luminescent System

Selection of clones containing genes from the bioluminescence operons is generally based on the detection of light emission. Only the luciferase genes need be present since the aldehyde substrate can be supplied exogenously to *E. coli* and sufficient $FMNH_2$ and O_2 are present intracellularly to allow generation of light. Screening may be accomplished by the following approaches.

Visual Scanning

Although this procedure is the least sensitive, it has the advantage of being the most direct technique for the identification of single luminescent colonies. Decanal (Aldrich) is applied to the inside of a Petri dish cover (20–50 μl of neat liquid) and the bacterial colonies are inverted over the

---

[6] R. F. Schleif and P. C. Wensink, "Practical Methods in Molecular Biology." Springer-Verlag, Berlin and New York, 1981.

aldehyde. While tetradecanal is believed to be the preferred aldehyde *in vivo*,[7] decanal is superior for screening because of its much greater volatility and possibly a greater ability to cross the bacterial membrane. For maximum sensitivity, the individual's eyes should be dark-adapted for at least 10 min although highly expressing clones can be observed in less than 1 min. Light-emitting transformed *E. coli* are still viable and apparently unaltered after 24 hr of exposure to aldehyde vapor, however, the Petri dish cover is slowly attacked by direct contact with aldehyde and should be replaced within a day to maintain sterile conditions. Colonies should be screened initially for light emission in the absence of aldehyde to determine if all the polypeptides of the luminescent system are present and expressed.

Exposure of Colonies to Film

The most sensitive method for detection of light emission by colonies involves exposure to Kodak XAR-5 film. The bacterial plates (plus or minus aldehyde) are inverted with the lid beneath them onto a support and secured to each other with transparent tape. Radioactive ink, applied to the tape, or a plate of light-emitting marine bacteria secured to the test plates, is used as a marker. The set of bacterial plates is transferred into an empty film container and the film placed on top of the plates. The closed container, with a weight placed on top, is left at room temperature for 5 sec to 1 day before the film is developed.

Scintillation Counter Readings

A sensitive and practical method for detecting light-emitting bacteria makes use of scintillation counters, which are readily accessible to most laboratories. Libraries of *E. coli* containing recombinant DNA from luminescent bacteria are replicated several times onto nitrocellulose filters and incubated colony-side-up on LB agar plates containing the appropriate antibiotic for several hours at 30°. One replica is cut into quadrants, and the cells from each quadrant are scraped into vials containing 1 ml of broth and 1 μl of decanal. Using a scintillation counter to detect luminescence, positive quadrants can easily be selected. Individual colonies from a replica of the positive quadrant can then be transferred into broth for analysis or, alternatively, the replica can be further subdivided and tested before screening for single colonies.

This method is particularly advantageous to confirm the existence of light-emitting clones after detection of positive signals on film. Often, it is

---

[7] S. Ulitzur and J. W. Hastings, *Proc. Natl. Acad. Sci. U.S.A.* **75**, 266 (1978).

difficult to pinpoint the exact colony responsible for a positive film signal among a cluster of colonies. In this case, individual colonies can be picked and dispersed in broth in sterilized glass scintillation vials. If required, 1 μl of aldehyde is applied to the lining of the vial cap. After approximately a minute, a positive response (10^3–10^5 cpm) can be detected in the scintillation counter. The cap is then replaced by a fresh sterile cap and the culture can continue to grow.

Hybridization Probes

On initial screening of *E. coli* containing recombinant DNA from *Photobacterium phosphoreum* (NCMB 844), light-emitting clones (plus aldehyde) could not be detected in the scintillation counter. Since the *lux* genes from different luminescent bacteria appear to have some conserved sequences,[8] DNA probes from the cloned luminescent system of *V. fischeri* (ATCC 7744, neotype strain) were used to screen the library containing the *P. phosphoreum* DNA. The luminescent system from the neotype *V. fischeri* strain is similar to the *V. fischeri* MJ1 system cloned by Engebrecht *et al.*,[1] except that the latter system is enclosed in a 9 kbp fragment between two *Sal*I sites whereas both *Sal*I sites are missing in the neotype strain and it is necessary to recombine a 16 kbp *Bam*HI fragment with pBR322 to encompass the luminescence system. However, most of the other restriction sites in the two *V. fischeri* luminescence systems are identical and their expression and regulation in *E. coli* appear to be similar.[9]

First, in order to prove that *V. fischeri* DNA could cross-hybridize with *P. phosphoreum* DNA, restriction fragments of *V. fischeri* DNA were nick-translated and used as probes to a Southern blot of *P. phosphoreum* DNA which had been restricted with various enzymes and electrophoresed on an agarose gel.[5] The blot was preincubated in heat-sealed bags for 2 hr at 50° with a 0.015% solution of sonicated denatured calf thymus DNA in 0.6 *M* NaCl, 0.12 *M* NaH$_2$PO$_4$, 3 m*M* EDTA, 2.5% dextran sulfate, 1% Sarkosyl, pH 6.2 (20 μl of solution per cm^2 of filter). Hybridization was then conducted for 6–16 hr at 50° with the denatured ^{32}P-labeled *V. fischeri* probe in fresh solution containing the same concentration of calf thymus DNA. Approximately 50 ng of probe DNA (~2.5 × 10^6 Cerenkov counts) was used per 300 cm^2 of filter. After extensive washing at 50° with 0.3 *M* NaCl, 30 m*M* Na citrate, 0.1% SDS, 0.1% Na

[8] K. H. Nealson and R. Cassin, *in* "Analytical Applications of Bioluminescence and Chemiluminescence" (L. J. Kricka, P. E. Stanley, G. H. G. Thorpe, and T. P. Whitehead, eds.), p. 87. Academic Press, Orlando, 1984.

[9] M. Boylan, A. F. Graham, and E. A. Meighen, *J. Bacteriol.* **163**, 1186 (1985).

pyrophosphate, pH 7.6 (0.5–1.0 ml per cm² of filter for each wash) until less than 100 Cerenkov counts could be detected per 20 ml of wash solution, the filter was dried and exposed to Cronex film (Dupont) at −70°.

Hybridization of the *V. fischeri* DNA probes to restricted *P. phosphoreum* DNA was obtained; results indicated that a 9 kbp fragment from a *Bgl*II digestion would encompass both the luciferase genes whereas other restriction enzymes tested appeared either to cut into the luciferase genes or to produce very large restriction fragments (≥20 kbp). Thus a *Bgl*II library of *P. phosphoreum* DNA was constructed by recombining the restricted DNA at the *Bam*HI site of pBR322. After transformation and growth of *E. coli* on LB agar plates containing ampicillin, the colonies were transferred to New England Nuclear (NEN) Colony/Plaque screen membranes and incubated colony-side-up on LB agar plates containing chloramphenicol (100 μg/ml) for 6 hr at 30° in order to amplify the plasmid. The colonies were then lysed with 10% SDS followed by 0.5 *M* NaOH, 1.5 *M* NaCl, neutralized, and prepared for *in situ* colony hybridization exactly as described elsewhere.[5] Hybridization with a probe encompassing the 5'-region of the *lux*A gene of *V. fischeri,* under the same conditions as described above for hybridization to Southern blots, yielded 6 clones out of 5000 colonies screened containing the 9 kbp *Bgl*II fragment of *P. phosphoreum* DNA. By using a 1.1 kbp fragment from the cloned *P. phosphoreum* DNA as probe, a clone containing a larger fragment (20–30 kbp) of *P. phosphoreum* DNA could be detected in a *Bam*HI library in *E. coli.* Experimental conditions were identical to those described for cross-hydridization between different DNAs, except that a temperature of 65° rather than 50° was used.

Thus cross-hybridization should prove applicable for screening libraries in *E. coli* containing luminescent systems from other marine bacteria. Depending on the homology of the probe of interest and on its size, different temperatures (45–65°) may prove optimal in other systems. Since Southern blots can serve as relatively permanent templates, it is recommended that various conditions for hybridization be tested with the same blot of restricted DNA prior to screening a library. The probe can be washed off with 2 ml of boiling water per cm² of filter and the filter reused several times.[10] Further applications of cross-hybridization include studies of the distribution of *lux* genes in nonluminescent marine bacteria and in other light-emitting systems.[11–13]

[10] P. S. Thomas, this series, Vol. 100, p. 255.
[11] C. J. Potrikus, E. P. Greenberg, N. V. Hamlett, S. Gupta and J. W. Hastings, *Soc. Microbiol., Abstr. Annu. Meet.* p. 84A-1 (1984).
[12] D. H. Cohn, Ph.D. Thesis, University of California, San Diego (1983).
[13] M. G. Haygood, Ph.D. Thesis, University of California, San Diego (1984).

Other Screening Techniques

Potentially the methods outlined in subsequent sections, such as fatty acylation of proteins, can be used for screening the genes from the luminescent system. However, the high sensitivity of the screening processes just described should preclude the necessity for using these latter techniques in most cases.

Light Emission in *E. coli*

Of the three luminescent systems studied (*V. fischeri, V. harveyi,* and *P. phosphoreum*), only the cloned *V. fischeri* system emits light in *E. coli* in a manner similar to the native strain. The *Bam*HI clone from the neotype *V. fischeri* strain can easily be detected by visual scanning in the absence of aldehyde as has been reported for the *Sal*I clone from MJ1 *V. fischeri*.[1,9] When exposed to film, the clone gives a strong signal within 5 sec. In liquid cultures, the maximum light intensity reaches the same levels as the native *V. fischeri* strain (150–300 LU/ml).[14]

The luciferase genes from *V. harveyi* (B392) have been cloned in a variety of restriction fragments into *E. coli*.[15,16] The nucleotide sequence of an 1.8 kbp fragment of DNA from *V. harveyi* including the whole *luxA* gene has recently been published.[17] A restriction map of a stretch of *V. harveyi* DNA including the *lux* genes is shown in Fig. 1.[15] Transformed cells can be detected by a positive response in the scintillation counter or by exposure of film for 5 min in the presence of decanal. The larger fragments of *V. harveyi* DNA (Vh *Sau*1 and Vh *Sau*2, Fig. 1) were generated by partial digestion with *Sau*3A1 and recombined with pBR322 at the *Bam*HI site. These clones give a positive signal on film after about 1 hr exposure in the absence of aldehyde suggesting that they may encompass the entire luminescence system. However, the maximum light intensity in liquid cultures for the *V. harveyi Sau*3A1 clones is still very low both in the absence (0.005 LU/ml) and presence (1.0 LU/ml) of aldehyde compared to the native *V. harveyi* strain (~400 LU/ml). Recently, much higher light has been obtained with *V. harveyi* clones by transformation of genetically altered *E. coli* indicating that the host/plasmid relationship is important for efficient expression of bacterial bioluminescence.

[14] One light unit (LU) corresponds to 5×10^9 quanta/sec based on the standard of J. W. Hastings and G. Weber [*J. Opt. Soc. Am.* **53**, 1410 (1963)].

[15] C. M. Miyamoto, A. D. Graham, M. Boylan, J. F. Evans, K. W. Hasel, E. A. Meighen, and A. F. Graham, *J. Bacteriol.* **161**, 995 (1985).

[16] T. O. Baldwin, T. Berends, T. A. Bunch, T. F. Holzman, S. K. Rausch, L. Shamansky, M. L. Treat, and M. M. Ziegler, *Biochemistry* **23**, 3663 (1984).

[17] D. H. Cohn, A. J. Mileham, M. I. Simon, K. H. Nealson, S. K. Rausch, D. Bonam, and T. O. Baldwin, *J. Biol. Chem.* **260**, 6139 (1985).

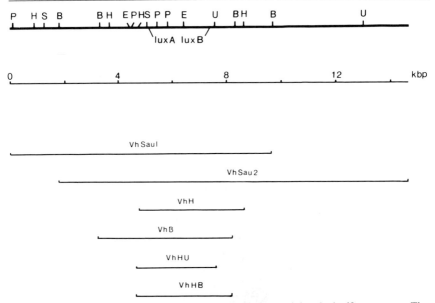

FIG. 1. Restriction map of 18 kbp of *V. harveyi* DNA containing the luciferase genes. The positions of *luxA* and *B* genes are indicated below the restriction sites along the DNA. E, *Eco*RI; B, *Bam*HI; H, *Hin*dIII; P, *Pst*I; S, *Sal*I; U, *Pvu*II. The locations and sizes of various cloned inserts of *V. harveyi* are presented at the bottom.

The maximum expression of light by the *P. phosphoreum* clones (*Bgl*II and *Bam*HI) in liquid culture is over 10^5-fold lower (0.02 LU/ml) than for the native strain (~4000 LU/ml) and explains why these clones were not detected on initial visual screening of the library for light emission. Interestingly, the *Bam*HI clone does not require aldehyde for expression indicating that the total luminescence system may be present. The reasons for the different expression of these luminescent systems in *E. coli* remain to be elucidated.

Optimization of Expression of the Cloned Genes

Experimental Conditions

The detection of luminescence in *E. coli* is dependent on the experimental conditions for cellular growth. Although the MJ1 *V. fischeri* luminescent system expressed very well in *E. coli* at 30°,[1] light emission by the cloned ATCC 7744 *V. fischeri* system is almost 100-fold higher at 20 than at 30°.[9] Similarly, the *P. phosphoreum* system will express light only at 20° in *E. coli,* although the *V. harveyi* system expresses well at both 20 and 30°. Optimal light expression of colonies is generally observed after 2 to 4 days of growth at room temperature. It is recommended that the colonies be grown at 30–37° for 1 day to obtain a high growth rate and then incubated at room temperature and screened for light emission over the next 2 days. Decanal is the preferred aldehyde for screening as it appears to give the maximum response under these experimental conditions. However, the possibility that other aldehydes (e.g., dodecanal or tetradecanal) could give higher light with other strains of luminescent bacteria must always be considered.

In liquid culture, the level of oxygenation may also affect the light emission in the cloned systems. For example, the large *V. harveyi* clones give higher light if grown under conditions where there is greater contact with air. The possibility that the concentration of salt in the LB broth is too low for high expression of the enzymes in the luminescent system should also be considered since some luminescent marine bacteria (e.g., *P. phosphoreum*) emit light very poorly at salt concentrations less than 3%. Finally, it should be noted that if the entire luminescent system is present and under regulation, then high levels of bioluminescence will only be detected after induction of the system.

Use of Promoters

A variety of promoters have been used to increase the expression of the *V. harveyi lux* genes including the P_L/P_R promoters of lambda[18] and the P1 promoter of pBR322.[16] A plasmid (pOP95-15) containing the *E. coli lac*-UV5 promoter, originally constructed by F. Fuller,[19] has been modified to include a polylinker and a kanamycin gene at the *Eco*RI site (pOP95-15K).[20] Figure 2 illustrates the construction of pOPVhSU in which the restriction fragment from pOP95-15K containing the *lac*UV5 promoter is inserted in front of the *V. harveyi lux* genes. The plasmid is maintained in *E. coli* DH21 cells which carry the *lac*I^sq gene on the F' plasmid.[21] When the recombinant *E. coli* is grown in LB broth the production of luciferase is stimulated 40-fold by the addition of isopropylthiogalactoside (IPTG) as shown in Fig. 3. The low level of light emitted in the control may be due to

[18] R. Belas, A. Mileham, D. Cohn, M. Hilmen, M. Simon, and M. Silverman, *Science* **218**, 791 (1982).

[19] F. Fuller, *Gene* **19**, 43 (1982).

[20] The plasmid pOP95-15K was generously given to us by Michael Dubow, McGill University, Montreal. It was constructed by J. Harel in his laboratory by inserting the polylinker-flanked *Kan*^R gene of TM903 from pUC71K into pOP95-15 [J. Vieira and J. Messing, *Gene* **19**, 259 (1982)].

[21] D. Hanahan, *J. Mol. Biol.* **166**, 557 (1983).

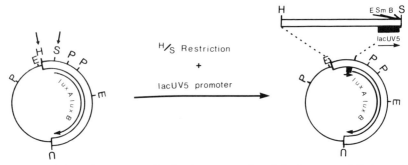

Fig. 2. Schematic representation for the construction of the *lux* genes of *V. harveyi* in pBR322 preceded by the *lac*UV5 promoter of *E. coli*. The plasmid pVhHU on the left was restricted with *Sal*I (S) and *Hin*dIII (H) and then ligated with the similarly restricted insert from pOP95-15K which contains the *lac*UV5 promoter. The resultant pOPVhSU plasmid on the right contains *V. harveyi* DNA now starting from *Sal*I (S) and terminating at *Pvu*II (U) on the map (Fig. 1), and encompassing the *lux* genes. The insert containing the *lac*UV5 (0.45 kbp) is recombined in a clockwise direction ahead of the *lux* genes.

incomplete repression and/or the presence of promoters internal to the *lac*UV5 promoter.

The high expression of the pVhH clone of *V. harveyi* appears to be due to a plasmid promoter in pBR322 between the *Hin*dIII and *Bam*HI sites oriented in the opposite direction to the tetracycline gene.[22] Much lower light emission results when the fragment is reversed (pVhH' in Fig. 4). Similarly, the 15-fold higher level of luminescence with the pVhB' clone compared to pVhB (Fig. 4) can be explained by facilitative expression by the P2 promoter for the tetracycline gene. In pBR322, *Hin*dIII cuts between the P1 promoter (one of two promoters for the β-lactamase gene required for ampicillin resistance) and the P2 promoter for the tetracycline gene, which are overlapping and oriented in opposite directions. After *Hin*dIII restriction, the tetracycline gene product is no longer made, while β-lactamase can still be produced due to the presence of the other natural promoter (P3).[22]

Analysis of mRNA

Hybridization of specific probes from cloned genes of the luminescent system to Northern blots can be used to detect the mRNA from the luminescent system as well as to determine the position and size of the mRNA within a stretch of DNA coding for the bioluminescence genes.[15] Although probing for minor mRNA species by this technique can prove to

[22] J. Brosius, R. L. Cate, and A. P. Perlmutter, *J. Biol. Chem.* **257,** 9205 (1982).

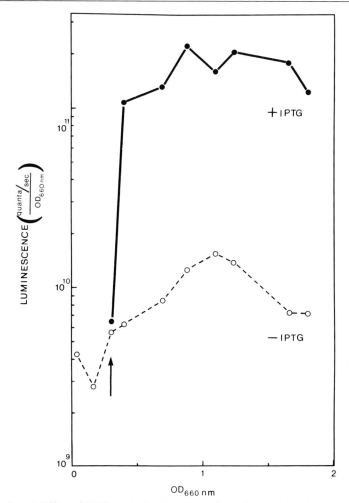

FIG. 3. Effect of IPTG on the luminescence of *E. coli* containing the plasmid pOPVhSU. The recombinant *E. coli* DH21 cells were grown in LB broth at 30°. When the $OD_{660\ nm} = 0.3$, one-half of the culture was treated with IPTG to a final concentration of $10^{-3}\ M$ (arrow) while the other half was left untreated. The light emitted with decanal (LU/ml) at various concentrations of cells was monitored. The luminescence converted to a quanta/sec per $OD_{660\ nm}$ is plotted as a function of concentration ($OD_{660\ nm}$).

be relatively difficult, fortunately the mRNAs from the induced bioluminescent system are usually present in relatively large amounts.

Bacteria are grown in 1% NaCl complex medium[23] at 28°. (In the case

[23] E. A. Meighen and I. Bartlett, *J. Biol. Chem.* **255**, 11181 (1980).

Clone	Insert	LU/OD_{660}

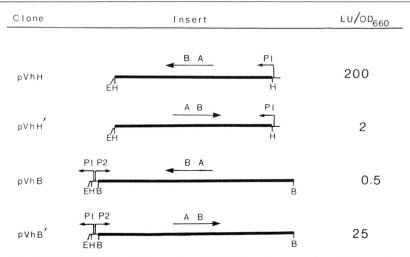

FIG. 4. Effect of the P1 and P2 promoters of pBR322 on the expression of *V. harveyi luxA* and *luxB* genes. The *V. harveyi Hin*dIII fragment containing the *lux* genes (Fig. 1) was ligated with pBR322 at the *Hin*dIII site in both directions resulting in the clones pVhH and pVhH'. Similarly, the *Bam*HI fragment of *V. harveyi,* also containing the *lux* genes (Fig. 1), recombined with pBR322 at the *Bam*HI site in both directions, yielded pVhB and pVhB' plasmids. The orientation of the *lux* genes (AB) is indicated. The two promoter regions of pBR322 (P1 of β*la* gene and P2 of the *tet* gene) are shown in each construction; pBR restriction sites indicated are *Eco*RI (E), *Hin*dIII (H), and *Bam*HI (B). On the right is listed the light emitted at $OD_{660\ nm} = 1$ of liquid cultures of *E. coli* carrying each plasmid.

of recombinant *E. coli,* ampicillin at 100 μg/ml is added.) RNA labeled with [³H]uracil can be extracted from bacteria by lysozyme treatment and phenol:SDS extraction.[24] RNA is electrophoresed at 40 V (50 mA) on 1.2% agarose containing 5 mM methylmercury (Alfa) in a borate–sulfate buffer[25] modified to contain 64 mM boric acid, 5 mM Na borate, 10 mM Na_2SO_4, pH 8.19. RNA is transferred to Pall membrane filters (0.2 μm, Biodyne) by electrophoresis in 25 mM sodium phosphate (pH 5.5) at 20 V (150 mA) in a Bio-Rad trans-blot apparatus overnight. Under the conditions described, other membranes and pore sizes did not give good transfer of RNA. The filters are baked *in vacuo* for 2 hr. Hybridization is carried out exactly as described for cross-hybridization of DNA probes to Southern blots (see above) except the incubation time at 50° was 16 to 24 hr.

Northern blots can be reused by washing the filters with 2 ml/cm² of

[24] J. F. Evans, S. McCracken, C. M. Miyamoto, E. A. Meighen, and A. F. Graham, *J. Bacteriol.* **153,** 543 (1983).
[25] J. M. Bailey and N. Davidson, *Anal. Biochem.* **70,** 75 (1976).

filter of 8 mM Tris–chloride (pH 7.6) and 0.4 mM EDTA for 10 min at 50°. The washing is repeated three times, and the filter is then shaken vigorously with distilled water for 1 min at room temperature. After 30 min *in vacuo* at 80°, the filter is ready for use.

Four mRNAs (8, 7, 4, and 2.6 kb) could be identified on hybridization of a ^{32}P-labeled DNA probe from the *luxA* gene of *V. harveyi* to RNA extracted from *E. coli* containing the pVhSau2 DNA or from the native *V. harveyi* strain after induction of bioluminescence (Fig. 5). Hybridization of the same DNA probe to Southern blots of restricted *V. harveyi* DNA indicated the existence of one gene for the α subunit of luciferase (thus eliminating the possibility that these are multiple genes for luciferase) and suggested that the various polycistronic mRNAs are all transcribed from the same segment of DNA. These overlapping mRNAs may have arisen from multiple promoters, transcription readthrough, and/or specific degradation. The detection of the same set of mRNA in recombinant *E. coli* indicates that a similar control of the mRNA size classes exists in both *E. coli* and *V. harveyi* and raises the question of why the levels of light are so low in the recombinant *E. coli*.

Analysis of Enzyme Function

A number of *in vitro* assays for specific enzymes can be applied to analyze the gene products from the cloned luminescent systems.

Luciferase

The presence of luciferase in sonicated bacterial extracts can be measured by injection of 50 μM FMNH$_2$ into 0.05 M phosphate, pH 7.0, containing the extract and aldehyde.[26] Maximum response is dependent on the aldehyde (usually decanal or tetradecanal) which is added to a final concentration of about 0.001%. Aldehyde stocks (0.1% v/v) can be conveniently prepared in isopropanol or dimethylformamide.

Enzymes of the Fatty Acid Reductase System

Reaction of cell extracts with labeled fatty acid can result in the specific identification of the polypeptides involved in fatty acid reduction.[27–29] The acyltransferase from both *V. harveyi* and *V. fischeri* has been spe-

[26] A. Gunsalus-Miguel, E. A. Meighen, M. Z. Nicoli, K. H. Nealson, and J. W. Hastings, *J. Biol. Chem.* **247**, 398 (1972).

[27] A. Rodriguez, L. Wall, D. Riendeau, and E. Meighen, *Biochemistry* **22**, 5604 (1983).

[28] L. Wall, A. Rodriguez, and E. Meighen, *J. Biol. Chem.* **259**, 1409 (1984).

[29] L. A. Wall, D. M. Byers, and E. A. Meighen, *J. Bacteriol.* **159**, 720 (1984).

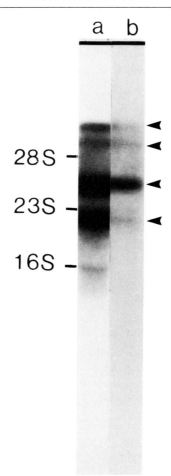

FIG. 5. Hybridization of ^{32}P-labeled *luxA* DNA of *V. harveyi* to RNA extracted from *E. coli* containing pVhSau2 (Fig. 1) and from native *V. harveyi*. The two bacterial strains were grown in 1% NaCl complex medium at 28° (in the case of the recombinant *E. coli*, ampicillin at 100 μg/ml was added) until $OD_{660\ nm}$ = 0.9. The culture of *V. harveyi* gave 78 LU/ml while that of the recombinant *E. coli* gave ≥0.001 LU/ml in the absence of decanal and 0.1 LU/ml, in the presence. Equal amounts of both RNAs (5 μg) were electrophoresed in 1.2% agarose containing methylmercury and a Northern blot of the gel was obtained. After ^{32}P labeling by nick translation of the *V. harveyi luxA* DNA,[5] and hybridization to the Northern blot, the blot was exposed to Cronex film at $-70°$. The pattern of hybridization is shown for *E. coli* (containing pVhSau2) RNA in (b) and for *V. harveyi* RNA in (a).

cifically identified in transformed *E. coli* by reaction of extracts with [3H]tetradecanoyl-CoA in 50 m*M* phosphate pH 7.0.[9,15] Labeling with [3H]tetradecanoic acid (plus ATP) *in vitro* has resulted in identification of all three polypeptides present in the cloned *V. fischeri* system.[9] Alternatively, incubation of the cells *in vivo* with [³H]tetradecanoic acid can be used specifically to label the polypeptides prior to extraction of the cells, and to identify the labeled polypeptides by autoradiography after SDS–gel electrophoresis.[9]

Acknowledgments

   This work was supported by Grant MA-7672 from the Medical Research Council of Canada. Michael Boylan was supported by a studentship from the Fonds de la Récherche en Santé du Québec. We thank Rose Szittner for her expert technical assistance, J. Mancini for obtaining the *P. phosphoreum Bam*HI clone, and David M. Byers for his valuable advice in the preparation of the manuscript.

[8] Techniques for Cloning and Analyzing Bioluminescence Genes from Marine Bacteria

By JoAnne Engebrecht and Michael Silverman

   The isolation by recombinant DNA techniques of genes for bioluminescence (*lux*) from marine bacteria has resulted in a rapid expansion of knowledge of the biochemical activities necessary for light production and of the regulatory mechanisms which govern the expression of these functions. We describe here a variety of genetic methodologies for cloning DNA fragments encoding luminescence functions, eliciting expression of luminescence genes, defining individual *lux* genes and transcriptional units containing *lux* genes, identifying the products of cloned *lux* genes, exploring the regulatory control of *lux* genes, and using *lux* gene fusions to measure transcriptional control of other gene systems. Most genetic analysis has been performed with luminescent *Vibrio,* and we will confine our attention to bacteria of this genus. Some of these topics as well as the biochemical analysis of the products of cloned *lux* genes and the study of mRNA transcripts of *lux* genes are discussed in another report in this volume.[1]

Cloning *lux* Genes

   Light production *in vitro* has been shown to require reduced flavin (FMNH₂), long chain fatty aldehyde, bacterial luciferase, and oxygen.

[1] C. Miyamoto, M. Boylan, A. Graham, and E. Meighen, this volume [7].

Luminescent marine *Vibrio* encode a NAD(P)H dehydrogenase FMN [NAD(P)H:flavin oxidoreductase] for the generation of reduced flavin, enzymes for the conversion of long chain fatty acid to long chain fatty aldehyde (tetradecanal), and the α and β subunits of bacterial luciferase.[2] Initial cloning strategies focused on the isolation of genes for the subunits of luciferase because of the central role these gene products have in bioluminescence and because simple *in vitro* methods were available for detecting these gene products. However, it is now apparent that, like many prokaryotic gene systems, genes encoding luminescence functions are organized into multigene transcriptional units (operons), and a DNA fragment encoding the luciferase subunits can also contain genes encoding other luminescence functions including genes for aldehyde biosynthesis and those with regulatory functions. Thus, cloning luciferase genes can yield other *lux* genes as well. By cloning and expressing *lux* genes in a nonluminous host such as *Escherichia coli* it is possible to deduce what functions are specific for bioluminescence and what functions are common to nonluminous bacteria. In addition, cloning genes for bioluminescence opens may avenues of genetic analysis which were not possible with conventional genetic techniques alone. Because it could not be assumed a priori that luciferase genes from marine bacteria would be expressed in *E. coli,* the cloning of luciferase genes was first attempted by two methods which did not rely on light production for the detection of *lux* gene recombinants. These two methods were the cloning of *lux* genes from transposon mutants and the cloning of *lux* genes by using radioactive oligonucleotide sequences.

By developing a procedure for transposon mutagenesis it was possible to collect mutants of *Vibrio harveyi* which contained transposon insertions in *lux* genes.[3,4] These mutants were easily recognizable. They produced little or no light because insertion of the transposon into a *lux* gene sequence interrupted the integrity of the gene. Furthermore, cloning was greatly facilitated because the transposon insertion physically linked a selectable marker, the transposon Tn5–*132* encoded tetracyline resistance, to the *lux* gene (see Fig. 1A). A restriction enzyme which did not cleave transposon sequences was used to fragment the *Vibrio* genome, and these DNA fragments were then ligated into a plasmid cloning vector. Among the population of recombinant molecules those containing the transposon and flanking *lux* gene sequences had the property of encoding a unique drug-resistance phenotype. Recombinant molecules were trans-

[2] M. M. Ziegler and T. O. Baldwin, *Curr. Top. Bioenerg.* **12,** 65 (1981).

[3] R. Belas, A. Mileham, M. Simon, and M. Silverman, *J. Bacteriol.* **158,** 890 (1984).

[4] R. Belas, A. Mileham, D. Cohn, M. Hilmen, M. Simon, and M. Silverman, *Science* **218,** 791 (1982).

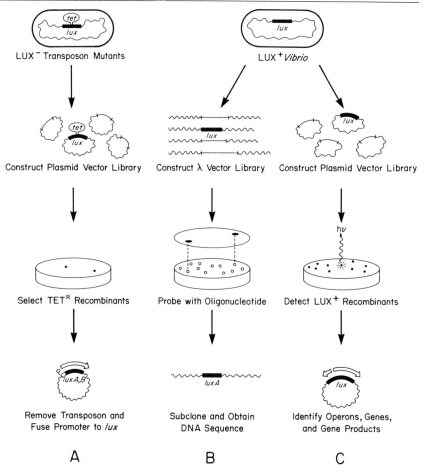

FIG. 1. Strategies used to clone *lux* genes from marine *Vibrio*. (A) Cloning from a transposon mutant; (B) the use of synthetic oligonucleotide DNA for *lux* gene cloning; and (C) the direct selection of *lux* recombinants.

ferred to *E. coli* by the calcium shock transformation procedure and then plated on a medium containing tetracycline. Tetracycline-resistant *E. coli* recombinants were obtained, and subsequent analysis showed that these recombinants did contain *lux* gene sequences corresponding to *luxA* (α subunit) and *luxB* (β subunit). As expected these *lux* sequences were interrupted by insertion of the transposon, but the continuity of the *lux* gene region was restored by "stitching together" a nondefective copy from two recombinant plasmids with transposon insertions located in different regions of the cloned *lux* gene fragment. *E. coli* with this reconstructed *lux* plasmid contained the linked *luxA* and *luxB* genes, but it produced relatively little light and required the exogenous addition of aldehyde for light production. Light production in the range observed for the parent *Vibrio harveyi* was obtained by constructing recombinants which contained transcriptional control elements, the P_R and P_L promoters of bacteriophage lambda,[5] inserted in close proximity with the *lux* genes. Induction of the lambda promoters by temperature shift increased transcription of the *luxA* and *luxB* (luciferase) genes and resulted in greatly amplified production of light by the recombinant *E. coli*. Furthermore, the products of the *luxA* and *luxB* genes were then identified by methods discussed in the next section.

A second approach for cloning *lux* genes, which also did not rely on obtaining expression in recombinants, employed a radioactively labeled DNA probe which was homologous to a *luxA* gene sequence[6] (see Fig. 1B). This probe consisted of a synthetic oligonucleotide corresponding to the DNA sequence which encodes the α subunit of the *V. harveyi* luciferase and was designed from amino acid sequence data. Due to the redundancy in codon usage for most amino acids, a mixture of oligonucleotides was actually synthesized. Labeled oligonucleotides were then used to detect the *luxA* gene by DNA–DNA hybridization. Recombinant molecules in a phage lambda vector library were probed by the plaque hybridization method,[7] and recombinant phage which gave positive hybridization were purified. These molecules contained a 1.85 kb *Eco*RI fragment which hybridized to the oligonucleotide probe. Production of light was not observed with recombinants containing the 1.85 kb insert. However, analysis of the DNA sequence revealed a complete *luxA* gene and the 5' end of the *luxB* gene. Extensive matches between DNA sequence of the cloned fragment and the amino acid sequence of peptides from the α and β subunits of *V. harveyi* luciferase unequivocally established the presence of these genes. The dideoxynucleotide method was used for DNA sequencing,[8] and this necessitated the subcloning of parts of the 1.85 kb fragment into M13 sequencing vectors.[9] Subsequent cloning attempts, guided by the precise restriction site map derived from the DNA sequence, resulted in the recovery of fragments with the entire *luxA* and *luxB* genes. Addition of long chain fatty aldehyde was required for light production by these *luxA, luxB* recombinants. Transcription of these *lux*

[5] J. Hedgpeth, M. Ballivet, and H. Eisen, *Mol. Gen. Genet.* **163,** 197 (1978).
[6] D. H. Cohn, R. C. Ogden, J. N. Abelson, T. O. Baldwin, K. H. Nealson, M. I. Simon, and A. J. Mileham, *Proc. Natl. Acad. Sci. U.S.A.* **80,** 120 (1983).
[7] W. D. Benton and R. W. Davis, *Science* **196,** 180 (1977).
[8] F. Sanger, S. Nicklen, and A. R. Coulson, *Proc. Natl. Acad. Sci. U.S.A.* **74,** 5463 (1977).
[9] J. Messing, R. Crea, and P. H. Seeburg, *Nucleic Acids Res.* **9,** 309 (1981).

genes, like the *V. harveyi* recombinants discussed earlier, probably was initiated from a promoter element located outside the cloned fragment. Apparently neither the regulatory control region nor the genes for synthesis of the aldehyde was contained on the cloned DNA. The *E. coli* host must have been capable of providing the reduced flavin for the light reaction because several of the luminescent *luxA, luxB* recombinants did not have sufficient coding capacity to direct the synthesis of an additional gene product which could have satisfied the requirement for a reductant.

The two cloning methods just described are applicable to the cloning of most genes. In the former case a method for transposon mutagenesis was required. In the latter case synthesis of the oligonucleotide probe required amino acid sequence information. Six to ten residues of amino acid sequence probably would be adequate for the design of a probe. Encouraged perhaps by the success in producing light in recombinant *E. coli,* a simpler method was subsequently used to clone *lux* genes. This method relied on light production to indicate the presence of *lux* recombinants. Cloning *lux* genes from *Vibrio fischeri* was achieved by constructing recombinant plasmid libraries and inspecting the clones for light production in the presence and absence of exogenously added aldehyde[10] (see Fig. 1C).

The problem of cloning incomplete *lux* genes was minimized by using a variety of restriction endonucleases to fragment the *Vibrio fischeri* genome. Luminescent recombinants were found in a genomic library which contained *Bam*HI restriction fragments. The use of *Bam*HI was particularly fortuitous since the luminescent recombinants, all of which contained one unique 16 kb *Bam*HI fragment, encoded the genes for the luciferase subunits, the genes for aldehyde biosynthesis, and also the control elements responsible for regulation of *lux* gene expression. Methods for genetic dissection of this DNA fragment will be discussed in the next section. A similar "shotgun" cloning approach has also been used to clone *V. harveyi* luciferase genes.[11] The DNA fragment cloned by this method, like those cloned by the two methods described earlier, apparently did not contain a full complement of *lux* genes because exogenous addition of aldehyde was required and the autoregulatory control characteristic of luminescent *Vibrio* was not manifested in the recombinant *E. coli*. To obtain the full complement of *lux* genes, sequential cloning might be productive. Luciferase genes might be cloned first, and aldehyde and regulatory genes added in subsequent cloning attempts. However, the

[10] J. Engebrecht, K. H. Nealson, and M. R. Silverman, *Cell* **32,** 773 (1983).
[11] T. O. Baldwin, T. Berends, T. A. Bunch, T. F. Holzman, S. T. Rausch, L. Shamansky, M. L. Treat, and M. M. Ziegler, *Biochemistry* **23,** 3663 (1984).

aldehyde and regulatory genes might not be capable of expression in *E. coli* due, for example, to failure of *E. coli* transcription enzymes to recognize *V. harveyi* promoters, or to protein instability in *E. coli,* or to failure of *E. coli* to furnish precursor molecules. In this case the use of a related nonluminous *Vibrio* as a cloning host might be advantageous. Remarkably, aldehyde biosynthesis functions have been demonstrated in extracts from *E. coli* recombinants containing *V. fischeri lux* genes, but the same activities could not be demonstrated in extracts from the parent *V. fischeri*.[12] It is clear that it is difficult to predict the consequences of cloning foreign genes into a recombinant host. Other cloning strategies could have been used to clone *lux* genes. For example, antibody to luciferase subunits might have been used to identify recombinant bacteria containing the *luxA* or *luxB* gene since detection of even relatively small amounts of a luciferase subunit was possible in principle by filter immunoassay methods.[13]

Identification of *lux* Genes and Gene Products

What methods can be used to identify the functions encoded by cloned DNA? As discussed earlier, if cloned DNA directs the synthesis of luciferase or if decoding the DNA sequence reveals the amino acid sequence of luciferase, the DNA fragment must contain the *luxA* and *luxB* genes. If, however, the identity of the gene products is not known, genetic techniques can be used to identify genes, gene products, and operons and to investigate regulation of expression of these genes. Generally, a functional phenotype is required as a starting point for genetic dissection, and cloned DNA is an ideal subject for the application of conventional genetic techniques. Extensive genetic analysis of recombinant molecules containing *lux* DNA from *V. fischeri* has been performed, and this study illustrates the application of genetic methods to the study of bacterial bioluminescence.

The DNA fragment cloned from *V. fischeri* (see previous section) encoded luciferase, aldehyde, and regulatory functions in *E. coli*.[10] Expression of light production in the recombinant closely mirrored that observed for the parent bacterium. An estimate of the amount of DNA devoted to *lux* genes was obtained by subcloning DNA with a variety of restriction endonucleases from the original 16 kb insert, and a 9 kb *Sal*I fragment with identical activity was isolated. This recombinant was used for further genetic investigation. Transposon mutagenesis, performed with Tn*5* and mini-Mu, was used to locate the *lux* genes and to generate a

[12] M. Boylan, A. F. Graham, and E. A. Meighen, *J. Bacteriol.* **163,** 1186 (1985).
[13] U. Henning, H. Schwartz, and R. Chen, *Anal. Biochem.* **97,** 153 (1979).

collection of mutations useful for identifying *lux* gene operons and for determining the direction of transcription of the *lux* operons. Transposon insertion results in a null phenotype (complete loss of function) and also prevents expression of genes transcribed later (downstream) in an operon. Thus, transposon mutations are polar and can negate the activity of several genes in an operon. Transposon insertion mutations have other useful properties including gross perturbation of DNA structure, which simplified mapping mutations, and the capacity to generate gene fusions (see next section).

Definition of *lux* genes and operons on cloned DNA was accomplished by application of the classical complementation test. Basically, pairs of mutant recombinant plasmids (with compatible replicons) were introduced into the same *E. coli* strain. The host cell was also recombination deficient to prevent genetic exchange between the plasmids. The resulting recombinant contained copies of two mutant plasmids neither of which was capable of encoding a complete set of *lux* gene functions. Production of light in these recombinant strains indicated that defects in both plasmids were mutually "complemented" by the other partner. Therefore, the defect in each recombinant plasmid must have been in a different functional component. Failure to obtain light production indicated that mutant defects in both plasmids affected the same functional component. Complementation tests are used routinely to define genes by constructing many combinations of nonpolar mutations situated *in trans*. Noncomplementing pairs of defects generally indicate that the mutations are located in the same gene. However, since transposon mutations are polar, all insertions in a given operon negate one or more genes in that operon. Therefore, pairs of recombinant plasmids each containing a transposon insertion in the same operon will not complement, and application of the complementation test to such mutations results in the definition of operons, not genes.

A large number of strains containing one recombinant plasmid mutated by transposon Tn*5* and one mutated with mini-Mu were constructed. Noncomplementing defects were located in two separate regions of the 9 kb *lux* fragment indicating the presence of two operons (see Fig. 2). A qualitative measure of light production was obtained by visual inspection. For a more quantitative determination, assays of recombinant *E. coli* were photographed by their own light as shown in Fig. 3 (one of the complementation tests described above), or used to expose X-ray film placed in close proximity to the bacterial cultures. When precise quantitation was required a photometer or scintillation counter (in chemiluminescence mode) was used. Identification of individual *V. fischeri* genes required the isolation of several hundred nonpolar point mutations for use in

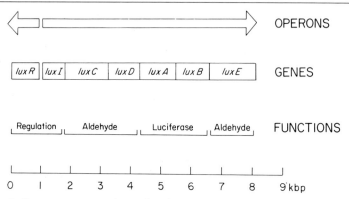

FIG. 2. Genes, operons, and gene functions encoded by a 9 kb fragment of DNA cloned from *Vibrio fischeri*. Arrows show the extent and direction of transcription of the two *lux* operons.

complementation tests.[14] All *lux* mutations were originally isolated on one recombinant plasmid by chemical mutagenesis *in vitro* with hydroxylamine and were subsequently transferred by reciprocal recombination to a second recombinant plasmid with a compatible replicon. Again "diploid" strains with pairs of mutated recombinant plasmids were constructed and assayed for light production. The observation of 7 noncomplementing groups of mutations indicated the presence of 7 *lux* genes. The location of *lux* gene defects was mapped by another complementation test which employed point mutations in combination with previously mapped transposon mutations. The locations and designations of the *lux* genes and operons is shown are Fig. 2.

The phenotypes of the mutants were helpful in assigning functions to the *lux* genes.[10,14] For example, exogenous addition of aldehyde (decanal) to recombinant strains with *luxC, luxD,* or *luxE* mutations restored light production. Therefore, these genes functioned to provide the aldehye substrate (see Fig. 2). No luciferase, as assayed by *in vitro* methods, was found in mutants with *luxA* or *luxB* defects. By using the *luxA* and *luxB* genes from *V. harveyi* as radioactive hybridization probes it was established that *V. fischeri luxA* encoded the α subunit of luciferase and *luxB* the β subunit. Mutations in *luxR* and *luxI* prevented expression of the operon containing aldehyde and luciferase functions. Strains with defects in *luxI* could produce light if the sensory molecule for regulatory control of bioluminescence, autoinducer, was provided exogenously. So, synthesis of autoinducer was assigned to this gene. A more complete description of the analysis of regulatory functions is made in the following section.

[14] J. Engebrecht and M. Silverman, *Proc. Natl. Acad. Sci. U.S.A.* **81,** 4154 (1984).

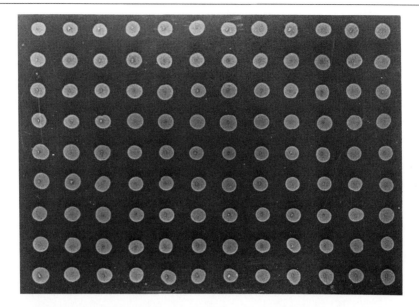

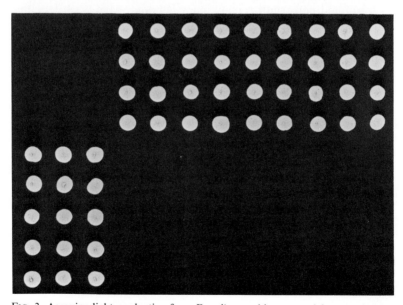

FIG. 3. Assaying light production from *E. coli* recombinants used for complementation analysis. Top photograph shows recombinant strains containing pairs of plasmids with transposon-generated *lux* mutations. The bottom photograph was exposed in the dark with bacterial light. Strains with particular combinations of *lux* mutations produced light indicating the presence of two *lux* operons (see text).

A variety of methods have been developed for the identification of gene products encoded by cloned genes. The minicell technique[15] was used to identify the products of the *V. fischeri lux* genes. Particular septation mutants of *E. coli* bud off "minicells." Recombinant plasmids often segregate with the minicells which are competent for transcription and protein synthesis. Since the host chromosome does not enter the minicells, protein synthesis in the minicells is directed exclusively by the recombinant plasmid. Consequently, minicells with *lux* plasmids incorporated radioactive amino acids (i.e. [^{35}S]methionine) only into plasmid and *lux* gene-encoded proteins. Protein products were assigned to specific *lux* genes on the basis of the correspondences between labeled protein bands on an SDS–polyacrylamide gel and the presence of specific *lux* genes on the recombinant plasmids which directed protein synthesis in minicells.[14] It was thus possible to make the following assignments of gene products to *lux* genes: *luxR,* 27,000 MW protein; *luxI,* 25,000 MW; *luxC,* 53,000 MW; *luxD,* 33,000 MW; *luxA,* 40,000 MW; *luxB,* 38,000 MW; and *luxE,* 42,000 MW.

Other methods for differentially programming the synthesis of proteins encoded by cloned genes include (1) infection of ultraviolet light (UV) irradiated cells with recombinant lambda phage,[16] (2) labeling the products of UV-irradiated maxicells containing recombinant plasmids,[17] (3) induction of strong promoters linked to cloned genes,[18] and (4) addition of cloned genes to transcriptionally coupled, cell-free translation systems.[19] The *luxA* and *luxB* genes from *V. harveyi* were linked to inducible promoters from phage lambda.[4] Induction of the promoters resulted in greatly amplified transcription of the *lux* genes which resided on a multicopy plasmid vector, and the disproportionate synthesis of *lux* gene products which resulted allowed the detection of radioactive *lux* proteins on SDS–polyacrylamide gels despite the presence of a background of labeled host proteins. The molecular weights of the *luxA* and *luxB* gene products were in agreement with those determined for the subunits of luciferase isolated from the parent *V. harveyi*. The ability to radioactively label specific gene products has another application. The cellular location of *lux* gene products could be determined by separating bacteria labeled in spe-

[15] P. Matsumura, M. Silverman, and M. Simon, *J. Bacteriol.* **132,** 996 (1977).

[16] S. R. Jaskunas, L. Lindahl, M. Nomura, and R. Burgess, *Nature (London)* **257,** 458 (1975).

[17] A. Sancar, A. M. Hack, and W. D. Rupp, *J. Bacteriol.* **137,** 692 (1979).

[18] T. Maniatis, E. F. Fritsch, and J. Sambrook, "Molecular Cloning—A Laboratory Manual," p. 405. Cold Spring Harbor Lab., Cold Spring Harbor, New York, 1982.

[19] J. F. Evans, S. McCracken, C. M. Miyamoto, E. A. Meighen, and A. F. Graham, *J. Bacteriol.* **153,** 543 (1983).

cific *lux* proteins into cytoplasmic, periplasmic, inner membrane, and outer membrane fractions. Luciferase is known to reside in the cytoplasm, but the cellular location of other *lux* gene products such as those involved in regulation of gene expression is not known. The efficacy of this approach has been demonstrated with other cloned genes.[20] Alternatively, the specific gene products can be labeled and used to guide the purification of large amounts of unlabeled proteins.

Use of Gene Fusions

Expression of the *luxA* and *luxB* genes cloned from *V. haveyi* was greatly stimulated by coupling or "fusing" promoter control elements from bacteriophage lambda to these luciferase genes. The native promoter for these *lux* genes apparently was not cloned or did not function well in *E. coli*. Linkage of such transcriptional control elements to the luciferase-coding region was accomplished by recombinant methods; a DNA fragment with a phage lambda promoter was inserted into a site proximal to the *lux* genes in the recombinant plasmid.[4] The purpose of this construction was to elicit expression of the *V. harveyi lux* genes, but the gene fusion technology has also been used for studying transcriptional regulation of other *lux* genes.

It was possible to deduce the direction of transcription of the *lux* operons cloned from *V. fischeri* by inserting the *lac* genes from *E. coli* into these operons. Because the native promoter for the *lac* genes was not present, expression of *lac* was dependent upon transcription initiated at the *lux* operon promoter. Furthermore, *lac* expression required correct alignment of the inserted *lac* genes relative to transcription from the *lux* promoter; only one of the two possible orientations of the inserted DNA could support the transcription of the *lac* genes. Thus, the direction of transcription of a *lux* operon could be found by determining which orientation of insertion gave a Lac⁺ phenotype. In practice, insertion of the *lac* genes was achieved by using a particular derivative of transposon Mu, i.e., mini-Mu, for *in vivo* mutagenesis.[21] Transposon mini-Mu contains, in addition to a drug-resistance gene and several functions required for transposition, the *lacZ* and *lacY* genes from *E. coli*. The *lac* genes, without the *lac* promoter, are situated near one terminus of the mini-Mu, and insertion of this transposon into a target gene in an operon can result in gene fusions which link transcription of the target gene to the *lac* genes. To obtain *lac* expression the transposon must insert in the orientation which aligns target gene transcription with *lac*, and this orientation can

[20] H. F. Ridgway, M. Silverman, and M. Simon, *J. Bacteriol.* **132,** 657 (1977).
[21] B. A. de Castilho, P. Olfsen, and M. J. Casadaban, *J. Bacteriol.* **158,** 488 (1984).

easily be determined by restriction mapping. *Lac* expression is convenient to measure by a variety of methods which detect β-galactosidase activity which is the product of the *lacZ* gene.

Transposon mini-Mu insertion mutants (discussed in another context in the previous section) had Lac$^+$ and Lac$^-$ phenotypes,[10] and mapping the orientation of insertion of the transposon in the Lac$^+$ strains gave the direction of transcription of the *lux* operons shown in Fig. 2. The polar effect of transposon insertion could also be used to deduce the direction of transcription. For example, since transposon insertion negates expression of genes distal to or downstream from the target gene, transposon insertion in the *luxD* gene would also negate expression of *luxA*, *luxB*, and *luxE*, but insertion in *luxE* would negate only *luxE*. This in fact was observed, thus confirming the result deduced from the properties of the gene fusions.

Light production in marine *Vibrio* is controlled by a mechanism called autoinduction. The extracellular concentration of a molecule, autoinducer, which is produced by the bacteria positively regulates the production of light. Each *Vibrio* species produces a specific autoinducer. In the case of *V. fischeri* it is *N*-(β-ketocaproyl)homoserine lactone. At low cell density little autoinducer has accumulated and little light is produced, whereas at high cell density a threshold concentration of autoinducer has been reached and 100 to 1000 times more light is produced. Strains with *lux–lac* gene fusions resulting from mini-Mu insertion were used to show that this control acts at the level of gene transcription.[10] Production of light in the recombinant strains with *V. fischeri lux* genes was replaced by the synthesis of β-galactosidase. In strains with *lac* fused to *luxC, D, A, B*, or *E* β-galactosidase synthesis was "autoregulated." It was thus apparent that autoinducer influences light production by controlling transcription of a *lux* operon rather than by affecting the biochemistry of luminescence or the translation of *lux* mRNA. Quantitation of *lux* specific mRNA transcripts using cloned *lux* DNA as a hybridization probe in northern blot analysis also suggested that autoinduction operates at the level of *lux* gene transcription.[22]

By integrating the results of genetic analysis a model for genetic control of bioluminescence was proposed (see Fig. 4). The *luxR* and *luxI* genes were shown to be necessary for transcription of the operon containing *lux* genes for the aldehyde and luciferase functions. The *luxI* gene is necessary for autoinducer production, and *luxR* encodes a function necessary for the transcriptional response to autoinducer. The *luxI* gene is also

[22] C. M. Miyamoto, A. D. Graham, M. Boylan, J. F. Evans, K. W. Hasel, E. A. Meighen, and A. F. Graham, *J. Bacteriol.* **161,** 995 (1985).

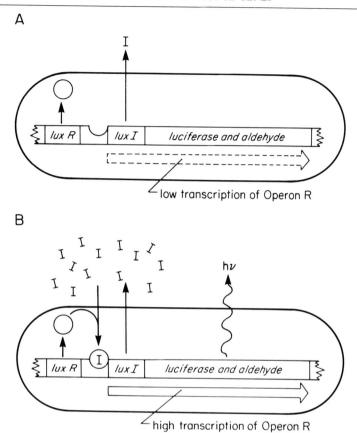

FIG. 4. Model for genetic control of bioluminescence of *Vibrio fischeri*. (A) Dilute cell culture; (B) dense cell culture. A diffusable sensory molecule, autoinducer (I), and a regulatory protein interact to increase transcription of operon R which contains genes for enzymes catalyzing light production. The *luxI* gene encodes a protein required for synthesis of autoinducer, and *luxR* encodes the regulatory protein. Initially, the concentration of autoinducer increases as a function of cell density. However, as a threshold concentration of autoinducer is reached transcription of operon R is stimulated. Because *luxI* is a part of operon R, this stimulation amplifies the synthesis of autoinducer. The accumulation of autoinducer is now influenced by positive feedback control, and transcription of operon R and synthesis of enzymes for light production increases exponentially.

the first gene in the above operon, and it is apparent that the *luxI* gene is positively regulated by its own product. A consequence of such a positive feedback circuit is an exponential increase in *lux* gene expression once a critical concentration of autoinducer is reached, and this prediction does in fact agree with actual measurements.

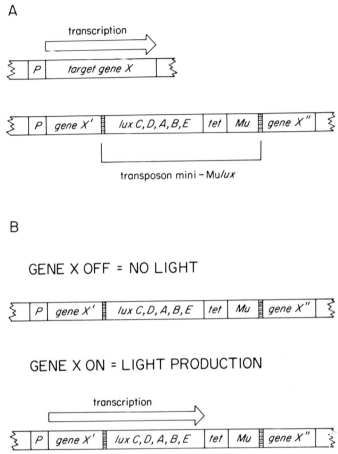

FIG. 5. Constructing *lux* gene fusions *in vivo* with transposon mini-Mu*lux*. Transposition of mini-Mu*lux* results in insertion of *lux* genes into the coding region of target gene X. Two orientations of insertion are possible, and the orientation of insertion shown in A aligns transcription of the target gene with that of the *lux* genes. When transcription of gene X occurs *lux* genes are also transcribed, and light production results (B).

The *lacZ* gene has been particularly useful for constructing gene fusions primarily because many sensitive methods have been devised for measuring β-galactosidase. As mentioned above the *lacZ* gene has been used in the construction of a transposon which generates gene fusion *in vivo*. Because luminescence is extremely convenient to measure and detection does not require disturbing the cells, *lux* genes were used as an indicator of the transcription of other genes. The *lux* genes encoding the enzymes for light production were used to construct a transposon which

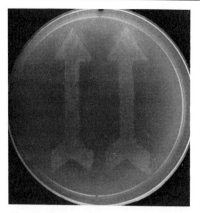 

FIG. 6. Measuring gene expression with light. The petri plate shown at the left contains a gradient of calcium from approximately 0.1 mM (bottom) to 10 mM (top). Strains of *Vibrio parahaemolyticus* which contain transposon-generated *lux* gene fusions have been innoculated across the surface of this plate. These strains were photographed by their own light (right panel). One strain (right arrow) only produced light in the presence of calcium, while light production by the other strain (left arrow) is not influenced by calcium. The *lux* genes in the strain on the right are fused to a gene which is regulated by the presence of calcium.

would generate gene fusions upon integration in a target gene. This transposon, mini-Mu*lux*, was constructed by inserting a restriction fragment containing the *luxC, D, A, B,* and *E* genes into a mini-Mu transposon (see Fig. 5A). The resulting transposon, like the mini-Mu containing *lac,* was capable of transposition, encoded a drug-resistance phenotype and could generate *lux* gene fusions.[23] Strains with such gene fusions expressed light as a function of target gene transcription (see Fig. 5B). For example, *E. coli* strains with *lux* genes fused to the chromosomal *lacZ* gene produced light when an inducer of the *lac* operon was added.

Vibrio parahaemolyticus was also mutagenized with mini-Mu*lux*. The isolation of a mutant bank of Lux[+] strains containing mini-Mu*lux* fusions in most nonessential genes required mobilization of the transposon into *V. parahaemolyticus* with transducing phage P1 followed by selection of transductants expressing the transposon-encoded drug resistance phenotype.[3] This mutant bank was exploited for a variety of purposes. Collection of strains with insertions in known gene systems was guided by inspection for particular mutant phenotypes. Subsequent studies included cloning of genes and analysis of transcriptional regulation of the target genes. However, the bank was also a valuable resource for identifying genes regulated by a particular physiological condition or metabolite but

[23] J. Engebrecht, M. Simon, and M. Silverman, *Science* **227**, 1345 (1985).

for which genes the mutant phenotypes were not known. For example, transposon mini-Mu*lux* mutants containing insertions in genes regulated by calcium were isolated by replica plating the bank onto nutrient plates with and without added calcium and selecting those strains which only produced light in response to addition of calcium. The regulation of light production in such a strain is shown in Fig. 6. The target gene, of unknown function, was fused to the *lux* genes which were expressed when calcium induced transcription of the target gene. Genetic study can now be commenced to define the function of this previously unrecognized gene. A similar approach could be used to identify genes regulated by cAMP, nutrient limitation, or a variety of other factors. Furthermore, now that the *lux* genes have been made mobile by inserting them in a transposon, many kinds of nonluminous bacteria could be tagged with the distinctive Lux phenotype. Strains marked with luminescence could be useful for following the fate of bacteria released into the environment.

Acknowledgment

The authors acknowledge support of their work at The Agouron Institute by a contract from the Office of Naval Research.

[9] Purification of Bacterial Luciferase by Affinity Methods

By Thomas O. Baldwin, Thomas F. Holzman, Rita B. Holzman, and Vicki A. Riddle

Any purification method requires an accurate and reproducible assay, and the various reported approaches to the purification of bacterial luciferase all benefit from the rapid, reproducible, and precise assays that are available. Bacterial luciferase catalyzes the reaction shown in Fig. 1.

The various commonly used assays, which have been summarized in a previous volume of this series,[1] differ only in the methods used to supply the substrates $FMNH_2$, aldehyde, and O_2. The conditions of each type of assay, as well as the intensity of the light detected, reflect specific aspects of the interactions of the luciferase with its substrates.[2] Understanding of these interactions is important to an understanding of the rationale of the various methods; we therefore review several important considerations here.

---

[1] J. W. Hastings, T. O. Baldwin, and M. Ziegler-Nicoli, this series, Vol. 57, p. 135.
[2] M. M. Ziegler and T. O. Baldwin, *Curr. Top. Bioenerg.* **12,** 65 (1981).

METHODS IN ENZYMOLOGY, VOL. 133

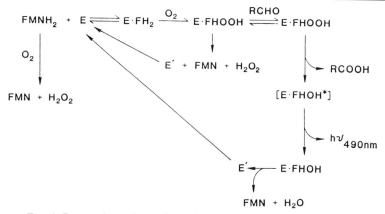

FIG. 1. Proposed reaction pathway for the bacterial luciferase reaction.

The FMNH₂ Injection Assay

The most commonly used assay for routine measurements is the $FMNH_2$ injection assay. The reaction is initiated by injection of reduced flavin (generally catalytically reduced by H_2 over platinized asbestos) into a vial containing enzyme, aldehyde, and O_2. The $FMNH_2$ binds to the luciferase (forming intermediate I) and reacts with O_2 to form the 4a-peroxydihydroflavin intermediate (intermediate II),[3] which in the presence of bound aldehyde substrate (intermediate IIA), then reacts with the aldehyde to form an excited flavin species and the carboxylic acid product. The excited flavin species emits a photon of blue-green light; ground state FMN is the flavin product of the reaction. Any reduced flavin that does not bind to the enzyme to form intermediate II by reaction with O_2 will react nonenzymically with O_2 to yield FMN and H_2O_2. The substrate is thereby removed from the reaction much more rapidly than the rate-limiting step in the catalytic cycle, so no catalytic turnover is possible in this assay. The activity of the enzyme is measured as the peak intensity of light emission reached following the injection of $FMNH_2$ (Fig. 2). In the absence of further $FMNH_2$ substrate, the emission intensity then decays in a first-order fashion, the rate of decay reflecting the rate-limiting step in the pathway after formation of intermediate IIA.

The peak intensity with saturating substrate concentrations is proportional to the amount of active enzyme present in the reaction over at least six orders of magnitude in enzyme concentration, but the decay rate of

[3] J. W. Hastings, C. Balny, C. LePeuch, and P. Douzou, *Proc. Natl. Acad. Sci. U.S.A.* **70,** 3468 (1973).

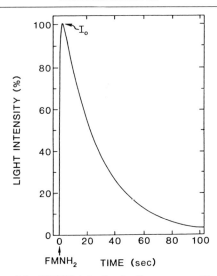

FIG. 2. Time course of the $FMNH_2$ injection luciferase assay. The assay was initiated by injection of 1.0 ml of catalytically reduced flavin mononucleotide into a vial containing luciferase and n-decanal [20 μl of a 0.01% (v/v) sonicated suspension] in 1.0 ml of air-equilibrated buffer (pH 7.0) containing 0.2% BSA (w/v). Light intensity was monitored as a function of time following injection of $FMNH_2$.

the light emission is independent of the concentration of enzyme[4]; rather, as indicated above, the decay of the luminescence reflects the rate-determining step in the catalytic pathway. While the decay rate is strongly affected by the chain length of the aldehyde substrate, the quantum yield of the reaction (proportional to the area under the emission curve) is independent of the chain length of the aldehyde for chain lengths greater than 6 carbons: the initial maximum light intensity is high if the decay rate is rapid and low if the decay rate is slow.[5]

The concentration of aldehyde in the assay mixture is of critical importance, since the enzyme from *Vibrio harveyi* is subject to aldehyde substrate inhibition.[5,6] The peak light emission is not only dependent on the concentration of aldehyde, but on the buffer composition. In buffers containing high concentrations of phosphate ($\sim 0.2\ M$), the luciferase is much less sensitive to aldehyde inhibition than in low phosphate ($\sim 0.02\ M$) buffers or cationic buffers such as Bis-Tris. Addition of bovine serum

[4] J. W. Hastings, Q. H. Gibson, J. Friedland, and J. Spudich, *In* "Bioluminescence in Progress" (F. H. Johnson and Y. Haneda, eds.), pp. 151–186. Princeton University Press, Princeton, N.J., 1966.

[5] J. W. Hastings, K. Weber, J. Friedland, A. Eberhard, G. W. Mitchell, and A. Gunsalus, *Biochemistry* **8**, 4681 (1969).

[6] T. F. Holzman and T. O. Baldwin, *Biochemistry* **22**, 2838 (1983).

albumin (BSA) to the assay buffer also protects the enzyme from aldehyde inhibition, possibly due to the aldehyde buffering effect of the protein. The enzyme–$FMNH_2$ complex does not appear to be susceptible to
aldehyde inhibition.

The affinity purification method that we describe here is based on the
cooperative interactions of the enzyme luciferase with the inhibitor 2,2-
diphenylpropylamine (DΦPA) and phosphate.[7,8] The inhibitor is competitive with the substrate aldehyde, but binds more tightly to the enzyme–
$FMNH_2$ complex. Likewise, the inhibitor binds more tightly to the
enzyme–phosphate complex than to the "free" enzyme. The impure enzyme in a high phosphate buffer can be bound to the inhibitor attached to
an insoluble support material (Sepharose); impurities can be removed by
washing with the high phosphate buffer. Subsequently switching to a
buffer without phosphate reduces the affinity between the enzyme and the
inhibitor, releasing the enzyme from the support. This selective ternary
complex formation allows for a very high level of purification.[8] We give
the detailed procedures below.

Reagents and Materials Required for DΦPA-Sepharose Synthesis

Sepharose (either 4B or 6B)
1,4-Butanediol diglycidyl ether (Aldrich)
2,2-Diphenylpropylamine (Aldrich; 10 mg/ml settled volume of
   Sepharose)
0.6 M NaOH (1 ml/ml settled volume of Sepharose)
$NaBH_4$
Dioxane/0.2 M sodium carbonate, pH 10.5 (1 : 1, v/v)
Dioxane/water (1 : 1)
Dioxane/0.20 M phosphate, pH 7.0 (1 : 1)
95% ethanol
Sintered glass funnel
Rotary evaporator with regulated water bath (Buchi model Rotava-
   por R, or equivalent)
Round-bottomed flask (at least four times the settled volume of
   Sepharose)

Synthesis of DΦPA-Sepharose

To 100 ml (settled volume) of Sepharose 6B in a 500-ml round-bottomed flask was added 100 ml of 1,4-butanediol diglycidyl ether, 100 ml
0.6 M NaOH, and 200 mg sodium borohydride. The flask was attached to
a rotary evaporator and the slurry mixed by slow rotation overnight at

[7] T. F. Holzman and T. O. Baldwin, *Biochemistry* **20,** 5524 (1981).
[8] T. F. Holzman and T. O. Baldwin, *Biochemistry* **24,** 6194 (1982).

room temperature. The resulting epoxy-activated Sepharose was collected on a sintered glass funnel, washed to neutrality with water, and suction dried.

The ligand 2,2-diphenylpropylamine was dissolved at 10 mg/ml in dioxane/0.2 M sodium carbonate, pH 10.5 (1 : 1, v/v). The coupling of the ligand to the epoxy-activated Sepharose was initiated by mixing 100 ml of the 2,2-diphenylpropylamine solution with 100 ml (settled volume) of the epoxy-activated Sepharose in a 500-ml round-bottomed flask. The flask was again attached to the rotary evaporator and the slurry mixed by slow rotation at 60° for about 24 hr. As the reaction proceeded, the pH of the solution dropped. To maintain a reasonable rate of reaction, the pH was checked and adjusted to 10.5 with 0.1 M NaOH every 3 hr for the first 12 hr, and again after 18 hr. The pH did not change between 18 and 24 hr, indicating that the reaction was complete.

After the 24-hr coupling reaction, the DϕPA-Sepharose (~100 ml) was collected on a sintered glass funnel and washed successively with 600 ml of 1 : 1 dioxane/water, 600 ml of 1 : 1 dioxane/0.20 M phosphate, pH 7.0, and 1200 ml of 95% ethanol. The DϕPA-Sepharose was then either used directly or stored in 50% ethanol/water at 4° (Fig. 3).

Preparation of Luciferase for Binding to the DϕPA-Sepharose

Lysis of bacterial cells was accomplished as described in detail previously.[1] There are several key points to note regarding cell lysis. First, freezing cells in liquid nitrogen or in a low temperature (−70° or lower) freezer results in poor lysis, probably because ice crystal formation is reduced by rapid freezing. Rather, slow freezing at −20° and storage in a non-frost-free freezer results in the best lysis. Storage and thawing are best accomplished in thin pads of cell paste in freezer bags. On rare occasions in which we have thawed and refrozen cells at −20°, we have obtained the most rapid and complete lysis.

Cell Lysis

The general procedure to lyse the cells by osmotic shock is presented below. The procedure is carried out at 4°.

1. Remove the bag of cells from the freezer and allow to thaw for several hours at room temperature or overnight at 4°. Add a small volume of 3% NaCl to the thawed cell paste and stir well, such that the cell paste becomes sufficiently fluid that it may be easily poured.

2. Slowly pour the cell paste into rapidly stirred deionized H$_2$O that is about 4°. The volume of water should be about 5 ml/g of original cell

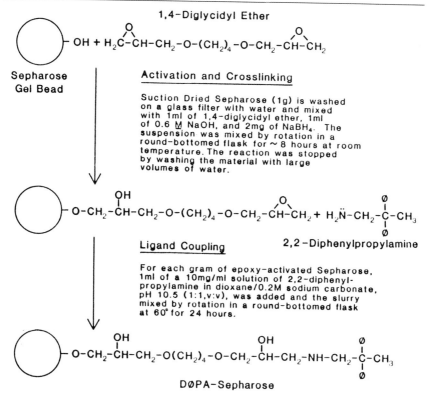

FIG. 3. Flowchart for the synthesis of DφPA-Sepharose. The details of the procedure are given in the text.

paste, and it is critical that the water be stirred as rapidly as can be accomplished with a magnetic stirrer, so that the cells can be dispersed quickly. Failure to stir the lysis mixture will result in poor lysis, apparently due to equilibration of the internal and external salt concentration without cell lysis. Addition of the cell paste to the stirred water should be done in small aliquots such that the complete addition of 500 g of cell paste to 2.5 liters would be complete in about 20 min. The pH of the lysate should be checked regularly during the addition of cell paste; the pH normally drops and should be maintained at about pH 7 by addition in small increments of Tris base to as much as 1 g/liter.

3. Lysis is usually complete within 15 min of the final addition of cells to the water. To check for complete lysis, it is important to remove the cell debris from the solution. Luciferase in unlysed cells will react and emit light in the flavin injection assay. Lysis is complete if the luciferase activity in 10 μl of supernatant following centrifugation for 1 min in an

Eppendorf microcentrifuge is within 10% of the activity of 10 μl of the uncentrifuged lysate.

4. If complete lysis does not occur, one must resort to sonication. Since the enzyme from all species studied is sensitive to oxidation, DTE should be added to about 0.5 mM prior to sonication. The only times that we have had serious trouble with cell lysis were when we tried to lyse fresh cells directly from the fermenter or cells that had only been frozen overnight. It appears that after a week or more in the freezer, the cells become more fragile and subject to lysis by osmotic shock.

Ammonium Sulfate Fractionation

If the standard purification method is to be employed,[1,9] DEAE-cellulose batch extraction is performed at this stage. The affinity purification methods described here, using either the column approach or the batch approach, require NH_4SO_4 fractionation of the lysate at this stage. The initial volume of the uncentrifuged lysate is measured and recorded. Solid ammonium sulfate is added to the lysate (at 0–4°) slowly, such that addition of 100 g of ammonium sulfate requires about 5 min. For the *V. harveyi* lysates, the fraction precipitating between 40 and 80% saturation of ammonium sulfate is collected; for the *Vibrio fischeri* and *Photobacterium phosphoreum,* the fractions precipitating between 35 and 80% ammonium sulfate are collected. Ammonium sulfate (242 g/1000 ml of initial volume for *V. harveyi* and 209 g/1000 ml of initial volume for *V. fischeri* and *P. phosphoreum*) is added to the uncentrifuged lysates and stirred slowly with a magnetic stir bar for about 15 min at 0–4°. The suspension is then clarified by centrifugation at 16,000 g (9500 rpm in a Beckman JA-10 rotor) for 30 min at 4°. The supernatants are collected and solid ammonium sulfate (318 g/1000 ml *initial* volume for *V. harveyi* and 351 g/1000 ml *initial* volume for *V. fischeri* and *P. phosphoreum*) is added as before. After about 15 min of slow stirring at 0–4°, the precipitated protein is collected by centrifugation as before. The precipitated protein is removed from the centrifuge bottles with a stainless-steel iced tea spoon and delivered to dialysis tubing through a small glass powder funnel. The remaining precipitated protein is washed out of the bottles and delivered to the dialysis tubing with a minimal amount of 0.1 M phosphate, pH 7.0, such that the protein concentration following dialysis will be 50 mg/ml or greater.

If the sample is to be stored for some time prior to purification of the luciferase, it is best to store it in the presence of ~0.5 mM DTE as a

[9] A. Gunsalus-Miguel, E. A. Meighen, M. Z. Nicoli, K. H. Nealson, and J. W. Hastings, *J. Biol. Chem.* **247,** 398 (1972).

APPLICATION AND ELUTION BUFFERS FOR AFFINITY CHROMATOGRAPHY OF
LUCIFERASES FROM *V. harveyi*, *V. fischeri*, AND *P. phosphoreum*

Bacterial strain	Application buffer	Elution buffer
V. harveyi	0.10 M phosphate, pH 8.5, 0.50 M NaCl, 0.50 M KCl, 0.5 mM DTE	25 mM ethanolamine, 5.0 mM Tris, pH 9.1, 0.5 mM DTE
V. fischeri	0.10 M phosphate, pH 7.0, 0.5 mM DTE	0.30 M Tris, pH 8.5, 0.5 mM DTE
P. phosphoreum	0.35 M phosphate, pH 7.0, 0.5 mM DTE	0.10 M Tris, pH 8.1, 0.5 mM DTE

frozen ammonium sulfate paste (for *V. harveyi* and *V. fischeri*). The luciferases from *V. harveyi* and *V. fischeri* are stable for over a year as ammonium sulfate pastes at $-20°$. The luciferase from *P. phosphoreum* is unstable to freezing whether in ammonium sulfate or in buffer, and is best stored for long times in 0.2 M phosphate, pH 7.0, 0.5 mM DTE, 40% glycerol at $-20°$.

Prior to application to the affinity matrix, the ammonium sulfate is removed from the protein by dialysis against sample application buffer (see the table). The sample is dialyzed against three changes of a 10-fold volume excess of the application buffer and the buffer from the last dialysis is used to equilibrate the affinity matrix prior to sample application. This latter point is of critical importance. We have found that equilibration of the affinity matrix with the buffer with which the sample was last equilibrated results in significantly higher yields and higher binding capacity than are observed when the affinity matrix is equilibrated with "fresh" buffer.

Use of DϕPA-Sepharose in a Column Mode

The approach to affinity purification of bacterial luciferases comprises a simple "on–off" strategy that is based on the observation that the binding of luciferases to 2,2-diphenylpropylamine is enhanced by binding of multivalent anions, especially phosphate, sulfate, and arsenate.[7,8] The procedure for all bacterial luciferases is essentially the same, with variation only in the composition of the application and elution buffers. The sample in application buffer (which contains phosphate) is loaded onto a column of the affinity matrix until the column is nearly saturated with the luciferase. The addition of luciferase is then stopped and the column washed with application buffer (the same buffer with which the column was equilibrated, i.e., the buffer used in the final dialysis equilibration) until the absorbance at 280 nm and luciferase activity in the effluent have dropped to near-baseline values. The column is then washed with elution buffer to remove the luciferase. The luciferases are significantly less stable in the elution buffers than in the application buffers; to circumvent the problem of the luciferase denaturing in the test tubes in the fraction collector used to collect the effluent from the column, a small amount (about 0.1 ml, depending on fraction size) of 1 M phosphate, pH 7.0, is added to each tube in the fraction collector so that the pH of the effluent will drop quickly to near neutrality and the enzyme will be afforded anion stabilization.

Affinity Purification of *V. harveyi* Luciferase

In a typical purification using a column with a 30 ml bed volume, the column is preequilibrated with about 100 ml of sample application buffer (used in the final dialysis of the sample) at a flowrate of 40 ml/hr maintained with a peristaltic pump. A typical sample would have an absorbance of 40–50 at 280 nm and 25,000 to 50,000 LU/ml luciferase activity (assayed with n-decanal; 1 LU (light unit) = 9.8×10^9 quanta/sec). The centrifuged (to remove material precipitating during dialysis) sample is applied to the column through the peristaltic pump at a flow rate of about 40 ml/hr. Fractions are collected from the effluent and assayed for protein (absorbance at 280 nm) and for luciferase. When the luciferase activity in the effluent reaches about 5–10% of the activity in the sample, application is terminated. A typical column of 30 ml bed volume can accommodate about 4.0×10^6 total light units, and loading of the sample usually requires between 1.5 and 2 hr. After sample loading, the column is washed with the same application buffer with which it had been equilibrated, also at about 40 ml/hr, until the absorbance and luciferase activity have returned to near-baseline values, usually about 4 hr. Luciferase elution is then effected by application of the elution buffer (see the table). The luciferase is eluted at a slower flow rate, about 5–10 ml/hr, to maintain a sharp elution profile. The luciferase elutes from the column in a sharp peak centered about 2–3 column volumes from the initiation of application of the elution buffer. Elution of the luciferase requires about 10–15 hr. While the elution can be carried out much more rapidly, both yields and purity suffer. By the procedure described here, we routinely achieve, by conservative pooling of fractions from the column, luciferase from *V. harveyi* of about 95% purity with overall yields of 50–75%.

Affinity Purification of *V. fischeri* Luciferase

The affinity method works well for both large scale and small scale purifications. To demonstrate the point, we will describe the purification

[10] Purification of Bacterial Luciferase by High-Performance Liquid Chromatography[1]

By Dennis J. O'Kane, Mushtaq Ahmad, Iain B. C. Matheson, and John Lee

High-performance liquid chromatography (HPLC)[2] has been used for nearly a decade to perform difficult separations of small organic compounds. Only recently, however, has HPLC become practical for use in the preparative chromatographic separation of proteins.[3,4] This application has been made feasible by the introduction of inert chromatography matrices that have minimized surface areas, but have high degrees of porosity, and have large pore radii which permit the penetration of macromolecules with large hydrodynamic radii. Furthermore, the availability of preparative columns permits the application in many cases of "gram quantities" of protein with little loss in resolution compared to smaller, analytical columns.

Large quantities of bacterial luciferase (Lase) can be rapidly prepared by HPLC using three chromatography procedures which have their respective counterparts in low-pressure liquid chromatography: high-performance size exclusion chromatography (HPSEC), high-performance anion exchange chromatography (HPAEC), and high-performance hydrophobic interaction chromatography (HPHIC). In particular, HPHIC and its low-pressure counterpart are most useful and provide purifications of Lase that equal or exceed those obtained by "affinity" chromatography on diphenylpropylamine-Sepharose.[5]

[1] Supported by grants from NSF DMB-85 12361 and ONR N00014-83-C-0614.

[2] Abbreviations used: HPLC, high-performance liquid chromatography; Lase, luciferase; HPSEC, high-performance size exclusion chromatography; HPAEC, high-performance anion exchange chromatography; HPHIC, high-performance hydrophobic interaction chromatography; DMSO, dimethyl sulfoxide; LU(s), light units(s); LumP, lumazine protein; 2-ME, 2-mercaptoethanol; ϕ-Sepharose, phenyl-Sepharose; DϕPA-Sepharose, diphenylpropylamine-Sepharose.

[3] W. W. Yau, J. J. Kirkland, and D. D. Bly, "Modern Size-Exclusion Chromatography," p. 180. Wiley (Interscience), New York, 1979.

[4] J. D. Pearson, W. C. Mahoney, M. A. Hemodsen, and F. E. Regnier, *J. Chromatogr.* **207**, 325 (1981).

[5] T. F. Holzman and T. Baldwin, *Biochemistry* **21**, 6194 (1982).

Brief Introduction to HPLC

It is beyond the scope and intent of this chapter to describe the theory and practice of HPLC. The reader is referred elsewhere for this.[6] However, a brief description of the chromatography system and sample preparation is warranted.

HPLC System

The chromatography system we use is manufactured by LKB, Bromma, Sweden, and can be operated unattended for long periods. Most operations are programmed and microprocessor controlled. The "controller" operates two high-pressure pumps which can be programmed to turn on, turn off, or form a gradient at selected times and with selected flow rates, and also allows the columns to be washed and regenerated automatically. The "controller" also sends an enabling signal to a fraction collector which in turn initiates its own set of programmed instructions. The sample is loaded into the loop of an injection valve using a syringe. (Other manufacturers, such as Pharmacia, offer another type of filling device for large volumes which also works quite well for large sample application). The injection valve is located between the pumps and column assembly and consequently the valve must be able to withstand the moderate pressure changes (26 kg cm^{-2}) that can occur when switching the sample loop in stream. The column assembly consists in all cases of a small, comparatively inexpensive precolumn and an expensive preparative column. The purpose of the precolumn is to filter the sample thus preventing the catastrophic clogging of the preparative HPLC column by the proteinaceous sample. The eluted protein passes through an 8 μl flow cell in an absorbance detector (278 nm) and the signal output is recorded by one channel of a two-channel chart recorder. The second channel is used to record the gradient employed, if any. The eluted protein is collected using a fraction collector which can be programmed for the number of fractions to be collected, to delay a set interval of time before starting collecting fractions, to collect fractions of a certain size, and to delay a set interval of time to allow the eluant in the flow cell to reach the fraction collector. Consequently, fractions can be reproducibly collected in the same set of tubes from several HPLC runs on the same column.

Many types of HPLC columns are available which differ in the type and form of the polymer matrix material used, the type of bonded phase applied and the type of functional group used if any. Some of these columns have recently been summarized by Regnier.[6] The preparative

---

[6] F. E. Regnier, this series, Vol. 91, Part I, p. 137.

TABLE I
PREPARATIVE HPLC COLUMN PACKINGS EMPLOYED IN LUCIFERASE PURIFICATIONS

Matrix	Manufacturer[a]	Column designation	Particle radius[b,c]	Functional group	Procedure	Mean pore radius (Å)[c]
Silica	Toya Soda (LKB)	TSK G3000SWG	6.5 ± 1	None	HPSEC	120[d]–150
Silica	Toya Soda (LKB)	TSK-DEAE 3SW	6.5 ± 1	Diethyl-amino-ethyl	HPAEC	120[d]
Silica	Pharmacia	Polyanion S1	8.5	Chiefly tertiary amine	HPAEC	150
Hydrophilic polymer	Toya Soda (Bio-Rad)	TSK Phenyl-5-PW	5	Phenyl	HPHIC	500

[a] Suppliers of the packed columns are listed in parentheses if different from the manufacturer.
[b] Radius ± standard deviation (where available), μm.
[c] Manufacturers' or suppliers' specifications unless otherwise noted.
[d] From Regnier.[6]

HPLC columns that have been employed in the purification of Lase are described in Table I. With one exception, the HPLC column matrices are all silica-based beads which have been modified with proprietary hydrophilic coatings to allow use with proteins and nucleic acid samples. The beads have very narrow size distributions, in contrast to conventional chromatography media, and have large "mean pore radius" (<120 Å). (Although single numbers are obtained by the mercury intrusion and nitrogen desorption techniques used to determine mean pore radius, a distribution of pore radii is detected by freeze-etching and direct examination of beads by electron microscopy.)[7] Other HPLC column types that may prove useful for protein separations are hydroxyapatite (Bio-Rad Laboratories) and chromatofocusing (Pharmacia) matrices. These have not been used to purify bacterial Lase to date. It is not recommended that reverse-phase HPLC columns be used in the preparation of Lase, since it is doubtful that the labile enzymatic activity would survive elution with the organic solvents and acids typically used to elute these HPLC columns.[8,9]

[7] W. E. Rigsby, W. L. Lingle, D. J. O'Kane, and J. Lee, *Proc.—Annu. Meet., Electron Microsc. Soc. Am.* **43**, 386 (1985).
[8] M. van Der Rest, H. P. J. Bennett, S. Soloman, and F. H. Glorieux, *Biochem. J.* **191**, 253 (1981).
[9] M. Hernodson and W. C. Mahoney, this series, Vol. 91, Part I, p. 352.

Preparation of Solvents and Samples for HPLC

The HPLC system contains several components that can be damaged or clogged by particulates present in samples and in buffers. These include the piston and piston seals of the pumps, the precolumns and HPLC columns, and any solvent filter in the system. The bed supports in the precolumns and columns are particularly susceptible to clogging.

The preparation of solvents for HPLC of proteinaceous extracts is not as tedious, nor is it as rigorous, as that required for the HPLC of small organic molecules and peptides by reverse-phase techniques. These latter applications require extensive degassing of solvents and/or purging with helium to prevent bubble formation and require, as well, the use of high quality, expensive, specially purified solvents. For example, there is little need to degas aqueous buffers providing these are prepared at room temperature and used at a lower temperature such as 4°. Second, bubble formation in the absorbance detector can be prevented by having a length of tubing connected to the exit of the flow cell, or by elevating the detector above the column outlet.

The major concern in the HPLC of biological extracts is that the columns will become clogged by particulates which possibly could ruin a costly HPLC column. Particulates are removed from the solvents by filtration using a membrane filter with a pore size of 0.45 μm or less (Millipore or Nucleopore). The filtered solvents are stored covered to prevent contamination by dust or other debris. The HPLC pumps and columns are secondarily protected from particulates by solvent filters placed on the solvent intake lines. It is recommended that all solvents be filtered on a daily basis to remove any bacteria and microcrystals of salts that form in concentrated buffer solutions. Buffers containing high concentrations of sodium phosphates are not recommended for HPLC of proteins at 4°. Microcrystals will form in filtered 0.5 M sodium phosphate buffer, at pH 7.0 and 4°, that will rapidly clog HPLC frits and columns. (Columns clogged by microcrystals of salt can frequently be unclogged by slowly warming to room temperature and washing with distilled water.)

Particulates must also be removed from the proteinaceous samples prior to application to the HPLC column. This is generally performed in two stages. The extract is first centrifuged in 1.5 ml capacity tubes for 15 min (Eppendorf Model 5412) to remove the bulk of the precipitates. The supernates are carefully withdrawn and 2-ml aliquots are filtered during sample application to the HPLC columns using a membrane filter (Spartan-3, 3 mm diameter, 0.45 μm pore size; Schleicher and Schuell) attached to the loading syringe. The centrifugation step removes most of the particulates which would otherwise clog the small surface area of the syringe filter. Consequently, the filter can be employed to make more than one sample application, when multiple applications are required, and this minimizes sample loss due to a hold-up volume in the filter assembly. Although this might seem to be a trivial point, sample loss due to hold-up volumes and surface films becomes significant when the samples may contain 200 mg of protein per ml.

Pump and Column Maintenance

The second major concern in HPLC is that the high-pressure pumps and columns may be destroyed by the solvents and samples used. The newer HPLC columns and pump heads are now designed of virtually noncorrosive materials such as titanium or ceramic materials and the newest columns are glass or glass lined. However, the HPLC pumps and columns manufactured up until very recently have stainless-steel surfaces that are wetted by buffers and these will corrode. The rate of corrosion can be minimized by flushing the HPLC system with distilled water between runs and storing the system and columns in distilled water, plus a bacteriostatic compound such as sodium azide, during idle periods.

Another component susceptible to damage by buffer salts is the piston–piston seal assembly. It is ironic that the piston of the HPLC pump, made from sapphire in the model we use, is tough enough to produce a pressure of 420 kg cm^{-2}, and yet is so delicate that it can be scored by a crystal of salt. Buffer that collects on the piston can crystallize by evaporation during periods in which the pump is not in use. Consequently, the piston seal and/or the piston will be scored by the salt when the pump is restarted. The HPLC pumps designed by LKB have a unique feature that allows the back edge of the piston to be continuously flushed with filtered distilled water. Consequently any buffer that passes the piston seal cannot crystallize on the piston, and is removed in the rinse water.

Assay of Luciferase Activity

Very different numbers are found in the literature of bacterial bioluminescence, each of which supposedly represents the maximum light emitting activity of a given type of purified bacterial Lase. These various numbers have arisen due to several factors including (1) using different definitions of the unit used to describe light emission, the light unit (LU), (2) employing different light standards which do not yield light measurements that agree with each other, (3) assaying Lase using photodetectors that have different spectral sensitivities, and (4) using photometers that have different geometries and therefore different light capturing efficiencies. These differences, and means to correct for them, are considered below. The corrections necessary will be simplified presently when a portable photodiode photometer, calibrated at the National Bureau of Standards, and new, portable light standards, become available.

Photometer Calibration

Absolute measurement of the number of photons emitted from a bioluminescent organism or from an *in vitro* bioluminescence reaction requires knowledge of both the absolute photometric sensitivity of the detector as a function of wavelength as well as the photon transfer efficiency from the sample to the detector.

Detector calibration requires the use of a standard lamp, optical bench, monochromators, attenuators, and a second detector, usually a thermopile. This instrumentation is generally not available in laboratories performing bioluminescence assays. Consequently secondary standardization based upon either a chemiluminescent[10] or radioactive standard[11] has been proposed and used.

The chemiluminescence calibration most often employed is the luminol reaction,[10] the reaction of 3-aminophthalhydrazide with H_2O_2 to produce light. The absolute quantum yield of chemiluminescence (Q) is the number of photons produced per luminol molecule reacted. Lee and Seliger[12] measured Q for the reaction of luminol in aqueous solution and in dimethyl sulfoxide (DMSO) and obtained values of 1.24% for Q in both cases. Q has since been remeasured by other workers using several different experimental techniques and these are collected in Table II. Photometer calibration from luminol chemiluminescence in water and in DMSO and the cross calibration of different photometers is described below.

The Luminol Reaction

Assay Solutions Required: Water Reaction

Luminol stock solutions: recrystallized 3-aminophthalhydrazide is dissolved in 0.1 M K_2CO_3 to an absorbance of ~20 at 347 nm (~47 mg/100 ml).

Luminol assay solutions: the luminol stock solution is accurately diluted with 0.1 M K_2CO_3 to an absorbance between 0.0001 and 0.01 at 347 nm.

Peroxide stock solution: 30% by volume H_2O_2 (~12.6 M).

Peroxidase stock solution: 10 mg/ml horseradish peroxidase in 0.1 M K_2CO_3.

---

[10] J. Lee and H. H. Seliger, *Photochem. Photobiol.* **4,** 1015 (1965).
[11] J. W. Hastings and G. Weber, *J. Opt. Soc. Am.* **53,** 1410 (1963).
[12] J. Lee and H. H. Seliger, *Photochem. Photobiol.* **15,** 227 (1972).

TABLE II
LUMINOL CHEMILUMINESCENCE QUANTUM YIELDS[a]

Authors	Date	Detector	Photons[b]	Q (%)[c]	Reference
Lee and Seliger	1965	PMT	9.75	1.25	d
Heller et al.	1971	PMT	6.3	0.83	e
Kalinichenko et al.	1974	Actinometer	9.0	1.18	f
Michael and Faulkner	1976	Integrating sphere	8.8	1.12	g
Lind and Merenyi	1981		9.6	1.23	h
Zalewski and Matheson	1985	Silicon photodiode	8.1	1.06	i

[a] All measurements are for the basic aqueous reaction.
[b] Units are 10^{14} photons per unit A_{347} per ml luminol.
[c] Q is the chemiluminescence quantum efficiency. Average value of $A_{(347)} = 1$ solution (with rejection of Heller et al.[e]) is 9×10^{14} photons ml^{-1}.
[d] J. Lee and H. H. Seliger, Photochem. Photobiol. **4**, 1015 (1965).
[e] C. A. Heller, D. T. Carlisle, and R. A. Harvey, J. Lumin. **4**, 81 (1971).
[f] I. E. Kalinichenko, A. T. Pilipenko, and G. Angeloud, Ukr. Khim. Zh. **40**, 859 (1974).
[g] P. R. Michael and L. R. Faulkner, Anal. Chem. **48**, 1188 (1976).
[h] J. Lind and G. Merenyi, Chem. Phys. Lett. **82**, 331 (1981).
[i] E. F. Zalewski and I. B. C. Matheson (unpublished).

DMSO Reaction

Reagent grade DMSO: due to the hygroscopic nature of DMSO it is preferable to use an unopened bottle of reagent grade.

Luminol stock solution: luminol dissolved in dry DMSO and diluted with dry DMSO to an absorbance between $A_{359} = 0.0001$ and 0.01.

Initiator solution: dry t-butanol saturated with potassium t-butoxide.

Assay Procedures

Reaction in Water. To 1 ml of luminol ($A_{347} = 10^{-4}$ to 10^{-2}) is added a small amount of H_2O_2 and 10 μl of peroxidase. The peroxide concentration (and hence added volume) is chosen to be 300 to 3000 times the luminol concentration. This corresponds to μl/ml amounts of peroxide. Since μl/ml amounts of peroxidase are also required care should be taken to ensure that the optical density does not exceed 0.02 cm^{-1} in order to eliminate the need for self-absorption correction.

Light signal measurement should start upon the peroxide addition since a small amount of light may be produced. The reaction proper is initiated by peroxidase addition. When the light level has dropped to below 5% of the initial maximum intensity a second aliquot of peroxidase should be added. This usually will produce increased light emission. Further peroxidase additions may be tried but are usually ineffective. Upon reaction completion, the integrated light signal is measured and recorded. If the absorbance at 347 nm and solution volume are known then the total number of photons emitted may be obtained since 1 ml of a solution of luminol ($A_{347} = 1$) produces on the average 9×10^{14} (Table II) photons.[12] The calibration factor will be the number of photons/integration unit.

Reaction in DMSO. One milliliter of luminol is placed in an assay vial and the reaction is started by the addition of 10 μl of the initiator solution. As with the aqueous reaction, light measurement should start with the first butoxide addition. Further additions should be made as necessary when the signal level drops to 5% of peak. Also as previously noted, 1 ml of $A_{359} = 1$ luminol yields 9.75×10^{14} photons, and the calibration is as defined above.

Comments

It is recommended that as nearly as possible the same volume of luminol calibration solution be used as in the bioluminescence assay measurements. Additions of reagents should be of minimum volume so that geometry changes are minimized. It has been observed[12] that more reproducible results are obtained when the quantum yield is referenced to the optical density at the luminol absorption maximum rather than volumetrically. The absorption maximum for basic aqueous luminol is at 347 nm with an extinction coefficient of 7640 M^{-1} cm^{-1}. The corresponding values for luminol in DMSO are 359 nm and an extinction coefficient of 7900 M^{-1} cm^{-1}. It is necessary that some means of integrating the photon signal output be available. This might take the form of some digital technique or a chart recorder output allowing area measurement.

Cross-Calibration of Photometers or Spectrophotometers

Problems arise when photodetectors of different geometries and spectral sensitivities are used to measure bioluminescence. This is the case when a photometer is used to assay luciferase activity and a spectrofluorimeter is used to obtain an *in vitro* spectrum of the light emitted. Different photodetectors can be compared to each other following a systematic correction for the different geometries and solvents (a transfer efficiency correction) and a correction for the detector wavelength response differences.

Transfer Efficiency. At first sight it may seem that no correction for the transfer efficiency need be applied if the same cell and geometry are used in calibration and normal use. If the solvent used for the calibration differs from that used for the normal measurements, a correction may be required for the different refractive indicies of the two solvents.[13] Lee and Seliger[12] used a relatively remote source and detector approximating point source geometry. Although the experimental detail is sketchy they report that a relative[12] correction is required in comparing the aqueous and dimethyl sulfoxide results. Matheson et al.[14] have redetermined Q for the aqueous luminol reaction using a photodiode (PD) detector. An approximate cube of luminol solution was used, 1 ml in a 1-cm^2 clear bottomed fluorescence cuvette. Part of the bottom face area was selected by a circular aperture. A similar aperture was placed at a measured distance in front of the PD detector. A fraction of the total light emerging from the near cubic rectangular parallelapiped sample volume is passed by the first aperture, and the fraction of light emerging from the upper aperture reaching the detector may be obtained from solutions to radiative transfer problems found in thermal engineering texts.[15] Thus the transfer efficiency may be calculated, the essential assumptions being that the radiation emerging from the cell is uniform, isotropic and diffuse. E. F. Zalewski and I. B. C. Matheson (unpublished) carried out experiments that demonstrated that the radiation was diffuse and uniform and deviated only slightly from isotropic, the radiation emission rate not being exactly the same from all faces. The effect was minor and easily corrected. They further verified the observation that Q from luminol was the same in aqueous and DMSO solutions, thus demonstrating that no correction for refractive index is required for "close up" geometry.

Detector Wavelength Response. Photometric measurements must be corrected for the wavelength-dependent sensitivities of the photodetector. If the emission spectrum of interest has a wavelength distribution of $E(\lambda)$ and the wavelength response of the detector is $D(\lambda)$ then the photometer response may be represented by

$$S = E(\lambda)D(\lambda)d \tag{1}$$

Thus where $D(\lambda)$ is known, the instrumental response for a second wavelength distribution $E(\lambda)$ may be calculated as

$$S1 = E1(\lambda)D(\lambda)d \tag{2}$$

[13] J. J. Hermans and S. Levinson, *J. Opt. Soc. Am.* **41**, 460 (1951).
[14] I. B. C. Matheson, J. Lee, and E. F. Zalewski, *Proc. Soc. Photo-Opt. Instrum. Eng.* **489**, 380 (1984).
[15] E. M. Sparrow, *in* "Handbook of Heat Transfer" (W. M. Rohsenow and J. P. Hartnett, eds.), Sects. 15–44 and 15–45. McGraw-Hill, New York, 1973.

The problem is that while the absolute emission spectra $E(\lambda)$ and $E1(\lambda)$ may be obtained relatively easily by comparison with a known fluorescence standard, $D(\lambda)$ is not readily accessible, most bioluminescence laboratories not having the requisite optical hardware outlined above. If such measurements are not possible the next best choice is to use the wavelength response supplied by the detector manufacturer. It should be noted that the manufacturer's data are for a "typical" PM and that there may be significant individual variations. If one is only interested in calibration for bacterial bioluminescence a third choice is possible. The emission band envelopes of luminol chemiluminescence in DMSO and the *in vitro* Lase spectrum of *Photobacterium leiognathi* S1 are almost coincident. Therefore in this case the luminol reaction in DMSO may be used for calibration without any consideration of adjustments for wavelength.

Luciferase Assay

Solutions Required[16,17]

Assay buffer: 50 mM phosphate, pH 7.0, containing 1 mg/ml bovine serum albumin and 0.3 mM EDTA

FMNH$_2$: Prepared by photoreducing 0.5 ml FMN (80 μM, in 50 mM phosphate buffer, pH 7.0, containing 20 mM EDTA) in 1 ml disposable plastic syringes with 23-gauge needles

Aldehyde stock solutions: for Lase isolated from bacteria other than *Vibrio harveyi* and *Xenorhabdus*, a saturated solution of tetradecanal in 95% ethanol; for *V. harveyi* Lase, in addition to the solution of tetradecanal a saturated solution of decanal in absolute methanol is also used, both prepared fresh daily.

Assay Procedure.[16,17] One to 10 μl of column fraction containing Lase is added, using a disposable glass micropipet, to 1 ml of assay buffer contained in a homeopathic vial (1 dram capacity). For most Lases, 10 μl of the aldehyde stock solution is added. In the case of *V. harveyi* Lase an optimized amount, usually 0.5–2 μl of the aldehyde solution is used. The vial is placed in the sample compartment of the photometer and the photoreduced FMN is rapidly injected. The bioluminescence signal is recorded using a chart recorder and a scaler/counter. The photometer is calibrated as described above.

Definition of Light Unit. In this laboratory we define a light unit (LU) as 10^{12} photons ml^{-1} sec^{-1}. Total light units are LU multiplied by total volume (ml) of a sample. It should be noted that this definition of LU corresponds to substantially more light than that employed by other labo-

[16] R. Gast and J. Lee, *Proc. Natl. Acad. Sci. U.S.A.* **75**, 833 (1978).

[17] J. Lee, *Photochem. Photobiol.* **36**, 698 (1982).

ratories, and consequently 1 mg of purified Lase will have fewer LUs than reported from other laboratories.[5,18-20]

Purification of Luciferase by HPLC

Luciferase from Photobacterium leiognathi DD17

This inducible, luminescent bacterium strain was isolated and identified by Dr. Reuben Levisohn (Tel Aviv University) from a sample of seawater from the Mediterranean Sea. The isolation of Lase from *P. leiognathi* is challenging and previous experience indicated that the Lase, which constitutes ~1% of the soluble protein, could not be purified to homogeneity by conventional liquid chromatography procedures. This isolate also serves as a source not only for Lase but is used to isolate lumazine protein (LumP) as well.[21]

Following ion-exchange chromatography on DEAE-Sepharose, the *P. leiognathi* DD17 extract is separated into low-molecular weight and Lase fractions by preparative gel filtration (see Hastings and Gibson[18] for complete details). The Lase fraction obtained is the starting material for the HPLC purification procedure (Table III) which requires three steps.

Preparative High-Performance Size Exclusion Chromatography (HPSEC)

The Lase starting material (96,000 total LU, ~1.7 g protein) was concentrated to ~20 ml by ultrafiltration (Amicon Diaflo ultrafiltration, PM30 membrane). The ultrafiltration cell was washed with an additional 4 ml of standard buffer [50 mM potassium phosphate, pH 7.0, containing 10 mM 2-mercaptoethanol (2-ME) and 1 mM EDTA]. An aliquot (2 ml, ~140 mg protein total) was loaded on a preparative HPSEC column (TSK G-3000 SWG, 2.15 × 7.5-cm precolumn plus 2.15 × 60-cm preparative column). The column was eluted at 2 ml min^{-1} using standard buffer at 4° (19.6 kg cm^{-2} back pressure). Fractions were collected at 1 min intervals following a programmed 40 min delay. A total of 12 samples was applied and collected in the same set of tubes, with samples being applied at 1.25 hr intervals. The elution profile of the protein is shown in Fig. 1. The Lase elutes as a resolvable protein peak which is contaminated on the smaller molecular size side by a low fluorescence flavoprotein which is similar to

[18] J. W. Hastings and Q. H. Gibson, *J. Biol. Chem.* **238**, 2537 (1963).
[19] D. Riendeau and E. Meighen, *Can. J. Biochem.* **59**, 440 (1981).
[20] R. A. Rosson and K. H. Nealson, *Arch. Microbiol.* **129**, 299 (1981).
[21] D. J. O'Kane and J. Lee, this volume [13].

TABLE III
PURIFICATION OF LUCIFERASE FROM *P. leiognathi*[a]

Stage	A_{280}	A_{450}	A_{280}/A_{450}	F_{470}	F_{520}[b]	Lase[c]	Lase activity[d]
G-75	1696	185	9.2	3.8×10^4	1.3×10^4	9.6×10^4	57
HPSEC	459.3	9.3	49.3	1.5×10^4	2.9×10^3	11.1×10^4	241
HPAEC	180.2	0.44	408.3	935	1.1×10^3	10.3×10^4	573
HPHIC	74.4	0.06	1181	66	21	8.8×10^4	1187

[a] A, Absorbance × total volume (ml); F, fluorescence intensity × total volume, in arbitrary units referenced to a sodium fluoresceine standard; F_{470}, 420 → 470 nm and F_{520}, 470 → 520 nm (excitation → fluorescence). Excitation path 1 cm, emission path 0.3 cm.

[b] A new photomultiplier tube and fluorescence standard have been employed herein which are more sensitive than those previously employed (D. J. O'Kane, V. A. Karle, and J. Lee, *Biochemistry*, **24**, 1461, 1985). One F_{470} unit is approximately equivalent to the fluorescence obtained from 0.6 μg of LumP. One F_{520} unit is equivalent to the fluorescence obtained from 55 nM FMN, pH 7.0.

[c] Units are 10^{12} photons sec^{-1} ml^{-1} × volume, using the luminol standard.

[d] Units are 10^{12} photons sec^{-1} per A_{280} (tetradecanal).

that isolated originally from *P. leiognathi* (Visser *et al.* strain S1[22]) (Stokes radius Lase, 34.5 Å; Stokes radius, low fluorescence flavoprotein ~27.5 Å) (D. J. O'Kane and J. Lee, unpublished). The indicated Lase peak was pooled (5 fractions total). There is a 3.7-fold reduction in A_{280} recovered and nearly a 20-fold reduction in A_{450}. The Lase specific activity has increased from 57 LU A_{280}^{-1} to 241 LU A_{280}^{-1} (Table III). The Lase pool is concentrated by ultrafiltration to approximately 10 ml.

Preparative HPAEC. The concentrated Lase pool was filtered and the entire amount was applied to a 2.15 × 15.0-cm column of silica-based TSK DEAE-3SW plus a small guard column. The column was eluted overnight at 0.2 ml min^{-1} (<1 kg cm^{-2} back pressure) using a linear gradient from standard buffer (pump A) to 0.5 M phosphate buffer, containing 10 mM 2-ME and 1 mM EDTA, pH 7.0 (pump B). Fractions were collected every 6 min after a programmed 4 hr delay. A portion of the chromatogram of eluted protein is shown in Fig. 2. The histogram shows Lase activity centered on the midpoint of each fraction collected. The shaded portion of the histogram was combined for Lase. The A_{280} recovered is only 39% of that applied, the A_{450} has been reduced an additional 21-fold, and the specific activity of the Lase increased an additional 2.4-fold.

Preparative HPHIC. The pooled Lase fraction was concentrated by

[22] A. J. W. G. Visser, J. Vervoort, D. J. O'Kane, J. Lee, and L. A. Carreira, *Eur. J. Biochem.* **131**, 639 (1983).

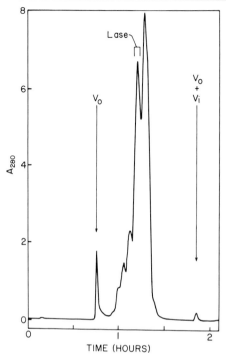

FIG. 1. Preparative high-performance size exclusion chromatography of *P. leiognathi* DD17 Lase. A 2 ml sample of partially purified Lase ($\sim$140 mg protein) was applied to an Ultrapac TSK G3000 SWG column and was eluted at 2 ml min^{-1}. The void volume (V_0) and included volume ($V_0 + V_i$) are indicated. The peak to the right of Lase is the low fluorescence flavoprotein which contributes the bulk of the A_{450} to the sample.

ultrafiltration and solid $(NH_4)_2SO_4$ was added to 25% saturation at 4° (136 g/liter). A slight precipitate was removed by centrifugation. One-half of the protein was applied to a preparative, polymer-based TSK-phenyl 5PW column (0.75 × 7.5-cm precolumn plus 2.15 × 15-cm column) which was previously equilibrated in 1M $(NH_4)_2SO_4$ in standard buffer. The column was eluted, as shown in Fig. 3, using decreasing concentrations of $(NH_4)_2SO_4$ at a flow rate of 1.75 ml min^{-1} (25.9 kg cm^{-2} back pressure). The column was automatically regenerated [0 to 1 M $(NH_4)_2SO_4$] and the remaining portion of protein was applied. The Lase eluted between 0.3 and 0.25 M $(NH_4)_2SO_4$. The best fractions were combined, reapplied to the same column and rechromatographed to yield Lase of constant specific activity (1180 LU A_{280}^{-1}) (Table II). The overall purification is 21-fold based on protein in the starting material, a 21-fold increase in specific activity of the Lase, and a 3100-fold reduction in A_{450}. Approximately 1 μmol of Lase was purified in 3 days time: 1.5 days devoted to HPSEC, 0.5 days for HPAEC, and 1 day for HPHIC.

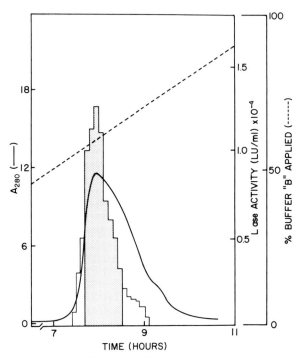

FIG. 2. Preparative high-performance anion-exchange chromatography of *P. leiognathi* DD17 Lase. The pooled sample from HPSEC was applied to an Ultrapak TSK DEAE-3SW column which was eluted at 0.2 ml min⁻¹ using a programmed gradient. Buffer "B" contained 0.5 *M* phosphate, pH 7.0, with 10 m*M* 2-ME and 1 m*M* EDTA. The Lase activity of collected fractions is indicated by the histogram. The shaded portion of the histogram was pooled for HPHIC.

Luciferase from an Aldehyde-Requiring Mutant of Vibrio harveyi

N-Methyl-*N'*-nitro-*N*-nitrosoguanidine was used to induce aldehyde-requiring mutations in cells grown from a bright strain of *V. harveyi*, strain MAV (from J. W. Hastings), by described methods.[23,24] The mutants are not completely dark but have a low light emission rate usually of the order of 10^6 photons sec⁻¹ per colony grown on a standard plate. The mutant selected is the one having the greatest ratio of the tetradecanal-induced maximum bioluminescence level to the low emission level in the absence of added aldehyde, typically 3×10^4.

[23] E. A. Adelberg, M. Mandel, and G. C. C. Chen, *Biochem. Biophys. Res. Commun.* **18**, 788 (1965).
[24] T. Cline and J. W. Hastings, *J. Bacteriol.* **118**, 1059 (1976).

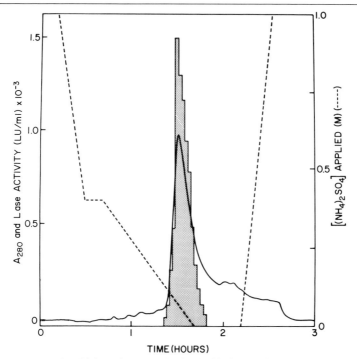

FIG. 3. Preparative high-performance hydrophobic interaction chromatography of *P. leiognathi* DD17 Lase. The pooled fractions from HPAEC were applied to a TSK-Phenyl 5 PW column equilibrated in 1 *M* (NH₄)₂SO₄. The column was eluted with a decreasing gradient of (NH₄)₂SO₄ as indicated. The Lase activity is indicated by the histogram. The fractions with high specific activity were combined and rechromatographed to obtain purified Lase.

Purification

In Table IV the bioluminescence activities obtained in processing extracts of the luminous strain MAV and the dark mutant MAVA are compared through each stage of purification.[25] The total bioluminescence activity in the lysate of MAVA is seen to be 4 times that in MAV and this ratio persists through all stages of purification. The total amount of soluble protein in the lysate determined as absorbance at 280 nm or assayed by the Biuret method is the same for both types of cells.

The cell extract cannot be applied directly to HPLC columns due to large quantities of particulates which would clog the columns. The extract is first precipitated with ammonium sulfate, desalted, and partially purified by ion exchange prior to application to a preparative HPAEC column.

[25] M. Ahmad and J. Lee, *Fed. Proc., Fed. Am. Soc. Exp. Biol.* **44,** 1216 (1985).

TABLE IV
PURIFICATION OF *V. harveyi* LUCIFERASES[a]

Step	Bioluminescence activity[b] LU A_{280}^{-1}		Total absorbance[c] at 280 nm		Total bioluminescence activity[d] 10^4 LU	
	MAV[e]	MAVA[f]	MAV	MAVA	MAV	MAVA
Cell lysate[g]	0.23	0.92	1.6×10^{5h}	1.63×10^{5h}	3.73	15.0[i]
30% ammonium sulfate	ND	ND	ND	ND	3.15	14.0
80% ammonium sulfate	ND	ND	ND	ND	2.88	14.0
Sephadex G-75	ND	ND	—	—	3.6	14.5
DEAE-Sephacel	21	68	1900	2000	3.3	14.5
DEAE-Sephadex A-50	68	136	—	—	3.3	14.5
Fractogel TSK	76	150	—	—	—	—
HPAEC[j]	—	150	—	360	—	5.4

[a] The extraction and purification were performed at 5° from 500 g of wet cell cake. All the buffers contained 2-ME (10 mM) and were at pH 7.0.

[b] Fractions of best activity were pooled and the bioluminescence activity was determined by the standard method as described in the text using decanal. ND, not determined.

[c] Total absorbance = $A_{280} \times$ volume (ml). The A_{280} unit refers to a 1 ml volume having an absorbance 1.0 at 280 nm.

[d] Total bioluminescence activity is LU ml$^{-1} \times$ volume (ml) and is given for 500 g of wet cell cake.

[e] MAV is luciferase from *V. harveyi*, wild type.

[f] MAVA is luciferase from a dark, aldehyde-requiring mutant.

[g] Cells were lysed on a French press.

[h] Total soluble protein in the lysates was determined by the Biuret method and converted to A_{280} units.

[i] Some batches of the mutant gave 2×10^5 total LU per 500 g of cells but in most of the cases the total bioluminescence activity was decreased to 1.5×10^5 LU after the first ammonium sulfate precipitation.

[j] From DEAE-Sephacel. Results are for a single pass on HPLC, average of pooled activities. The recovery is improved to 80% on reapplying side fractions.

The cell extract from the French Press is made 30% saturated with ammonium sulfate (pH 7) over a period of 30 min and then immediately centrifuged (17,000 g, 30 min). The supernatant is then separated and will contain at least 90% of the total bioluminescence activity if the procedure has been carried out in the times specified. The supernatant is made 80% saturation in ammonium sulfate and stirred for at least 2 hr or preferably overnight before the precipitated luciferase is removed by centifugation (17,000 g, 30 min). This crude luciferase can be frozen for storage or redissolved in a minimum amount (300 ml) of standard buffer for desalting in the following step. The solution should contain no more than about

1.5×10^5 total LU using decanal for MAVA or three times less than this for MAV, otherwise the high protein concentration will cause channeling of the following Sephadex column.

The solution is centrifuged (17,000 g, 30 min) and applied to a gel filtration column (Sephadex G-75 coarse, 10×60 cm) and eluted with standard buffer as rapidly as the column back pressure will allow. All fractions with bioluminescence activity above 10% of the maximum are pooled and loaded to a DEAE-Sephacel column (3.5×20 cm) equilibrated with standard buffer. The column is then washed with standard buffer (500 ml), then with 120 mM phosphate buffer (all solvents contain 10 mM 2-mercaptoethanol, pH 7.0) which removes some yellow proteins and NADH : FMN oxidoreductase activity. The luciferase is eluted with 250 mM phosphate buffer (500 ml). The specific activity of the luciferase at this juncture should be above 15 LU A_{280}^{-1} (decanal) for MAV or 40 for MAVA if the preceding steps have been properly executed. For preparations below this activity the following purification steps yield correspondingly lower activities. The luciferase is concentrated to 50 ml if it is to be applied to the Sephadex A-50 column or to 5–10 ml for HPLC.

The results for chromatography on the Sephadex A-50 are given here for comparison with the more efficient HPLC method. The luciferase is diluted with distilled water to a conductivity equivalent to 150 mM phosphate buffer and then applied to a Sephadex A-50 column (3.5×18 cm) equilibrated with 150 mM phosphate buffer. The pressure head is maintained at all times at less than 18 cm above the gel bed. The column is washed with 180 mM phosphate buffer (1.5 liters) followed by 200 mM phosphate buffer (2 liters). The latter concentration must be carefully adjusted to a value just below where the luciferase starts to elute. At this point a number of yellow proteins along with about 1% of the luciferase elute. The phosphate buffer concentration is increased to 220 mM (0.5 liter) followed by 230 mM which elutes all the luciferase. The major disadvantages of this procedure are that it takes about 5 days and has a variable success rate. The specific activity of the MAV luciferase produced in Table IV is seen to be a little higher than that published from this laboratory about 10 years ago.[26] It is clear that the aldehyde mutant itself is producing luciferase of a much higher intrinsic activity.[25]

Preparative HPAEC

For the HPLC column the luciferase solution is adjusted to a phosphate buffer concentration of 50 mM. For the best resolution no more than 700 A_{280} units in a volume less than 20 ml (using a 2-ml injection loop)

[26] J. Lee and C. L. Murphy, *Biochemistry* **14**, 2259 (1975).

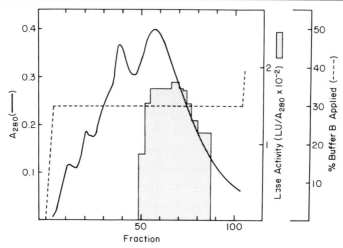

FIG. 4. Preparative high-performance anion-exchange chromatography of *V. harveyi* MAVA Lase. The pooled fractions from DEAE-Sephacel (4 ml, about 300 mg total protein, specific activity 60 LU A_{280}^{-1}) were applied to an Ultrapak TSK DEAE-25W column, and eluted at 1 ml min^{-1}. The histogram shows the Lase activity.

should be applied to the DEAE-TSK column (LKB Ultrapak, 21.5 × 150 nm). The elution conditions and profile are shown in Fig. 4. Elution is essentially isocratic (193 mM phosphate buffer) and the luciferase takes about 12 hr to come off. The recovery of highest activity luciferase, the constant activity fractions (54–75 in Fig. 4) is only about 30% in a single pass. However, the side fractions of lesser activity can be reapplied or combined with a following batch. The overall yield on HPAEC of top activity is 80% with a specific activity of 150 LU/A_{280}.

When the data of Holzman and Baldwin[5] are corrected for the different light standard (2.74-fold)[27] and the different light unit employed (100-fold), affinity chromatography purified Lase and that obtained by HPAEC are comparable in specific activity (150 LU/A_{280} by HPAEC; 120 LU/A_{280} from Holzman and Baldwin[5]). However, the HPLC method allows a larger throughput than the affinity method.

The activity of the preparation from the HPAEC is the same as off Sephadex A-50. The type MAV Lase has not yet been run on HPLC. The disadvantage of the HPLC method is that the amount to be loaded to the column is about 3 times less than can be applied to Sephadex A-50. However the HPLC is more reliable and reproducible.

[27] J. W. Hastings and G. T. Reynolds, *in* "Bioluminescence in Progress" (F. H. Johnson and Y. Haneda, eds.), p 45. Princeton Univ. Press, Princeton, New Jersey, 1966.

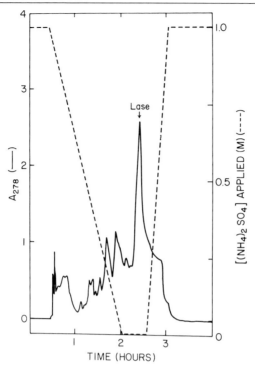

Fig. 5. Preparative high-performance hydrophobic interaction chromatography of *P. phosphoreum* A13 Lase. Partially purified *P. phosphoreum* Lase (300 mg protein) was applied to a TSK-Phenyl 5 PW column in 1 *M* (NH₄)₂SO₄. The column was eluted with a decreasing gradient of (NH₄)₂SO₄ as indicated. The Lase elutes with the indicated peak in A_{278}.

Hydrophobic Interaction Chromatography of Lase

Chromatography of Lase on an HPHIC column, or simply on phenyl-Sepharose[24] (φ-Sepharose), is an excellent first step in the purification of Lase and has some advantages over the "affinity" chromatographic purification of Lase on diphenylpropylamine-Sepharose (DφPA-Sepharose).[5] As pointed out by Holzman and Baldwin,[5] the basic conditions used to elute DφPA-Sepharose are particularly deleterious to *P. phosphoreum* Lase activity. The *Photobacterium* Lases can be readily purified from partially purified extracts, or from crude cell-free extracts, by HPIC or by chromatography on φ-Sepharose, respectively. Figure 5 shows the elution of *P. phosphoreum* Lase from a preparative HPHIC column employing a decreasing concentration gradient of (NH₄)₂SO₄ in standard buffer. There is no loss of total Lase activity under these conditions.

The purification of Lase on φ-Sepharose compares favorably with that obtained using DφPA-Sepharose. A lysate of *V. harveyi* MAVA Lase was precipitated with ammonium sulfate, as described by Holzman and Baldwin,[5] and applied to a 1 × 15-cm column of φ-Sepharose as a 1 *M* $(NH_4)_2SO_4$ solution. The recovered Lase was purified 74-fold based on A_{280}, with a recovery of 63% of the applied Lase Activity (43,000 LU total). This chromatography step required 4 hr. This should be compared to the published results with DφPA-Sepharose[5] where a 34.5-fold purification with an 88% recovery of activity was obtained and this required 32 hr.

[11] Bacterial Luciferase 4a-Hydroperoxyflavin Intermediates: Stabilization, Isolation, and Properties

By SHIAO-CHUN TU

All monooxygenases activate molecular oxygen to insert one oxygen atom into a substrate and reduce the other oxygen atom to water. For flavin-dependent external monooxygenases (also referred to as hydroxylases), the activated oxygen has been demonstrated to be in the form of 4a-hydroperoxyflavin[1-4] (abbreviated 4a-FH-OOH[5]). This flavin peroxide intermediate has been particularly well characterized in the case of reduced FMN-dependent bacterial luciferase.

The luciferase 4a-hydroperoxyFMN intermediate (4a-FMNH-OOH) can be formed in three ways, by reacting luciferase-bound reduced flavin with oxygen,[6] flavin neutral semiquinone with superoxide anion,[7] or oxi-

---

[1] V. Massey and P. Hemmerich, *in* "The Enzymes" (P. D. Boyer, ed.), 3rd ed., Vol. 12, p. 191. Academic Press, New York, 1975.

[2] J. W. Hastings, C. Balny, C. LePeuch, and P. Douzou, *Proc. Natl. Acad. Sci. U.S.A.* **70,** 3468 (1973).

[3] L. L. Poulsen and D. M. Ziegler, *J. Biol. Chem.* **254,** 6449 (1979).

[4] N. B. Beaty and D. P. Ballou, *J. Biol. Chem.* **256,** 4619 (1981).

[5] The abbreviations used are isoFMN, 6,7-dimethyl-10-(1'-D-ribityl-5'-phosphate)isoalloxazine; 2S-FMN, 2-thio-7,8-dimethyl-10-(1'-D-ribityl-5'-phosphate)isoalloxazine; RF, riboflavin; CM-FMN, 3-carboxymethyl-7,8-dimethyl-10-(1'-D-ribityl-5'-phosphate)isoalloxazine; DA-FMN, 7,8-dimethyl-10-(1'-D-ribityl-2',3'-diacetyl-5'-phosphate)isoalloxazine; CP-F, 7,8-dimethyl-10-(ω-carboxypentyl)isoalloxazine; FH₂ and 4a-FH-OOH, reduced flavin and 4a-hydroperoxyflavin, respectively, in which F can be substituted with any of the flavin abbreviations listed above.

[6] J. W. Hastings and Q. H. Gibson, *J. Biol. Chem.* **238,** 2537 (1963).

[7] M. Kurfürst, S. Ghisla, and J. W. Hastings, *Biochemistry* **22,** 1521 (1983).

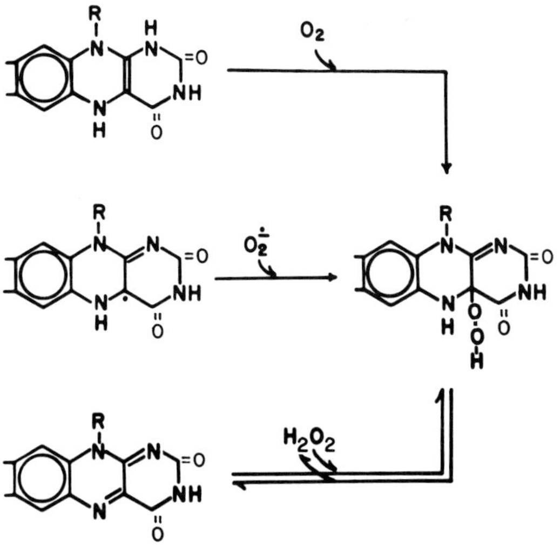

FIG. 1. The formation of luciferase 4a-FMNH-OOH by three pathways.

dized flavin with H_2O_2[8] (Fig. 1). The yields of 4a-FMNH-OOH by the second and third pathways are much lower than that by the first route. Therefore, only the first method for the formation of flavin peroxide was used in this work. Principles and procedures for the stabilization and isolation of luciferase 4a-FH-OOH intermediates formed from various flavins are described, and their properties are summarized.

Materials and General Methods

Reagents and Materials

Na/K phosphate buffer, 1 M, pH 7.0: 106.25 g of K_2HPO_4 and 53.82 g of $NaH_2PO_4 \cdot H_2O$ dissolved in water to final volume of 1 liter; pH is checked after 20-fold dilution and adjusted for the stock if necessary; used with appropriate dilutions

Decanal, 0.01% suspension: mix 1 ml of 1% decanal in 95% ethanol with 100 ml of 0.05 M phosphate, pH 7, under vigorous stirring

Long-chain aliphatic alcohols: 10 mM of dodecanol or tetradecanol in 95% ethanol

Ethylene glycol: mix 2 volumes of ethylene glycol with 3 volumes of 0.165 M phosphate, pH 7

Solid sodium hydrosulfite (dithionite)

Riboflavin 5'-phosphate (FMN)

[8] J. W. Hastings, S.-C. Tu, J. E. Becvar, and R. P. Presswood, *Photochem. Photobiol.* **29**, 383 (1979).

Sephadex G-25 and Sephadex LH-20
Vibrio harveyi luciferase[9]

Assays

For the intermediate assay, a 10- to 50-μl aliquot of isolated intermediate is withdrawn using a Hamilton syringe, precooled to 0° for aqueous samples or −20° for samples in 40% ethylene glycol, and injected into 1 ml of the 0.01% decanal suspension kept at 23° to initiate the bioluminescence reaction. Luciferase activity is, in some cases, also determined by the dithionite assay.[9]

Stabilization

Low Temperature

Certain luciferase reaction intermediates, including the 4a-FH-OOH species, have unusually long lifetimes. For example, the half-life of luciferase 4a-FMNH-OOH is about 10 sec in 0.1 M phosphate, pH 7, at 23°.[10] Furthermore, the decay of this flavin peroxide is associated with a rather large positive enthalpy of activation.[10] As indicated by the following relationship

$$\ln \frac{k_2}{k_1} = \frac{\Delta H\ddagger}{R} \left(\frac{1}{T_1} - \frac{1}{T_2} \right) + \ln \frac{T_2}{T_1} \tag{1}$$

where k_1 and k_2 are the decay rate constants at absolute temperatures T_1 and T_2, respectively, R is the gas constant, and $\Delta H\ddagger$ is the activation enthalpy for decay, the lifetime of luciferase 4a-FMNH-OOH can be markedly lengthened at low temperatures. Hastings *et al.*[2] have capitalized on such a temperature effect to demonstrate in their elegant work that successful stabilization and quantitative isolation of 4a-FMNH-OOH can be achieved at −20°.

High Salt Conditions

High salt conditions appear to affect at least three aspects of luciferase property. Luciferase is known to resist inactivation by heat[11,12] and proteases[12] in the presence of high concentrations of phosphate. High phosphate or sulfate media also significantly enhance the bioluminescence activities of neutral flavin substrates.[13] High salt conditions effectively

[9] J. W. Hastings, T. O. Baldwin, and M. Z. Nicoli, this series, Vol. 57, p. 135.
[10] S.-C. Tu, *Biochemistry* **18**, 5940 (1979).
[11] S.-C. Tu and J. W. Hastings, unpublished results (1975).
[12] T. F. Holzman and T. O. Baldwin, *Biochem. Biophys. Res. Commun.* **94**, 1199 (1980).
[13] E. A. Meighen and R. E. MacKenzie, *Biochemistry* **12**, 1482 (1973).

stabilize the 4a-FMNH-OOH intermediate.[14] However, the basis for these effects of high salt is not well understood.

Nonaldehyde Long-Chain Aliphatic Compounds

Stabilities of 4a-FH-OOH chemical model compounds have been shown to be enhanced in more hydrophobic media.[15] It is thus conceivable that luciferase 4a-FH-OOH may be stabilized if the hydrophobicity of the surrounding microenvironment can be augmented. Several types of nonaldehyde long-chain aliphatic compounds are known to be competitive inhibitors for luciferase.[16,17] On the assumption that the binding of these aliphatic inhibitors will increase the active site hydrophobicity, the effects of long-chain alcohols, carboxylic acids, and methyl ester of the acids on the stability of luciferase 4a-FMNH-OOH have been examined (Table I). Indeed, marked stabilization is obtained in all cases. In general, long-chain alcohols are somewhat more effective. Values of K_d for the binding of alcohols to luciferase 4a-FMNH-OOH have been determined to be 2.4, 1.2, 0.04, and 0.05 μM for octanol, decanol, dodecanol, and tetradecanol, respectively, at 23°.[10] Using dodecanol, such a stabilization effect has also been demonstrated for 4a-FH-OOH intermediates formed from a number of reduced flavin derivatives (Table II). The extents of stabilization by long-chain alcohols, e.g., dodecanol, are generally superior than that by high salt conditions.[10,14,18]

Additional Considerations

For exceedingly unstable intermediates, the three singly cited parameters can be used in combinations (e.g., high salt plus long-chain alcohol or subzero temperature plus long-chain alcohol) to achieve the desired stabilization. Variations of pH could conceivably also be used to our advantage.

Isolation by Sephadex Column Chromatography

Methods for the isolation of luciferase 4a-FMNH-OOH in a high salt medium (e.g., 0.5 M NaCl in 0.3 M phosphate, pH 7) at 0° and in 50% ethylene glycol–phosphate buffer cosolvent at −20° by Sephadex column

[14] J. E. Becvar, S.-C. Tu, and J. W. Hastings, *Biochemistry* **17**, 1807 (1978).
[15] C. Kemal and T. C. Bruice, *Proc. Natl. Acad. Sci. U.S.A.* **73**, 995 (1976).
[16] J. W. Hastings, Q. H. Gibson, J. Friedland, and J. Spudich, *in* "Bioluminescence in Progress" (F. H. Johnson and Y. Haneda, eds.), p. 151. Princeton Univ. Press, Princeton, New Jersey, 1966.
[17] A. L. Baumstark, T. W. Cline, and J. W. Hastings, *Arch. Biochem. Biophys.* **193**, 449 (1979).
[18] S.-C. Tu, *J. Biol. Chem.* **257**, 3719 (1982).

TABLE I
HALF-LIFE OF 4a-FH-OOH : ALIPHATIC
COMPOUND COMPLEXES AT 23°

Aliphatic compound	Half-life[a] (min)
None	0.16
$CH_3(CH_2)_7OH$	5.8
$CH_3(CH_2)_9OH$	3.3
$CH_3(CH_2)_{11}OH$	17.3
$CH_3(CH_2)_{13}OH$	7.7
$CH_3(CH_2)_8COOH$	0.4
$CH_3(CH_2)_{10}COOH$	0.4
$CH_3(CH_2)_{12}COOH$	2.7
$CH_3(CH_2)_8COOCH_3$	5.0
$CH_3(CH_2)_{10}COOCH_3$	1.1
$CH_3(CH_2)_{12}COOCH_3$	0.54

[a] Determined in 0.1 M phosphate, pH 7, containing saturating levels of the aliphatic compounds. From Tu.[10] [Reprinted with permission from *Biochemistry*. Copyright (1979) American Chemical Society.]

TABLE II
K_m OF REDUCED FLAVINS AND EFFECT OF DODECANOL ON THE
STABILITIES OF VARIOUS 4a-FH-OOH

Flavin	K_m for FH_2[a] (μM)	Half-life of 4a-FH-OOH (min)[b]	
		−Dodecanol	+Dodecanol
FMN	0.3	11	320
CP-F	3	2[c]	75
DA-FMN	29	25[c]	138
CM-FMN	—	14	138
RF	22	15[c]	45
2S-FMN	0.5	3	1400
IsoFMN	17	2	68

[a] Determined at 23° in 0.05 M phosphate, pH 7.
[b] Determined at 0° in 0.05 M phosphate, pH 7.
[c] These intermediates exhibited, in the absence of dodecanol, biphasic decays. The initial decay was brief and the second phase was a slow monoexponential decay. Values of half-life for the slow phase are shown. Adapted from Tu.[18]

chromatography have been described previously.[19] A similar procedure is described here for the preparation and isolation of 4a-FH-OOH : dodecanol complex in aqueous media. Operations can be carried out at either 0 or 4° throughout or by using tetradecanol with few differences in results.

Approximately 10 μl of 10 mM dodecanol is mixed with 0.5 ml of 0.1 M phosphate, pH 7, containing 2.5 mg of luciferase and 50 μM FMN. Some precipitates of dodecanol may be formed but no precipitation of luciferase has been encountered. A few milligrams of sodium dithionite is added to reduce the flavin, and the solution is quickly applied to a Sephadex G-25 column (1 × 16 cm) preequilibrated with 0.1 M phosphate, pH 7, containing saturating dodecanol. The dodecanol-saturated buffer is prepared by stepwise additions of the dodecanol stock solution to the buffer under constant stirring until a faint turbidity is visible. The buffer is then clarified by centrifugation or filtration. The column is eluted with the same dodecanol-containing buffer under aerobic conditions and the 4a-FMNH-OOH : dodecanol complex is obtained in the void volume well separated from FMN, dithionite, and its oxidized products.

Since exposure of luciferase 4a-FMNH-OOH to 370 nm light can transform it from a weakly to a strongly fluorescent species (described later), the isolation should be carried out under red light if the fluorescence property of the original intermediate form is to be characterized. Otherwise, operations under room light produce no known adverse effect.

The same procedure can be used for the isolation of luciferase peroxide intermediates formed from various reduced flavin derivatives. In all cases, K_d (or K_m as an approximation) for the binding of reduced flavins should be preevaluated. This information is useful for the estimation of the amounts of flavins required to saturate luciferase at the step of dithionite reduction. Values of K_m for several flavins are shown in Table II.

Isolation by Sephadex/Centrifugation

A previously reported method of centrifugation on Sephadex matrix for desalting[20] has been adapted for the isolation of luciferase 4a-FH-OOH. In one application, the entire operation is carried out at 4 ± 1° in a cold room. A small Sephadex G-25 column (1 × 8.5 cm) is equilibrated in 0.1 M phosphate, pH 7, containing 10 μM dodecanol and the buffer is drained by gravity. The column is then centrifuged at the top speed of a tabletop centrifuge (e.g., Fisher Centrific Model 225 equipped with a four-place horizontal rotor and 3 × 10-cm shields) for 30 sec in an arrangement

[19] J. W. Hastings and J. E. Becvar, this series, Vol. 57, p. 194.
[20] M. W. Neal and J. R. Florini, *Anal. Biochem.* **55**, 328 (1973).

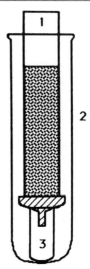

FIG. 2. Sephadex/centrifugation setup. (1) A Bio-Rad glass barrel Econo column (1 cm i.d. cut to the length of 10 cm) containing preequilibrated Sephadex G-25 or Sephadex LH-20. (2) A Nalgene brand polycarbonate centrifuge tube (28.5 × 103 mm, o.d. × length). (3) An extra heavy-wall plastic tube (1.2 cm o.d.) cut to the length of 3 cm.

shown in Fig. 2 and the filtrate discarded. The volume of the semidry Sephadex cake should be kept at 5 to 6 ml/ml of protein sample to be applied subsequently. To 1 ml 0.1 M phosphate buffer, pH 7, containing 2.5 mg luciferase, 50 μM FMN, and 0.1 mM dodecanol, a few milligrams of solid dithionite is added. The sample is immediately applied onto the semidry Sephadex column and centrifuged at top speed for about 30 sec. The filtrate contains the 4a-FMNH-OOH intermediate with ≤10% change in protein concentration and volume. Peroxide intermediates formed from flavin derivatives can be similarly isolated and care should be taken to keep the initial reduced flavin concentration at a saturating level.

The same method can be modified to isolate the peroxide intermediate at subzero temperatures, such as at −20° carried out in a chest freezer (e.g., Hotpoint Model FH25CW). A 40% ethylene glycol–0.1 M phosphate, pH 7, cosolvent is used throughout as the medium for luciferase sample and column equilibration. The luciferase sample is prepared by the addition of a small aliquot (e.g., 10 to 50 μl) of a concentrated luciferase stock in aqueous buffer to 1 ml of −20° mixed solvent containing flavin. No long-chain alcohol is needed for additional stabilization of 4a-hydroperoxyFMN. The centrifugation is set at the top speed for 10–15 min. The column material is Sephadex LH-20. All other procedures are the same as those described above. Again, the same method can be used for the isolation of derivatized flavin peroxide intermediates. For the more labile intermediates, the inclusion of dodecanol may provide further stabilization.

The Sephadex/centrifugation method has three advantages. The final intermediate samples are obtained with very little dilution of protein concentration and very little change in the sample volume. Second, only simple instrumentation (i.e., a cooling system and a tabletop centrifuge) is required for the operation. Most importantly, this method is much faster than the corresponding column chromatography method at either 0 to 4° or subzero temperatures and, therefore, is suitable for the isolation of more labile intermediates. With appropriate modifications for sample preparations, the same Sephadex/centrifugation method should be quite applicable to the isolation of other relatively stable luciferase intermediates such as neutral flavin semiquinone[7] and 4a-hydroxyFMN.[21]

Properties

Absorption and Fluorescence

The absorption spectrum of luciferase 4a-FMNH-OOH : tetradecanol complex at 0° in a neutral aqueous phosphate buffer is shown in Fig. 3. The spectrum is indistinguishable from that of the 4a-FMNH-OOH isolated at −20° in ethylene glycol–phosphate buffer mixed solvent.[2] The change from one type of long-chain alcohol to another also exhibits no detectable effect on the absorption. Essentially identical absorption spectra are observed with 4a-FH-OOH formed from side-chain derivatized flavins such as reduced CP-F and DA-FMN at 0° in the presence of dodecanol and riboflavin at −20° in mixed solvent. 4a-FH-OOH : dodecanol species have also been isolated using reduced 2S-FMN (Fig. 4) and isoFMN (Fig. 5) each as a starting substrate. In these cases, significantly different absorption spectra are observed. In all cases, a gradual decay of 4a-FH-OOH intermediates occurs upon standing, leading to the formation of the corresponding oxidized flavins. Such a process is first order in kinetics and associated with well-defined absorption isosbestic point(s).

The isolated 4a-FMNH-OOH : tetradecanol complex initially exhibits weak fluorescence (emission max near 535 nm). However, exposure to 370 nm excitation light transforms it to a species with higher fluorescence intensity (13-fold increase at 480 nm) and a blue shift of the spectrum (emission max 505 nm) (Fig. 6). However, neither the absorption spec-

[21] M. Kurfürst, S. Ghisla, and J. W. Hastings, *Proc. Natl. Acad. Sci. U.S.A* **81**, 2990 (1984).

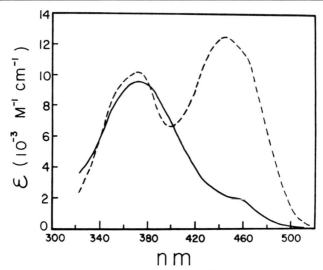

FIG. 3. Absorption spectra of luciferase 4a-FMNH-OOH : tetradecanol complex and the decay product. The flavin peroxide spectrum (——) was taken at 0° in 0.1 M phosphate, pH 7, containing saturating tetradecanol immediately after isolation. The sample was then warmed to and kept at 23° for 3 hr for a complete decay and subsequently cooled to 0° for spectrum determination (————). The latter is identical to the spectrum of an authentic FMN sample. Reproduced from Tu.[10] [Reprinted with permission from *Biochemistry*. Copyright (1979) American Chemical Society.]

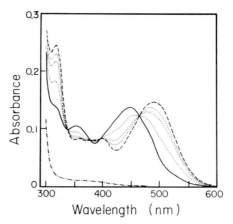

FIG. 4. Absorption spectra of 4a-2S-FMNH-OOH : dodecanol (——) and the final flavin decay product (————). The former was taken at 0° and then the temperature was raised to 10° allowing the intermediate to decay. Transient spectra during the decay are also shown (·········). The background absorption due to luciferase alone (–·–) was not subtracted. Reproduced from Tu.[18]

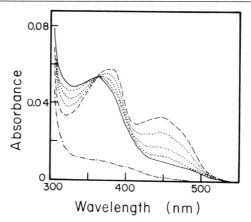

FIG. 5. Absorption spectra of 4a-isoFMNH-OOH : dodecanol (———) and the final flavin decay product (– – – –). Transient spectra during the decay (·······) are also shown. The background absorption due to luciferase alone (–·–) was not subtracted. Other conditions are the same as described in Fig. 4. Reproduced from Tu.[18]

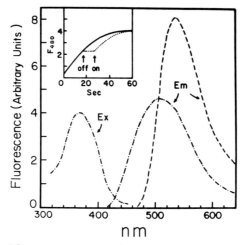

FIG. 6. Corrected fluorescence spectra of 4a-FMNH-OOH : tetradecanol and FMN. Inset: The isolated intermediate was immediately subject to 370-nm excitation and fluorescence emission at 480 nm was followed (———). The increase in emission can be stopped by cutting off the excitation and resumed later by reexposing to 370 nm light (······). Main figure: After 1 min of irradiation at 370 nm, the transformed sample was used for the determination of excitation and emission spectra (–·–) with emission detected at 480 nm for the former and excitation set at 370 nm for the latter measurement. For comparison, the emission spectrum of FMN (– – – –) upon 370 nm excitation is also shown. The flavin peroxide intermediate and the FMN sample were adjusted to have the same absorbance at 370 nm. Reproduced from Tu.[10] [Reprinted with permission from *Biochemistry*. Copyright (1979) American Chemical Society.]

TABLE III
REACTION PRODUCTS OF 4a-FH-OOH IN THE
PRESENCE AND ABSENCE OF ALDEHYDE

Intermediate[a]	Decanal addition	H_2O_2/Flavin
4a-FMNH-OOH	−	0.97 ± 0.09
	+	0.17 ± 0.04
4a-CP-FH-OOH	−	0.99 ± 0.10
	+	0.16 ± 0.05
4a-DA-FMNH-OOH	−	1.18 ± 0.04
	+	0.59 ± 0.20
4a-2S-FMNH-OOH	−	0.99 ± 0.06
	+	0.15 ± 0.01
4a-Iso-FMNH-OOH	−	0.96 ± 0.01
	+	0.71 ± 0.06

[a] All initially isolated as dodecanol complexes.
Adapted from Tu.[18]

trum nor the bioluminescence activity (upon reacting with aldehyde) of the intermediate shows any significant changes after the phototransformation.[10] These properties are essentially the same as those reported for 4a-FMNH-OOH isolated at −20° in 50% ethylene glycol.[22] Both the spectrum and fluorescence quantum yield (0.17) of the photoinduced 4a-FMNH-OOH species closely resemble those of the bioluminescence emission. The fluorescence emission properties of 4a-FH-OOH: dodecanol formed from reduced 2S-FMN and isoFMN have also been reported.[23] In both cases, the fluorescence spectra of flavin peroxides correlate well with the corresponding bioluminescence emissions.

Reaction Products

In the absence of aldehyde substrate, all isolated 4a-FH-OOH species decay exponentially to yield about equal molar quantities of oxidized flavin and H_2O_2 (Table III). In the presence of saturating decanal, reduced but significant levels of H_2O_2 formation are still observed (Table III). Bioluminescence quantum yields of the 4a-FH-OOH species, based on the flavin content of each isolated intermediate, in the absence of aldehyde are quite low but are much enhanced on reacting with decanal (Table IV). Assuming that the excited emitter is formed from the fraction of 4a-

[22] C. Balny and J. W. Hastings, *Biochemistry* **14**, 4719 (1975).
[23] J. W. Hastings, S. Ghisla, M. Kurfürst, and P. Hemmerich, *in* "Bioluminescence and Chemiluminescence: Basic Chemistry and Analytical Applications" (M. A. DeLuca and W. D. McElroy, eds.), p. 97. Academic Press, New York, 1981.

TABLE IV
BIOLUMINESCENCE QUANTUM YIELDS^a (ϕ)

| Flavin | ϕ_{FH_2} | $\phi_{4a\text{-}FH\text{-}OOH}$ | | $\phi_{Emitter}$ | $\phi_{4a\text{-}FH\text{-}OOH}/\phi_{FH_2}$ |
		$-$Decanal	$+$Decanal		
FMN	0.16	0.008	0.15	0.18	0.94
Riboflavin	0.0006	—	0.029	—	48.3
	0.008^b				
CP-F	0.03	<0.001	0.077	0.09	2.6
DA-FMN	0.0004	<0.001	0.021	0.05	52.5
2S-FMN	0.003	<0.001	0.009	0.01	3.0
IsoFMN	0.001	<0.001	0.012	0.04	12.0

^{a} Determined at 23°.
^{b} Determined at 10°. Adapted from Tu.[18]

FH-OOH that does not generated H_2O_2 upon reacting with aldehyde, the quantum yields of emitter species are also determined. Furthermore, luciferase is titrated with various reduced flavins using the dithionite activity assay and, for each flavin, the bioluminescence quantum yield is determined based on the amount of luciferase-bound reduced flavin (Table IV).

Three aspects of the above described results are particularly important. (1) Once the values of emitter quantum yields are determined, light intensity measurements can be directly correlated with the absolute quantities of excited emitter formed in the bioluminescence reaction. (2) With the exception of 4a-FMNH-OOH, all other isolated peroxide intermediates exhibit significantly higher quantum yields than those determined based on the amounts of bound reduced flavins. This indicates that certain luciferase-bound reduced flavins can react with O_2 in two pathways: one leads to the formation of 4a-FH-OOH and the other produces oxidized flavin and H_2O_2 without involving 4a-FH-OOH as an intermediate. Assuming that these two pathways have different temperature dependence, one would expect to detect changes in quantum yields of bound flavin at different temperatures. This is indeed the case with reduced riboflavin. (3) Even in the presence of saturating decanal, significant amounts of H_2O_2 are still formed from 4a-FH-OOH. A dark pathway is thus also indicated at the stage of flavin peroxide reacting with aldehyde.

Acknowledgments

The work reported here was supported by a GM 25953 grant from the National Institute of General Medical Sciences and a Welch Foundation grant E-1030. The author also acknowledges the support of a Research Career Development Award (ES00088) from the National Institute of Environmental Health Sciences.

[12] Bacterial Luciferase Intermediates: The Neutral Flavin Semiquinone, Its Reaction with Superoxide, and the Flavin 4a-Hydroxide

By M. Kurfürst, S. Ghisla, and J. W. Hastings

The luciferase from bacteria is a flavin monooxygenase that cleaves the oxygen molecule during the light reaction, with one oxygen atom being incorporated into the long-chain aliphatic aldehyde substrate to give the corresponding acid, and the other in water.[1] The reaction consists of two experimentally separable steps: the oxidation of luciferase-bound reduced flavin mononucleotide to give a peroxide intermediate, and the reaction of this intermediate with the aldehyde to give light emission peaking at about 490 nm, fatty acid, oxidized flavin, and H_2O.

The molecular structure of the peroxide intermediate was proposed to be the flavin C(4a)-hydroperoxide[2]; this was confirmed[3] [however, see Vervoort et al.[3a]] by means of ^{13}C NMR spectroscopy, where the C(4a) position of oxidized FMN was enriched with ^{13}C. In the original report the isolated luciferase-flavin hydroperoxide exhibited an absorption peaking at about 375 nm with no absorption above 520 nm.[4,5] In later work, however, preparations were obtained with appreciable absorption in the 550–650 nm range.[6] This blue species was later isolated and characterized, and shown to be a neutral flavin semiquinone,[7] the formation and properties of which are described below. While this luciferase-bound radical does not itself give significant light emission it will emit light upon reaction with superoxide ion.[8]

From a mechanistic point of view, it is important to know whether the semiquinone recombination as such might generate an excited state or

---

[1] J. W. Hastings, C. J. Potrikus, S. C. Gupta, M. Kurfürst, and J. Makemson, *Adv. Microb. Physiol.* **26,** 235 (1985).

[2] A. Eberhard and J. W. Hastings, *Biochem. Biophys. Res. Comm.* **47,** 348 (1972).

[3] S. Ghisla, J. W. Hastings, V. Favaudon, and J. M. Lhoste, *Proc. Natl. Acad. Sci. U.S.A.* **73,** 995 (1978).

[3a] J. Vervoort, F. Müller, J. Lee, W. A. M. van den Berg, and C. T. W. Moonen, *Biochemistry,* in press.

[4] J. W. Hastings, C. Balny, C. Le Peuch, and P. Douzou, *Proc. Natl. Acad. Sci. U.S.A.* **70,** 3468 (1973).

[5] J. E. Becvar, S. C. Tu, and J. W. Hastings, *Biochemistry* **17,** 1807 (1978).

[6] R. P. Presswood and J. W. Hastings, *Photochem. Photobiol.* **30,** 93 (1979).

[7] M. Kurfürst, S. Ghisla, and J. W. Hastings, *Eur. J. Biochem.* **123,** 355 (1982).

[8] M. Kurfürst, S. Ghisla, and J. W. Hastings, *Biochemistry* **22,** 1521 (1983).

whether light is being produced subsequent to this process, for example, via the formation of the flavin hydroperoxide. The latter appears to be the case.

In the reaction of the peroxide with long-chain aldehyde, the luciferase-bound flavin 4a-hydroxide chromophore has been proposed to be the emitter.[9-10] Kinetic measurements[11] showed that in the reaction of purified luciferase-bound flavin hydroperoxide with aldehyde, the decay of bioluminescence occurs faster than the appearance of the oxidized flavin. This indicates the existence of an intermediate species formed after the light-emitting step but prior to the formation of oxidized flavin. From absorbance and fluorescence studies of this intermediate and the comparison with other enzyme systems, it may be inferred that it is indeed the luciferase-bound flavin C(4a)-hydroxide (see Scheme 1).

Formation and Properties of the Luciferase Neutral Flavin Radical

Materials

Luciferase was isolated and purified from the luminous bacterium *Vibrio (Beneckea) harveyi* mutant strain M-17[12] according to procedures previously described[13]

Commercial FMN was from Sigma and pure FMN was obtained by affinity chromatography with an apoflavodoxin column[14]

Sephadex G-25 (medium) column (1 × 10 cm; 5 ml void volume)

In the EPR measurements the semiquinone of 5-ethyl riboflavin[15] was used as a calibration standard

Lamp for irradiation (projection lamp or equivalent)

Procedure

The luciferase neutral flavin radical can be formed by comproportionation of equimolar concentrations of oxidized and reduced FMN in the presence of luciferase in the absence of O_2. A solution (~0.5 ml) containing 0.1 mM luciferase and 0.2 mM commercial FMN in 0.35 M phosphate

[9] J. W. Hastings and T. Wilson, *Photochem. Photobiol.* **23,** 461 (1976).

[10] J. W. Hastings and K. H. Nealson, *Annu. Rev. Microbiol.* **31,** 549 (1977).

[11] M. Kurfürst, S. Ghisla, and J. W. Hastings, *Proc. Natl. Acad. Sci. U.S.A.* **81,** 2990 (1984).

[12] C. A. Waters and J. W. Hastings, *J. Bacteriol.* **131,** 519 (1977).

[13] J. W. Hastings, T. O. Baldwin, and M. Z. Nicoli, this series, Vol. 57, p. 135.

[14] S. G. Mayhew and M. J. J. Strating, *Eur. J. Biochem.* **59,** 539 (1975).

[15] H. Michel and P. Hemmerich, *J. Membr. Biol.* **60,** 143 (1981).

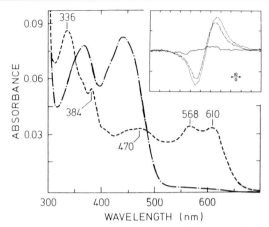

FIG. 1. Absorption spectrum of the luciferase neutral flavin semiquinone (dashed line) along with the spectrum after its aerobic decay to form oxidized enzyme–FMN. The semiquinone was purified by chromatography on Sephadex G-25. Inset: EPR spectra for the semiquinone (dotted line), a standard flavin radical (dashed line), and the sample after warming to 20° for 10 min (solid line), also measured at liquid N_2 temperature.

buffer, pH 7.0, and 0.25 M NaCl is made anaerobic by repeated evacuation and flushing with N_2. The same volume of FMN (0.2 mM) in the same buffer containing 3 mM EDTA is reduced photochemically in a side arm of the vessel.

After mixing, the formation of the radical will be complete within 5–10 min, and is stable enough to allow its purification in the presence of oxygen. Upon chromatography at 2° on a short Sephadex G-25 (5 ml void volume) the nearly homogeneous blue flavin species elutes with the protein. Its spectrum exhibits maxima at 610, 568, 470, 384, and 336 nm (Fig. 1). This resembles the spectrum of a neutral flavin radical in apolar solvents[16] and is closely similar to the radical spectrum of *Azotobacter*,[17] which is considered typical for a flavin semiquinone in an apolar environment. The extinction coefficient at 610 nm for the luciferase neutral flavin radical was estimated to be about 4500 M^{-1} cm^{-1}.[7] The rates of formation and decay of the semiquinone depend strongly on the temperature. A value of $\Delta H = 88$ kJ mol^{-1} was estimated for the temperature-dependent equilibrium between the semiquinone and the oxidized and reduced flavin couple.

A characteristic property of the luciferase neutral flavin semiquinone is its stability in the presence of oxygen ($t_{1/2} = 20$ hr at 0°), and its highly

[16] F. Müeller, M. Brüstlein, P. Hemmerich, V. Massey, and W. H. Walker, *Eur. J. Biochem.* **25,** 573 (1972).
[17] V. Massey and G. Palmer, *Biochemistry* **5,** 3181 (1966).

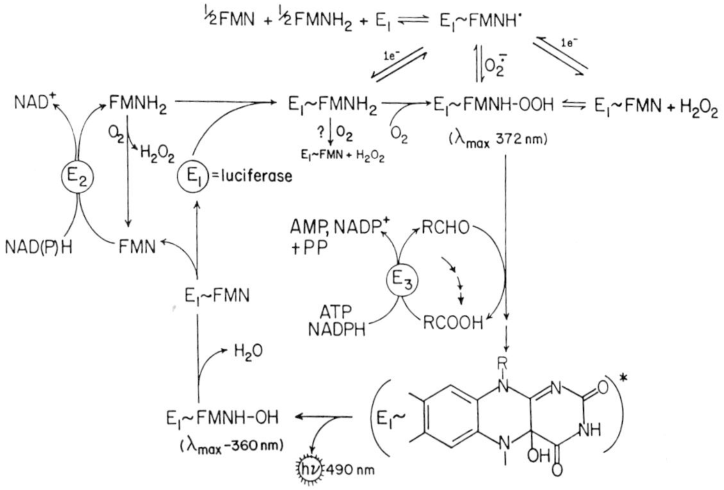

$$\tfrac{1}{2}FMN + \tfrac{1}{2}FMNH_2 + E_I \rightleftharpoons E_I{\sim}FMNH^{\cdot}$$

$$NAD^+ \quad FMNH_2 \longrightarrow E_I{\sim}FMNH_2 \longrightarrow E_I{\sim}FMNH\text{-}OOH \rightleftharpoons E_I{\sim}FMN + H_2O_2$$

$$E_I{\sim}FMN + H_2O_2 \quad (\lambda_{max}\ 372\ nm)$$

(E_2)

(E_I) = luciferase

$$NAD(P)H \quad FMN$$

AMP, NADP$^+$ + PP   RCHO

(E_3)

ATP NADPH   RCOOH

$$E_I{\sim}FMN$$

$$H_2O$$

$$E_I{\sim}FMNH\text{-}OH \longleftarrow (E_I{\sim})$$

$(\lambda_{max}\ {\sim}360\ nm)$

$h\nu\ {\sim}490\ nm$

SCHEME 1. In the proposed mechanism, luciferase (E_1) binds reduced flavin mononucleotide ($FMNH_2$) to form a reduced enzyme–flavin complex which can either be oxidized in a one electron step to give the luciferase-bound flavin semiquinone (E-FMNH) or oxidized by molecular oxygen to give the flavin 4a-hydroperoxide. In the presence of a long-chain aldehyde this peroxide reacts to form a peroxihemiacetal, which, after breaking the peroxide bond, gives the carboxylic acid and the flavin 4a-hydroxide. In the absence of accessory fluorescent proteins the flavin 4a-hydroxide is formed in its excited singlet state. After light emission, the flavin 4a-hydroxide is in its ground state; oxidized flavin is formed by eliminating water.

temperature-dependent rate of decay (Arrhenius activation energy is 160 kJ/mol). Addition of decanal (a substrate of bacterial luciferase) generates no light emission; rather, the stability of the semiquinone is increased approximately 3-fold, even though the activation energy for the decay remains the same. The stability of the luciferase flavin semiquinone reflects the general concept of flavin–luciferase interaction which must be involved also in the specific stabilization of the flavin C(4a)-hydroperoxide. The high activation energy reflects a strong interaction of a protein function with the flavin, most probably the N(5) hydrogen, which prevents elimination of H—O—O$^-$ and the formation of oxidized flavin. Furthermore, the high activation energy suggests that a conformational change might also play a role in the decay of the enzyme-bound flavin semiquinone to the products oxidized flavin and luciferase; the latter remains enzymatically active after the decay.

The EPR spectrum of a sample (0.25 ml) of the luciferase semiquinone frozen in liquid nitrogen is shown in Fig. 1 (inset); there is a direct correlation between the absorbance at 610 nm and the EPR signal.

As indicated in Scheme 1, the luciferase neutral flavin semiquinone can also be formed by a 1-electron oxidation of the $FMNH_2$–luciferase complex or by a 1-electron reduction of the oxidized complex. The latter may be carried out under anaerobic conditions by the stepwise addition of dithionite in small aliquots to a deoxygenated solution (typically 3 ml) of FMN ($\sim 10 \ \mu M$) and luciferase (10 μM) at 2°. The course of the reaction is conveniently followed by recording the absorbance at 610 nm after each addition. The absorbance at 610 nm first increases to a maximum; further additions of dithionite result in the loss of absorbance in this range, which indicates complete reduction to E-$FMNH_2$.

To achieve a 1-electron oxidation of the reduced complex, commercial FMN ($2 \times 10^{-4} \ M$ in 0.04 M phosphate buffer, pH 7) is reduced at 0° in the presence of luciferase ($8 \times 10^{-5} \ M$) with a small excess of dithionite in the same buffer. The mixture is then slowly oxygenated with molecular oxygen using a Pasteur pipet, while the 600 nm absorbance is followed spectroscopically; the latter increases over the next 20–30 min. The solution is then chromatographed on a Sephadex G-25 column as described above; essentially homogeneous semiquinone is found in the protein fraction. In this procedure it should be pointed out that the total flavin must be in excess of luciferase; alternatively, some electron transfer mediator other than FMN itself must be present. This is fulfilled when commercial FMN is used since it contains up to 25% flavin impurities.[14] When pure FMN and a large excess of luciferase are used, no flavin semiquinone is formed; instead, the luciferase flavin C(4a)-hydroperoxide is formed quantitatively.

The Reaction of Luciferase Flavin Radical with Superoxide

Materials

> Luciferase flavin semiquinone, prepared and purified as described above
> Xanthine oxidase (Boehringer), xanthine (Merck)
> Superoxide dismutase (SOD; Sigma)
> Hydroxylamine (Merck)
> KO_2 (Fluka), dissolved in Me_2SO in the presence of 18-crown-6 ether (Aldrich)[18]

Procedure

To chromatographically purified luciferase flavin semiquinone (10^{-7}) M in 0.5 ml of 0.01 M phosphate buffer, pH 7.0 (Fig. 2 at A), decanal and xanthine ($10^{-4} \ M$) are added (at B). The addition of xanthine oxidase (0.05 units) initiates luminescence (at C). Light emission is suppressed if the

[18] J. S. Valentine and A. B. Curtis, *J. Am. Chem. Soc.* **97,** 224 (1975).

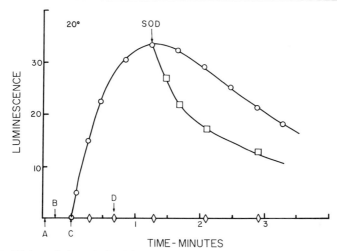

FIG. 2. Light emission obtained from the reaction of the luciferase semiquinone with superoxide (generated from xanthine and xanthine oxidase) with decanal. Superoxide dismutase added prior to initiation prevented light emission; added secondarily it caused it to decay more rapidly.

incubation mixture contains hydroxylamine ($1.5 \times 10^{-2} M$), but the subsequent addition of excess decanal results in light emission (not shown). When SOD is present in the assay from the beginning, no light emission is detected at all (D; baseline), and the secondary addition of SOD (at arrow) can cause the light emission to decay more rapidly.

The formation of the luciferase flavin C(4a)-hydroperoxide by recombination of the luciferase flavin semiquinone and $O_2^{\bar{}}$ can also be demonstrated by chromatographic isolation of the peroxide. A reaction mixture containing 0.5 ml of the luciferase flavin semiquinone (7 μM), 0.1 ml of a saturated xanthine solution, and 15 μl of a xanthine oxidase suspension (10 mg/ml) is incubated for 10 min at 0°. This mixture is then applied at 2° to a Sephadex G-25 column (void volume 5.5 ml; preequilibrated with 0.35 M phosphate buffer, pH 7.0), and eluted with the same buffer. Samples of 0.5 ml each are collected and kept at 0°. Aliquots in 0.5 ml of 0.01 M phosphate buffer are assayed for light emission by the addition of 0.5 ml of 0.1% v/v decanal. The amount of eluted protein correlates well with the capability of the fractions to emit light upon mixing with decanal, and the kinetics of this light emission are the same as observed with authentic luciferase flavin hydroperoxide. The decay of the activity of this preparation, which should reflect the decay of the luciferase hydroperoxide, occurs with a half-time of about 50 min at 0°, the same as that observed with authentic hydroperoxide.[7]

The same type of light emission occurs when the luciferase semi-quinone is reacted with different O_2^--generating systems such as xanthine, acetaldehyde or purine with xanthine oxidase, or the system O_2^- + 18-crown-6 (K^+, in Me_2SO).[19] The effects of systematic variation of the different reagents, trapping agents, and their sequence of addition,[8] support the postulate that L-FMNH⁻, O_2^- and long-chain aldehyde are requirements for light emission.

Luciferase-Bound Flavin 4a-Hydroxide

Materials

Luciferase purified from *Vibrio harveyi* M-17[13]
Sephadex G-25 column (medium) (1 × 15 cm; void volume 5 ml)
FMN (Sigma)
Sodium dithionite (Mallinckrodt)
Decanal (Aldrich)
0.1 *M* phosphate or Tris buffer, pH 8.5

Procedure

For the purification of C(4a)-hydroxide, a Sephadex G-25 column is equilibrated at 2° with 0.01% v/v decanal solution at pH 8.5. A decanal suspension is prepared by mixing 100 μl of aldehyde solution (10 μl decanal in 90 μl ethanol) with 100 ml of 0.1 *M* phosphate, pH 8.5. The solution is sonicated for about 1–2 min, cooled to 2°, and then filtered before use. A mixture of 100 μl luciferase (4.6 × 10^{-4} *M*), 25 μl FMN (4.5 × 10^{-3} *M*), and 175 μl equilibration buffer is reduced by adding a few grains of solid sodium dithionite, and promptly applied to the aldehyde-equilibrated column, then eluted with the same buffer under aerobic conditions. After about 10 sec, a bright band of luminescence is observed in the upper part of the column, due to the light-emitting reaction of the formed luciferase flavin hydroperoxide with aldehyde. The light emission is complete prior to elution (elution time, ~15 min) of the main protein fraction (0.5 ml). This fraction is collected and transferred to a cuvette for measurement of fluorescence and absorbance spectra. No residual luciferase-bound flavin hydroperoxide activity can be detected by injecting aliquots of the flavin hydroxide fraction into a freshly prepared decanal solution. The absorption and the fluorescence spectra of the preparation, before and after decay of the luciferase-bound flavin hydroxide decay, are shown in Fig. 3; the kinetics of fluorescence and absorbance changes

---

[19] R. L. Arudi, A. O. Allen, and B. H. J. Bielski, *FEBS Lett.* **135**, 265 (1981).

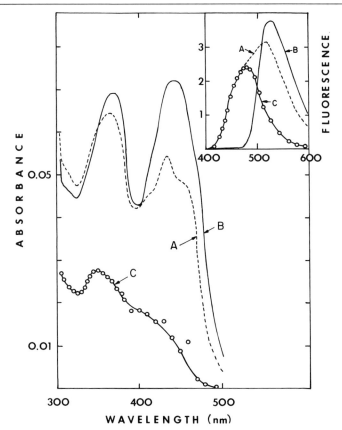

FIG. 3. The luciferase flavin 4a-hydroxide. Illustrated are absorbance and (inset) fluorescence emission spectra, both taken with the same sample of a preparation immediately (dashed lines, A) after its elution from the "aldehyde column," at which time it contained a mixture of flavin 4a-hydroxide and oxidized FMN. The solid lines (B) give the spectra for oxidized flavin taken after complete decay and the open circles (C) show absorption and fluorescence spectra calculated for the hydroxide.

during this are plotted in Fig. 4. The spectra for the hydroxide itself are obtained by subtraction.

The enzyme-bound flavin C(4a)-hydroxide intermediate possesses properties that are essential requirements for the emitting chromophore. Moreover, from a mechanistic point of view, the formation of the flavin hydroxide as a product and emitter would rule out proposed flavin structural rearrangements,[20–22] since such arrangements could not lead to the

[20] H. J. Mager and R. Addink, *Tetrahedron Lett.* p. 3545 (1979).
[21] F. McCapra and D. W. Hysert, *Biochem. Biophys. Res. Commun.* **52,** 298 (1973).
[22] A. Wessiak, G. E. Trout, and P. Hemmerich, *Tetrahedron Lett.* p. 739 (1980).

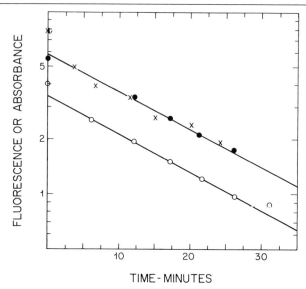

FIG. 4. Correlation of the kinetics of decay of fluorescence at 460 nm (○), increase of fluorescence at 520 nm (●), and increase in absorbance at 440 nm (×) taken from the preparation of luciferase flavin 4a-hydroxide in Fig. 3. Half-time for decay, 5°: 15 min.

formation of this flavin C(4a)-hydroxide. Scheme 1 shows the postulated pathways of bacterial luciferase, which includes the proposed flavin intermediates to which roles can be assigned unambiguously. In some cases, a separate fluorescent protein is involved in the emission and possibly also the reaction.[23,24]

The postulated reaction mechanism does not demand the formation of an excited chromophore per se, but it does account for the energetic requirements of light emission. A peroxide bond is broken in the reaction step involved in, or immediately preceding the formation of the luciferase-bound flavin 4a-hydroxide in its excited singlet state. The mechanism of excited state formation is not known. One possible route is patterned on the chemically induced electron exchange luminescence (CIEEL) mechanism.[25,26] Such an intermolecular electron exchange mechanism, in which

[23] P. Koka and J. Lee, *Proc. Natl. Acad. Sci. U.S.A.* **76,** 3068 (1979).
[24] G. Leisman and K. H. Nealson, *in* "Flavins and Flavoproteins" (V. Massey and C. Williams, eds.), p. 383. Elsevier, Amsterdam, 1982.
[25] J.-Y. Koo, S. P. Schmidt, and G. B. Schuster, *Proc. Natl. Acad. Sci. U.S.A.* **75,** 30 (1978).
[26] T. Wilson, *in* "Singlet Oxygen" (A. Frimer, ed.), Vol. 2, p. 37. CRC Press, Boca Raton, Florida, 1985.

the 4a, 5-dihydroflavin plays the role of an internal activator, has been proposed for the bacterial luciferase reaction,[1,27] and may be represented as follows:

$$E—FHOO—CHOH—R \rightarrow E—^{\dagger}FHO^{-}\,O—CHOH—R \rightarrow E—^{\dagger}FHOH^{-} + RCOOH$$

$$F + H_2O \leftarrow h\nu + FHOH \leftarrow [E—\overset{\downarrow}{F}HOH]^*$$

[27] P. Macheroux, S. Ghisla, M. Kurfürst, and J. W. Hastings, in "Flavins and Flavoproteins" (R. Bray, P. Engel, and S. G. Mayhew, eds.), p. 669. de Gruyter, Berlin, 1984.

[13] Purification and Properties of Lumazine Proteins from *Photobacterium* Strains

By Dennis J. O'Kane and John Lee[1]

The bioluminescence emission spectra of bacterial luciferases (Lases)[2] *in vitro* have single emission maxima ranging from approximately 487 nm (*Vibrio harveyi* strain MAV[3,4]) to approximately 500 nm (unclassified bacterium designated SQ−1[5]) with the emission maxima of most Lases clustering in the 495 nm region. The *in vivo* and *in vitro* bioluminescence spectral maxima of *Vibrio* species generally, but not always,[6–8] are quite similar, but parameters describing the spectral distributions (i.e., mean

[1] Supported by grants from the National Institutes of Health GM 28139 and National Science Foundation DMB 8512361.

[2] Abbreviations used: Lase, bacterial luciferase; CARS, coherent anti-Stokes Raman spectroscopy; Lum, 6,7-dimethyl-8-(1′-D-ribityl)lumazine; LumP, lumazine protein; PMT, photomultiplier tube; PD, photodiode detector; 2-ME, 2-mercaptoethanol; CHAPS, 3-[(3-cholamidopropyl)dimethylammonio]-1-propane sulfonate; HPLC, high-performance liquid chromatography.

[3] The taxonomy of luminescent bacteria follows that recommended by Baumann and co-workers: P. Baumann, A. L. Furniss, and J. E. Lee, in "Bergey's Manual of Systematic Bacteriology" (N. R. Krieg, ed.), 9th ed., Vol. 1, p. 518. Williams & Wilkins, Baltimore, Maryland, 1984; P. Baumann, *ibid.*, p. 539.

[4] J. W. Hastings, *Annu. Rev. Biochem.* **37,** 597 (1968).

[5] H. H. Seliger and R. A. Morton, in "Photophysiology" (A. G. Giese, ed.), Vol. 4, p. 253. Academic Press, New York, 1968.

[6] J. M. Fitzgerald, Ph.D. Thesis, Monash University, Melbourne, Australia (1978).

[7] E. G. Ruby and K. H. Nealson, *Science* **196,** 432 (1977).

[8] G. Leisman and K. H. Nealson, in "Flavins and Flavoproteins" (V. Massey and C. H. Williams, eds.), p. 383. Elsevier/North Holland, New York, 1982.

emission wavelength, variance, skewness, and kurtosis) may not be identical. Many isolates of *Photobacterium phosphoreum* and *P. leiognathi,* particularly the brighter strains, have *in vivo* bioluminescence spectra and emission maxima substantially shifted to shorter wavelengths than those of their Lases *in vitro,*[5-11] with peak emission wavelengths as low as 472 nm reported.[12]

In 1978, Gast and Lee[9] reported the isolation of a "blue fluorescence protein" from *P. phosphoreum* that stimulated light emission from the *in vitro* reaction of purified Lase *in vitro* and which provided the first clue to understanding the origin of these spectral shifts. They demonstrated that the fluorophore was protein bound, that its fluorescence emission spectrum closely matched the *in vivo* bioluminescence of *P. phosphoreum,* that addition of the blue fluorescence protein to Lase *in vitro* resulted in a 3-fold increase in the relative quantum efficiency of light emission with a dramatic alteration of the bioluminescence kinetics, and that the *in vitro* bioluminescence spectrum was shifted to shorter wavelengths to closely match the *in vivo* spectrum.[9] This blue fluorescence protein was subsequently demonstrated to bind a single mole[11,13] of noncovalently bound fluorescent ligand which was identified by several techniques including NMR and coherent anti-Stokes Raman spectroscopy (CARS), as 6,7-dimethyl-8-(1'-D-ribityl)lumazine[14] (Lum, Fig. 1), the immediate enzymatic precursor to riboflavin.[15-17] The holoprotein has been named generically for the ligand: lumazine protein (LumP).

Lumazine proteins have been identified in six strains of *Photobacterium* in which the *in vivo* bioluminescence spectra are shifted to shorter wavelengths than the corresponding *in vitro* spectra.[9-11,18,19] The amounts of LumP obtained per kilogram wet cell paste vary greatly with less LumP obtained from strains which have *in vivo* and *in vitro* emission maxima closer together (e.g., *P. leiognathi* B477; *in vivo* λ_{max} 483 nm; *in vitro* λ_{max} 496 nm). The purification protocol for each LumP is different. However, the procedures for preparing LumP from two of the Lum overproducers,

[9] R. Gast and J. Lee, *Proc. Natl. Acad. Sci. U.S.A.* **75,** 833 (1978).
[10] J. Lee, *Photochem. Photobiol.* **36,** 689 (1982).
[11] D. J. O'Kane, V. A. Karle, and J. Lee, *Biochemistry* **24,** 1461 (1985).
[12] A. Spruit-Van Der Burg, *Biochim. Biophys. Acta* **5,** 175 (1950).
[13] E. D. Small, P. Koka, and J. Lee, *J. Biol. Chem.* **255,** 8804, (1980).
[14] P. Koka and J. Lee, *Proc. Natl. Acad. Sci. U.S.A.* **76,** 3068 (1979).
[15] G. W. E. Plaut, *Annu. Rev. Biochem.* **30,** 409 (1961).
[16] C. H. Winestock and G. W. E. Plaut, *in* "Plant Biochemistry" (J. Bonner and J. E. Varner, eds.), p. 391. Academic Press, New York, 1965.
[17] G. W. E. Plaut and R. A. Harvey, this series, Vol. 18, p. 515.
[18] J. Vervoort, D. J. O'Kane, L. A. Carreira, and J. Lee, *Photochem. Photobiol.* **37,** 117 (1983).
[19] J. Lee and P. Koka, this series, Vol. 18, p. 226.

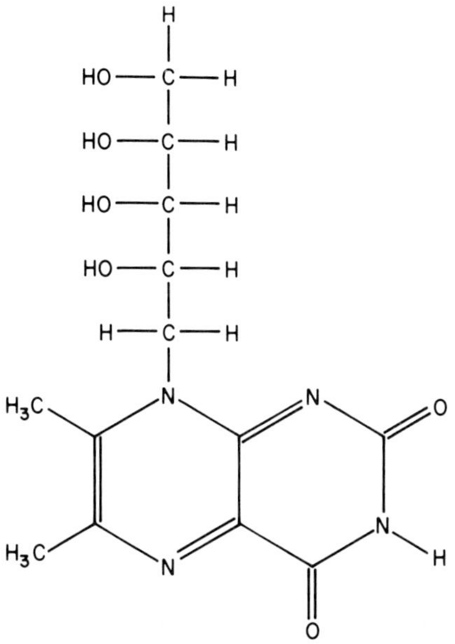

FIG. 1. Structure of 6,7-dimethyl-8-(1'-D-ribityl)lumazine.

P. phosphoreum strain A13 and *P. leiognathi* A2D, are sufficiently similar to be described together.

Assay Methods

Lumazine Protein Sensitization of Luciferase Activity

The addition of LumP to bacterial luciferases alters the kinetics of light emission and blue shifts the spectral maximum of the *in vitro* bioluminescence.[9-11,20] Two types of nonturnover assays are routinely employed to assess all or some of these changes "sensitized" by LumP. The simplest is the "squirt and flash" photometer assay where the stimulation of maximum light emission (I_0), the increase in total photons emitted (I_Σ), and changes in kinetics of light emission all can be observed. The assay[9] contains in a total volume of 1 ml of 50 mM phosphate buffer, pH 7.0, 22°: bovine serum albumin, 1 mg; Lase, 10–15 μM; 10 μl of a long-chain aldehyde solution, either a 5% saturated solution of decanal in methanol for *V. harveyi* Lase or a saturated solution of tetradecanal in ethanol for the other Lases: and 0–30 μM LumP. FMNH$_2$ (0.5 ml) photoreduced from FMN (80 μM) in 20 mM EDTA in standard buffer is injected and the

---

[20] D. J. O'Kane, I. B. C. Matheson, and J. Lee, *Photochem. Photobiol.* **41**, Suppl., 425 (1985).

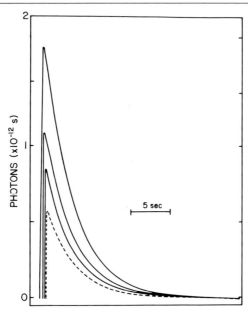

FIG. 2. Stimulation of *P. phosphoreum* Lase activity by LumP. Lase, either alone (dashed curve) or with increasing concentrations of *P. phosphoreum* LumP (12, 24, and 56 μM), was assayed by the injection of $FMNH_2$.

bioluminescence intensity is recorded over time using a photomultiplier (PMT) or photodiode (PD) detector (Fig. 2). A variation of the above "squirt and flash" assay is the generation of $FMNH_2$ from FMN and NAD(P)H by using a coupled turnover assay.[10,20] In this case the sample size is generally scaled down to 0.2 ml volume. The maximum bioluminescence intensity, in this case, is not directly related to the "squirt and flash" I_0, but the effects of LumP on I_Σ can be conveniently recorded using a coulombmeter to integrate the bioluminescence output (Fig. 3).[20]

It is possible to simultaneously determine the LumP-sensitized spectral shifting, the change in bioluminescence emission kinetics, and the increases in light output. The "squirt and flash" assay recipe described above is scaled down to 0.3 ml total volume and the reaction is started by the injection of 50 μl $FMNH_2$ using a fluorimeter cuvette having an emission path of 0.3 cm in order to minimize self-absorption of bioluminescence and to conserve on reagents. The assay is performed at 0–4° in a computer-controlled scanning spectrofluorimeter[21,22] where the bilumi-

[21] J. E. Wampler, M. G. Mulkerrin, and E. S. Rich, Jr., *Clin. Chem.* (*Winston-Salem, N.C.*) **25**, 1628 (1979).

[22] J. E. Wampler, *Appl. Spectrosc. Rev.* **171**, 407 (1981).

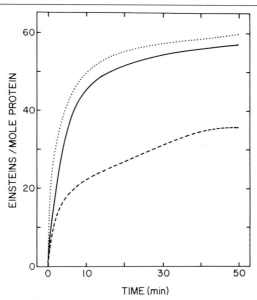

TIME (min)

FIG. 3. Stimulation of *P. leiognathi* S1 Lase activity by LumP in a turnover assay. Lase, either alone (dashed curve) or with 30 μM of either *P. phosphoreum* (solid curve) or *P. leiognathi* (dotted curve) LumP was assayed with FMN reduction linked to NADH oxidation, with NADH being regenerated from glucose 6-phosphate with the appropriate dehydrogenase from *Leuconostoc*. The signal output was integrated with a coulombmeter.

nescence is viewed simultaneously by two PMTs.[10] The reference PMT views a constant fraction of undispersed bioluminescence while the second PMT views the bioluminescence at discrete wavelengths through a grating monochromator. The monochromator is stepped through a predetermined wavelength range, 400–600 nm for example, and the monochromatized signal is ratioed to the reference PMT to correct for the change of total bioluminescence intensity with time.[23] This produces the technical bioluminescence emission spectrum. The reference PMT signal output over time is also recorded so that the kinetics of bioluminescence can be determined. Providing that the spectrofluorimeter has been properly calibrated (see below), and the number of photons to which 1 mV of reference signal corresponds is known (by calibration with luminol), then the reference PMT signal can be integrated and the absolute number of photons emitted can be determined following several correction procedures.

[23] H. H. Seliger, *Anal. Biochem.* **1**, 60 (1960).

Correcting Spectra and Spectral Analyses

It is regrettable that much of the literature in the field of bacterial bioluminescence presents technical data which are not corrected for the artifacts introduced by the measuring system. Although the value of proper corrections of results has been clearly pointed out,[5,24] it has largely been ignored in the literature. A number of sources can be consulted regarding the determination of appropriate correction factors[5,24-28] and their implementation should be greatly facilitated by the general availability of computer-interface spectrophotometers and fluorimeters.

Correction for Self-Absorption. Technical bacterial bioluminescence spectra must be corrected for trivial self-absorption due mostly to oxidized FMN and LumP present in the *in vitro* assays. Immediately following the filing of the technical emission spectrum, the absorption spectrum of the reaction mixture is determined over the same wavelength interval and employing the same optical path length. This must be performed on a calibrated, photometrically accurate spectrophotometer. A multiplying factor, F_{λ_i}, is calculated for each wavelength λ_i from the relationship[25,26]

$$F_{\lambda_i} = 2.30259\ OD_{\lambda_i}/(1 - 10^{-OD_{\lambda_i}})\qquad(1)$$

where OD is the product of absorbance and path length. The bioluminescence (fluorescence) intensity at each λ_i in the technical spectrum is multiplied by F_{λ_i} to yield the corrected intensity at each wavelength. This spectrum is subsequently stored for further corrections (below).

Although it may be tempting, or expedient, to "overlook" this correction step, we have found that even under conditions specifically chosen to minimize this artifact (i.e., 0.3 cm optical path and less than 0.1 OD averaged over the emission band), a 1- to 2-nm red shift results when self-absorption correction is not performed.[10]

Corrections for Emission Monochromator and PMT. Following correction for self-absorption, the spectra are corrected for the wavelength-dependent sensitivities of the monochromator/PMT pair. This is performed by calculating a set of correction factors,[28] [C_{λ_i} (quinine)], for the primary fluorescence standard quinine hemisulfate dihydrate, which is available from the National Bureau of Standards, Washington, D.C., that

[24] H. H. Seliger, this series, Vol. 57, p. 560.
[25] C. A. Parker, "Photoluminescence of Solutions." Elsevier, Amsterdam, 1968.
[26] J. E. Wampler, *in* "Bioluminescence in Action" (P. J. Herring, ed.), p. 1. Academic Press, London, 1978.
[27] J. R. Lakowicz, "Principles of Fluorescence Spectroscopy," Chapter 2. Plenum, New York, 1983.
[28] R. A. Velapoldi and K. D. Mielenz, *NBS Spec. Publ. (U.S.)* **260-264** (1980).

corrects the technical emission spectrum of the primary standard $[T_{\lambda_i}$ (quinine)] to the normalized emission spectrum of quinine $[A_{\lambda_i}$ (quinine)], such that

$$A_{\lambda_i} \text{ (quinine)} = [T_{\lambda_i} \text{ (quinine)}] \times [C_{\lambda_i} \text{ (quinine)}] \qquad (2)$$

The correction factors, C_{λ_i} (quinine), in fact represent the inverse of the spectral sensitivities of the monochromator/PMT pair and any technical spectrum T_{λ_i} can be corrected by multiplying the C_{λ_i} (quinine) in the range 375 to 675 nm. Since the absolute emission of quinine was originally determined at 5 nm intervals,[28] while most emission spectra are obtained at 1 nm increments, linear interpolation is employed to obtain the corrected emission spectra in 1 nm intervals.

It should be noted that this procedure employs a relative sensitivity correction function C_{λ_i} (quinine) which is normalized to the wavelength of peak emission for quinine. Absolute photometric sensitivity of the spectrofluorimeter can be determined by using luminol chemiluminescence as described in detail in a previous chapter.[29]

Determining Spectral Parameters and Comparing Spectra

The question frequently arises in bacterial bioluminescence as to whether two spectra are identical or whether they are different. Generally, the criteria used to answer this question are the "coincidence" of the peak emission wavelengths and the similarity of the two spectral shapes. Furthermore, this question is generally answered by visual inspection. If two corrected spectra are identical, then they will have not only the same peak emission wavelength but will also the same mean emission wavelength, $\bar{\lambda}$, and the same skewness (S) and kurtosis (K). These last two parameters describe the shape of a spectrum and are determined mathematically by the methods of reduced moments[21,26.]

$$U_n = \left[\sum_i (\lambda_i - \bar{\lambda})^n I_i \right] \bigg/ \sum_i I_i \qquad (3)$$

where U_n is the nth reduced moment, I_i is the intensity at wavelength λ_i, and

$$\bar{\lambda} = \left[\sum_i \lambda_i I_i \right] \bigg/ \sum_i I_i \qquad (4)$$

---

[29] D. J. O'Kane, M. Ahmad, I. B. C. Matheson, and J. Lee, this volume [10].

Skewness and kurtosis are calculated respectively from U_3 and U_4 normalized to the variance, U_2; K is further normalized to the Gaussian distribution, such that

$$S = U_3/(U_2)^{3/2} \tag{5}$$

and

$$K = [U_4/(U_2)^2] - 3 \tag{6}$$

The Gaussian distribution has $K = 0$ according to this definition. The reproducibility of the determined spectral parameters is excellent, the standard deviations obtained with fully corrected *in vivo* bioluminescence spectra are ± 5 Å ($\bar{\lambda}$), $\pm 1\%$ (S), and $\pm 1\%$ ($K + 3$).[11] Consequently, parameters obtained from two spectra which differ by approximately 3 times these deviations (95% confidence for small sample size) would indicate that the two spectra are not identical.

Spectra can also be compared by assuming that the differences between two spectra are only a scaling factor, F, and a wavelength position displacement D such that

$$\text{spectrum 1} = F \, (\text{spectrum 2}) + D \tag{7}$$

where perfect matching is indicated by values of 1 and 0 for F and D, respectively.

The peak emission wavelength, λ_{max}, should not be determined simply by eye. Emission spectra are always complicated by noise which is more noticeable around λ_{max} than on either side of the peak. The effect of noise is reduced if λ_{max} is found by linear regression analysis of the first derivative of a portion of the emission spectrum around the maximum. The point at which $dI/d\lambda = 0$ is λ_{max} which is reproducible to ± 1 nm.[11]

Selection and Growth of Bacteria

Selection

Not all bacteria are created equal, not even those isolated originally from a single colony. This is born out by finding a distribution of *in vivo* bioluminescence emission maxima rather than a single common λ_{max}, for clones isolated after subculturing a single colony in liquid media. This is particularly true for *P. leiognathi* A2D and DD17. The range of λ_{max} for clones of *P. leiognathi* A2D after one experiment was 477–489 nm while with DD17, the range was 477–482 nm. An additional problem encountered with some strains (e.g., *P. leiognathi* B477) is the spontaneous formation of dark variants. Clones of *P. phosphoreum* A13, however, have shown no statistically significant variation in λ_{max} over a 10-year period, and dark variants occur at low frequency compared to the *P. leiognathi* strains, but the level of bioluminescence intensity can vary in strain A13 by a factor of 5, clone to clone. Consequently single colony isolates should be routinely examined and selected for the level of bioluminescence, the position (λ_{max}) of the *in vivo* spectral maximum, and the ability to grow to high cell densities without the loss of bioluminescence potential.[11] With the exception of one dim clone of *P. leiognathi* DD17, that apparently had some defect in bioluminescence, the brighter colonies yield more LumP, and this is also well correlated with the position of λ_{max} *in vivo*.

Growth of Bacteria

The optimum conditions for growth of the selected clone must be determined prior to a large-scale, expensive fermentation. Generally for *P. phosphoreum* the conditions are low aeration, pH 7.2, 20° for initial growth and, once bioluminescence has been induced, the temperature is decreased to 12°. The condition for *P. leiognathi* are quite different: maximum aeration and 27–28°. An additional factor to monitor is the rate at which the bacteria acidify the medium. These studies are performed on 200 ml cultures in 500-ml flasks.

Once the optimum conditions for growth have been determined, and it has been established that the level of bioluminescence and λ_{max} *in vivo* are at the desired values, the growths can be scaled up to 400 liter fermentor volume. *P. phosphoreum* A13, the best studied LumP-producing strain, is inoculated (~8000 Klett units total; 100 Klett units = A_{660} = 2) at 20° with 100 liter min^{-1} aeration, pH 7.2, using a minimal salt medium supplemented with peptone and glycerol.[19] Eighteen to 20 hr later, the cells reach a density of 60 to 70 Klett units and bioluminescence has been induced. At this time the aeration is decreased (<30 liter min^{-1}) and the fermentor is cooled to 12°. After a further 20 to 24 hr, the cells reach a Klett of approximately 150 and the fermentor is harvested while cooling to 4°, using a cooled Sharples centrifuge. The yield of wet cell paste is usually between 1 and 1.5 kg per 400 liters.

P. leiognathi strains[11] are grown at 28° with 350 liter min^{-1} aeration and 300 rpm stirring rate. Approximately 1000 total Klett units (Klett × volume) are used as an inoculum. After 13 to 14 hr of growth, the Klett has increased to approximately 200 and the cells are cooled at 4° and harvested. The yield of wet cell paste ranges from 1.6 to 2.6 kg depending on the strain employed.[11] All cell pastes are stored at −60° until processing begins.

Purification for Lumazine Proteins[30]

Cell Breakage

Several procedures have been employed in this laboratory to disrupt cells, the easiest to use with large volumes being a continuous flow French press. Cell paste (500–750 g) is thawed in 2 volumes (per weight of paste) of 50 mM phosphate buffer, pH 7.0, containing 1 mM EDTA and the following concentrations of protease inhibitors: 0.5 mM phenylmethanesulfonyl fluoride, 0.5 mM benzamidine, and 0.1 mM diphenylcarbamyl chloride. After 45 min stirring the cells are dispersed and 2-mercaptoethanol (2-ME) is added to a final concentration of 10 mM (2-ME cannot be added initially since it rapidly inactivates phenylmethanesulfonyl fluoride).

Following stirring overnight at 4° the cells are disrupted by two passages through the thoroughly cooled French press at 600 kg cm^{-2}. Between passages, the extract and the press are cooled with ice. (Under no circumstances should osmotic lysis[31] by employed for the extraction. The resulting low ionic strength will cause dissociation of Lum and result in formation of apoprotein!)

Cell debris is removed by centrifugation twice for 3 hr at 23,300 G_{max} at 2° (DuPont-Sorvall, GSA rotor). The conductivity of the supernatant is generally between 80 and 90 mS and is adjusted to 60 mS with distilled H$_2$O.

The next three purification steps are identical for *P. phosphoreum* A13 and *P. leiognathi* A2D LumPs and are described together. The final purification steps for these two proteins are quite different and each is discussed separately below. Two preparations of the same cell type can be combined after the preparative gel filtration procedure described below to increased productivity.

Ion-Exchange Chromatography on DEAE-Sepharose with Step-Elution

The centrifuged cell extract (1.5–2 liters) is applied to a 5 × 50-cm column of DEAE-Sepharose equilibrated with 50 mM phosphate buffer, pH 7.0, containing 10 mM 2-ME and 1 mM EDTA (referred to below as standard buffer) plus the protease inhibitors minus PMSF. It takes 12 to 18 hr to load the column, and care must be taken to avoid channeling. The column is eluted at a flow rate of 60 ml hr^{-1} with 1 liter each of 50, 75, 100,

---

[30] Adapted with permission from supplementary materials submitted with D. J. O'Kane, V. A. Karle, and J. Lee, *Biochemistry*, **24**, 1461–1467. Copyright (1985) American Chemical Society.

[31] J. W. Hastings, T. O. Baldwin, and M. Z. Nicoli, this series, Vol. 57, p. 135.

125, 150, and 200 mM phosphate buffer, pH 7.0, containing the above additions. Fractions of 400 ml are collected in 800 ml beakers on a fraction collector. *P. leiognathi* LumP is easily visualized at this stage with a long wavelength UV lamp, as a blue fluorescent band running between Lase (green fluorescence) and a dark band (a low fluorescence flavoprotein[32]). The beakers are assayed for Lase activity, A_{280}, A_{420}, and for fluorescence at 470, excited at 420 nm (F_{470}) for LumP, and for fluorescence at 520, excited at 470 nm (F_{520}) for flavin. *P. leiognathi* LumP is well separated from the Lase at this stage while *P. phosphoreum* LumP coelutes with the bulk of the Lase (Tables I and II).

Preparative Gel Filtration Chromatography

The pooled LumP fractions are concentrated by ultrafiltration (Amicon YM10 membrane), centrifuged to remove precipitated protein and 3-[(3-cholamidopropyl)dimethylammonio]-1-propane sulfonate (CHAPS) a zwitterionic detergent, is added to a concentration of 2.5 mM. The extract (25 ml) is applied to a 9.6 × 65-cm column of Sephadex G-75 equilibrated with standard buffer containing CHAPS as above and 12 ml fractions are collected at 60 ml hr^{-1}. More than 97% of the Lase activity is separated from the LumP using this procedure (Tables I, II, and Fig. 4). This was not previously possible using gel filtration without CHAPS. The Lase fractions are pooled, glycerin is added to 30% (v/v), and these are stored frozen until further purification.

Ion-Exchange Chromatography on DEAE-Sepharose

The pooled LumP fractions are applied to a 4 × 60-cm column of DEAE-Sepharose equilibrated with standard buffer. The column is eluted with a gradient of 1 liter of standard buffer and 1 liter of 350 mM phosphate buffer, pH 7.0, containing EDTA and 2-ME, as above, at a flow rate of 32 ml hr^{-1}. Fractions of 8.5 ml are collected and assayed as above. *P. phosphoreum* LumP elutes at approximately 140 mM phosphate while *P. leiognathi* LumP elutes at 310 mM phosphate (at the peaks).

At this stage the absorption ratio A_{280}/A_{420} for the LumPs has improved considerably over that of the original extract. The *P. leiognathi* has in fact been purified to approximately 85% of homogeneity and the peak fractions (those with $A_{280}/A_{420} < 2.2$) can be combined to yield homogeneous LumP (Table II). On the other hand, *P. phosphoreum* LumP is only purified to 8% of homogeneity and is severely contaminated by fluo-

---

[32] A. J. W. G. Visser, J. Vervoort, D. J. O'Kane, J. Lee, and L. A. Carreira, *Eur. J. Biochem.* **131,** 639 (1983).

TABLE I

PURIFICATION OF LUMAZINE PROTEIN FROM *P. phosphoreum* STRAIN A13[a,b]

Step	A_{280}	A_{420}	A_{280}/A_{420}	F_{470}	F_{520}	F_{470}/F_{520}	Luciferase (10^{12} photons sec^{-1})
1. Cell-free extract[c]	37450	330	2113	5260	1956	2.7	8.0×10^5
2. DEAE-Sepharose: step-elution	22380	225	100	4020	1340	3.0	6.9×10^5
3. Gel filtration	3044	192	16	3504	1142	3.1	3.4×10^3
4. DEAE-Sepharose: gradient elution	2238	174	13	3230	1124	2.9	ND[e]
5. Blue Sepharose[d]	233	104	2.24	2570	88	29	ND
6. DEAE-Sepharose: gradient elution	175.6	40	2.18	1974	67	29	ND
Recovery (%)	0.5	12	—	37.5	3.4	—	$<2 \times 10^{-5}$

[a] Adapted with permission from supplementary materials submitted with O'Kane *et al.*[11] Copyright (1985) American Chemical Society.
[b] All procedures at 5° in standard buffer. A, Absorbance × total volume (ml); F, fluorescence intensity × total volume, in arbitrary units referenced to a sodium fluorescein standard; F_{470}, $420 \to 470$ nm; F_{520}, $470 \to 520$ nm (excitation → fluorescence). The purifications are standarized to 1 kg wet weight of starting materials. Luciferase units are bioluminescence activity/ml × total volume (ml).
[c] Turbidity in the cell extract prevents an accurate measurement of fluorescence and absorbance. Therefore the values in this row are estimated by summing all fractions from step 2.
[d] Combined pools of several columns. Tables I and II, three columns. Average data listed.
[e] ND, Not detectable.

160

TABLE II

PURIFICATION OF LUMAZINE PROTEIN FROM *P. leiognathi* STRAIN A2D[a,b]

Step	A_{280}	A_{420}	A_{280}/A_{420}	F_{470}	F_{520}	F_{470}/F_{520}	Luciferase (10^{12} photons sec^{-1})
1. Cell-free extract[c]	25540	766	33	2516	613	4	5.1×10^6
2. DEAE-Sepharose step-elution	6330	257	25	2334	220	11	5.6×10^5
3. Gel filtration	268	84	3.2	1762	129	14	1.4×10^3
4. DEAE-Sepharose gradient elution applied to step 5	115	40	2.9	1050	99	11	ND[e]
Pooled	66	30	2.2	642	39	16	ND
5. Gel filtration AcA-54[d]	90	42	2.2	870	46	19	ND
Final pool	153	70	2.2	1510	85	18	ND
Recovery (%)	0.6	9	—	60	14	—	$<10^{-7}$

[a–e] Footnotes as in Table I. Reprinted with permission from O'Kane *et al.*[11] Copyright (1985) American Chemical Society.

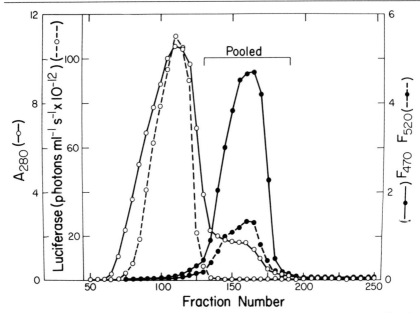

FIG. 4. Separation of *P. phosphoreum* LumP from Lase by preparative gel filtration chromatography on Sephadex G-75 with 2.5 mM CHAPS. Adapted with permission from supplementary materials submitted with O'Kane *et al.*[11] Copyright (1985) American Chemical Society.

rescent and low-fluorescent flavoproteins. The two separate final purification procedures are described below.

Final Purification of P. leiognathi LumP on Ultragel AcA 54

The fractions of LumP from DEAE-Sepharose which have not been previously pooled ($2.2 < A_{280}/A_{420} < 6.0$) are concentrated to 12 ml by ultrafiltration and are applied to a 4×110-cm column of Ultrogel AcA 54 equilibrated with standard buffer. The column is eluted at a linear flow rate of 3 cm hr^{-1} and fractions of 5 ml are collected (Fig. 5). Those fractions with $A_{280}/A_{420} < 2.2$) are combined, and the high- and low-molecular-weight tails are concentrated and reapplied twice to the same column, each time removing the pure fractions ($A_{280}/A_{420} < 2.2$). The final pool (Table II, steps 4 and 5) represents a recovery of 60% of the original F_{470} units and corresponds to approximately 145 mg of protein.

Final Purification of P. phosphoreum A13 LumP

Chromatography on Blue Sepharose. The pooled fractions of *P. phosphoreum* LumP from the DEAE-Sepharose step are concentrated by ul-

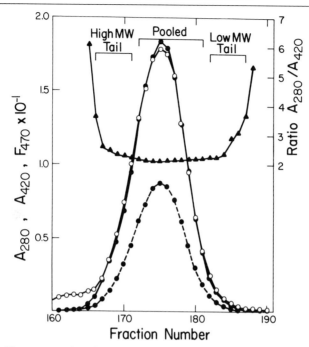

FIG. 5. Chromatography of *P. leiognathi* LumP on Ultragel AcA 54. Adapted with permission from supplementary materials submitted with O'Kane *et al.*[11] Copyright (1985) American Chemical Society.

trafiltration, centrifuged, and adjusted to a conductivity of 32 mS with cold distilled H_2O. The concentrated extract (15 ml) is applied to a 6.5 × 96-cm column of Blue Sepharose (prepared as described by Small *et al.*[13]) equilibrated with 40 mM phosphate buffer, pH 7.0, containing EDTA and 2-ME as above. The column is isocratically eluted at a linear flow rate of 2.7 cm hr^{-1} and fractions of 4.5 ml are collected. The early LumP fractions are contaminated by a low fluorescence flavoprotein and the tailing fractions are contaminated with fluorescent flavoprotein(s). The pooled LumP fractions are reconcentrated and reapplied to the same column twice. Between applications, the column is washed with 500 mM phosphate buffer containing 6 M urea to elute tightly bound proteins. The pooled LumP after three applications to Blue Sepharose has an A_{280}/A_{420} of 2.24, indicating that it is nearly homogeneous (approximately 3% contaminants remaining) and the ratio of F_{470}/F_{520} has improved dramatically indicating the removal of fluorescent flavoproteins.

Ion-Exchange Chromatography on DEAE-Sepharose. The residual contaminants are removed from the *P. phosphoreum* LumP by ion-exchange chromatography on a 2.5 × 60-cm column of DEAE-Sepharose, equilibrated with standard buffer. The column is eluted with a gradient formed from 500 ml each of standard buffer and 300 m*M* phosphate buffer at a flow rate of 20 ml h^{-1}. Fractions with $A_{280}/A_{420} < 2.2$ are pooled, concentrated by ultrafiltration, and stored at $-60°$.

Summary of Lumazine Protein Purification and General Comments

Lumazine proteins have been purified in good yields (> 30% recovery of F_{470}) from 6 strains of *Photobacterium* (Table III). The two LumP overproducers *P. phosphoreum* A13 and *P. leiognathi* A2D have the bluest *in vivo* emission spectra and have substantially greater amounts of LumP than do other strains. *P. leiognathi* B477 contains little LumP (~5 mg/kg wet weight) and has a spectrum intermediate between *P. phosphoreum* A13 and Lase *in vitro* with λ_{max} 483 nm.

The best yields of LumP were obtained from *P. leiognathi* A2D (145 mg/kg wet weight) with a recovery of 60%. *P. phosphoreum* A13 produces more LumP than *P. leiognathi* A2D, but there are greater losses during the *P. phosphoreum* A13 purification and a subsequent lower re-

TABLE III
SUMMARY OF PURIFICATION OF LUMAZINE PROTEINS FROM *Photobacterium* STRAINS

Species and strain	LumP recovered[a] (mg/kg wet weight cells)	λ_{max} (nm)	Maximum bioluminescence[b] (10^{12} photons sec^{-1} ml^{-1})
P. phosphoreum A13	83	475	5.5
P. phosphoreum NCMB844	18	476	2.0
P. leiognathi A2D	145	476	40
P. leiognathi DD17	20	478	0.85[c]
			(15)
P. leiognathi S1	13.2	481	20
P. leiognathi B477	~5	483	2.5

[a] Determined from A_{420}, with the exception of *P. leiognathi* B477 which is estimated from fluorescence.

[b] Maximum bioluminescence observed *in vivo* with our laboratory strains, at 22°.

[c] This particular isolate showed low bioluminescence *in vivo* but had Lase activity *in vitro* comparable to *P. leiognathi* S1. Other isolates of *P. leiognathi* DD17 have maximum bioluminescence of up to 15 × 10^{12} photons sec^{-1} ml^{-1}.

covery of LumP. The LumP from *P. leiognathi* B477 has not been purified to homogeneity.[10,11] The best preparations of this still contain a fluorescent flavoprotein which has charge properties and an effective hydrodynamic (Stokes) radius similar to those of the LumP, making the separation of these components difficult. The flavoprotein contaminant(s) also fails to separate from LumP by chromatography on Blue Sepharose and on Procion Red HE-3B-Sepharose (D. J. O'Kane and J. Lee, unpublished). The LumP, however, can be demonstrated by its fluorescence and by its shifting of the bioluminescence spectrum of Lase *in vitro*.[11,12]

LumP from *P. leiognathi* DD17 has recently been purified in part by high-performance liquid chromatography (HPLC) in which proprietary silica-based matrices replace the conventional column packing materials (Fig. 6). Two major difficulties have been encountered with the HPLC purification procedure. The first is that Lum dissociates from LumP during HPLC, at least on the silica-based columns, and results in the formation of apoprotein. This can be reconverted to LumP by addition of authentic Lum following ultrafiltration. The second difficulty is the LumP cannot be quantitatively separated from Lase during size exclusion chromatography in the presence of CHAPS (see below) which limits the usefulness of this procedure.

Effects of CHAPS

The zwitterionic detergent CHAPS is essential for separating LumP from Lase in *P. phosphoreum* preparations.[11] In the absence of CHAPS some 30 to 40% of the LumP remains "associated" with the Lase[13,19] while quantitative separations of these proteins is achieved by incorporation of only 2.5 mM CHAPS under the same conditions. CHAPS is a relatively innocuous detergent which allows the solublization of proteins under gentle conditions.[33-37] In addition, it does not interfere with subsequent chromatographic steps on ion-exchange columns and it can be removed from the preparations by ultrafiltration since the critical micelle concentration is ~8 mM and it has a small aggregation number.[37] The addition of CHAPS (9 mM) to *P. phosphoreum* LumP does not alter the fluorescence lifetime of bound Lum (14.5 nsec, 2°; 14.4 nsec reported for

[33] W. R. Simonds, G. Kosi, R. A. Streaty, L. M. Hjelmeland, and W. A. Kleen, *Proc. Natl. Acad. Sci. U.S.A.* **77**, 4623 (1980).
[34] L. M. Hjelmeland, *Proc. Natl. Acad. Sci. U.S.A.* **77**, 6368 (1980).
[35] D. S. Liscia, T. Alhadi, and B. K. Vanderhaar, *J. Biol. Chem.* **257**, 9401 (1982).
[36] A. J. Bitonti, J. Moss, L. Hjelmeland, and M. Vaughan, *Biochemistry* **21**, 3650 (1982).
[37] L. M. Hjelmeland, D. W. Nebert, and J. C. Osborne, Jr., *Anal. Biochem.* **130**, 72 (1983).

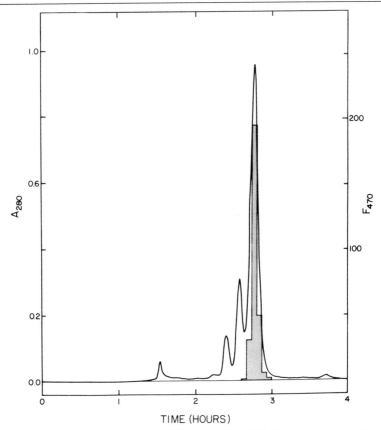

TIME (HOURS)

FIG. 6. Purification of *P. leiognathi* DD17 LumP by HPLC. A crude preparation was applied to a preparative (2.15 × 60-cm) column of a silica-based size exclusion chromatography column (TSK G3000 SWG) and eluted at 1 ml min⁻¹. The histogram shows the fluorescence of LumP at 470 nm centered on the midpoint of each fraction collected. The solid curve is the absorbance profile at 280 nm. Approximately 4 mg LumP was recovered from this one column.

LumP[38]; 10 nsec reported for free Lum[38]). Similarly, the ability of *P. phosphoreum* LumP to blue-shift the emission spectrum of Lase *in vitro* is not changed (D. J. O'Kane and J. Lee, unpublished). (We have observed an inhibition of FMN reduction by NADH and a partially purified extract of *P. leiognathi* S1 oxidoreductase by CHAPS however.)

The mechanism of separation of Lase and LumP by CHAPS is enigmatic since it fails to separate these proteins when this is attempted by

[38] A. J. W. G. Visser and J. Lee, *Biochemistry* **19**, 4366 (1980).

HPLC (D. J. O'Kane and J. Lee, unpublished). It is possible that CHAPS affects the separation of these proteins by both detergent action as well as by decreasing the viscosity of the protein solution and that substantial dilution of the extract (as obtained by size exclusion chromatography on Sephadex G-75) is required to affect the actual separation. This has not been investigated further.

Properties of Lumazine Proteins

Physicochemical Properties

Photobacterium LumPs are acidic, hydrophilic, and highly hydrated proteins with anhydrous molecular weights of 21,200 ± 300 (*P. leiognathi* A2D) and 21,300 ± 500 (*P. phosphoreum* A13) by several direct techniques.[39] The isoelectric points and amino acid compositions of the two proteins are slightly different.[40] The minimum molecular weights from the amino acid composition and Lum content are 21,220 (*P. leiognathi*) and 19,650 (*P. phosphoreum*). Single bands of protein are observed on SDS–polyacrylamide gels with Coomassie Blue staining or fluorescence detection of fluorescamine-derivatized proteins.[11,39] The two LumPs have statistically identical sedimentation coefficients (2.18 S *P. leiognathi* and 2.16 S *P. phosphoreum*) and Stokes radii (22.9 Å, *P. leiognathi*; 23.0 Å, *P. phosphoreum*).[39] Several estimations of hydration all indicate that *P. phosphoreum* LumP is more highly hydrated than *P. leiognathi* LumP (0.52–0.53 g H_2O/g protein, versus 0.43–0.44 g/g protein).[39] The two LumPs have similar maximum axial ratios of approximately 1.25 using a hydrated prolate ellipsoid model. Emission anisotropy decay measurements[41,42] of the bound Lum are in agreement with both the hydration and axial ratio estimates and indicate that both LumPs behave as near spherical molecules with hydrated molecular weights of 30,000 to 31,000.[38,39]

The stoichiometries of Lum to protein have been shown to be 1 : 1 by several methods.[13,40] The proteins bind Lum noncovalently and reversibly[13,40,42] and apparently in very similar ways since the CARS spectra of *P. phosphoreum* and *P. leiognathi* S1 LumP are quite similar to each other in the 1100 to 1600 cm^{-1} region.[18] The CARS spectra of the LumPs are also similar to that of free Lum with the exception of the 1242 cm^{-1} band and this may reflect a difference in hydrogen bonding to Lum when in solution and when bound to protein. *P. leiognathi* LumP binds Lum an

[39] D. J. O'Kane and J. Lee, *Biochemistry* **24,** 1484 (1985).
[40] D. J. O'Kane and J. Lee, *Biochemistry* **24,** 1467 (1985).
[41] A. J. W. G. Visser and J. Lee, *Biochemistry* **21,** 2218 (1982).
[42] J. Lee, D. J. O'Kane, and A. J. W. G. Visser, *Biochemistry* **24,** 1476 (1985).

order of magnitude more tightly than *P. phosphoreum* LumP. The dissociation constant, K_d (20°; 50 mM phosphate), is 16 nM (compared to 160 nM for *P. phosphoreum* LumP[42]). The temperature dependence of K_d results in an estimate of 6 kJ mol^{-1} for $\Delta H°$ for both LumPs. The difference in K_d for the two LumPs is believed to result from entropic or non-temperature-dependent interactions between Lum and the proteins requiring approximately 2–4 kJ mol^{-1}.[42]

Biological Properties of Lumazine Proteins

The proposed biological function of LumP is to shift the emission spectrum of Lase from the 490–495 nm region, which is suitable for light transmission is shallow waters, to bluer wavelengths suitable for bioluminescence in deeper waters. The color of light emitted by Lase *in vitro* is unsuitable for countershading by leiognathid fishes at depths greater than a few meters since the transmission spectrum of sunlight in sea water shifts to shorter wavelengths as depth increases. Similarly, if bright bioluminescence is an advantage because reception of this light over a distance greater than several meters is important, then the *in vitro* Lase bioluminescence should be shifted to shorter wavelengths *in vivo* which would facilitate greater light transmission. The maximum transmission of sunlight by sea water is at 476 nm, which is virtually identical to the peak fluorescence emission of LumP.[9–11,13]

The ability of LumP to induce a spectral shift, and the degree of spectral shifting, is concentration dependent.[9–11] The effect of LumP on the bioluminescence spectrum from lase is observable with as low as 1 μM LumP while >30 μM LumP may be required for the maximum effect.[10] An analysis of the spectral parameters for the *in vitro* bioluminescence spectra obtained with increasing concentrations of LumP is quite revealing. The values for λ_{max}, $\bar{\lambda}$, S, and K are identical, within statistical error, to those obtained from the *in vivo* bioluminescence of different bacteria.[11] An isoemissive point is observed in the *in vivo* and *in vitro* spectra at 503 nm.[12] The coincidence of the spectral parameters and the isoemissive point for the *in vivo* spectra and *in vitro* spectra with LumP indicate that (at least) two emitters are involved *in vivo*: the first is LumP excited to fluorescence by the Lase reaction; the second presumably is Lase *in vitro* emission alone. This is supported by spectral modeling studies with the *in vitro* bioluminescence spectrum of Lase and the fluorescence emission spectrum of LumP. The coincidence of the isoemissive point in the *in vivo* and *in vitro* spectra is consistent with the proposal that the same two emitters are active *in vivo* as well as *in vitro*.

TABLE IV
STIMULATION OF BIOLUMINESCENCE ACTIVITY LASE BY LumP[a]

Lase	LumP stimulation (fold)	
	P. phosphoreum	P. leiognathi
P. phosphoreum A13	1.6	1.7
P. leiognathi A2D	2.4	2.6
V. harveyi MAV	0.2	<0.1
V. harveyi MAVA[b]	1.9	<0.1
V. fischeri 7744	<0.1	<0.1

[a] Optimized tetradecanal, 22°.
[b] Aldehyde-deficient mutant isolated by Ahmad and Lee.[43]

The addition of LumP to Lase *in vitro* increases the overall quantum efficiency of the reaction (Table IV). A stimulation of light emission as low as 15% is observed with *V. harveyi* (strain MAV) Lase. This slight stimulation is real since it has been observed by kinetic experiments as well (I.B.C. Matheson, unpublished). LumP stimulates Lase purified from an aldehyde-deficient mutant derived from *V. harveyi* MAV[43] by a factor of approximately 2-fold, however. The increased stimulation of the Lase from the mutant compared to that from the wild-type is attributed to having less inactive Lase in the nonluminescent mutant strain. The only Lase which has not been found to be stimulated by LumP is from *V. fischeri* 7744. The reasons for this failure are unknown.

Spectral Properties

The absorption spectra of the two types of LumP are slightly different from each other with absorbance maxima at 262, 275, and 417.5 nm (*P. phosphoreum* LumP) and 262 and 419.5 nm with a shoulder in the 275–280 nm region (*P. leiognathi* A2D LumP).[40] The differences in the UV absorption are due to differences in the number of aromatic amino acids in the two proteins. *P. phosphoreum* LumP has 3 tyrosine residues while *P. leiognathi* LumP has only 2.[40] The difference in the position of the visible absorbance maxima for the two LumPs can be accounted for by having identical ground state environments for Lum in both proteins, but having different excited state interactions between Lum, the protein, and sol-

[43] M. Ahmad and J. Lee, *Fed. Proc., Fed. Am. Soc. Exp. Biol.* **44**, 1216 (1985).

vent. The fluorescence emission spectra of the two LumPs are virtually identical with an emission maximum at 475 nm which also coincides with the position of the *in vivo* bioluminescence maximum of *P. phosphoreum* A13.[9,11]

Mechanism of Bioluminescence Excitation

The transformation of the chemical energy released by the luciferase reaction into excitation of the LumP to its fluorescence state occurs by an unknown mechanism. However several mechanisms can be eliminated. A trivial absorption of the luciferase emission by the LumP and reemission of fluorescence does not occur since the optical density is too low and, moreover, the bioluminescence quantum efficiency increases and the kinetics of light emission change. Energy transfer of the long-range Förster type or short-range Dexter type, from the luciferase *in vitro* emitter as a donor to the LumP acting as a secondary emitter, is improbable. The 0–0' transition for LumP is approximately 440 nm determined by correlating line-shape changes in the bands of the CARS spectra with excitation wavelength (J. Vervoort, A. J. W. G. Visser, and L. A. Carreira, unpublished). At least 270 kJ mol^{-1} is therefore required to excite LumP. This is 30 kJ mol^{-1} higher than the energy available from the luciferase emitter.[44] Förster energy transfer is not favored by the additional fact that the overlap of the absorption spectrum of LumP with the luciferase bioluminescence is only 5–10% of the maximum possible (Fig. 7). Some unpublished observations are available that do not favor an exciton coupling mechanism. FMNH$_2$ can be replaced in the *in vitro* reaction by 2-thioFMNH$_2$ and the bioluminescence spectral maximum is around 560 nm.[45,46] There is no spectral shift or kinetic effect of including LumP in this 2-thioFMNH$_2$ reaction mixture. Furthermore, Lum can be replaced on the apolumazine protein by analogs such as riboflavin. This fluorescent riboflavin protein does not shift the bioluminescence spectrum toward the riboflavin fluorescence (maximum 530 nm) when it is included in the *in vitro* reaction. Apolumazine protein reconstituted with the authentic Lum recovers all the bioluminescence properties of the original LumP.

There are two processes that can be postulated for the excitation of LumP by the bioluminescence reaction. The first is an electronic energy

[44] F. Muller, *Photochem. Photobiol.* **34,** 753 (1981).
[45] G. Mitchell and J. W. Hastings, *J. Biol. Chem.* **244,** 2572 (1969).
[46] I. B. C. Matheson, J. Lee, and F. Muller, *Proc. Natl. Acad. Sci. U.S.A.* **78,** 948 (1981).

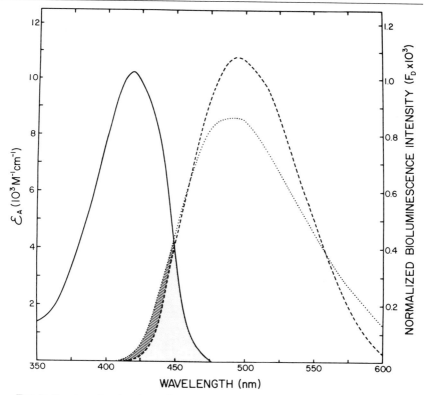

FIG. 7. Overlap of molar absorption spectrum of *P. leiognathi* LumP (solid curve) with the normalized bioluminescence of *P. leiognathi* (dashed curve) and *V. harveyi* (dotted curve) Lases in the absence of LumP.

transfer[47–49] from an otherwise weakly emitting donor species of energy around 300 kJ mol⁻¹, to the LumP, either long range, or short range within a protein–protein collision. The second process is one involving electron transfer such as "chemically induced electron exchange luminescence."[50] Kosower has calculated an overall enthalpy of -347 kJ mol⁻¹ for the steps leading to light emission from a proposed excited FMNH+.[51]

[47] I. B. C. Matheson and J. Lee, *Photochem. Photobiol.* **38,** 231 (1983).
[48] A. J. W. G. Visser and J. Lee, *in* "Excited State Probes in Biochemistry and Biology" (A. G. Szabo and L. Masotti, eds.). Plenum, New York, 1987 (in press).
[49] J. Lee, *in* "Chemi- and Bioluminescence" (J. G. Burr, ed.), p. 401. Dekker, New York, 1985.
[50] H. I. X. Mager and R. Addink, *in* "Flavins and Flavoproteins" (R. C. Bray, P. C. Engel, and S. G. Mayhew, eds.), p. 37. de Gruyter, Berlin, 1984.
[51] E. M. Kosower, *Biochem. Biophys. Res. Commun.* **92,** 356 (1980).

The development of protocols for purifying the relevant proteins in quantity and of top activity will make more practical the systematic physical studies that will need to be made in order to piece together the workings of this intriguing energy transformation phenomenon.

Acknowledgments

We wish to thank Dr. J. E. Wampler, Department of Biochemistry, University of Georgia, for providing the SPECOS computer programs used to obtain and to process the spectral data.

[14] Fatty Acid Reductase from *Photobacterium phosphoreum*

By Angel Rodriguez, Lee Wall, Denis Riendeau, and Edward Meighen

Fatty acid reductases from bioluminescent bacteria are responsible for the synthesis of long-chain aldehydes used as substrates in the luciferase-catalyzed luminescent reaction.[1-5] Both ATP and NADPH are required for the reduction of fatty acids as shown below:

$$RCOO^- + ATP + NADPH \longrightarrow R\overset{\overset{\displaystyle O}{\displaystyle \|}}{C}H + AMP + PP_i + NADP^+$$

The purified fatty acid reductase from *Photobacterium phosphoreum* is a multienzyme complex consisting of three distinct subunits (designated as 58K, 50K, 34K) which are synthesized in concert with luciferase during luminescence induction in the late stages of bacterial growth.[6,7] Two of the components, an acyl-protein synthetase (50K) and a reductase (58K), are essential for fatty acid reductase activity, while the third component

[1] S. Ulitzur and J. W. Hastings, *Proc. Natl. Acad. Sci. U.S.A.* **75**, 266 (1978).
[2] S. Ulitzur and J. W. Hastings, J. Bacteriol. **132**, 854 (1979).
[3] E. A. Meighen, *Biochem. Biophys. Res. Commun.* **87**, 1080 (1979).
[4] D. Riendeau and E. Meighen, *J. Biol. Chem.* **254**, 7488 (1979).
[5] S. Ulitzur and J. W. Hastings, *Curr. Microbiol.* **3**, 295 (1980).
[6] D. Riendeau, A. Rodriguez, and E. Meighen, *J. Biol. Chem.* **257**, 6908 (1982).
[7] L. Wall, A. Rodriguez, and E. Meighen, *J. Biol. Chem.* **259**, 1409 (1984).

(34K), an acyltransferase, appears to be involved in channeling of fatty acids into this reaction.[8–10] The acyl-protein synthetase catalyzes the ATP-dependent activation of fatty acids to form a tightly bound acyl-AMP (carboxyphosphate anhydride) intermediate that can acylate the synthetase (50K) component.[8,11]

$$50K + ATP + RCOO^- \xrightarrow[PP_i]{} 50K\ (R-\overset{\overset{\textstyle O}{\|}}{C}-AMP) \xrightarrow[AMP]{} 50K-\overset{\overset{\textstyle O}{\|}}{C}-R$$

The reduction of acylated 50K polypeptide with NADPH is catalyzed by the reductase (58K) component of the complex, converting the activated fatty acid to aldehyde. The 58K component can be assayed independently using acyl-CoA as a substrate and has thus been referred to as acyl-CoA reductase.[12]

$$R-\overset{\overset{\textstyle O}{\|}}{C}-X + NADPH + H^+ \longrightarrow R-\overset{\overset{\textstyle O}{\|}}{C}-H + NADP^+ + HX$$

Assay of Fatty Acid Reductase Activity

Principle. The amount of aldehyde produced can be determined quantitatively by the addition of luciferase and measurement of the light produced upon injection of $FMNH_2$.[13]

Procedure. The fatty acid reductase sample (≤50 μl) is incubated at room temperature in 1 ml of 50 mM phosphate, pH 7.0, 20 mM 2-mercaptoethanol, 10 mM $MgSO_4$, containing 10 μl each of 50 mM ATP, 10 mM NADPH, and 0.5 mM tetradecanoic acid (in isopropanol). Luciferase (10 μl of a 0.5 mg/ml solution of purified *P. phosphoreum* luciferase) is added and the sample placed in the chamber of a luminometer. At 4 min, 1 ml of 50 μM $FMNH_2$, catalytically reduced under H_2 in the presence of platinum, is injected through a syringe and the maximum light intensity recorded. Although the time for luciferase addition is not crucial, it is most convenient to add the luciferase just before (~15 sec) $FMNH_2$ injection. The production of tetradecanal is linear with both time and enzyme concentration up to 0.8 nmol [equivalent to 200 light units (LU) under the described conditions where 1 LU = 6×10^9 quanta/sec based on the

[8] A. Rodriguez, L. Wall, D. Riendeau, and E. Meighen, *Biochemistry* **22**, 5604 (1983).
[9] L. M. Carey, A. Rodriguez, and E. Meighen, *J. Biol. Chem.* **259**, 10216 (1984).
[10] D. Byers, A. Rodriguez, L. Carey, and E. Meighen, this volume [15].
[11] A. Rodriguez and E. Meighen, *J. Biol. Chem.* **260**, 771 (1985).
[12] A. Rodriguez, D. Riendeau, and E. Meighen, *J. Biol. Chem.* **258**, 5233 (1983).
[13] D. Riendeau and E. Meighen, *J. Biol. Chem.* **255**, 12060 (1980).

standard of Hastings and Weber[14]]. All phosphate buffers (pH 7.0) were made from 1 M stock solutions of K_2HPO_4 and NaH_2PO_4.

Assay for the Acyl-Protein Synthetase Component

Acyl-Protein Synthetase Assay[6,8]

Principle. Formation of acylated protein is measured by the incorporation of [^{3}H]tetradecanoic acid into material insoluble in organic solvent.[15]

Procedure. An aliquot (10–50 μl) of the enzyme preparation is mixed at room temperature in 100 μl final volume of 50 mM phosphate, pH 7.0, 20 mM 2-mercaptoethanol, containing 5 μl of 50 mM ATP and 5 μl of 0.25 mM [^{3}H]tetradecanoic acid (~5 Ci/mmol in ethanol). The reaction can be initiated either by addition of the enzyme or the radiolabeled fatty acid. At 10 min, 75 μl of the reaction mixture is applied to 3 cm^2 of filter paper (Whatman 3) and washed 3 times in 100 ml of $CHCl_3/CH_3OH/CH_3CO_2H$ (3/6/1). The filter papers are dried and the amount of acyl-protein determined by counting in 10 ml of Econofluor (NEN) with a 15% efficiency.

The incorporation of fatty acid at levels less than 20 pmol is proportional to enzyme concentration.[6] Under optimal conditions incorporation of fatty acid by the fatty acid reductase complex reaches a plateau within 5 min corresponding to the steady-state level of the acyl–enzyme intermediate.[6,8] Once separated from the complex, the acyl-protein synthetase subunit (50K) shows only a low level of incorporation which could be stimulated (~6-fold) by the addition of an excess of the reductase subunit (58K). The acyl-protein synthetase activity of the 50K can also be stimulated by the acyltransferase subunit (34K). However, the rate of acyl-protein formation in this case is lower and does not reach a plateau since the acyl group does not turnover in the absence of the reductase subunit.[8]

PP_i–ATP Exchange Assay[11]

Principle. The acyl-protein synthetase (50K) catalyzes the exchange of PP_i between ATP and free PP_i through formation of an enzyme-bound acyl–AMP intermediate. Incorporation of $^{32}PP_i$ into ATP is measured by adsorption on charcoal under conditions where PP_i remains soluble, followed by release of the radioactivity by hydrolysis of the ATP.[16]

$$R\overset{\overset{\text{O}}{\|}}{-}\text{COH} + \text{ATP} + \text{50K} \rightleftharpoons \text{50K (Acyl—AMP)} + PP_i$$

[14] J. W. Hastings and G. Weber, *J. Opt. Soc. Am.* **53**, 1410 (1963).
[15] T. K. Ray and J. E. Cronan, *Proc. Natl. Acad. Sci. U.S.A.* **73**, 4374 (1976).
[16] R. K. Crane and F. Lipmann, *J. Biol. Chem.* **201**, 235 (1953).

Procedure. The enzyme preparation (10–25 μl) is incubated at room temperature in a final volume of 100 μl of 50 mM phosphate, pH 7.0, 20 mM 2-mercaptoethanol, 0.5 mM MgSO$_4$ containing 5 μl of 100 mM ATP, 2 μl of 0.5 mM tetradecanoic acid (in isopropanol) in a 1.5 ml Eppendorf tube. Five microliters of 100 mM ^{32}PP$_i$ (~1.0 Ci/mmol) is added to start the assay. At 30 min, 0.6 ml of charcoal suspension (35 mg/ml in 5% TCA) is added and the tube centrifuged for 2 min in a microfuge. The charcoal is washed 3 times with 0.7 ml of 5% TCA and 1 time with 0.7 ml of H$_2$O by resuspension and centrifugation. The washed charcoal is then boiled 10 min in 0.5 ml of 1 N HCl and a 250-μl aliquot of the supernatant counted in 10 ml of Aquasol-2 (NEN).

This activity is not affected by the presence of the reductase (58K) subunit. Maximum activity is found at a fatty acid concentration of 10 μM and formation of labeled ATP is linear with time and enzyme concentration until 20% of the PP$_i$ has been exchanged.

Assay for the Reductase Subunit[6,12]

Principle and Procedure. The activity of the reductase subunit (58K) is measured as described for the fatty acid reductase–luciferase coupled assay, except that ATP and tetradecanoic acid are substituted by 5 μM tetradecanoyl-CoA and the activity is only dependent on the reductase component. The assay is optimal between 5 and 8 μM tetradecanoyl-CoA and is linear with enzyme concentration until a concentration of 80 nM of aldehyde has been attained.

Purification of Fatty Acid Reductase Components[6,8,12]

Growth and Extraction of Bacteria. *Photobacterium phosphoreum* (NCMB844) from the National Collection of Marine Bacteria, Aberdeen Scotland, is grown at 19–20° in 10 liters of 3% NaCl complex medium[12] to peak luminescence (4–6 × 10^3 LU/ml). The cells are harvested by centrifugation and stored frozen before being lysed by osmotic shock for 1 hr at 4° in 1 mM phosphate (pH 7.0), 10 mM EDTA, 1 mM 2-mercaptoethanol (15 ml/g wet cells) followed by sonication. Cellular debris is removed by centrifugation and, in most cases, the lysis procedure is repeated using 10 ml of lysis buffer per g of wet cell paste. The pooled supernatants are made 50 mM in phosphate and 2-mercaptoethanol by addition of 1 M phosphate (pH 7.0) containing 1 M 2-mercaptoethanol. All subsequent steps, summarized in the table, are performed at 4° in phosphate buffer at pH 7.0, containing 50 mM 2-mercaptoethanol.

Anion-Exchange Chromatography. The lysate is applied to a DEAE-cellulose column (4.4 × 40 cm) preequilibrated in 50 mM phosphate and

PURIFICATION OF FATTY ACID REDUCTASE COMPONENTS[a]

Purification step	Protein (mg)	Fatty acid reductase (nmol/ min/mg)	Acyl-CoA reductase (nmol/ min/mg)	Acyl protein synthetase (nmol/mg)
Extract	2350	0.53	0.50	0.09
DEAE-cellulose	256	6.6	6.1	0.65
$(NH_4)_2SO_4$ fractionation	130	9.9	10.0	0.95
Gel filtration	51	20	19.4	2.15
Aminohexyl-Sepharose	32	25	23.6	2.50
Blue Sepharose				
58K	11	—	33	—
50K	7	$(24)^b$	—	$(2.4)^b$
DEAE-Sepharose (58K)	8	—	33	—
DEAE-Sepharose (50K)	5	$(29)^b$	—	$(2.5)^b$

[a] From 44 g wet weight of *P. phosphoreum*.

[b] After complementation with an excess of the reductase subunit (58K).

eluted with 400 ml of the same buffer followed by a 2.2 liter linear gradient of 0.05 to 0.55 M phosphate. Fractions containing fatty acid reductase activity (eluting at ~0.4 M phosphate) are dialyzed against 50 mM phosphate and concentrated by adsorption onto a DEAE-Sepharose column (7 × 2.6 cm) followed by elution with this same buffer containing 0.5 M NaCl.

Ammonium Sulfate Fractionation and Gel Filtration. The DEAE-Sepharose pool is made 30% saturated in ammonium sulfate by addition of a saturated solution (pH 7.0), stirred for 30 min, and the precipitate removed by centrifugation at 27,000 g for 15 min. The supernatant is then made 55% saturated in ammonium sulfate, stirred for 30 min, and the precipitate collected by centrifugation. The pellet is redissolved in 5 ml of 50 mM phosphate, 0.2 M NaCl, and applied on a Sephacryl S-300 column (1.6 × 90 cm), equilibrated and eluted in the same buffer. Fatty acid reductase fractions are pooled and then dialyzed against 50 mM phosphate.

Aminohexyl-Sepharose Chromatography. The dialyzed Sephacryl pool is applied to an aminohexyl-Sepharose column (2.5 × 8.5 cm) and washed with 50 ml of 50 mM phosphate. The enzyme activity is eluted near the end of a 400 ml linear gradient of 0.05 to 0.6 M phosphate.

At this stage of purification, the fatty acid reductase constitutes approximately 80% of the protein and includes both required subunits (50K and 58K) as well as the acyltransferase subunit (34K).[12] The enzyme can be stored for months at −20° in 50 mM phosphate, pH 7.0 (with 50 mM 2-

mercaptoethanol) containing 30% glycerol and 0.2 M NaCl. Alternatively, the individual subunits can be dissociated and separated by dye binding chromatography and purified further.

Resolution of the Subunits on Blue Sepharose CL-6B. The fatty acid reductase sample is dialyzed 4 hr against 1 liter (2 changes) of 20 mM HEPES buffer, pH 7.0, containing 50 mM 2-mercaptoethanol. The dialyzed sample is applied to a Blue Sepharose column (1.5 × 20 cm) and the acyl-CoA reductase activity (58K component) eluted with 100 ml of the HEPES buffer. The column is then washed with 100 ml of 50 mM phosphate containing 0.5 M NaCl followed by 100 ml of 50 mM phosphate containing 0.2 M NaSCN which removes the acyltransferase subunit.[10] The acyl-protein synthetase enzyme (50K) is then eluted with 150 ml of 0.5 M NaSCN in phosphate buffer.

The reductase component (58K) as eluted in HEPES buffer from the Blue Sepharose column is greater than 90% pure with only a few minor contaminants.[12] If desired, these can be removed by aminohexyl chromatography or, alternatively, by gel filtration on Sephacryl S-300. The final preparation of enzyme can be precipitated by dialysis against saturated ammonium sulfate in 50 mM phosphate and redissolved in 50 mM phosphate, pH 7.0, 50 mM mercaptoethanol, 30% glycerol, and 0.2 M NaCl and stored at −20° for several months with no loss of activity.

The acyl protein synthetase (50K) component, which is eluted with 0.5 M NaSCN from Blue Sepharose, is dialyzed against 50 mM phosphate, 5% glycerol, applied to a DEAE-Sepharose column (1.5 × 4 cm), and eluted with a 50 ml linear gradient of 0 to 0.5 M NaCl in the same buffer.[8] The synthetase pool is dialyzed against the 50 mM phosphate containing 50 mM mercaptoethanol, 30% glycerol, and 0.2 M NaCl, and can be stored at −20° for several months without loss of activity. Attempts to recover the synthetase activity after precipitation with ammonium sulfate have been unsuccessful, since the protein would only redissolve in the presence of denaturants such as 8 M urea.

Properties of Fatty Acid Reductase Components

Fatty acid reductase from *P. phosphoreum* is the first reported NADPH fatty acid reductase to be purified and characterized. Most NADPH-dependent acyl-CoA reductases are membrane bound and have not been solubilized and resolved from other proteins.

Stability. The fatty acid reductase is sensitive to thermal inactivation, primarily due to loss of acyl protein synthetase activity. Incubation at 35° for 15 min results in the loss of 70% of fatty acid reductase activity and

50% of acyl protein synthetase activity, while over 90% of the acyl-CoA reductase activity remains even after 1 hr.[6]

The fatty acid reductase enzyme requires thiol reducing agents for both stability and activity. Removal of reducing agent results in a fast inactivation of the enzyme. N-Ethylmaleimide (NEM) at 3 mM inactivates the enzyme completely within 10 min. Both the synthetase (50K) and reductase (58K) subunits react with NEM, however, the synthetase protein is much more sensitive to inactivation.

The stability of the enzyme as a complex is also dependent on protein concentration.[6] After a 10-fold dilution of a 1 mg/ml preparation, for instance, only 25% of the fatty acid reductase activity remains after 1–2 hr incubation at 0° in phosphate buffer, 20 mM mercaptoethanol. This inactivation appears to be due to dissociation of the synthetase subunit from the complex and its subsequent inactivation as dilution of the purified synthetase subunit into buffers containing no glycerol also results in its inactivation. No dilution effect has ever been observed for the acyl-CoA reductase activity. Dissociation of the complex occurs much more readily in low ionic strength buffers such as the HEPES buffer used in separation of the synthetase and reductase subunits on Blue Sepharose. The dependence of the stability of fatty acid reductase on protein concentration, salt concentration, and reducing agent must always be considered, particularly if only low amounts of material are available and/or the procedure for purification must be modified.

Mg^{2+} Requirement. Only the PP_i–ATP exchange activity has been shown to be dependent on Mg^{2+} for optimal activity[11] and this requirement appears to be related to the binding of ATP. By contrast, a dependence on Mg^{2+} has not yet been found for the fatty acid reductase and acyl protein synthetase activities, although both activities are dependent on ATP and inhibited by 10 mM EDTA. The requirement of Mg^{2+} (10 mM) in the acyl-CoA reductase assay is not specific since it can be substituted by 0.2 M NaCl.

Structural Features. The fatty acid reductase enzyme migrates on gel filtration at a molecular weight of about 5×10^5. However, since the diluting effect of the column results in partial separation of the subunits,[6] the exact molecular weight of the complex is not known. An active complex can also be reconstituted by mixing of the synthetase and reductase subunits prepared by separation on Blue Sepharose.[6] The stoichiometry and arrangement of the subunits in this multienzyme complex is currently under investigation. The properties of the enzyme subunits, particularly the synthetase subunit, indicate the presence of strong hydrophobic domains raising the possibility that these proteins are associated with the membrane and are part of a more highly organized structure *in vivo.*

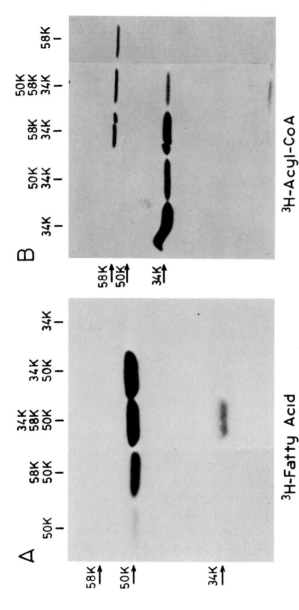

Fig. 1. Acylation of fatty acid reductase related polypeptides from *P. phosphoreum*. Mixtures of the three polypeptides in 100 μl (at about 60 μg/ml each) were incubated for 10 min at room temperature with 12 μM [³H]tetradecanoic acid (21 Ci/mmol) and 5 mM ATP (A) or with 10 μM [³H]tetradecanoyl-CoA (1.1 Ci/mmol) (B) in a 50 mM phosphate buffer (pH 7) containing 25 mM 2-mercaptoethanol. The reaction was stopped by addition of SDS–electrophoresis sample buffer and aliquots containing 3 μg (A) or 5 μg (B) of each polypeptide were boiled and run on a 10% SDS–polyacrylamide gel using the method of U. K. Laemmli [*Nature (London)* **227**, 680 (1970)]. The gels were fixed, soaked in En³Hance (NEN) and in 5% glycerol, dried, and exposed to Kodak film (XAR-5) for 1 day (A) or 3 days (B). From Rodriguez et al.[8]

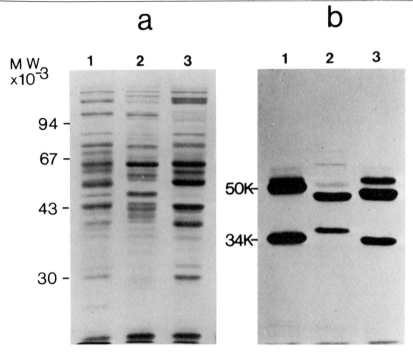

FIG. 2. Acylation pattern of protein extracts from different strains. Protein extracts from three different bacteria strains: *P. phosphoreum* (1), *P. leiognathi* (2), and *P. phosphoreum-A13* (3) were incubated for 10 min with 12 μM [³H]tetradecanoic acid (21 Ci/mmol) and 5 mM ATP. Aliquots containing 25 μg of total protein were run on a 10% SDS–polyacrylamide gel and fluorographed for 7 days, as described for Fig. 1. (a) Coomassie blue staining, (b) fluorography. In part, from Rodriguez *et al.*[8]

Kinetic Parameters. Fatty acid reductase has a high affinity for ATP with a K_m of 20 nM. The other substrates are characterized by apparent K_m values of 5 μM for NADPH and 1 μM for tetradecanoic acid and tetradecanoyl-CoA in the fatty acid reductase and acyl-CoA reductase assays, respectively.[17] Concentrations of acyl-CoA above 10 μM result in enzyme inhibition.[12] At saturating concentrations of fatty acid, the enzyme has a preference for tetradecanoic acid showing the same chain length specificity as luciferase.[17] The enzyme is highly specific for NADPH with only an activity of 3% with NADH.[6] The fatty acid reductase has a maximal activity of about 25 nmol/min/mg and a similar rate is obtained when the tetradecanoic acid is substituted by tetradecanoyl-CoA,[12] suggesting that the reductase subunit is involved in the rate limiting step.

[17] A. Rodriguez, I. R. Nabi, and E. Meighen, *Can. J. Biochem. Cell Biol.* **63,** 1106 (1985).

a   b

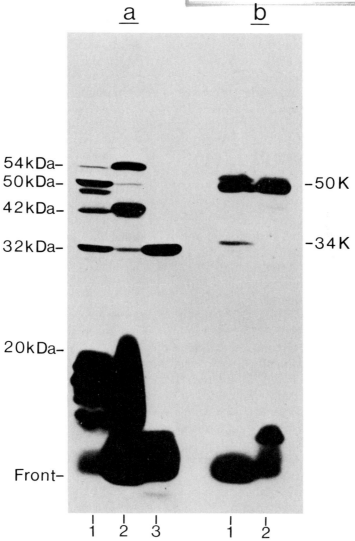

FIG. 3. *In vitro* and *in vivo* acylation of *V. harveyi* and *P. phosphoreum*. Cells were grown to near peak luminescence. *In vitro* acylation of bacterial extracts, followed by separation of the polypeptides by SDS–polyacrylamide gel electrophoresis and fluorography, was performed as described for Figs. 1 and 2. *In vivo* acylation was achieved by incubating an aliquot of growing cells with 2.5 μM [³H]tetradecanoic acid (21 Ci/mmol) for 5 min at room temperature. After 2 washes with cold medium, the cells were lysed by sonication in 1 mM phosphate (pH 7) containing 1 mM 2-mercaptoethanol, the cellular debris removed by centrifugation, and the protein extract resolved by SDS–polyacrylamide electrophoresis. (a) *V. harveyi*, 50 μg of protein per lane. (b) *P. phosphoreum*, 25 μg of protein per lane. Lane 1, *in vitro* acylation with [³H]tetradecanoic acid (+ATP); lane 2, *in vivo* acylation with [³H]tetradecanoic acid; lane 3, *in vitro* acylation with [³H]tetradecanoyl-CoA. From Wall *et al.*[18]

Acyl Transferase Activity. The reductase subunit (58K) can transfer activated fatty acids (i.e., acyl-50K, acyl-CoA) to thiol compounds such as mercaptoethanol present in the buffer. Addition of NADPH completely blocks the acyltransferase activity as the fatty acid is converted to aldehyde.[12] This activity causes the turnover of the activated acyl-50K intermediate, with concomitant cleavage of ATP by the fatty acid reductase even in the absence of NADPH.[17] The rate of formation of acyl-*S*-mercaptoethanol and aldehyde products is equivalent, suggesting the existence of a common intermediate for both reactions[12] (probably the acylated form of the 58K subunit).

Protein Acylation. Fatty acid activation catalyzed by the fatty acid reductase results in the formation of acyl-protein products that are sensitive to neutral NH_2OH treatment but are stable enough to allow analysis by SDS–gel electrophoresis.[6,8] Incubation of a reconstituted enzyme complex with ^3H-labeled fatty acid and ATP in a buffer containing 20 mM 2-mercaptoethanol results in the acylation of the synthetase (50K) and acyltransferase (34K) subunits (Fig. 1A). The 50K subunit is responsible for the acylation reaction and it is acylated, although at a much lower level, when incubated alone.

The labeling of the 34K subunit in *P. phosphoreum* can be blocked by the addition of NADPH or the removal of the 2-mercaptoethanol or the reductase (58K) subunit. Acylation of this subunit is due to reaction with the acyl-*S*-mercaptoethanol produced by the acyltransferase activity of the reductase subunit.[9] In this regard, specific acylation of the acyltransferase (34K) subunit, and to a lesser degree the reductase (58K) subunit, can be achieved by incubation with $[^3H]$acyl-CoA (Fig. 1B).

The *in vitro* identification of the acylated polypeptides of luminescent bacteria appears to be directly related to the presence of functional fatty acid reductase activity in extracts. Characteristic acylation patterns are obtained for extracts of *Photobacterium* strains that contain fatty acid reductase activity (Fig. 2). In contrast, fatty acid reductase activity has not yet been extracted from *Vibrio harveyi* or *V. fischeri* and only the acyltransferase subunit can be identified *in vitro* by acylation with tetradecanoyl CoA.[18] In these cases, however, the polypeptides of the fatty acid reductase complex can be specifically labeled *in vivo* by incubating ^3H-labeled fatty acid with the cells before extraction (Fig. 3).

Acknowledgments

We acknowledge the valuable technical assistance of Rose Szittner. This work was supported by Grant MT-4314 from the Medical Research Council of Canada.

[18] L. A. Wall, D. M. Byers, and E. A. Meighen, *J. Bacteriol.* **159,** 720 (1984).

[15] Bioluminescence-Related Acyltransferases from *Photobacterium phosphoreum* and *Vibrio harveyi*

By David Byers, Angel Rodriguez, Luc Carey, and Edward Meighen

The long-chain aldehyde substrate of bacterial luciferase is synthesized via the reduction of fatty acids.[1,2] The enzymes responsible for supplying endogenous fatty acids for this process have recently been isolated from *Photobacterium phosphoreum*[3] and *Vibrio harveyi*[4] (Fig. 1, reaction A). These enzymes have subunit molecular weights of 34 and 32 kDa, respectively, and possess acyl-CoA acyltransferase activity *in vitro*.[5] Recent evidence has indicated that acyl-acyl-carrier protein (acyl-ACP) might be the precursor involved in generating fatty acyl groups for bioluminescence.[6] Both acyltransferases appear to be induced during the midexponential growth phase characteristic of bacterial luminescence *in vivo*.[7,8]

Assay

Reagents. [3H]Tetradecanoyl-CoA is synthesized from [3H]tetradecanoic acid (Amersham)[9,10] and stored frozen under N_2 at $-20°$ in 50 mM phosphate (pH 6). Silica gel N-HR/UV$_{254}$ plates and En^{3}Hance spray are obtained from Fisher and New England Nuclear, respectively. Organic solvents are reagent grade and are used without further purification. Phosphate buffers are prepared by mixing NaH_2PO_4 and K_2HPO_4 in the appropriate ratio.

Method. The assay is based on the cleavage of [3H]tetradecanoyl-CoA to form a labeled hexane-soluble product.[10] For the *P. phosphoreum* acyl-

[1] S. Ulitzur and J. W. Hastings, *Proc. Natl. Acad. Sci. U.S.A.* **75**, 266 (1978).

[2] D. Riendeau and E. Meighen, *J. Biol. Chem.* **254**, 7488 (1979).

[3] L. M. Carey, A. Rodriguez, and E. Meighen, *J. Biol. Chem.* **259**, 10216 (1984).

[4] D. Byers and E. Meighen, *J. Biol. Chem.* **260**, 6938 (1985).

[5] The term "acyltransferase" has been used for these enzymes, although they exhibit acylhydrolase (transfer to water) activity under certain conditions. "Acyl-CoA cleavage" refers to the cleavage of acyl derivatives to form hexane-soluble products, without regard to the final acceptor.

[6] D. Byers and E. Meighen, *Proc. Natl. Acad. Sci. U.S.A.* **82**, 6085 (1985).

[7] L. Wall, A. Rodriguez, and E. Meighen, *J. Biol. Chem.* **259**, 1409 (1984).

[8] L. Wall, D. M. Byers, and E. A. Meighen, *J. Bacteriol.* **159**, 720 (1984).

[9] J. E. Bishop and A. K. Hajra, *Anal. Biochem.* **106**, 344 (1980).

[10] A. Rodriguez, D. Riendeau, and E. Meighen, *J. Biol. Chem.* **258**, 5233 (1983).

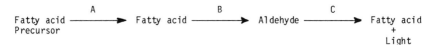

FIG. 1. A scheme outlining the acyltransferase (A), reductase (B), and luciferase (C) reactions of bacterial bioluminescence.

transferase, the enzyme preparation is incubated with 8 μM [³H]tetradec-anoyl-CoA (100 Ci/mol) in 1 M phosphate, 50 mM 2-mercaptoethanol, pH 7.0, in a total volume of 100 μl at room temperature (22°).[3] The reaction is stopped by adding 10 μl of glacial acetic acid and the solution is extracted twice with 1 ml hexane. The hexane washes are combined and counted directly in 10 ml Econofluor (New England Nuclear).

The assay for the *V. harveyi* enzyme is essentially identical, except that the solvent used is 50% ethylene glycol, 50 mM phosphate (pH 7).[4] For both acyltransferases, enzyme concentration and incubation time are chosen such that less than 60% of the substrate is cleaved.

The hexane-extractable product of [³H]acyl-CoA cleavage can be ana-lyzed by thin-layer chromatography and fluorography.[10] Hexane extracts are concentrated under nitrogen to a volume of ~10 μl and spotted on Silica gel plates. The chromatogram is developed in benzene/diethyl ether/acetic acid (90/10/2), dried, sprayed with En³Hance, and exposed to Kodak XAR-5 film at −70°.

Comments. The relatively low specific activities of the *P. phospho-reum* and *V. harveyi* acyltransferases in aqueous buffers and the presence of interfering acyl-CoA thioesterases in crude extracts are the major prob-lems faced in the assay for these bioluminescence-related enzymes. The activity of the *P. phosphoreum* enzyme in 1 M phosphate, 50 mM 2-mercaptoethanol (120 nmol/min/mg) is sufficient to allow its resolution at all preparative stages. However, the *V. harveyi* acyltransferase repre-sents less than 1% of the total acyl-CoA cleavage activity in *V. harveyi* extracts under similar conditions and high concentrations of ethylene glycol or glycerol must be included in the assay mixture, at least during the initial purification steps.[4] For example, the acyl-CoA cleavage rate of the 32 kDa enzyme in 50 mM phosphate is stimulated 100-fold (to 1500 nmol/min/mg) when 50% ethylene glycol is present; consequently, over 50% of the activity in *V. harveyi* lysates is attributable to this enzyme.

If [³H]tetradecanoyl-CoA is not available, labeled acyl-CoAs of simi-lar chain length can also be used as substrates, although their effects on the activities of interfering enzymes are not known. Another potential substrate is acyl-ACP: it has been found that the cleavage of [³H]tetra-decanoyl-ACP to a hexane-soluble product is dependent upon the pres-ence of these acyltransferases in bacterial extracts.[6] Thus, acyl-ACP may in fact be the substrate of choice in studies of enzyme expression under various conditions. If no labeled substrates are available, acyltransferase activities can be monitored with spectrophotometric assays based on the release of thiol groups or on the hydrolysis of tetradecanoyl-*p*-nitrophenol. Measurement of fatty acid production by a coupled *P. phosphoreum* fatty acid reductase-luciferase assay has also been used to demonstrate the hydrolysis of tetradecanoyl-*S*-mercaptoethanol and tetradecanoyl-l-glycerol by the 34 kDa enzyme.[3]

Purification

Reagents. DEAE-Sepharose CL-6B and Blue Sepharose CL-6B are purchased from Pharmacia. BioGel HT and Ultrogel AcA 44 are obtained from Bio-Rad and LKB Instruments, Inc., respectively.

P. phosphoreum acyltransferase. The 34 kDa acyltransferase can be isolated from the partially purified *P. phosphoreum* fatty acid reductase complex, with which it copurifies during ion-exchange, gel filtration, and aminohexyl Sepharose chromatography.[11] No change in the acyl-CoA cleavage/fatty acid reductase activity ratio is observed through these steps, starting from the initial Cellex D pool.[3] The acyltransferase is separated from the 50 and 58 kDa fatty acid reductase subunits on a Blue Sepharose column, using batch elution with 0.2 *M* NaSCN.[11] The fractions containing 34 kDa acyl-CoA cleavage activity are pooled and dialyzed vs 50 m*M* phosphate, 20 m*M* 2-mercaptoethanol (pH 7) containing 15% glycerol. If desired, the enzyme can be stored at this stage at $-20°$ without significant loss of activity after 1 month. Further purification to remove minor contaminants is carried out on an Ultrogel AcA 44 gel filtration column (80 × 1.5 cm) in 0.1 *M* NaCl, 50 m*M* phosphate, 20 m*M* 2-mercaptoethanol (pH 7). The activity is pooled, dialyzed vs 15% glycerol, 50 m*M* phosphate, 20 m*M* 2-mercaptoethanol, and stored at $-20°$.

Cellex D chromatography of *P. phosphoreum* cell-free extracts separates three peaks of acyl-CoA cleavage activity (Fig. 2, I–III)[3]; Peak III corresponds to the activity associated with the aforementioned fatty acid reductase complex. However, Peak II activity is also characteristically stimulated by high concentrations of phosphate (Fig. 2) and is attributed to free 34 kDa acyltransferase, which has been resolved from the other fatty acid reductase subunits. The relative amount of activity in Peak II is variable with different preparations, but the acyltransferase can be purified from this source using Blue Sepharose and gel filtration. A typical yield of the 34 kDa enzyme from Peak III alone is 2–4 mg per 40 g cells.

V. harveyi acyltransferase. Although a corresponding fatty acid reductase complex has not been isolated from extracts of *V. harveyi*, the 32

---

[11] The initial preparative stages for purifying the fatty acid reductase complex are described by A. Rodriguez, L. Wall, D. Riendeau, and E. Meighen, this volume [14].

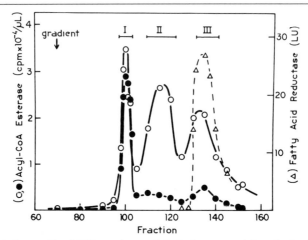

FIG. 2. Elution profile of [³H]tetradecanoyl-CoA cleavage activity of a *P. phosphoreum* extract after Cellex D chromatography.³ The column was eluted with a linear 0.05–0.5 *M* phosphate gradient, pH 7, and assays were conducted in 50 m*M* (●) or 1 *M* (○) phosphate containing 50 m*M* 2-mercaptoethanol as described in the text.

kDa acyltransferase can be separated from luciferase and the majority of the soluble protein by $(NH_4)_2SO_4$ fractionation.[4] Proteins associated with aldehyde metabolism in *V. harveyi,* including the acyltransferase, aldehyde dehydrogenase, as well as the 57 and 42 kDa fatty acid-labeled polypeptides thought to be involved in fatty acid reduction,[8] are coprecipitated in 30–50% saturated $(NH_4)_2SO_4$. This is a possible indication of a complex similar to that found in *P. phosphoreum.*

V. *harveyi* B392 cells are grown to maximum luminescence (A_{660} = 1.5–2) at 27° in 10 liters of 1% complete medium.[2] The cells are harvested by centrifugation and lysed in 700 ml of 50 m*M* phosphate, 10 m*M* 2-mercaptoethanol (pH 7) by sonication (3 × 45 sec) in small batches. A cell-free supernatant is obtained by centrifugation (17,000 *g,* 25 min). Lysis and subsequent steps are carried out at 4° (see Table I).

The *V. harveyi* lysate is made 30% saturated with solid $(NH_4)_2SO_4$, stirred for 25 min, and centrifuged to remove the precipitate. The resulting supernatant is made 50% saturated with ammonium sulfate, stirred for 30 min, and the precipitate is collected by centrifugation (17,000 *g,* 30 min). The 30–50% precipitate is dissolved in 100 ml of 50 m*M* phosphate, 10 m*M* 2-mercaptoethanol (pH 7) and dialyzed overnight. The dialyzed fraction is applied to a DEAE-Sepharose CL-6B column (2.5 × 20 cm), which is washed with 50 m*M* phosphate, 10 m*M* 2-mercaptoethanol at pH 7 (150 ml) followed by a linear 0–0.5 *M* NaCl gradient (1 liter total) in the same buffer. Fractions with acyl-CoA cleavage activity in 50% ethylene glycol are pooled and concentrated by precipitation in 75% saturated

TABLE I
PURIFICATION OF THE *V. harveyi* 32 kDa ACYLTRANSFERASE[a]

Purification step	Volume (ml)	Total protein (mg)	Acyl-CoA cleavage activity[b]	
			Total (nmol/min)	Specific (nmol/min/mg)
Lysate supernatant	700	970	43,000[c]	45
(NH₄)SO₄ 30–50% fraction	100	125	12,000[c]	100
DEAE-Sepharose pool	32	12.1	8,000	660
Gel filtration pool	10	4.5	5,300	1,170
Hydroxylapatite pool	2.7	2.6	3,800	1,460

[a] From Byers and Meighen.[4]

[b] [³H]Tetradecanoyl-CoA cleavage assay was performed in 50% ethylene glycol, 50 mM phosphate, pH 7.

[c] Not all of the activity at these stages is due to the 32 kDa acyltransferase.

$(NH_4)_2SO_4$. This precipitate is dissolved in 50 mM phosphate, 10 mM 2-mercaptoethanol, pH 7 (2 ml) and applied to an Ultrogel AcA 44 column (1.5 × 46 cm). Following elution of the column with the same buffer, the appropriate fractions are pooled and concentrated on a small (1 ml) DEAE-Sepharose CL-6B column. The concentrated protein is eluted with 0.5 M NaCl and dialyzed vs 10 mM phosphate, 20 mM 2-mercapto-ethanol. The dialyzed sample is applied to a hydroxylapatite column (BioGel HT, 1 ml) and the 32 kDa acyltransferase is eluted in the void volume with 10 mM phosphate, 20 mM 2-mercaptoethanol (pH 7). The purified enzyme is made 10% in glycerol and stored at −20° with no appreciable loss of activity after 1 month.

Properties

The 32 kDa acyltransferase from *V. harveyi* appears to be monomeric in the phosphate buffers used for its purification, whereas the 34 kDa *P. phosphoreum* enzyme tends to aggregate unless NaCl (at least 0.1 M) is included in the gel filtration eluant. With both acyltransferases, the acyl-CoA cleavage activity in phosphate buffer is maximal at pH values above pH 7 and the apparent K_m for [³H]tetradecanoyl-CoA is ~1 μM.[3,4] The *P. phosphoreum* enzyme exhibits substrate inhibition at tetradecanoyl-CoA concentrations above 10 μM, but no such effect is observed for the *V. harveyi* enzyme (up to 40 μM). Both acyltransferases are inhibited by the sulfhydryl reagent N-ethylmaleimide, although the *V. harveyi* enzyme appears to be protected from NEM inactivation by high concentrations of phosphate buffer.[4] This observation, together with the phosphate-induced

TABLE II
[³H]TETRADECANOYL-CoA CLEAVAGE CATALYZED BY *P. phosphoreum* AND
V. harveyi ACYLTRANSFERASES: EFFECT OF SOLVENT COMPOSITION

Solvent composition	Enzyme activity (nmol/min/mg)	
	P. phosphoreum	*V. harveyi*
50 mM phosphate, pH 7	0	15
+50 mM 2-mercaptoethanol	40	40
+40% ethylene glycol	1000	1500
+50% glycerol	2000	800
1 M phosphate, pH 7	0	20
+50 mM 2-mercaptoethanol	120	—

stimulation of the activities of the 32 and 34 kDa enzymes (Table II), suggests that the effect of this anion could have physiological relevance. Neither enzyme is appreciably affected by similar concentrations of other salts (i.e., NaCl) or buffers.

One of the more interesting properties of the bioluminescence-related acyltransferases is the dramatic stimulation of the acyl-CoA cleavage activity observed in the presence of low molecular weight thiol and alcohol acceptors (Table II).[4] In fact the *P. phosphoreum* enzyme appears to be essentially inactive in phosphate buffer alone, while the *V. harveyi* enzyme exhibits a low acylhydrolase activity. The large increase in acyl-CoA cleavage rate by such compounds as glycerol, ethylene glycol, or 2-mercaptoethanol is accompanied by the transfer of the fatty acyl moiety to form the *O*- or *S*-acyl ester derivative, as observed by TLC.[3,4] Other organic compounds, either without or with poor acceptor capabilities, still produce a significant increase in the acyl-CoA cleavage rate, probably due to effects on the enzyme environment or on the effective substrate concentration. Thus, these enzymes do not function optimally in a purely aqueous environment, with water as an acceptor, suggesting perhaps that their true role is to channel fatty acyl groups to and from specific acceptors (i.e., lipids, other enzymes, etc.). Indeed, it has been demonstrated that the *P. phosphoreum* 50 kDa acyl-protein synthetase subunit can modify both the rate and final product of acyl-CoA cleavage catalyzed by the 34 kDa acyltransferase.[3]

Acknowledgments

This work was supported by a Postdoctoral Fellowship (D.B.), a Graduate Studentship (L.C.), and a Research Grant (MT-4314) from the Medical Research Council of Canada.

[16] Bioluminescent Analysis of Insect Pheromones

By DAVID MORSE, ROSE SZITTNER, GARY GRANT, and
EDWARD MEIGHEN

Light emission catalyzed by bacterial luciferases requires the presence of a long-chain fatty aldehyde[1] as illustrated in the reaction below. Under conditions of constant enzyme, $FMNH_2$ and O_2, the bioluminescence response is

$$FMNH_2 + O_2 + CH_3(CH_2)_n\overset{O}{\overset{\|}{C}}-H \rightarrow FMN + H_2O + CH_3(CH_2)_n\overset{O}{\overset{\|}{C}}-OH + h\nu_{490}$$

dependent only on the amount and type of fatty aldehyde with low amounts of saturated and unsaturated aldehydes of 13 to 16 carbons in chain length giving high responses.[2] A number of economically important insect pests, including the corn earworm, the tobacco budworm, the spruce budworm, the navel orangeworm, and the dermestid beetles, have unsaturated aldehydes of 14 to 16 carbons in chain length as their major pheromone component.[3–6] Consequently, the bioluminescence assay provides a highly sensitive, direct, and very rapid method of analysis for these pheromones. Moreover, the sex pheromones or attractants identified for over 350 lepidopteran species are primarily composed of long-chain unsaturated acetate esters, alcohols, and aldehydes.[7] By conversion of fatty alcohols and acetate esters to the corresponding fatty aldehydes, the bioluminescence assay can be extended to analyze pheromones from a very large number of insect species.

[1] J. W. Hastings, this series, Vol. 57, p. 125.
[2] E. A. Meighen, K. N. Slessor, and G. G. Grant, *J. Chem. Ecol.* **8**, 911 (1982).
[3] J. A. Klun, J. R. Plimmer, B. A. Bierl-Leonhardt, A. N. Sparks, and O. L. Chapman, *Science* **204**, 1328 (1979).
[4] C. J. Sanders and J. Weatherston, *Can. Entomol.* **108**, 1285 (1976).
[5] J. A. Coffelt, K. W. Vick, P. E. Sonnet, and R. E. Doolittle, *J. Chem. Ecol.* **5**, 955 (1979).
[6] J. H. Cross, R. C. Byler, R. F. Cassidy, Jr., R. M. Silverstein, R. E. Greenblatt, W. E. Burkholder, A. R. Levinson, and H. Z. Levinson, *J. Chem. Ecol.* **2**, 457 (1976).
[7] M. N. Inscoe, *in* "Insect Suppression with Controlled Release Pheromone Systems" (A. F. Kydonieus and M. Beroza, eds.). CRC Press, Boca Raton, Florida, 1982.

Dithionite Assay for Aldehydes

Principle. Aldehyde is injected into a solution containing luciferase and $FMNH_2$, reduced with sodium dithionite.[8,9] Light emission, detected by a photomultiplier tube, is proportional to the amount of aldehyde at low concentrations.

Reagents. Aldehyde standards (in dimethylformamide) are diluted into water ($\leq 0.1\%$ v/v) and used within 30 min. Luciferases,[10] purified from *Vibrio harveyi* (B392) and *Photobacterium phosphoreum* (NCMB 844) and stored in 30–40% glycerol, 0.1 M dithiothreitol, 0.05 M phosphate, pH 7.0, at $-20°$, were diluted 1 : 20 into 0.001 M NH_2OH, 0.05 M mercaptoethanol, 0.05 M phosphate, pH 7.0, at 4° and used within 2 weeks. All phosphate buffers were composed of K_2HPO_4 and NaH_2PO_4.

Procedure. One milliliter of the test aldehyde in aqueous solution is injected into 1.0 ml of 0.001 M NH_2OH, 0.05 M mercaptoethanol, 0.05 M phosphate, pH 7.0, at room temperature, containing 10 μl of a 0.05% solution of luciferase and 50 μM $FMNH_2$ reduced just before analysis by addition of either a small amount ($\sim$0.3 mg) of solid sodium dithionite ($Na_2S_2O_4$) or 10–20 μl of a 1.5% $Na_2S_2O_4$ solution. The latter solution should be maintained under N_2 and used within 4 hr of preparation.[10] Light emission rises rapidly to a maximum (<1 sec) and then decays. The average of the maximum light intensities for 2 to 3 analyses is recorded in arbitrary light units (LU). In the present experiments, 1 LU equals 6×10^9 quanta/sec, based on the standard of Hastings and Weber.[11]

The presence of NH_2OH and $Na_2S_2O_4$ in the assay buffer serves to lower the background (endogenous) response by scavenging endogenous aldehydes. As a consequence, this system is about 10-fold more sensitive than the standard bioluminescence assay in which $FMNH_2$ is injected into a solution containing luciferase and aldehyde. Although an increase in luciferase concentration increases the bioluminescence response, no gain in sensitivity is achieved in the standard assay because the endogenous response also increases in parallel.

As little as 0.1 pmol of aldehyde can be quantitated in this assay. At low aldehyde concentrations, the light emission is directly proportioned to the amount of aldehyde injected into the reaction mixture. As shown in Fig. 1, the bioluminescence response of *V. harveyi* luciferase to (*E*)-11-tetradecenal, the major pheromone component of the spruce budworm,[4] is linear over a 5×10^4 range (0.1–5000 pmol) of aldehyde concentration.[2]

[8] E. A. Meighen and J. W. Hastings, *J. Biol. Chem.* **246**, 7666 (1971).

[9] E. A. Meighen and R. E. MacKenzie, *Biochemistry* **12**, 1482 (1973).

[10] J. W. Hastings, T. O. Baldwin, and M. Z. Nicoli, this series, Vol. 57, p. 135.

[11] J. W. Hastings and G. Weber, *J. Opt. Soc. Am.* **53**, 1410 (1963).

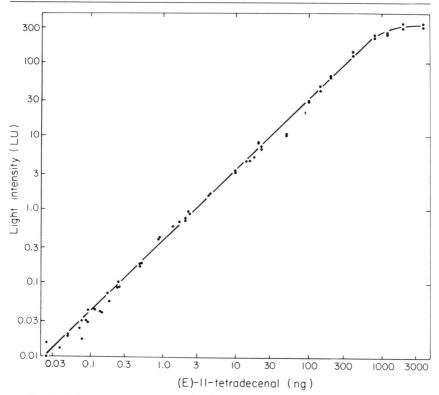

FIG. 1. Bioluminescence response of *V. harveyi* luciferase to different amounts of (*E*)-11-tetradecenal plotted on logarithmic scales. Each point represents the average of two independent assays after correction for the background response. From Meighen *et al.*[2]

At higher levels of (*E*)-11-tetradecenal (>5 nmol), the bioluminescence response remains relatively constant, due to either saturation of the aldehyde binding sites on the enzyme or saturation of the aqueous solution with the nonpolar lipid substrate.

Aldehyde Specificity

The luminous response at low aldehyde concentration is dependent primarily on the aldehyde chain length with the optimal response occurring with aldehydes having backbones of 13 to 16 carbons (Table I). Earlier studies on aldehyde specificity gave a somewhat different relationship because high saturating concentrations of aldehyde were used.[12–14]

[12] J. W. Hastings, J. Spudich, and G. Malnic, *J. Biol. Chem.* **238**, 3100 (1963).
[13] T. Watanabe and T. Nakamura, *J. Biochem. (Tokyo)* **72**, 647 (1972).
[14] E. A. Meighen and I. Bartlet, *J. Biol. Chem.* **255**, 11181 (1980).

TABLE I
BIOLUMINESCENT RESPONSE OF LUCIFERASE TO FATTY ALDEHYDES[a]

Aldehyde	Luciferase	
	V. harveyi	P. phosphoreum
None (background)	<0.1	<0.1
Dodecanal	6	6
Tridecanal	50	40
Tetradecanal	100	100
(E)-11-Tetradecenal	100	70
(Z)-11-Tetradecenal	70	30
(Z)-9 Tetradecenal	30	60
(Z,E)-9,11-Tetradecadienal	100	90
Pentadecanal	60	90
Hexadecanal	40	15
(Z)-11-Hexadecenal	120	70
(Z)-9-Hexadecenal	100	40
(Z)-7-Hexadecenal	25	3
(E)-14-Methyl-8-hexadecenal	30	2
(Z)-14-Methyl-8-hexadecenal	60	30
(Z,Z)-11,13-Hexadecadienal	160	20
Heptadecanal	15	1
Octadecanal	7	1

[a] Maximum light intensities in the dithionite assay for 100 pmol of aldehyde relative to 100 pmol of tetradecanal.

Although the presence of unsaturation, found in most aldehyde phero- mones, affects the kinetics of light emission to some degree, the activity still remains relatively high providing the double bonds are not in close proximity to the aldehyde functional group.[2,15] In this regard, the pres- ence of a double bond in the 16-carbon aldehydes at a position more distal than the 8-carbon, results in a bioluminescence response even greater than with hexadecanal (Table I). The specificity of P. phosphoreum luci- ferase for aldehydes is much more restrictive than V. harveyi luciferase. Only two aldehydes [(Z)-9-tetradecenal and pentadecanal] give higher light emission with P. phosphoreum luciferase than with V. harveyi luci- ferase in comparison to their respective responses to tetradecanal, the putative in vivo substrate.[16,17]

The differences in the relative bioluminescent responses of the two luciferases can be used to qualitatively distinguish between certain alde-

[15] J. Spudich and J. W. Hastings, J. Biol. Chem. 238, 3106 (1963).
[16] O. Shimomura, F. H. Johnson, and H. Morise, Proc. Natl. Acad. Sci. U.S.A. 71, 4666 (1974).
[17] S. Ulitzur and J. W. Hastings, Proc. Natl. Acad. Sci. U.S.A. 76, 265 (1979).

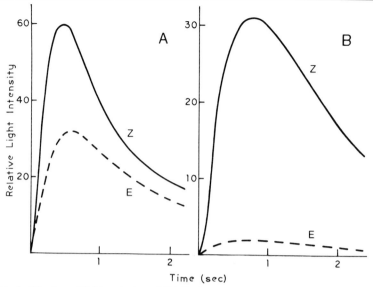

Fig. 2. Kinetics of light emission of (A) *V. harveyi* luciferase and (B) *P. phosphoreum* luciferase with the pheromone components of the dermestid beetles; 100 pmol of (*Z*)-14-methyl-8-hexadecenal (solid line); 100 pmol of (*E*)-14-methyl-8-hexadecenal (dashed line).

hyde pheromones. Figure 2 illustrates the kinetics of light emission of *V. harveyi* and *P. phosphoreum* luciferases with the isomers (*Z*)- and (*E*)-14-methyl-8-hexadecenal, the pheromone components of the dermestid beetles.[6] Both luciferases give similar responses to the *Z* isomer whereas *V. harveyi* luciferase gives a much higher light emission at its peak intensity (~15-fold) than *P. phosphoreum* luciferase with the *E* isomer. The pheromone of the dermestid beetle, *Trogoderma glabrum*, which has been identified as (*E*)-14-methyl-8-hexadecenal, would thus be characterized by a much higher bioluminescence response with *V. harveyi* luciferase compared to *P. phosphoreum* luciferase. In contrast, the dermestid beetles, *T. inclusum*, *T. variabile*, and *T. granarium*, with primarily the *Z* isomer as the main pheromone component (92–100%) would have a much lower ratio of activities with these luciferases. Consequently, the luminescence assays could be used not only to quantitate the amount of pheromone but also to determine the relative amounts of *Z* and *E* isomers in beetles using this aldehyde as a pheromone. Similarly, differences in the rate of decay of light emission after reaching peak intensity (Fig. 2) might also be used to distinguish qualitatively between aldehyde pheromones.

Analysis of Fatty Alcohols and Acetate Esters

Principle. Hydrolysis of long-chain acetate esters to alcohols followed by oxidation generate fatty aldehydes which can be analyzed in the bacterial bioluminescence assay.

Reagents. Horse liver alcohol dehydrogenase (HLAD) and porcine liver esterase Type I (carboxylic ester hydrolase) were purchased from Sigma. Stock solutions of long-chain acetate esters and alcohols are prepared in solvents miscible in water (dimethyl formamide, isopropanol) or in hexane. Hexane stocks (10 μl) can be added directly to the test tube or vial and the hexane removed by a controlled flow of N_2 (2 min) followed by equilibration in buffer (3.5 ml) for 5 to 10 min. This procedure is generally used if the ester or alcohol has limited solubility in the water miscible solvent since concentrations of dimethyl formamide or isopropanol greater than 0.1% will interfere with the assay.

Procedure. One milliliter of 0.05 M phosphate, pH 8.0, containing (E)-11-tetradecenol is vortexed for 5 sec with 10 μl of NAD (0.5 μmol) and 10 μl of a 0.1% HLAD solution and then analyzed by injection into the dithionite luminescence assay.[18] The luminescence response is dependent on the amount of alcohol from 5 to 500 pmol. The levels of HLAD and NAD and the incubation time were selected to give maximum luminescence responses upon conversion of a fixed amount of alcohol to aldehyde. The presence of protein in the luminescence assay at concentrations greater than 5 μg/ml will result in inhibition due in part to the interaction with the fatty aldehyde product.

Acetate esters were hydrolyzed to fatty alcohols by mixing the acetate ester with 0.1 units of esterase (in 5 μl of phosphate buffer) for 5 min in 1.0 ml of 0.05 M phosphate, pH 8.0. The fatty alcohol product was converted into aldehyde as described above and analyzed in the luminescence assay. Table II compares the relative sensitivities and the detection range for the analyses of a (E)-11-tetradecenyl acetate, (E)-11-tetradecenol, and (E)-11-tetradecenal. Although the sensitivity of the assay is about 50-fold less with the acetate ester and alcohol than with aldehyde, due to lower bioluminescence responses and higher endogenous (background) responses, concentrations in the range of 0.005–0.5 μM can still be measured for these compounds.

Analysis of Insect Extracts

Pheromone-producing glands (or other body parts) are excised and extracted in 10 μl of heptane in a small tube for 10 min. The extract is

---

[18] D. Morse and E. Meighen, *J. Biol. Chem.* **259**, 475 (1984).

TABLE II
COMPARISON OF BIOLUMINESCENCE ASSAYS FOR (E)-11-TETRADECENAL,
(E)-11-TETRADECENOL, AND (E)-11-TETRADECENYL ACETATE

Compound	Relative luminescent response (LU/pmol)	Range of detection (pmol)[a]
Aldehyde	0.07	0.1–5000
Alcohol	0.04[b]	5–500
Acetate ester	0.04[b]	5–500

[a] The lower limit of detection is limited by the endogenous response and will depend on the prior treatment of the sample.

[b] Measured for amounts less than 300 pmol as the assay may not be strictly linear above these levels with fatty alcohols and acetate esters.

transferred to a 20-ml glass vial and the heptane removed by evaporation for 2 min under a controlled flow of nitrogen. The residue is immediately dissolved in 10 ml of water, vortexed for 10 sec, and 1.0-ml aliquots assayed 10 min later by injection into the luminescence assay. Pheromone levels have been measured by analysis of individual glands excised from the spruce budworm (2.2 ng/gland), the western spruce budworm (2.8 ng/gland), the corn earworm (4.7 ng/gland), and the navel orangeworm (0.56 ng/gland).[19] The luminescence response is converted into nanograms based on standards for the major pheromone component carried through the same procedure. The aldehyde standards are (E)-11-tetradecenal for the two budworms, (Z)-11-hexadecenal for the corn earworm, and (Z,Z)-11,13-hexadecadienal for the navel orangeworm.[3–5] *V. harveyi* luciferase rather than *P. phosphoreum* luciferase is used for analyses due to its higher response with most aldehydes.

Analysis of Pheromone Release Rates from Insect Lures

Pheromone lures are routinely used as baits for attracting insects to traps not only for the purpose of monitoring the level of a specific insect population but, in some cases, to trap-out or eliminate the insect. Knowledge of the pheromone release rates of insect lures is necessary for their effective use.

Pheromone lures are placed in a 16-mm glass tube under constant flow (0.1–1.0 mph) of air, which has been prefiltered through an inert absorbent. The airborne pheromone is trapped by passage of the air through a 20-ml scintillation vial maintained at −20 to −70°. The inlet tube should

[19] G. G. Grant, K. N. Slessor, R. B. Szittner, D. Morse, and E. A. Meighen, *J. Chem. Ecol.* **8,** 923 (1982).

be situated about 3 mm from the bottom of the vial. Trapping can only be conducted for periods of 1 hr or less since ice will accumulate in the vial and eventually block the flow of air. Water (~9 ml), part of which is used to rinse the inlet tube, is added to the vial containing the pheromone and ice (~1 ml) to give a final volume of 10 ml. The samples are then vortexed for 10 sec and equilibrated at room temperature for 10 min before analyses of 1.0-ml aliquots in the bioluminescence assay. The luminescence responses are then converted into pheromone levels based on standards (in 10 μl hexane) placed in the release chamber and carried through the same trapping procedure.

The release rates of fiber, flake, and polyvinyl chloride lures containing (E)-11-tetradecenal, and fiber lures containing (Z)-9-hexadecenal have been analyzed by the luminescence assay and found to agree with values measured by other techniques.[20] Recoveries of (E)-11-tetradecenal and (Z)-9-hexadecenal standards with the trapping procedures described were 69 and 52%, respectively. Although a second trap (vial) placed in series increases the trapping efficiency of the system for (E)-11-tetradecenal to 85%, this procedure is not necessary as the amount of pheromone in the primary trap is directly proportional to the amount of pheromone placed in the airstream. Release rates as low as 2 ng/hr can be measured after 1 hr of cold trapping. Since most lures release greater amounts of pheromone, trapping can be conducted for as short a time as 5 min making this procedure an extremely rapid method for determining pheromone release rates.

Measurement of Pheromone Release Rates from Insects

Female eastern spruce budworm moths are placed in an airflow of 0.7 liter/min in a cylindrical glass tube ($r = 8$ mm). Porapak Q, 50–80 mesh (Waters Assoc.) is used to prefilter the air as well as to trap the released pheromone. The Porapak or other inert absorbent must be extensively washed with hexane before use to remove any materials inhibitory to the luminescence assay as well as to lower the endogenous response. The outlet Porapak trap (200 mg) is removed periodically and extracted for 30 min with 1.0 ml of hexane and 0.2 ml of the extract is transferred to a 6 ml glass vial. The hexane is removed by controlled evaporation under nitrogen (2 liters/min, 7 min) and the pheromone redissolved in 3.5 ml of water for 10 min before analysis of 1.0-ml aliquots in the luminescence assay. The light intensity is converted into nanograms of pheromone based on the response of (E)-11-tetradecenal standards and the recoveries of fixed

[20] E. A. Meighen, R. B. Szittner, and G. G. Grant, *Anal. Biochem.* **133**, 179 (1983).

amounts of this aldehyde placed in the airstream and trapped for the same time period. The recovery of (*E*)-11-tetradecenal decreases with time: 60% after 1 hr, 40% after 4 hr, and 30% after 8 hr. Release rates as low as 2 ng/hr can be measured. Thus, the determination of the temporal pattern of pheromone release by individual female budworm moths as affected by age and photoperiod[21] was well within the sensitivity of the technique.

Measurement of Enzyme Activities Involved in Pheromone Biosynthesis

A long-chain acetate esterase and a fatty alcohol oxidase have been measured in gland extracts of the spruce budworm moth by application of luminescence assays.[18] Glands are excised from female moths, homogenized for 2 min at 500 rpm with a motor-driven pestle in 0.05 *M* phosphate, pH 7.0 (10 glands/ml), and the extract clarified by centrifugation. Only low amounts of extract ($\leq$20 μl) can be used for enzyme analysis in order to maintain the protein concentration at less than 5 to 10 μg/ml in the assay.

Alcohol Oxidase. Reactions are initiated by addition of extract to 1.4 μM (*E*)-11-tetradecenol in 0.05 *M* phosphate, pH 7.0, and the luminescence response for 1.0-ml aliquots measured with time. The formation of the aldehyde product is linear with time ($\leq$20 min) and extract concentration ($\leq$5 μg/ml). The presence of fatty alcohol at a concentration of 1.4 μM does not inhibit the luminescence assay. Rates of alcohol oxidase activity as low as 0.1 pmol/min can be detected.

Acetate Esterase. Reactions are initiated by addition of extract to 1.4 μM (*E*)-11-tetradecenyl acetate in 0.05 *M* phosphate, pH 8.0. The alcohol product was measured with time in the luminescence assay after conversion into aldehyde with HLAD (10 μg/ml) and NAD (0.5 m*M*). The presence of 1.4 μM acetate ester in the luminescence assay causes an inhibition of less than 20%. Formation of product was linear with time ($\leq$20 min) and extract concentration ($\leq$5 μg/ml) with a lower limit of sensitivity of 4 pmol of acetate ester hydrolyzed per min.

Acknowledgment

This work was supported by Grant MT-4314 from the Medical Research Council of Canada.

---

[21] D. Morse, R. Szittner, G. G. Grant, and E. A. Meighen, *J. Insect Physiol.* **28,** 863 (1982).

[17] Bioluminescent Assays Using Coimmobilized Enzymes

By G. Wienhausen and M. DeLuca

Introduction

Luminescent marine bacteria contain enzymes that catalyze the following reactions:

$$NAD(P)H + FMN + H^+ \rightarrow NAD(P)^+ + FMNH_2 \tag{1}$$
$$FMNH_2 + RCHO + O_2 \rightarrow FMN + RCOOH + H_2O + h\nu \tag{2}$$

The first reaction is catalyzed by an NAD(P)H dehydrogenase (FMN) [NAD(P)H : FMN oxidoreductase] and results in the production of $FMNH_2$. The second reaction is catalyzed by bacterial luciferase. In this reaction, $FMNH_2$ and a long-chain aldehyde are oxidized and light is produced. It is possible to use the two enzymes in a coupled assay, and if NAD(P)H is the limiting component, the amount of light produced is proportional to its concentration.

In *Vibrio harveyi* there are two oxidoreductases present, one which is specific for NADH and the other one for NADPH.[1] It is therefore possible to assay either of the reduced pyridine nucleotides specifically.

Another enzyme, microbial diaphorase (dihydrolipoamide dehydrogenase) catalyzes the NAD(P)H-dependent reduction of FMN, and this enzyme can also be used in the coupled reaction with luciferase.

In principle then any compound XH_2 which can be oxidized according to the general reaction (3) can be assayed with the oxidoreductase–luciferase or diaphorase-luciferase system.

$$XH_2 + NAD(P)^+ \rightarrow X + NAD(P)H + H^+ \tag{3}$$

Similarly, if the substrate XH_2 is in excess and the enzyme-catalyzing reaction (3) is limiting, then it is possible to measure the amount of enzyme present. Since there are several hundred specific pyridine nucleotide-dependent dehydrogenases known and an equal number of substrates, the number of assays which are possible is quite large.

There are two general types of assays for various metabolites. For compounds where a suitable dehydrogenase is available, the enzyme is coimmobilized along with the oxidoreductase (diaphorase) and luciferase. For example, the assay for malate involves malate dehydrogenase which

---

[1] E. Jablonski and M. DeLuca, *Biochemistry* **16**, 2932 (1977).

catalyzes the following reaction:

$$\text{malate} + \text{NAD}^+ \rightarrow \text{oxalacetate} + \text{NADH} + \text{H}^+ \tag{4}$$

If one wants to measure a compound which cannot be directly linked to the production of NAD(P)H, it is possible to immobilize an additional enzyme as in the case of the assay for glucose.

$$\text{Glucose} + \text{ATP} \xrightarrow{\text{hexokinase}} \text{glucose 6-phosphate} + \text{ADP} \tag{5}$$

$$\text{Glucose 6-phosphate} + \text{NAD}^+ \xrightarrow{\text{G6P dehydrogenase}} \text{6-phosphogluconic acid} $$
$$+ \text{NADH} + \text{H}^+ \tag{6}$$

In this case hexokinase and glucose-6-phosphate dehydrogenase (G6PDH) are coimmobilized with the bioluminescent enzymes. It is possible to immobilize even more enzymes, but in general the sensitivity of the assays decreases with the increasing numbers of enzymes in the coupled reactions.[2]

Bioluminescent assays using a variety of enzymes have been well documented by many investigators.[3–6] Most of these assays previously described were done with soluble enzymes. What we will describe here are assays in which the various enzymes have been coimmobilized onto Sepharose beads. The advantages of using coimmobilized enzymes are an increased stability and sensitivity. In addition it is possible to put them into a flow cell where the assays can be automated and the enzymes can be reused. For additional discussions of the application of immobilized bioluminescent assays see Schoelmerich *et al.*, this volume [19], or De-Luca *et al.*[7–9] The various flow systems which have been designed are described by Vellom and Kricka [20] and Roda *et al.* [21] in this volume.

[2] M. DeLuca and L. J. Kricka, *Arch. Biochem. Biophys.* **226,** 285 (1983).

[3] M. DeLuca, ed., this series, Vol. 57.

[4] L. J. Kricka, P. E. Stanley, G. H. G. Thorpe, and T. P. Whitehead, eds., "Analytical Applications of Bioluminescence and Chemiluminescence." Academic Press, London, 1984.

[5] M. Serio and M. Pazzagli, eds., "Luminescent Assays: Perspectives in Endocrinology and Clinical Chemistry." Raven Press, New York, 1982.

[6] K. Van Dyke, ed., "Bioluminescence and Chemiluminescence: Instruments and Applications." CRC Press, Boca Raton, Florida, 1985.

[7] M. DeLuca, *in* "Analytical Applications of Bioluminescence and Chemiluminescence" (L. J. Kricka, P. E. Stanley, G. H. G. Thorpe, and T. P. Whitehead, eds.), p. 111. Academic Press, London, 1984.

[8] D. Slawinska and J. Slawinski, *in* "Chemi- and Bioluminescence" (J. G. Burr, ed.), p. 533. Dekker, New York, 1985.

[9] K. Kurkijarvi, R. Raunio, J. Lavi, and T. Lovgren, *in* "Bioluminescence and Chemiluminescence: Instruments and Applications" (K. Van Eyke, ed.), Vol. 2, p. 167. CRC Press, Boca Raton, Florida, 1985.

Materials and Methods

Enzymes

Luciferase was purified from *V. harveyi* according to the method of Baldwin *et al.*[10] The enzyme, after chromatography on Sephadex G-100, was stored at $-20°$ in 0.1 *M* phosphate buffer, pH 7.0 with 0.5 m*M* DTT.

The NADH and NADPH : FMN oxidoreductases were purified according to the method of Jablonski and DeLuca.[1] These enzymes were stored at $-70°$ in 0.1 *M* phosphate buffer, pH 7.0 containing 2 m*M* DTT.

Diaphorase ("NADH : dye oxidoreductase") from microorganism (Cat. No. 411558) was supplied by Boehringer Mannheim (Indianapolis, IN).

Yeast alcohol dehydrogenase and yeast hexokinase were purchased from Calbiochem-Behring Corp. (La Jolla, CA). Chicken liver malic enzyme was from Sigma Chemical Co. (St. Louis, MO). *Bacillus subtilis* L-alanine dehydrogenase, yeast, glucose-6-phosphate dehydrogenase, beef liver L-glutamate dehydrogenase, *B. subtilis* glutamate, pyruvate aminotransferase, beef heart lactate dehydrogenase, and yeast 6-phosphogluconate dehydrogenase were all obtained from Boehringer Mannheim.

Chemicals

FMN and all of the pyridine nucleotides were of highest purity and were purchased from Boehringer Mannheim. Substrates and buffers were from Calbiochem-Behring. DTT, decanal, and BSA were obtained from Sigma Chemical Co. Cyanogen bromide was supplied by Eastman Chemical Co. Sepharose 4B was purchased from Pharmacia Fine Chemicals.

Luminescent Assays

Bacterial Luciferase. The activity of bacterial luciferase can be assayed using any commercial luminometer. We used an Amicon Chem-Glo, or an ALL Monolight 401. The assay was performed by injecting 0.1 ml of 0.15 *M* $FMNH_2$, photoreduced in the presence of 5 m*M* EDTA, into 0.5 ml of a solution containing either soluble luciferase or an aliquot of the Sepharose-bound enzyme, 0.0005% decanal (v/v) and 0.1 *M* phosphate buffer, pH 7.0. Peak light intensity is proportional to enzyme activity.

Coupled Oxidoreductase or Diaphorase–Luciferase Assay. A decanal–ethanol mixture was prepared by vigorously shaking 5 μl decanal with 10 ml EtOH. One hundred microliters of this mixture, 200 μl aqueous

[10] T. O. Baldwin, M. Z. Nicoli, J. E. Becvar, and J. W. Hastings, *J. Biol. Chem.* **250,** 2763 (1975).

FMN (73 μM), and 10 ml of 0.1 M phosphate buffer, pH 7.0 were mixed and stored at room temperature in the dark. This mixture is stable for 3–4 hr. An appropriate aliquot of the immobilized enzyme suspension is added to 500 μl of the substrate mixture, NADH was added, and the tube was mixed twice by inversion. The peak light emission was measured in the photometer and the activity is expressed as relative light units. It is important to know if any of the enzymes contain traces of alcohol dehydrogenase since then the decanal mixture must be prepared in water not ethanol.

Spectrophotometric Assays

Oxidoreductases. The soluble enzymes are assayed by following the initial rate of oxidation of NADH or NADPH by the decrease in absorbance at 340 nm. The reaction is initiated by adding 0.1 ml of 2 mM NADH or NADPH in 0.1 M potassium phosphate, pH 7.0, to 0.9 ml of a solution containing the enzyme, 0.13 mM FMN and 0.1 M potassium phosphate, pH 7.0. The immobilized enzymes were assayed in the same way with constant stirring of the reaction mixture.

Diaphorase. This activity was also measured by the decrease in the absorbance at 340 nm. The reaction was started by the addition of enzyme to 1 ml of 0.1 M phosphate, pH 7.0, containing 0.2 mM NADH (or NADPH) and 0.3 mM FMN. A control reaction minus FMN was also measured and this rate, with oxygen as the electron acceptor, was subtracted from the FMN-dependent rate.

In the diaphorase reaction it is important that the reaction be started either by the addition of enzyme or NAD(P)H. If FMN is added last there is an apparent decrease in activity of about 30%. This is observed for both the soluble and immobilized enzyme and is possibly due to a reduction of the enzyme by the NAD(P)H.

Other Enzyme Assays

The activities of the various other dehydrogenases and coupling enzymes were measured according to published procedures.[11]

Immobilization of Enzymes: Bacterial Luciferase and Auxiliary Enzymes

The Sepharose 4B was activated according to the procedure of March *et al.*[12] In general between 12 and 16 mg of total protein was reacted with

[11] H. U. Bergmeyer, ed., "Methods of Enzymatic Analysis," 2nd Engl. ed., Vols. 1 and 2. Academic Press, New York, 1971.
[12] S. C. March, I. Parith, and P. Cuatrecasas, *Anal. Biochem.* **60,** 149 (1974).

1 g of activated Sepharose. When only two enzymes were coimmobilized, 4 mg of bacterial luciferase and 4 mg of either oxidoreductase or diaphorase were added to 8 mg of BSA in a final volume of 3 ml of 0.1 M pyrophosphate, pH 8.0. The enzymes were dialyzed overnight against 0.1 M sodium pyrophosphate, pH 8.0, to remove DTT. The protein mixture was added to 1 g of freshly activated Sepharose and stirred gently for 2 hr at 22° or 4° for 16 hr. Either procedure gave comparable results. The Sepharose was then washed with 200 ml of cold 0.1 M phosphate, 0.5 mM DTT, pH 7.0, followed by 500 ml of 1 M NaCl, 0.1 M phosphate, 0.5 mM DTT, pH 7.0, and finally 300 ml of 0.1 M phosphate, 0.5 mM DTT, pH 7.0. The oxidoreductase luciferase Sepharose bound enzymes were then suspended in 6 ml of 0.1 M phosphate, pH 7.0, containing 0.2% BSA, 0.2% sodium azide, and 2 mM DTT and 1 mM EDTA. There was a significant increase in the activity of the bound enzymes over the first 48 hr after which the activity remained constant. This was probably due to the reduction of some sulfhydryls by the DTT. Every week an additional 20 μl of 0.1 M DTT was added to the suspension.

The diaphorase–luciferase preparation was stored in the same buffer minus DTT since this compound produces background light.

For the preparation of three or more coimmobilized enzymes the following amounts of these enzymes (see tabulation below) were added to 5 mg of bacterial luciferase and 3 IU of either the NADH or NADPH oxidoreductase. Final protein concentration was brought to about 8 mg/ml with BSA.

Compound to be assayed	Oxido-reductase used	Amount of enzymes added to Sepharose	Active enzyme bound (%)
Glucose	NADPH	2 IU hexokinase	62
		2 IU G6P dehydrogenase	28
L-Lactate	NADH	3 IU lactate dehydrogenase	26
6-Phospho-gluconate	NADPH	7 IU 6-phosphogluconate dehy-drogenase	35
L-Malate	NADH	8 IU malate dehydrogenase	45
L-Alanine	NADH	7.5 IU alanine dehydrogenase	10
L-Glutamate	NADH	3.5 IU glutamate dehydrogenase	7

The recovery of the oxidoreductases was in the range of 50–90% and the luciferase was about 10%. The amounts of the auxiliary enzymes were arrived at empirically and it is possible that other combinations might increase the sensitivity of the assays. However, some of the commercial

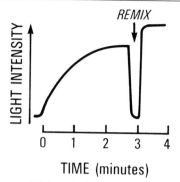

TIME (minutes)

FIG. 1. Typical time course of light emission obtained with NADH using coimmobilized bacterial luciferase–oxidoreductase. The slight increase of light observed after remixing is due to resuspension of all of the Sepharose beads which had settled out.

enzymes contain other dehydrogenases and these may contribute to background light, so it is not possible to use large amounts of these enzymes without further purification.

Stability of Immobilized Oxidoreductase–Luciferase[13]

Only the two enzyme systems were studied in detail and it is likely that storage conditions for other enzymes might vary. DTT in the storage buffer is important for maintaining activity. If this was added weekly the enzymes remained stable for several months when stored at 4°. Freezing at −20° (slow freezing) resulted in loss of activity. However, if the Sepharose enzymes were frozen rapidly in liquid nitrogen in the presence of 15% glycerol, there was no loss of activity and after 2 months they were fully active. Presumably they could be stored this way indefinitely.

Results

Assays of NAD(P)H

Figure 1 shows a typical time course for light emission after the addition of 50 pmol of NADH.[13] The small increase in light observed after remixing occurs because some of the Sepharose has settled to the bottom and is resuspended by the mixing. In general, the lower limit of detection for NAD(P)H is about 1 pmol and peak light is a linear function of the concentration up to about 1 nmol. This is essentially the same for either of the oxidoreductases or diaphorase and luciferase.

[13] G. K. Wienhausen, L. J. Kricka, J. E. Hinkley, and M. DeLuca, *Appl. Biochem. Biotechnol.* **7,** 463 (1982).

TABLE I
ASSAY CONDITIONS OF METABOLITES

Metabolite	Reagents	Concentration (mol/liter)[a]
D-Glucose	Buffer	TES: 2.5×10^{-2}; pH 7.4
	FMN	3×10^{-7}
	NADP	3.6×10^{-4}
	ATP	3.6×10^{-4}
	MgCl$_2$	5.5×10^{-4}
L-Lactate	Buffer	TAPS: 5×10^{-2}; pH 8.5
	FMN	2×10^{-6}
	NAD	3.6×10^{-4}
6-P-Gluconate	Buffer	Imidazole: 5×10^{-2}; pH 7.0
	FMN	2×10^{-6}
	NADP	3.6×10^{-4}
	Mg acetate	10^{-4}
	NH$_4$ acetate	10^{-2}
L-Malate	Buffer	TAPS: 5×10^{-2}; pH 8.5
	FMN	2×10^{-6}
	NAD	3.6×10^{-4}
L-Alanine	Buffer	TAPS: 5×10^{-2}; pH 8.5
	FMN	2×10^{-7}
	NAD	3.6×10^{-4}
L-Glutamate	Buffer	TAPS: 5×10^{-2}; pH 8.5
	FMN	2×10^{-7}
	NAD	1.8×10^{-5}
NAD	Buffer	TAPS: 5×10^{-2}; pH 8.5
	FMN	2×10^{-6}
	Lactate	2×10^{-4}
NADP	Buffer	Imidazole: 5×10^{-2}; pH 7.0
	FMN	2×10^{-6}
	6-P-Gluconate	2×10^{-4}

[a] All assays contained 0.0005% decanal.

Assays of Other Metabolites

Immediately prior to assay, the immobilized enzymes were diluted with buffer so that a 20- to 50-μl aliquot produced a good light signal with a small amount of the metabolite to be measured. The reaction is started by adding the immobilized enzymes to 0.5 ml volume containing all of the additional substrates in optimal concentrations. The exact buffers and substrate concentrations used are shown in Table I.[14] After the tube is mixed, the light emission is measured in the luminometer and recorded. In some instances peak light is measured and in others the rate of increase (slope) of light was used.

[14] G. Wienhausen and M. DeLuca, *Anal. Biochem.* **127,** 380 (1982).

For example, in the assay of glucose using the four enzyme system, the slope of increasing light intensity is proportional to glucose concentration. This is shown in Fig. 2.

In Fig. 3 the amount of light obtained with glucose is compared to that with glucose 6-phosphate or NADPH using the same immobilized enzymes.

Figure 4 shows an assay for NADP using 6-phosphogluconic acid and 6-P-gluconate dehydrogenase. The NADP is reduced to NADPH and in this case peak light is measured after about 1 min.

The various metabolites that were measured and the concentration range in which the assays are linear are shown in Table II.

The differences in the lower limits of detection are the result of both the amount of active enzymes bound as well as the equilibrium of the reaction. The greatest sensitivity is obtained when the equilibrium of the reaction catalyzed lies in the direction of NADH production. For example, the equilibrium constant for malate dehydrogenase is 6.4×10^{-13} M and the lowest detectable concentration of malate is 100 pmol.

Using the immobilized 6-phosphogluconate dehydrogenase system NADP could be measured at 0.2 pmol. With lactate dehydrogenase the lower limit of detection of NAD was 2 pmol. This is also probably due to the less favorable equilibrium constant of lactate dehydrogenase relative to 6-P-gluconate dehydrogenase.

Enzyme Assays

In addition to the metabolites discussed above, it is possible to assay various enzymes which catalyze the production of NAD(P)H. The assays are performed as described with the enzyme to be assayed as the limiting component using the coimmobilized oxidoreductase–luciferase to measure the NAD(P)H. Several typical examples are lactate dehydrogenase (3–700 fmol), alcohol dehydrogenase (15–3000 fmol), and glucose-6-phosphate dehydrogenase (1.5–100 fmol). The lower limit of detection of the enzymes is due to the turnover. Even greater sensitivity is possible if the enzyme is allowed to react for several minutes before the immobilized enzymes are added.

Limitations of the Assay and Interfering Substances

In addition to unfavorable equilibrium constants discussed previously there are several factors which presently limit the detection of smaller amounts of metabolites. There is always some background light associated with the immobilized enzymes. This appears to be determined by the purity of the reagents used. Although we routinely used analytical grade

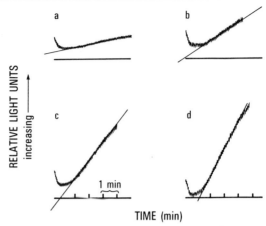

FIG. 2. Time course of light emission in response to various amounts of D-glucose: (a) background light, (b) 2, (c) 5, and (d) 10 pmol glucose. The slope of the line is proportional to glucose concentration after background has been subtracted.

chemicals, some of the substrates were apparently contaminated with traces of reduced pyridine nucleotides. Both NAD and NADP were found to contain some of the reduced forms.

The commercially available enzymes which were used are also frequently contaminated with other dehydrogenases or bound pyridine nu-

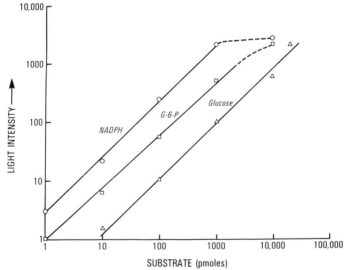

FIG. 3. Light obtained using varying concentrations of substrates with coimmobilized hexokinase, glucose-6-phosphate dehydrogenase, oxidoreductase, and bacterial luciferase.

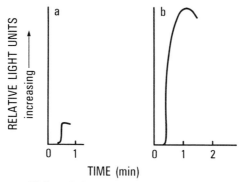

FIG. 4. Time course of light emission in the assay of NADP, an example where peak light is measured rather than the slope as in Fig. 2: (a) 0 pmol (background); (b) 2 pmol NADP.

cleotides. Also in some cases the immobilized oxidoreductase–luciferase was found to contain traces of lactate dehydrogenase and alcohol dehydrogenase. A more serious contaminant was malate dehydrogenase. When 100 μl of the immobilized oxidoreductase–luciferase was assayed spectrophotometrically for malate dehydrogenase activity it was found to contain 0.0025 units which is at the very lowest limits of the spectrophotometric assay. However, when malate dehydrogenase contamination of this coimmobilized enzyme system was determined in a bioluminescent assay the light obtained corresponded to 200 pmol of NADH produced in 1 min.

The various enzymes are also inhibited by certain buffers. For oxidoreductase–luciferase phosphate, TES and TAPS were the best buffers.

Because of the sensitivity of the bioluminescent assays the analytical purity of the various reagents and substrates is of critical importance.

TABLE II
CONCENTRATIONS OF COMPOUNDS ASSAYED

Compound	Linear range of assay (pmol–nmol)	Type of assay
D-Glucose	10–15	Slope
L-Lactate	100–100	Peak
6-P-Gluconate	10–100	Peak
L-Malate	100–10	Peak
L-Alanine	50–10	Slope
L-Glutamate	100–5	Peak
NAD	1–2 (LDH)	Peak
NADP	0.2–0.2	Peak

Since this will be variable with the source and batch these must be routinely checked.

Interfering Substances

Fresh human serum was found to inhibit light emission significantly. Similarly BSA exhibited a concentration-dependent inhibition of the NADH determinations. At 80 mg/ml a 40% inhibition was observed. These high concentrations of protein may bind some of the NADH thus making it unavailable for the immobilized enzymes. Various other dehydrogenases in serum may also contribute to the apparent inhibition by oxidizing the NADH. At least in the case of serum bile acid analysis both of these problems could be overcome by briefly heating the serum and then diluting it by 50-fold prior to assay.[15]

Reproducibility and Accuracy

The reproducibility of the assay seems to be most dependent upon the accuracy with which multiple aliquots of the Sepharose enzymes can be delivered into the assay tubes. In our experience, addition of 30–70 μl of the suspension gives adequate reproducibility. The precision of the assay based on 10 replicate samples during a 1-day period was 3.4% for 1×10^{-5} M NADH and 5.8% at 50 pmol NADH.[13] This can be improved by using the immobilized enzymes in the automated flow system (see Vellom and Kricka, this volume [20]).

There are now several reports in which hydroxysteroid dehydrogenases have been coimmobilized along with the bioluminescent enzymes.[15–17] In these publications the bioluminescent values agreed well with independent determinations by radioimmunoassay or gas–liquid chromatography.

Summary

In summary the use of immobilized luciferases along with other enzymes offers a method for measuring a wide variety of metabolites or enzymes. The assays are rapid, sensitive, and specific and can be automated. It is anticipated that many more assays for different compounds will be developed in the future.

[15] A. Roda, J. L. Kricka, M. DeLuca, and A. F. Hofmann, *J. Lipid Res.* **23,** 1354 (1982).
[16] J. Schoelmerich, G. P. van Berge Henegouwen, A. F. Hofmann, and M. DeLuca, *Clin. Chim. Acta* **137,** 21 (1984).
[17] J. Schoelmerich, J. E. Hinkley, I. A. Macdonald, A. F. Hofmann, and M. DeLuca, *Anal. Biochem.* **133,** 244 (1983).

[18] Bioluminescent Assays of Estrogens

By JEAN-CLAUDE NICOLAS, ANNE MARIE BOUSSIOUX, and
ANDRÉ CRASTES DE PAULET

Estrone and estradiol concentrations are measured using the trans-hydrogenase function of human placental estradiol 17β-dehydrogenase. The hydrogen, from NADPH continuously formed by a regenerating system, is transferred to NAD through the rapid cycling enzymatic conversion of estrone and estradiol.

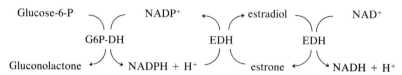

G6P-DH, Glucose-6-phosphate dehydrogenase
(glucose-6-phosphate : NADP$^+$ 1-oxidoreductase, EC 1.1.1.49)
EDH, Estradiol 17β-dehydrogenase [estradiol-17β : NAD(P)$^+$
17-oxidoreductase, EC 1.1.1.62]

Under appropriate conditions, there is a linear relationship between the concentration of estrogen (estrone + estradiol) and the rate of production of NADH.[1] However 5 to 16 hr of incubation is required to accumulate enough NADH (10^{-6} to 10^{-4} M) to be determined spectrophotometrically. To decrease this incubation time NADH must be determined at low concentration (10^{-8} to 10^{-6} M). This is possible using fluorescence measurement, but in this assay it is not suitable since NADPH is continuously generated at a concentration of 10^{-7} to 10^{-6} M. The bioluminescent measurement of NADH is a thousand times more sensitive than spectrophotometric determination. The NADH is determined specifically with FMN oxidoreductase and luciferase from *Beneckea harveyi* as described by Stanley[2]:

$$\text{NAD(P)H + FMN} \xrightarrow{\text{FMN reductase}} \text{FMNH}_2 + \text{NAD(P)}$$

$$\text{FMNH}_2 + \text{R—CHO} + O_2 \xrightarrow{\text{luciferase}} \text{FMN + R—COOH} + H_2O + h\nu$$

[1] J. C. Nicolas, A. M. Boussioux, B. Descomps, and A. Crastes de Paulet, *Clin. Chim. Acta* **92,** 1 (1979).
[2] P. Stanley, this series, Vol. 57, p. 135.

Thus we propose a two-step assay of estrogens. The first reaction is the transhydrogenase or amplification reaction in which NADH is accumulated (15 min to 2 hr) and the second reaction is the bioluminescent assay of NADH.

Equipment and Material

Enzymes. Glucose-6-phosphate dehydrogenase (G6P-DH, Boehringer) from yeast contained less than 0.001% NAD-dependent activity. Luciferase and reductase was prepared from *Beneckea harveyi* as described previously.[3] Estradiol dehydrogenase (EDH, 4 U/mg in phosphate buffer 2 M glycerol) must not contain estrogen or estrogen-independent transhydrogenase activity. It is commercially available (Biomerieux Marcy l'Etoile) but can also be prepared from fresh placenta.

Two placentas were homogenized in phosphate buffer 0.03 M, pH 7.2, glycerol 2 M (buffer A), centrifuged at 50,000 g for 30 min, and then the supernatant added to a DEAE-cellulose column (5 × 20 cm). The enzyme was eluted directly by phosphate buffer 0.1 M, pH 7.2, 2 M glycerol (buffer B). Fractions containing the EDH activity were pooled and precipitated by adding 1 volume of ammonium sulfate (saturated solution, pH 7). The precipitate was collected after centrifugation and dissolved in buffer A and stored at 4°. When at least 10 placentas had been processed, the glycerol concentration was increased (5 M) and diethylstilbestrol was added (10^{-4} M). This solution was incubated at 65° for 4 hr and then dialyzed twice against buffer A. The solution was centrifuged at 20,000 g and the supernatant added to a second DEAE-cellulose column (3 × 15 cm). The enzyme was eluted by a linear gradient formed by using 400 ml buffer A and B. The fractions containing high enzyme activity were pooled and diluted 3-fold with 2 M glycerol. EDH (50 IU) was then applied on a third DEAE-cellulose column (2 × 5 cm) and the column was washed thoroughly with buffer A containing 20% ethanol, in order to remove bound estrogens. The column was equilibrated with buffer A and EDH was eluted with buffer B. The enzyme must not be frozen but stored in small aliquots at 4° in buffer with 2 M glycerol.

Buffers and Solutions

Tris–HCl buffer 0.05 M; BSA 0.5 g/liter, pH 7
G6P-DH buffer (NAD, 0.2 mM; NADP, 0.2 μM; G-6-P, 0.5 mM;
   Tris–HCl, 0.05 M; G6P-DH, 500 U/liter, pH 7)
EDH buffer (NAD, 0.4 mM; NADP, 0.4 μM; G-6-P, 1 mM; Tris–
   HCl, 0.05 M; estradiol dehydrogenase 40 U/liter)

---

[3] J. C. Nicolas, Y. Chikhaoui, N. Bressot, A. Bonardet, B. Descomps, P. Cristol, and B. Hedon, *J. Gynecol. Obstet. Biol. Reprod.* **11**, 477 (1982).

Estradiol solution (400 ng/liter) in Tris–HCl buffer 0.05 M; BSA 0.5 g/liter, pH 7

FMN solution: FMN is light sensitive and the solution must be prepared under dim light and kept in a colored vial; dissolve 10 mg FMN in 10 ml water

Decanal solution: 10 μl decanal to 5 ml isopropanol

Bioluminescent reagent (BL reagent): 0.5 ml FMN solution, 0.5 ml decanal solution, 1.5 mg bacterial luciferase, and 0.3 U of NADH specific FMN reductase from *B. harveyi* were added to 100 ml phosphate buffer; this reagent must be stored in a colored bottle and kept at 4°

Sample Preparation

Urine. For hydrolyzing conjugates 1 drop glucuronidase solution (Pasteur Institute) was added to 1 ml of urine and incubated for 30 min at 37°. The reaction was stopped by adding 1 ml ethanol and the solution was centrifuged for 10 min at 3000 g; 5 to 50 μl of supernatant was diluted in 5 ml of Tris–HCl buffer.

Plasma. Plasma (0.5 ml) was extracted twice with 2 ml of diethyl ether. The organic phase was washed with 2 ml water and the ether was evaporated. The residue was dissolved in 1 ml of Tris–HCl buffer. Estrogen can also be extracted from plasma using an extrelut column: 0.5 ml plasma was diluted with 0.5 ml water, and passed through a 6-ml extrelut column (Merck). Estrogens were eluted with 6 ml isooctane/ethyl acetate (9/1, v/v), solvent was evaporated under a nitrogen stream in a 50° water bath, and the dry extract dissolved in with 1 ml Tris–HCl buffer.

Assay Procedure. The assay requires three tubes for each sample and two steps.

In the assay tube the sample is incubated with all the reagent and enzymes as indicated in the table. The incubation time is dependent on the sensitivity required (30 min to 2 hr). The accumulated NADH is determined by adding the bioluminescent reagent.

In the blank tube the sample is incubated with the reagents and enzymes except for G6P-DH. We have found that EDH interferes with the bioluminescent reagent and decreases the light signal. The G6P-DH buffer is prepared at the same time as EDH buffer and is added to the blank tube 20 sec before the addition of the bioluminescent reagent.

The standard tube contains the sample, all the reagents and enzymes, and a known mass of estrogen as internal standard. The mass of standard must be more than that in the assay tube and is dependent on the duration of the amplification reaction. Internal standards of 10, 20, or 50 pg were used when the incubation time was 2 hr, 1 hr, or 30 min, respectively.

BIOLUMINESCENT ASSAY OF ESTROGENS

Reagents	Blank (ml)	Assay (ml)	Standard (ml)
Sample	0.05	0.05	0.05
G6P-DH buffer		0.02	0.02
EDH buffer	0.10	0.10	0.10
Tris–HCl buffer	0.05	0.05	
Standard solution			0.05
Incubate 30 min to 2 hr			
G6P-DH buffer	0.02		

References containing 0, 10, or 20 pg of estradiol were measured every 48 tubes. They were used to determine estrogen contamination (0 pg), and to evaluate the transhydrogenase activity in the absence of sample (10 and 20 pg).

The table shows the volumes of each reagent added to each tube. All tubes were incubated at 25° and then loaded into the luminometer (LKB 1251). G6P-DH buffer was injected automatically to the blank tube 20 sec before injection of the bioluminescent reagent (0.2 to 0.4 ml).

Luminescence values were integrated over 30 sec after a delay time of 20 sec. The luminescent reagent is very sensitive to light and black tubing must be used to connect the automatic dispenser to the luminometer. A 50% inhibition of the signal is obtained if the luminescent reagent is exposed 30 sec to daylight.

Calculation. The concentration of total estrogen (estrone + estradiol) was calculated using the following formula:

$$C = \frac{LU(sample) - LU(blank)}{LU(standard) - LU(sample)} \times \frac{m}{v} \quad (\mu g/liter)$$

where m is the mass of estradiol added as standard (ng), v is the volume of the sample (ml), and C is the mass concentration (μg/liter).

Results and Discussion

Figure 1 shows the luminescence values obtained for various amounts of estradiol. A linear relationship between volts and mass of estradiol was obtained. The range of linearity is dependent on incubation time.

Specificity. The enzyme recognizes estradiol, estrone, and some other derivatives such as equilenine, 7-ketoestradiol, and 2-hydroxyestradiol.

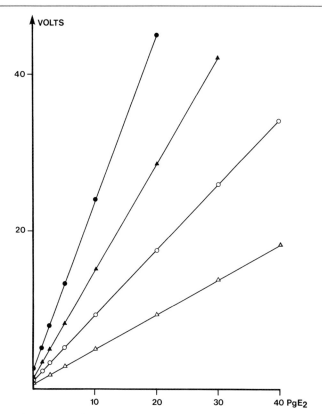

FIG. 1. Luminescence values as a function of estradiol concentration after 2 hr (●), 1 hr (▲), 30 min (○), and 15 min (△) of transhydrogenase reaction.

Estradiol 3-sulfate and 3-glucuronide were poor substrates (less than 3%) for EDH in the transhydrogenase reaction. Thus, the specificity of the transhydrogenase is higher than the dehydrogenase.[1]

Precision, Accuracy, and Detection Limits. For concentrations of estradiol within 50 to 1000 ng/liter plasma, the relative standard deviation is 5%. The accuracy has been proved by comparison with RIA.[3] The limit of detection is 0.1 pg of estrogen per assay. It is dependent on the amount of estrogen contained in the reagent, mainly EDH. This estrogen contamination can be determined precisely for each batch using the sample blank and can be subtracted from assay values.

Linearity. The estrogen concentrations are estimated by comparing the values obtained for the internal standard and the assay tube and assuming a linear relationship between absorbance and estrogen concentration. However, when the concentration of estrogen is too high, the rate of NADH formation is reduced, due to the accumulation of NADH. In this case, the use of the internal standard leads to an overestimation of the concentration of estrogen. For a linear relationship the concentration of the NADH produced must be less than 5×10^{-6} M. This is produced when 50 pg estrogen is incubated for 2 hr in 0.22 ml of transhydrogenase mixture.

Interferences with the Transhydrogenase Reaction. The amount of biological fluid which can be assayed is dependent on the content of the extract. Lipids, salts, or solvent in the medium can lead to an inhibition of the transhydrogenase reaction. This inhibition is easily quantitated by the ratio of the luminescence value of the internal standard run with the sample to the luminescence value of the internal standard run with the buffer. This ratio must be higher than 0.5. If human plasma is extracted with diethyl ether the maximum sample volume which can be used is 25 μl. However, using extraction on an extrelut column the assay can be performed on 1 ml plasma.

Interferences with the Bioluminescent Reaction. EDH causes diminution of light signal. Biological extract can also decrease the signal. Therefore the amount of EDH and sample must be the same in the three tubes.

Interferences with NADH Production. All systems producing or using NADH (enzymatic or chemical) can interfere. Samples able to produce or consume NADH must be inactivated by heating or solvent and all reagents must be prepared every week to avoid growth of microorganisms.

Preparation of the Blank. If high sensitivity is required the preparation of the blank tube is very important since G6P-DH and EDH in the presence of NAD can produce small amounts of NADH. All enzymes and components of the mixture must be contained in the blank tube. In this chapter the blank tube is constituted by adding the G6P-DH buffer after the amplification step. Since the nonspecific production of NADH by G6P-DH is dependent on the concentration of coenzymes, these must be identical in the assay tube and in the G6P-DH buffer added to the blank tube. G6P-DH also produces NADPH and in order to obtain the same NADPH concentration in the blank and in the assay the G6P-DH must be added to the blank at least 20 sec before the addition of the bioluminescent reagent.

We have described previously a different assay procedure[4] in which the blank tube was constituted by adding the sample to the complete reagent after the amplification reaction. In this case the luminescence background is produced by both G6P-DH and also from the transhydro-

[4] J. C. Nicolas, A. M. Boussioux, A. M. Boularan, B. Descomps, and A. Crastes de Paulet, *Anal. Biochem.* **135,** 141 (1983).

genase reaction due to trace amount of estrogen which contaminate the transhydrogenase reagent (EDH). Some errors can occur with this procedure because the sample partially inhibits the transhydrogenase reaction resulting in a higher blank value. This results in underestimation of the estrogen mass and an error of 0.1–0.4 pg. The blank can also be prepared by addition of EDH after the amplification; using this protocol the enzyme preparation must be very pure since trace amounts of other dehydrogenases can produce with the biological extract a small quantity of NADH.

For a very sensitive assay we recommend the blank be prepared as in this chapter and all reagents and solvents be checked for contamination with estrogen. For routine analyses of samples containing in 25 μl between 4 and 40 pg of estrogens (urine or plasma from women under ovarian stimulation) preparation of the blank is less critical and the sample can be replaced by an equivalent volume of buffer in the blank tube.

[19] Immobilized Enzymes for Assaying Bile Acids

By J. Schoelmerich, A. Roda, and M. DeLuca

Introduction

Although it is well-established that serum bile acid (SBA)[1] concentrations are elevated in patients with various liver diseases,[1a] their clinical utility is still debated. Several different types of SBA analysis have been used in an attempt to define the clinical utility of these determinations. Gas–liquid chromatography (GLC) combined with mass spectrometry is the currently available reference method.[2]

---

[1] Abbreviations used: BA, bile acid(s); BSA, bovine serum albumin; EIA, enzyme immunoassay; FMN, flavin mononucleotide; GLC, gas–liquid chromatography; HPLC, high-pressure liquid chromatography; HSD, hydroxysteroid dehydrogenase(s); NAD(H), nicotinamide-adenine dinucleotide (reduced form); NADP(H), nicotinamide-adenine dinucleotide phosphate (reduced form); RHPLC, reverse-phase high-pressure liquid chromatography; RIA, radioimmunoassay; SBA, serum bile acids.

[1a] S. Sherlock and V. Walshe, *Clin. Sci.* **6**, 223 (1984).

[2] K. D. R. Setchell and A. Matsui, *Clin. Chim. Acta* **127**, 1 (1983).

The enzymatic methods are based on the use of specific NAD-dependent hydroxysteroid dehydrogenases (HSD)[3] and on a spectrophotometric or fluorimetric detection of NADH production.[4]

More recently, the detection of NADH by bioluminescence resulted in an assay which was simple and as sensitive as any of the other published methods.

The reactions catalyzed are the following:

$$OH - \text{bile acid} + NAD(P)^+ \rightarrow O = \text{bile acid} + NAD(P)H + H^+ \qquad (1)$$
$$NAD(P)H + FMN + H^+ \rightarrow NAD(P)^+ + FMNH_2 \qquad (2)$$
$$FMNH_2 + RCHO + O_2 \rightarrow FMN + RCOOH + h\nu + H_2O \qquad (3)$$

The first reaction is catalyzed by a specific hydroxysteroid dehydrogenase. The second reaction is catalyzed by an NADH : FMN oxidoreductase or a diaphorase, and the third reaction is catalyzed by bacterial luciferase and results in the production of light. If the three enzymes are coimmobilized on Sepharose, the sensitivity of the assay is increased and the enzymes are very stable.

We describe here assays for 3α-, 7α-, and 12α-hydroxy bile acids in serum, urine, and bile.

Materials

Enzymes

Bacterial luciferase from *Vibrio harveyi,* Strain B 392, was used for all assays described. It was purified according to the procedure of Jablonski and DeLuca.[5] However, a preparation from *Vibrio fischerii* (Fa. Boehringer, Mannheim, FRG) has been used later with similar success (unpublished results). NADH : FMN-oxidoreductase from *Vibrio harveyi* was used initially and later replaced by bacterial disphorase (Fa. Boehringer, Mannheim, FRG). Several hydroxysteroid dehydrogenases (HSD) were used: 3α-HSD and 7α-HSD from Sigma Chemical Company, St. Louis, USA, 3α-HSD from Worthington Biochemical Cooperation, Freehold, USA, and from Fa. Nyegaard, Oslo, Norway. 12α-HSD and 7α-HSD were provided by the late Dr. I. A. Macdonald, Halifax, Canada,

---

[3] T. Iwata and K. Yamasaki, *J. Biochem.* (*Tokyo*) **56,** 424 (1964).
[4] F. Mashige, N. Tanaka, A. Maki, S. Kamei, and M. Yamanata, *Clin. Chem.* (*Winston-Salem, N.C.*) **27,** 1352 (1981).
[5] E. Jablonski and M. DeLuca, *Biochemistry* **16,** 2932 (1977).

TABLE I
BILE ACID NOMENCLATURE

Trivial name	Systematic nomenclature
Cholic acid	$3\alpha,7\alpha,12\alpha$-Trihydroxy-5β-cholanoic acid
β-Muricholic acid	$3\alpha,6\beta,7\beta$-Trihydroxy-5β-cholanoic acid
Ursocholic acid	$3\alpha,7\beta,12\alpha$-Trihydroxy-5β-cholanoic acid
Chenodeoxycholic acid	$3\alpha,7\alpha$-Dihydroxy-5β-cholanoic acid
Deoxycholic acid	$3\alpha,12\alpha$-Dihydroxy-5β-cholanoic acid
Ursodeoxycholic acid	$3\alpha,7\beta$-Dihydroxy-5β-cholanoic acid
Lithocholic acid	3α-Monohydroxy-5β-cholanoic acid

Taurine- and glycine-conjugated BA are termed by the prefixes tauro-
and glyco-. The salt form is depicted by the ending "cholate"

prepared from *Clostridium group P*[6-8] and from *Clostridium absonum.*[9]
The 3α-HSD from Worthington was contaminated with an aldehyde dehydrogenase and could only be used after purification by chromatography on
a Sephadex G-100 column.[10] The enzymes from Sigma were found to have
varying amounts of aldehyde dehydrogenase, depending on the lot used.
Some preparations could be utilized without further purification. The 3α-
HSD from Nyegaard did not contain aldehyde dehydrogenase.

Sepharose 4B was purchased from Pharmacia, Piscataway, USA; cyanogen bromide from Eastman Chemical Company, Rochester, USA;
Decanal, FMN, and glutathione from Sigma; and NAD, NADH, NADP,
NADPH grade I from Boehringer, Mannheim, FRG. All other chemicals
used were of analytical grade.

Bile Acids

Bile acids which were used as standards were obtained from a variety
of sources and had a purity of more than 98% as judged by thin-layer
chromatography[11] (Table I).

[6] I. A. Macdonald, E. C. Meier, D. E. Mahony, and G. A. Costain, *Biochem. Biophys. Acta*
450, 142 (1976).
[7] I. A. Macdonald, J. F. Jellett, and D. E. Mahony, *J. Lipid Res.* **20,** 234 (1979).
[8] I. A. Macdonald and Y. P. Rochon, *J. Chromatogr.* **259,** 154 (1983).
[9] I. A. Macdonald and J. D. Sutherland, *Biochim. Biophys. Acta* **750,** 397 (1983).
[10] J. Ford and M. DeLuca, *Anal. Biochem.* **110,** 43 (1981).
[11] A. Roda, A. F. Hofmann, and K. J. Mysels, *J. Biol. Chem.* **258,** 6362 (1983).

TABLE II

ENZYME COMPOSITION OF THE IMMOBILIZED ENZYME SYSTEMS USED FOR THE
ASSAYS DESCRIBED[a]

Enzyme	Hydroxysteroid dehydrogenase	Diaphorase	Luciferase
3α-IES	2.3 mg (10.9 U/mg)	6.0 mg (1 U/mg)	10.0 mg (6 × 10^7 LU/mg)
7α-IES	7.0 mg (7.0 U/mg)	6.0 mg (1 U/mg)	15.0 mg (6 × 10^7 LU/mg)
12α-IES	1.5 mg (0.1 U/mg)	10.0 mg (1 U/mg)	5.0 mg (6 × 10^7 LU/mg)

[a] From Schoelmerich et al.[23]

Sera, Urine, and Bile Samples

Sera and urine samples from patients with different hepatobiliary diseases were provided and analyzed by GLC and HPLC by Dr. G. van Berge Henegouwen, Arnheim, The Netherlands. Some sera were in addition analyzed by RIA as well.[12,13] Bile samples extracted in methanol from the National Cooperative Gallstone Dissolution Study[14,15] analyzed by GLC were provided by Dr. A. F. Hofmann, San Diego, CA.

Instruments

An Aminco Chem Glow photometer was used for measuring the bioluminescence. In principle, any luminometer could be used and the choice of the instrument depends on the needs of the individual laboratory. A continuous flow system utilizing immobilized enzymes has also been developed (see A. Roda, S. Girotti, and G. Carrea, this volume [21]).

Methods

Enzyme Immobilization

The combinations of enzymes shown in Table II were dissolved in 3.5 ml of 0.1 M sodium pyrophosphate, pH 8.0. BSA was added to give a final protein concentration of 8 mg/ml. The enzyme mixture was dialyzed

[12] E. Roda, R. Aldini, M. Capelli, D. Festi, C. Sama, G. Mazella, A. M. Rorselli, and L. Barbara, Clin. Chem. (Winston-Salem, N.C.) 26, 1647 (1980).
[13] A. Roda, L. J. Kricka, M. DeLuca, and A. F. Hofmann, J. Lipid Res. 23, 1354 (1982).
[14] J. J. Albers, S. M. Grundy, P. A. Cleary, D. M. Small, and J. M. Lachin, Gastroenterology 82, 638 (1982).
[15] A. F. Hofmann, S. M. Grundy, J. M. Lachin, S.-P. Lan, R. A. Baum, R. F. Hanson, T. Hersh, N. C. Hightower, J. W. Marks, H. Mekhijian, R. A. Shaefer, R. D. Soloway, J. L. Thistle, F. B. Thomas, and M. B. Tyor, Gastroenterology 83, 738 (1982).

against 4 liters of 0.1 M sodium pyrophosphate, pH 8.0, for 16 hr, in order to remove dithiothreitol (DTT) which is used for storage of luciferase but interferes with the immobilization procedure.

One gram of Sepharose 4B was activated with cyanogen bromide according to the method of March *et al.*[16] Three milliliters of the enzyme mixture was added to the Sepharose and the mixture was stirred for 2 hr at room temperature. The Sepharose-enzymes were then washed with 200 ml of 0.1 M potassium phosphate, pH 7.0, 500 ml of 0.1 M potassium phosphate, pH 7.0, with 1 M NaCl, and finally with 300 ml of 0.1 M potassium phosphate. For preparations containing diaphorase the washing buffers contained 0.5 mM glutathione.

The washed Sepharose-enzymes were stored in 10 ml of 0.1 M potassium phosphate, pH 7.0, containing 0.02 M sodium azide and 2 mM DTT. The latter was added weekly. If diaphorase was used the storage buffer contained 0.5 mM glutathione instead of DTT which interferes with diaphorase activity. For longer storage glycerol (10% v/v) was added. The immobilized systems were stored at 4° for up to 3 months or rapidly frozen in liquid nitrogen and stored indefinitely at −20°.

The recovery of active enzymes on the Sepharose was tested using standard spectrophotometric assays for HSD and diaphorase. The activity of luciferase was measured by injecting 100 μl of 0.15 mM FMNH$_2$–0.5 mM EDTA, which had been photoreduced, into a tube containing 400 μl potassium phosphate, pH 7.0, 0.0001% decanal, and 100 μl of the Sepharose-enzyme suspension (diluted 1 : 10–1 : 1000) in an Aminco Chem Glow photometer. The resulting light emission peak was used to calculate the relative light units/ml of the immobilized enzymes. The amounts of active immobilized enzymes obtained were between 30 and 40% for the different HSDs, 40–60% for oxidoreductase, 30–50% for the bacterial diaphorase, and 6–10% for the bacterial luciferase. The immobilization of these enzymes onto nylon tubes has been successfully performed as well for use in continuous flow systems (see A. Roda *et al.,* this volume [21]).

Assay Techniques

The basic assay mixture consists of 400 μl of 0.1 M potassium phosphate, pH 7.0, containing decanal, 0.0001%, 10 μl of 0.5 mM FMN, 10–50 μl of the Sepharose-enzyme suspension, and 10 μl of 5 mM NAD(P)$^+$. Standard or sample (10–50 μl) is added and the tube is vigorously shaken and placed into the luminometer. Light emission is recorded continuously until a peak is reached after 20–40 sec. This is the first peak method.

[16] S. C. March, I. Parikh, and P. Cuatrecasas, *Anal. Biochem.* **60,** 149 (1974).

Alternatively, the tube is removed every 20 sec, remixed, and light is again measured until a final peak is reached (final peak method).

A further technique is to initiate the reaction by injection of 100 μl of 0.5 mM NAD(P)$^+$ into the tube which is already placed in the luminometer and contains assay buffer, decanal, FMN, Sepharose-enzymes, and sample or standard. The injection starts the reaction and, in addition, mixes the contents of the tube. This method proved to be the easiest and had the best reproducibility.[17,18] Figure 1 shows a typical time course of light emission obtained with the different procedures. Background is obtained by measuring the light emission without sample or standard.

It is important to be aware that the decanal suspension is unstable and therefore has to be prepared freshly before use by dissolving 10 μl of decanal in 10 ml of ethanol and adding the appropriate amount to the assay buffer at 1–2 hr intervals.

Since the descending part of the kinetic curve of the reaction is caused by sedimentation of the Sepharose beads it is also of importance that the peak light emission is reached during the first 30 sec after injection.

Results obtained by the three different test techniques on 20 serum samples revealed a good correlation ($r = 0.99$ for each comparison).

Standard Curves

Standard curves for 3α, 7α-, and 12α-OH bile acids could be constructed for concentrations from 0.2 to 1000 μM. With the enzyme activities and the instruments used the light output was less than twice the background at concentrations below 0.2 μM (0.5 pmol/tube). Above 1000 μM the amount of light obtained was not linear with the concentration of bile acid. While standard curves for taurine and glycine conjugates and the corresponding unconjugated bile acid were identical (Fig. 2), significant differences were found among the different bile acids when standards were prepared without bovine serum albumin. These differences disappeared, however, when bovine serum albumin (BSA) was added to the standard solutions at a final concentration of 1.5 g/dl.

When 13 different bile acids or conjugates were analyzed at the same concentration a coefficient of variation of 5.7 was found in the presence of albumin (Table III). The exact amount of BSA necessary has not yet been determined. However, kinetic studies of some of the enzymes used revealed that the affinity of 3α-HSD and 7α-HSD for different bile acids is different and can be modified by the addition of albumin (Fig. 3). Since the binding affinity of albumin to bile acids parallels the affinity of the HSD

[17] I. A. Macdonald, C. N. Williams, and D. E. Mahony, *Anal. Biochem.* **57,** 127 (1974).
[18] I. A. Macdonald, C. N. Williams, and B. C. Musial, *J. Lipid Res.* **21,** 381 (1980).

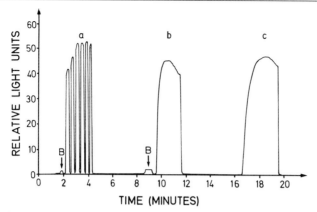

FIG. 1. Kinetics for the different test systems described. (a) Repeated shaking until a final peak height of light emission is reached = final peak method. (b) First peak method. (c) Injection assay. Taurocholate (20 μmol/liter) is used in each test. B, Background light emission without bile acid present. From Schoelmerich *et al.*[22]

the effect on the test systems can be explained. In addition the high affinity of the reaction product—an oxo bile acid—for albumin might be responsible for the increase of V_{max} and, thus, for the sensitivity of the assays. Therefore, albumin should be used in the preparation of standard solutions and for urine and bile samples. In the latter case it is easier to add the albumin to the assay buffer.

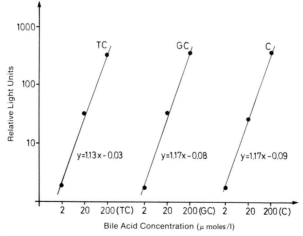

FIG. 2. Comparison of standard curves for conjugates and the unconjugated form of cholate using the 12α-HSD system.

TABLE III
LIGHT EMISSION WITH IDENTICAL STANDARD CONCENTRATION
OF DIFFERENT BILE ACIDS[a]

Bile acid	Concentration (μmol/liter)	Light units	Variation (%)
Taurocholate	100	348	+4.5
Glycocholate	100	335	+0.6
Cholate	100	342	+2.7
Taurochenodeoxycholate	100	345	+3.6
Glycochenodeoxycholate	100	290	−12.9
Chenodeoxycholate	100	320	−3.9
Taurodeoxycholate	100	320	−3.9
Deoxycholate	100	352	+5.7
Tauroursodeoxycholate	100	359	+7.8
Ursodeoxycholate	100	323	−3.0
Glycolithocholate	100	315	−5.4
Ursocholate	100	336	+0.9
β-Muricholate	100	343	+3.0
Mean		332.9 19.1	CV: 5.7%

[a] Light units represent mean of two analyses with the 3α-OH bile acid assay. Variation is calculated as percentage of the mean light emission for all measurements. From Schoelmerich et al.[23]

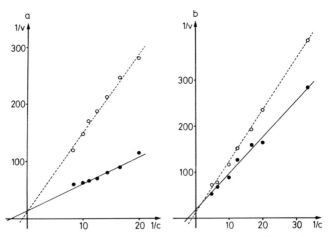

FIG. 3. Enzyme kinetics for 3α-HSD (a) and for 7α-HSD (b). Comparison with (open circles) and without (closed circles) albumin using taurochenodeoxycholate (a) and taurocholate (b). From Schoelmerich et al.[23]

Specificity of the Assays

None of the immobilized enzyme systems studied reacted with bile acids which did not have the specific hydroxy groups. Other compounds such as malate, lactate, and β-hydroxybutyrate did not produce any light emission up to concentrations of 10 mM. Androsterone did react with the 3α-system as described.[10] But this compound is present in biological fluids in much smaller concentrations and, therefore, the interference is negligible. Cholesterol did not react with the enzymes and the addition of a 3β-OH bile acid resulted in a light emission of 0.28% of that obtained with the corresponding α-OH compound, probably due to impurity of the BA.

Stability of the Systems

Sepharose-enzyme suspensions containing oxidoreductase were stable over a period of 3 months when stored at 4°. Preparations containing diaphorase were stable when stored in the presence of 0.5 mM glutathione.

Sample Preparation

Sera were diluted 1:4 with 0.1 M potassium phosphate, pH 7.0, and heated for 15 min at 68° in order to inactivate other NAD(P)H-generating enzymes. Urine samples were used undiluted after centrifugation (300 g, 5 min). Bile extracts were diluted 1:100–1:1000 and used without further preparation. Unmodified rat bile samples were studied after dilution (1:500). As stated above, it may be useful to add BSA to the samples or to the assay buffer in order to overcome the different affinity of the HSD for different bile acids. However, no detailed studies on sample preparation have so far been performed. No influence of the degree of dilution of the biological samples on the results obtained was found. The sample size could be changed from 10 up to 100 μl without significant influence on the concentrations measured.

Recovery and Precision

In order to determine the recovery of added standards, several standard solutions with different concentrations were added to different samples (serum, urine, and bile) with a large range of bile acid concentrations. The recovery in serum ranged from 87 to 113%; in urine and bile similar values were found. Intraassay precision was analyzed by 10 repeated analyses of the same sample in one day. The coefficient of variation ranged between 5.2 and 8.2 for all samples studied and was independent of the sample concentration.

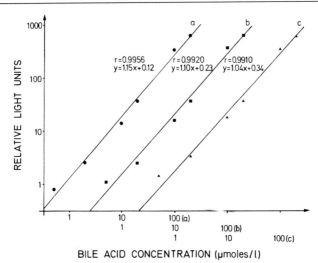

FIG. 4. Standard curves for taurocholate (a), taurochenodeoxycholate (b), and tauro-deoxycholate (c) using the 3α-OH bile acid assay. Observe different scales on the horizontal axis. From Schoelmerich *et al.*[21]

Interassay precision was determined by day to day analyses of the same sample and had the same range as intraassay precision.

Calculations

Using the equations given by Macdonald[17,18]

$$3\alpha\text{-OH bile acids} = \text{cholic acid} + \text{chenodeoxycholic acid} + \text{deoxycholic acid} \quad (4)$$
$$7\alpha\text{-OH bile acids} = \text{cholic acid} + \text{chenodeoxycholic acid} \quad (5)$$
$$12\alpha\text{-OH bile acids} = \text{cholic acid} + \text{deoxycholic acid} \quad (6)$$

the three major bile acids appearing in human serum and urine could be calculated.

Validation of the Assays

Results obtained with these bioluminescence assays were compared with those revealed by GLC and HPLC on the same sera, urines, or bile extracts.[19,20] Furthermore, the correlation between the mean of both mea-

[19] A. T. Ruben and G. P. van Berge Henegouwen, *Clin. Chim. Acta* **119**, 41 (1982).
[20] G. P. van Berge Henegouwen, A. Reuben, and K.-H. Brandt, *Clin. Chim. Acta* **54**, 249 (1974).

TABLE IV
ANALYSIS OF MIXED STANDARDS OF BILE ACIDS
WITH THREE BIOLUMINESCENCE ASSAYS[a]

Total bile acid concentration (μmol/liter)	3α assay		7α assay		12α assay	
	μmol/ liter	%	μmol/ liter	%	μmol/ liter	%
1000	902.5	90.2	600.0	90.9	571.0	86.3
100	100.1	100.1	59.0	89.4	68.1	102.2
10	8.8	88.2	6.3	95.4	7.1	107.0

[a] Standards were mixed with cholic acid : chenodeoxycholic acid : deoxycholic acid = 1 : 1 : 1. Results are given as μmol/liter and as percentage of the expected value as calculated from the mixture. From Schoelmerich et al.[23]

surements $M = C[GLC] + C[BIOLUMINESCENCE]/2$ and the difference between both values as percentage of M was calculated.

Results

Standards

Using BSA for the preparation of standards for different BA, each of the major human bile acids, either free or as taurine or glycine conjugates, could be used for the standard curve. Figure 4 gives as an example the curves obtained with the taurine conjugates of the three major human bile acids. As shown in Table IV, the use of other BA gave similar results. Therefore, cholyltaurine was used for all standards used in validation experiments.

Validation

Results obtained with the three bioluminescence assays were closely correlated with those determined by GLC or HPLC.[13,21,22] Figure 5 gives the comparison for 12α-OH BA in serum. Regardless of the sample or the assay, values obtained with either bioluminescence or GLC were always closely correlated ($r > 0.88$).

[21] J. Schoelmerich, G. P. van Berge Henegouwen, A. F. Hofmann, and M. DeLuca, Clin. Chim. Acta 137, 21 (1984).
[22] J. Schoelmerich, J. E. Hinkley, I. A. Macdonald, A. F. Hofmann, and M. DeLuca, Anal. Biochem. 133, 244 (1983).

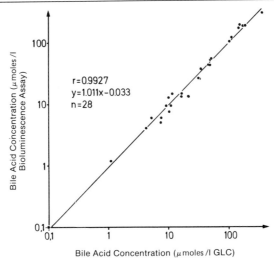

FIG. 5. Comparison of 12α-OH bile acid concentrations in serum obtained by the bioluminescence assay and those obtained by gas–liquid chromatography (GLC).

Using Eqs. (4)–(6), concentrations of chenodeoxycholic and cholic
acid were calculated for the 28 serum samples where all three assays had
been performed. The results were again closely correlated with those
measured by GLC ($r > 0.88$). However, due to multiple calculation steps,
the error was increased and, thus, deoxycholate present at a concentration of less than 10% of the total could not be calculated accurately.
Similar results were obtained for serum, urine, and bile extracts. However, the results with bile extracts were somewhat less satisfactory than
for the other samples. This might be due to the interference of methanol
with the enzyme system or to dilution problems. Preliminary experience
with extracted bile suggests better results and, in addition, makes the test
much easier.

No correlation was found between the error as percentage and the
total BA concentration (Fig. 6). This was true for all sample types and for
all three assays studied.

The use of rat serum or bile revealed a close correlation between
bioluminescence and HPLC values.[23]

Applications

In addition, some experimental studies on rats were performed using
these assays. SBA concentrations proved to be as effective as serum ALT

[23] J. Schoelmerich, A. Roda, and M. DeLuca, *in* "The Bile Acids" (K. Setchell, P. P. Nair,
and D. Kritchevsky, eds.), Vol. 4. Plenum, New York (in press).

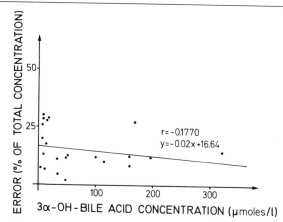

FIG. 6. Correlation between the total bile acid concentrations (mean of bioluminescence and GLC) and the difference between both values as percentage of the total concentration. From Schoelmerich *et al.*[23]

in detecting galactosamine-induced liver damage. Furthermore, SBA determinations could be used for sequential analysis during the course of the experiment, since in contrast to all other tests a very small sample size is required, which can be repeatedly taken from the animal.[24]

Furthermore, the assays have been used for studies on BA transport in the perfused rat liver. They proved to be of value in the determination of BA uptake, secretion, and metabolism.[25,26] Some studies on SBA in hamsters[23] and rats with selective biliary obstruction[27] have shown the possible usefulness of the method when only small samples are available.

Using a system of equations similar to Eqs. (4)–(6) bile acid groups could be estimated in rat serum and bile. However, since more than three major bile acids appear in this species,[25] an additional assay, i.e., for 7β-OH bile acids seems to be necessary in order to calculate individual bile acids. So far, such an assay has been developed but due to the characteristics of the 7β-HSD its detection limit is about 10 μmol/liter and, due to impurities of the enzyme, specificity is questionable.

[24] J. Schoelmerich, M. DeLuca, and M. Chojkier, *Hepatology* **4,** 639 (1984).

[25] J. Schoelmerich, S. Kitamura, and K. Miyai, *Biochem. Biophys. Res. Commun.* **115,** 518 (1983).

[26] J. Schoelmerich, S. Kitamura, and K. Miyai, *Hepatology* **4,** 763 (abstr.) (1984).

[27] S. Bellentani, W. G. M. Hardison, and F. Manenti, *J. Hepatol.* **2,** 525 (1985).

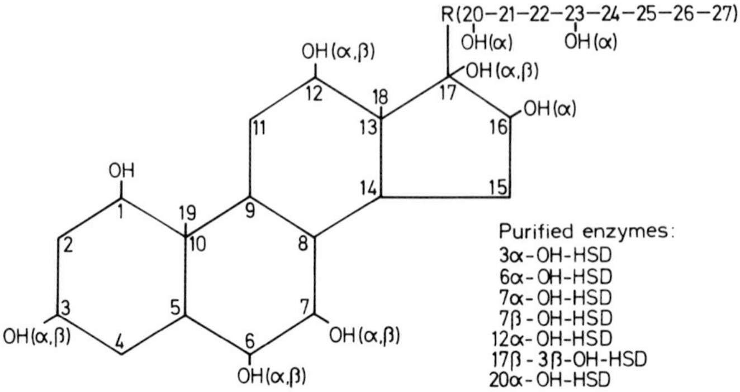

FIG. 7. Purified available hydroxysteroid dehydrogenases and the corresponding hydroxy group at the schematic steroid molecule, as present in bile acids and steroid hormones.

Discussion

Applications

Experiences with the clinical application of the bioluminescence assays are limited so far. Availability of this simple technique may well facilitate studies to assess the clinical utility of total bile acids or classes of hydroxy bile acids measurements. So far, the role of SBA determination in the assessment of liver disease has not been finally determined.[28] The extreme sensitivity of the bioluminescence assays suggests that it could be useful for the detection of BA malabsorption. In such patients, the postprandial increase in serum bile acid levels is lower than in healthy subjects, and the lower postprandial rise in such patients has heretofore been detectable only by radioimmunoassay and could not be detected by GLC or enzymatic methods.[29,30]

The use of the combination of all three assays developed, probably in the flow system described (see Roda et al., this volume [21]), allows the determination of the pattern of the human BA. The nylon-immobilized enzyme system can also be adapted as a detector for BA reverse-phase high-pressure liquid chromatography (RHPLC). Potentially, it will represent an extremely sensitive detector of column separated BA, offering the advantages of increased specificity and sensitivity, and permitting analysis of unconjugated, glycine conjugated, and taurine conjugated BA with

[28] A. F. Hofmann, Hepatology 2, 512 (1982).
[29] R. Aldini, A. Roda, D. Festi, G. Mazzella, A. M. Morselli, C. Sama, E. Roda, N. Scopinaro, and L. Barbara, Gut 23, 829 (1982).
[30] W. F. Balistreri, F. J. Suchy, and J. E. Heubi, J. Pediatr. 96, 582 (1980).

the same detector response. In addition, the determination of BA patterns in urine and bile is possible without much preparation. This may facilitate further clinical studies.

In summary, the bioluminescence assays described in either the standard form as injection assay or in an automated form using nylon immobilization or flow cells offer the possibility of studying the diagnostic value of the determination of total or individual BA in health or in hepatic or ileal disease. The bioluminescent assays are rapid, specific, and as sensitive as GLC or other binding assays, and it is possible to use commercially available diaphorase and luciferase to prepare the immobilized enzymes.

Finally, the principle may be used for the determination of other steroids in biological fluids, since a number of other HSD has been purified so far (Fig. 7).

[20] Continuous-Flow Bioluminescent Assays Employing Sepharose-Immobilized Enzymes

By DANIEL C. VELLOM and LARRY J. KRICKA

Biochemical reactions which produce light have in recent years been exploited as versatile analytical tools by a growing number of clinical and research laboratories.[1-4] Recent developments in enzyme immobilization and coupled enzymatic assays, as well as the continuing quest for sensitive nonradioactive methodologies, have contributed to a rapidly expanding repertoire of bioluminescent assays for metabolites, hormones, drugs, and even explosives. Many of these assays utilize luciferase isolated from luminous marine bacteria, coupled to NAD(P)H : FMN oxidoreductase, as the terminal detection system. Coimmobilization of these enzymes with a specific dehydrogenase, or a reactive pathway of enzymes which ultimately produces NAD(P)H, allows picomole detection of most substrates in a single step. The reactions involved may be represented as follows.

[1] L. Kricka, P. E. Stanley, G. H. G. Thorpe, and T. P. Whitehead, eds., "Analytical Applications of Bioluminescence and Chemiluminescence." Academic Press, New York, 1984.
[2] M. DeLuca, ed., this series, Vol. 57.
[3] L. J. Kricka and T. J. N. Carter, eds., "Clinical and Biochemical Luminescence." Dekker, New York, 1982.
[4] M. Serio and M. Pazzagli, eds., "Luminescent Assays." Raven Press, New York, 1982.

Dehydrogenase

$$\text{Reduced substrate} + \text{NAD(P)}^+ \rightarrow \text{Oxidized product} + \text{NAD(P)H} + \text{H}^+ \qquad (1)$$

Oxidoreductase

$$\text{NAD(P)H} + \text{H}^+ + \text{FMN} \rightarrow \text{NAD(P)}^+ + \text{FMNH}_2 \qquad (2)$$

Luciferase

$$\text{FMNH}_2 + \text{RCHO} + \text{O}_2 \rightarrow \text{FMN} + \text{RCOOH} + \text{H}_2\text{O} + h\nu \qquad (3)$$

Alternatively, the firefly luciferase, coupled with ATP : NMN adenylyl-transferase (NAD$^+$ pyrophosphorylase, nicotinamide-nucleotide adenylyltransferase) may be utilized as the detection system for substrates for which there is a specific NADH-dependent reductase. These reactions are as follows.

Reductase

$$\text{Oxidized substrate} + \text{NADH} \rightarrow \text{Reduced product} + \text{NAD}^+ \qquad (4)$$

Pyrophosphorylase

$$\text{NAD}^+ + \text{PP}_i \rightarrow \text{NMN} + \text{ATP} \qquad (5)$$

Luciferase

$$\text{ATP} + \text{Luciferin} + \text{O}_2 + \text{Mg}^{2+} \rightarrow \text{AMP} + \text{PP}_i + \text{CO}_2 + \text{Oxyluciferin} + h\nu \qquad (6)$$

One widely used support for immobilization of enzymes and other proteins is Sepharose. Assay systems utilizing Sepharose-immobilized luminescent enzymes have proven to be highly sensitive due to a very high enzyme concentration within the enclosed volume of the bead. It is thought that the proximity of coupled enzymes to each other within the pores of the bead reduces diffusional barriers, thus making coupled pathways more efficient. Since the final product, light, is not subject to diffusion, detection is accomplished instantly.

The protocol adopted in our laboratory[5,6] for metabolite assays using Sepharose-immobilized bioluminescent enzymes has proven to be rather cumbersome. Typically, the sample, buffer containing decanal and other cofactors, an FMN solution, and the Sepharose-enzyme suspension, must be separately pipetted into the assay cuvette. Many systems are sensitive to the order and timing of additions, and the subsequent mixing of these reagents to achieve reproducible light emission. For peak light intensity measurements, timing is critical as the peak can be generated very rap-

[5] J. Ford and M. DeLuca, *Anal. Biochem.* **110,** 43 (1981).
[6] G. Wienhausen and M. DeLuca, *Anal. Biochem.* **127,** 380 (1982).

idly. Initiation of the reaction by injection of the reagents into the sample positioned in the luminometer has not satisfactorily solved these problems. This is because the finite number of Sepharose beads in a measured volume leads to large fluctuations in the amount of enzyme contributing to the measured light emission. These fluctuations arise from physical processes such as uneven mixing and beads adhering to the walls of pipets and assay tubes out of range of the photomultiplier tube (PMT). Thus, distributing small volumes of Sepharose suspensions and making measurements with injection mixing are difficult to achieve reproducibly. In addition, the Sepharose beads settle rapidly, making sustained measurements of an assay suspension impractical. However, the use of immobilized enzyme systems is still desirable since they show increased stability as compared to soluble forms and they are reusable.

Kricka *et al.* established the feasibility of continuous-flow bioluminescent assays[7,8] using coimmobilized multienzyme systems. A flow cell was constructed by packing a small piece of Tygon or glass tubing with a few hundred microliters of coimmobilized Sepharose enzymes and, with suitable tubing for delivery of reagents, it was mounted in front of the PMT in an Aminco luminometer. Delivery of reagents was accomplished via a peristaltic pump with an AAI AutoAnalyzer to deliver automatically measured amounts of substrate. This system proved to be relatively reliable, with sensitivities for the detection of NADH, glucose 6-phosphate, and ATP in the picomole range. The reproducibility was improved over suspension type assays. Coefficients of variation were 3–5% for substrates measured at both 2.5×10^{-7} and 2×10^{-5} M. Thirty samples could be measured per hour. Moreover, a single flow cell could be used for up to 700 assays before background light emission increased to a level where sensitivity was compromised.

Ultimately, sensitivity was limited by the variable level of background light produced by the substrate/cofactor reaction buffer. In an effort to reduce this background light level and improve sensitivity, we modified the earlier design. Concomitant reductions of the flow cell bed volume and reagent flow rates, as well as separation of FMN and decanal reagents into individual reservoirs, reduced the background light to approximately 0.1% of its previous level. The result of these manipulations was a 1000-fold increase in sensitivity for the measurement of NADH. Similar increases in sensitivity were observed for ATP, NAD, and 6-phosphogluconic acid measurements.

[7] L. J. Kricka, G. K. Wienhausen, J. E. Hinkley, and M. DeLuca, *Anal. Biochem.* **129**, 392 (1983).
[8] M. DeLuca and L. J. Kricka, *Arch. Biochem. Biophys.* **226**, 285 (1983).

Construction of the Capillary Flow Cell

The flow cell, which contains the reactive bed of Sepharose-immobilized enzymes, is constructed from a 20-μl Drummond Microcap capillary pipet. These capillaries are thin wall glass with an inside diameter of 0.63 mm and an overall length of 64 mm. One end of the capillary is constricted by very brief heating in a flame such that the inside diameter of the tip is reduced by approximately one-half. Then a very small amount of glass beads (Superbrite, Type 100-5005, 3M Co.) is introduced into the other end of the capillary and tapped down to form a 3–4 mm bed at the constricted end. This is best accomplished with dry glass beads in a manner similar to loading sample into a melting point capillary. We have found it useful to tamp the glass beads down with the plunger of a 5-μl Hamilton syringe in order to permanently fix them in place as a bed support. The capillary column is then packed with Sepharose-immobilized enzymes with the help of a peristaltic pump connected downstream of the capillary. A small loading reservoir is constructed from a truncated pasteur pipet and connected to the unconstricted end of the capillary tube with a short piece of 1/16 in. i.d. Tygon tubing. With the pump running at a flow rate of 0.4 to 0.5 ml/min, 1 ml of storage buffer is placed in the reservoir and allowed to fill the capillary. Fifty to one-hundred microliters of a 1 g/10 ml suspension of Sepharose-immobilized enzymes is introduced into the storage buffer in the reservoir. The Sepharose will settle to the bottom of the capillary and form a packed bed 15–20 mm long. More storage buffer is added to the reservoir as necessary to prevent the capillary from running dry. The packed "flow cell" is now ready to be plumbed into the flow manifold.

The Flow Manifold

A schematic of the flow manifold is shown in Fig. 1A. Small test tubes are used for buffer and reagent reservoirs. Automatic sampling is accomplished with a Technicon AAI AutoAnalyzer. A 10-ml beaker is used to hold buffer wash and the samples are held in 0.5-ml disposable polystyrene cups. Three separate channels are employed: one to deliver sample and the other two to deliver buffer and cofactors. Pump tubing is 0.25-mm-i.d. silicon rubber or polyvinyl chloride. Manifold tubing is 0.4-mm-i.d. teflon. Microbore "T" fittings (MER Chromatographic) were employed to mix the three streams just prior to their entering the sample chamber of the luminometer (LKB model 1250). Twenty-two-gauge stainless-steel tubing pierces the rubber septum of the sample injection port of the luminometer to provide a light-tight seal. All tubing connections were silicone rubber or Tygon tubing. Each channel delivered 30 μl/min to give

A

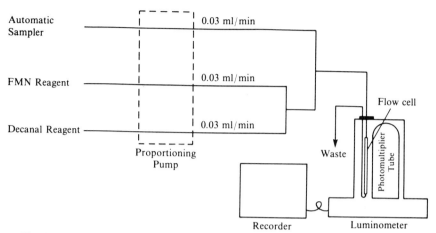

FIG. 1. (A) The flow manifold. (B) Design of immobilized enzyme packed capillary flow cell and its orientation in the luminometer: (a) glass beads bed support, (b) capillary tube flow cell (20 μl Drummond Microcap), (c) Sepharose-enzyme bed, (d) thick wall capillary for outflow, (e) silicone rubber tubing, (f) 22-gauge stainless-steel hypodermic needle, (g) rubber septum, (h) knurled screw port, (i) body of sample chamber of LKB 1250 luminometer, (j) window of sample chamber: 38 mm, (k) distance from flow cell to PMT ~5 mm, (l) Teflon tubing.

a total flow rate of 90 μl/min. A controller cam for the automatic sampler was constructed to give a 2 min sampling time followed by a 4 min wash. The capillary flow cell and its orientation in the luminometer are depicted in Fig. 1B. A small glass mirror is mounted behind the flow cell to increase the percentage of emitted light falling on the PMT. Light intensity was monitored continuously on a single channel chart recorder.

Immobilization of Enzymes

Sepharose 4B or CL-6B was activated by cyanogen bromide according to the method of Axén et al.[9] or by tresyl chloride according to the method of Nilsson and Mosbach.[10] Immobilization of enzymes was essentially as previously described except that 100 mM sodium pyrophosphate, pH 8.0, was used as buffer and BSA was omitted from the protein mixture. The immobilized enzymes could be stored as a 1 g/10 ml suspension in 50 mM Tris–HCl, pH 7.0, 2 mM 2-mercaptoethanol, 1 mM EDTA, and 0.02% NaN$_3$ at 4° for weeks, or indefinitely at −20°.

[9] R. Axén, J. Poráth, and S. Ernback, *Nature (London)* **214,** 1302 (1967).
[10] K. Nilsson and K. Mosbach, *Biochem. Biophys. Res. Commun* **102,** 449 (1981).

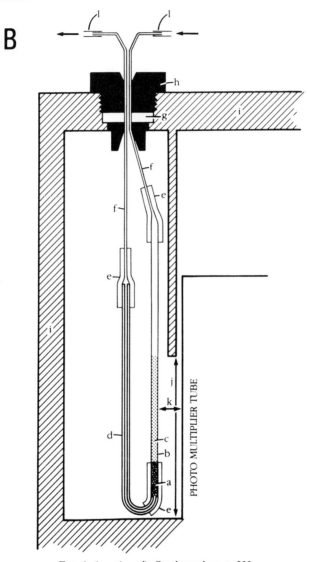

FIG. 1. (*continued*). See legend on p. 233.

Reagents and Enzymes

Reagent grade FMN, decylaldehyde and 2-mercaptoethanol were from Sigma Chemical Co. ATP, NAD, NADP, NADH, NADPH, and Tris were the highest purity available from Boehringer-Mannheim. Cyanogen bromide was from Eastman. Tresyl chloride (2,2,2-trifluoroethanesulfo-nate) was from Fluka. Sepharose 4B and CL-6B were from Pharmacia.

Bacterial luciferase was isolated from a frozen cell paste of *Vibrio harveyi* as described by Hastings *et al.*[11] NADPH and NADH : FMN ox-idoreductases were purified from *V. harveyi* according to Jablonski and DeLuca.[12] Diaphorase "from Microorganism" could be substituted for oxidoreductase and was from Boehringer-Mannheim as was 6-phos-phogluconate dehydrogenase and NAD^+ pyrophosphorylase. Firefly lu-ciferase was purified from isolated lanterns by the method of Green and McElroy.[13]

Assay and Wash Buffers

NADH Assay. Three channels were used: (1) 100 μl 0.15% decanal in ethanol was added to 4.9 ml 2 mM 2-mercaptoethanol, 2mM EDTA in 50 mM Tris, pH 7.0, (2) 5 ml 0.1 mM FMN in distilled H_2O, and (3) Tris buffer used to prepare samples and wash flow cell between samples.

6PGA Assay. Three channels were used: (1) 1.6 mM NADP, 0.4 mM magnesium acetate, 36 mM ammonium acetate, 0.1 mM FMN in H_2O, (2) decanal, 2-mercaptoethanol, EDTA (as for NADH assay) in 50 mM Tris–HCl, pH 7.9, and (3) sample wash in pH 7.9 Tris–HCl.

ATP Assay. Luciferin (1 mM) (0.8 ml), 200 mM $MgCl_2$ (0.25 ml), and 25 mM glycylglycine, pH 7.8 or 50 mM Tris–HCl (8.85 ml), in two chan-nels, sample in third channel.

NAD Assay. The same as the ATP assay but includes 0.1 ml 100 mM sodium pyrophosphate.

Results and Discussion

NADH and 6PGA Assays Using Bacterial Luciferase

The peak light intensity recorded for known concentrations of NADH and 6-phosphogluconic acid is shown in Fig. 2. Peak intensity increased linearly over the range 100 fmol/ml to 1 nmol/ml for NADH and 1 pmol/ml to 10 nmol/ml for 6PGA. Thus a 60-μl sample volume gives a minimum level of detection of 6 fmol of NADH and 60 fmol of 6PGA. An analytical trace for repetitive measurement of 100 pmol/ml NADH is shown in

[11] J. W. Hastings, T. D. Baldwin, and M. Z. Nicoli, this series, Vol. 57, p. 135.
[12] E. Jablonski and M. DeLuca, *Biochemistry* **16**, 2932 (1977).
[13] A. A. Green and W. D. McElroy, *Biochim. Biophys. Acta* **20**, 170 (1956).

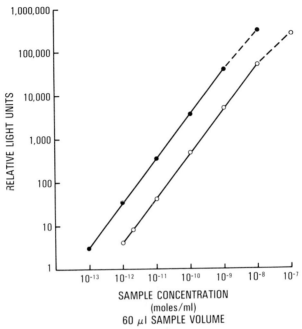

FIG. 2. Peak light intensity as a function of the amount of NADH (●) and GPGA (○). Two minute samples = 60 μl, 4 min wash.

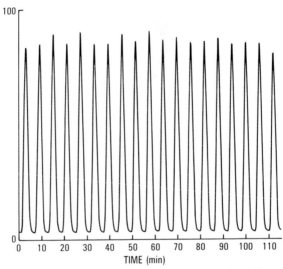

FIG. 3. Typical analytical traces obtained for repeated analysis of 10 pmol/ml of NADH: 2 min samples, 4 min wash.

Fig. 3. Intraassay coefficients of variation were typically 2–10% over the linear range of the assay.

NAD Assays Using Firefly Luciferase

NAD concentrations as low as 10 pmol/ml could be measured with the linear range to 10 nmol/ml (data not shown). Concentrations as low as 1 pmol/ml could be detected; however the assay was not linear below 10 pmol/ml.

Comparison of this flow system with assays performed utilizing a suspension of immobilized enzymes[5,6] or the previously described flow method[7] has shown a marked increase in reproducibility as well as a 10- to 1000-fold increase in sensitivity. Five factors are presumed to contribute to the increased sensitivity: (1) reduced bed volume in the flow cell has decreased the background (baseline) light emission; (2) reduced flow rates allow the sample to remain in contact with the reactive bed for a longer period of time; (3) the smaller diameter of the capillary tube allowed a greater proportion of the emitted light to reach the PMT; (4) the flow cell is positioned closer to the PMT, increasing the measured light intensity; (5) the small manifold tubing and fittings reduced the dead volume and turbulent mixing. Our sensitivity is limited by a small relatively constant baseline light emission which is noisy due to pulsation produced by the pump. We have found that light emission is strongly influenced by flow rate and propose that piston-type infusion pumps could be utilized to deliver reagents, thus eliminating pulsation. This type of pump would also allow for the incorporation of in-line filters and HPLC type sample injection valves to fine-tune the system. Background light emission gradually begins to increase after several hours of operation. This is most likely due to contamination of the Sepharose enzyme bed and we presume that filtration of reagents could reduce this problem significantly.

In conclusion, we have found this method to be both reliable and sensitive for the measurement of pmol/ml levels of NADH, 6-phosphogluconic acid, ATP, and NAD. In principle, these methods could be applied to any coupled enzyme system where immobilization on a particulate gel matrix is possible. DeLuca and Kricka[8] have shown that as many as 11 enzymes may be coimmobilized onto a single Sepharose matrix and a limiting substrate for the first enzyme in the sequence will result in light emission corresponding to its concentration. The method shows promise for both clinical and research applications.

[21] Flow Systems Utilizing Nylon-Immobilized Enzymes

By ALDO RODA, STEFANO GIROTTI, and GIACOMO CARREA

The assay of NAD(P)H and NAD(P)H-generating metabolites with bacterial bioluminescent enzymes is rapid, sensitive, and specific.[1,2]

In recent years many studies have been carried out in order to elucidate the properties of immobilized enzymes and their use for analytical purposes.[3] In fact, the stability of immobilized biocatalysts, that generally exceeds that of soluble ones, and their reusability make them desirable tools for the determination of substrates, activators, inhibitors, and other enzymes in biological fluids.

The use of Sepharose-immobilized bioluminescent enzymes for analytical purposes has been described by several investigators and attempts have also been made to automate these assays using Sepharose columns.[4,5]

The continuous-flow determination of NAD(P)H, ethanol, glycerol, aldehyde, and bile acids in biological samples (serum and saliva) with nylon tube-immobilized bioluminescent enzymes and specific dehydrogenases is described in the present chapter.

Experimental

Materials

Luciferase [EC 1.14.14.3, alkanal monooxygenase (FMN)] (specific activity 12 mU/mg) from *Photobacterium fischeri,* NAD(P)H : FMN oxidoreductase [EC 1.6.8.1, NAD(P)H dehydrogenase (FMN)] (specific activity 4 U/mg) from *Photobacterium fischeri,* alcohol dehydrogenase (EC 1.1.1.1) (specific activity 400 U/mg) from yeast, aldehyde dehydrogenase (EC 1.2.1.5) (specific activity 15 U/mg) from yeast, glycerol dehydrogenase (EC 1.1.1.6) (specific activity 25 U/mg) from *Enterobacter aerogens,* NAD (lithium salt), and FMN were purchased from Boehringer-

[1] M. DeLuca and W. D. McElroy, eds., "Bioluminescence and Chemiluminescence: Basic Chemistry and Analytical Applications." Academic Press, New York, 1981.
[2] L. J. Kricka and T. J. N. Carter, eds., "Clinical and Biochemical Luminescence." Dekker, New York, 1982.
[3] L. Goldstein, this series, Vol. 44, p. 397.
[4] K. Kurkijarvi, R. Raunio, and T. Korpela, *Anal. Biochem.* **125,** 415 (1982).
[5] L. J. Kricka, G. Wienhausen, J. E. Kinkley, and M. DeLuca, *Anal. Biochem.* **129,** 392 (1983).

Mannheim. 3α-Hydroxysteroid dehydrogenase (EC 1.1.1.50) (specific activity 2.5 U/mg) from *Pseudomonas testosteroni* chromatographically purified, 7α-hydroxysteroid dehydrogenase (EC 1.1.1.159) (specific activity 6 U/mg) from *Escherichia coli,* *n*-decyl aldehyde (decanal), and dithiothreitol (DTT) were obtained from Sigma and glutaraldehyde (25% aqueous solution) was obtained from Merck. 12α-Hydroxysteroid dehydrogenase (EC 1.1.1.176) (specific activity 1.5 U/mg) was extracted from *Clostridium* group P as described by Macdonald *et al.*[6] Bile acids were purchased from Calbiochem and were crystallized before use. All the solutions were made with apyrogenic reagent-grade water prepared with a Milli-Q System (Millipore). Nylon-6 tubes with 1 mm internal diameter were obtained from Snia Viscosa. All other reagents and compounds were of analytical grade.

Enzyme Assay

The activity of free enzymes was measured in a 3-ml cuvette by spectrophotometrically monitoring at 340 nm the formation or consumption of NADH.

The conditions for the various assays were as follows: NAD-(P)H : FMN oxidoreductase in 0.1 *M* potassium phosphate buffer, pH 7.0, containing 0.15 m*M* NADH, 0.2 m*M* FMN, and 1 m*M* dithiothreitol; 3α-hydroxysteroid dehydrogenase and 7α-hydroxysteroid dehydrogenase in 0.1 *M* potassium phosphate buffer, pH 9.0, containing 0.5 m*M* NAD and 1.5 m*M* cholic acid; 12α-hydroxysteroid dehydrogenase in 0.1 *M* potassium phosphate buffer, pH 8.0, containing 0.5 m*M* NADP and 1.5 m*M* cholic acid; alcohol dehydrogenase in 0.1 *M* potassium phosphate buffer, pH 9.0, containing 2 m*M* NAD and 0.1 *M* ethanol; glycerol dehydrogenase in 0.1 *M* potassium phosphate buffer, pH 9.0, containing 2 m*M* NAD and 0.1 *M* glycerol; and aldehyde dehydrogenase in 0.1 *M* potassium phosphate buffer, pH 9.0, containing 2 m*M* NAD and 0.2 *M* acetaldehyde. The activity of the immobilized enzymes was determinated by spectrophotometrically monitoring at 340 nm the eluate from the nylon tubes; 20–100 ml/hr flow rates and 25–100 cm tubes were used.[7] The composition of assay buffers was identical to that used with free enzymes.

Enzyme Immobilization

Nylon coils (1 cm diameter) were formed by heating tubes at 100° for 15 min. Nylon tubes (3–5 m) were then O-alkylated through triethylox-

[6] I. A. Macdonald, J. F. Jellet, and D. E. Mahony, *J. Lipid Res.* **20,** 234 (1979).
[7] A. Roda, S. Girotti, S. Ghini, and G. Carrea, this series, in press.

onium tetrafluoroborate[8] which was prepared as follows: 12 ml of 1-chloro-2,3-epoxypropane was slowly added to 150 ml of 15% (v/v) boron trifluoride in dry ether. The mixture was stirred under reflux for 1 hr and then the precipitated triethyloxonium tetrafluoroborate was washed three times with 100-ml aliquots of dry ether and finally dissolved in dry dichloromethane (final volume 200 ml). Within 24 hr, nylon tubes (3–5 m) were filled by suction with the triethyloxonium tetrafluoroborate solution and incubated at 25° for 10 min.

The O-alkylated tubes were washed with dichloromethane, filled immediately with a solution of 1,6-diaminohexane in methanol (10%, w/v), and incubated for 1 hr at 30°. After extensive washing with water the tubes were activated, within 48 hr, by perfusion with 5% (w/v) glutaraldehyde in 0.1 M borate buffer, pH 8.5, for 15 min at 20°. Thereafter, the tubes were washed with 0.1 M potassium phosphate buffer, pH 8.0, filled (1-meter portions) with solutions of enzyme in 0.1 M potassium phosphate buffer, pH 8.0, 0.2 mM dithiothreitol, and 0.5 mM NAD, and left overnight at 4°.

After removal of the enzyme solutions, the tubes were washed thoroughly with 0.1 M potassium phosphate buffer, pH 7.0, to remove protein that was not covalently linked. The proportion of enzyme immobilized was calculated by subtracting the unbound enzyme activity from the total added activity.

The immobilized enzymes were stored in 0.1 M potassium phosphate buffer, pH 6.9, 1% bovine serum albumin, 1 mM DTT, and 0.02% sodium azide, at 4°.[9,10]

Continuous-Flow Assay

Principles. The determination of NAD(P)H[9] is based on an enzymatic system consisting of an NAD(P)H:FMN oxidoreductase and a luciferase, which emits light in the presence of FMN, NAD(P)H, a long-chain aldehyde (RCHO), and molecular oxygen according to the following reactions:

$$\text{NAD(P)H} + \text{H}^+ + \text{FMN} \underset{\text{oxidoreductase}}{\rightleftharpoons} \text{NAD(P)}^+ + \text{FMNH}_2 \tag{1}$$

$$\text{FMNH}_2 + \text{RCHO} + \text{O}_2 \xrightarrow{\text{luciferase}} \text{FMN} + \text{RCOOH} + \text{H}_2\text{O} + \text{LIGHT} \tag{2}$$

[8] W. E. Hornby and L. Goldstein, this series, Vol. 44, p. 118.
[9] S. Girotti, A. Roda, S. Ghini, B. Grigolo, G. Carrea, and R. Bovara, *Anal. Lett.* **17,** 1 (1984).
[10] A. Roda, S. Girotti, S. Ghini, B. Grigolo, G. Carrea, and R. Bovara, *Clin. Chem. (Winston-Salem, N.C.)* **30,** 206 (1984).

The determination of bile acids,[10] ethanol, glycerol, and aldehyde is based on the coupling reactions (1) and (2) with the following NAD(P)H-generating reactions:

$$\text{Hydroxy bile acid} + \text{NAD(P)}^+ \overset{\text{HSD}}{\rightleftharpoons} \text{oxo bile acid} + \text{NAD(P)H} + \text{H}^+ \qquad (3)$$

where HSD is immobilized 3α-, 7α-, or 12α-hydroxysteroid dehydrogenase for the assay of 3α-, 7α-, or 12α-hydroxy bile acids, respectively.

$$\text{Ethanol} + \text{NAD}^+ \overset{\text{ADH}}{\rightleftharpoons} \text{NADH} + \text{H}^+ + \text{acetaldehyde} \qquad (4)$$

$$\text{Glycerol} + \text{NAD}^+ \overset{\text{GDH}}{\rightleftharpoons} \text{dihydroxyacetone} + \text{NADH} + \text{H}^+ \qquad (5)$$

$$\text{Aldehyde} + \text{NAD(P)}^+ + \text{H}_2\text{O} \overset{\text{AlDH}}{\longrightarrow} \text{acid} + \text{NAD(P)H} + \text{H}^+ \qquad (6)$$

where ADH is alcohol dehydrogenase, GDH is glycerol dehydrogenase, and AlDH is aldehyde dehydrogenase.

Apparatus. The manifold developed for bioluminescent continuous-flow assay is shown in Fig. 1. For the analysis by means of separately immobilized enzymes, the flow system involved four streams. The first supplied the bioluminescent enzymes with the working bioluminescent solution. The second and the third supplied the immobilized dehydrogenases (1 m coil), placed outside the luminometer, with 1 mM aqueous NAD solution and 0.02 M sodium pyrophosphate buffer, pH 9.0. The fourth was a continuous flow of air into which a known volume of sample was intermittently added. For NAD(P)H assay only two streams (working bioluminescent solution and sample air) were used (Fig. 1, dotted line). A multichannel peristaltic pump (Minipuls HP4, Gilson) and calibrated tubes of different diameters were used to produce different flow rates. The bioluminescent reactor—a 0.5–1 m coil of nylon tube containing coimmobilized luciferase, and NAD(P)H:FMN oxidoreductase—was wound around a Plexiglas support and positioned inside the luminometer in front of the photomultiplier window. Before reaching the reactor the stream passed through a stainless-steel coil (0.8 mm i.d.) which both mixed the stream and prevented a possible "optical fiber" light-diffusion effect[9]; a similar steel coil was also inserted after the reactor. The luminometer we used was the Model 1250 (LKB, Wallac), which required only slight modifications of the original light-recording system.

The sample (5–100 μl) was aspirated uniformly without fragmentation.[9,10] Operational steady state (stable background) was reached in about 5 min after a preliminary washing with 0.1 M potassium phosphate buffer, pH 6.9, containing 0.5 mM dithiothreitol.

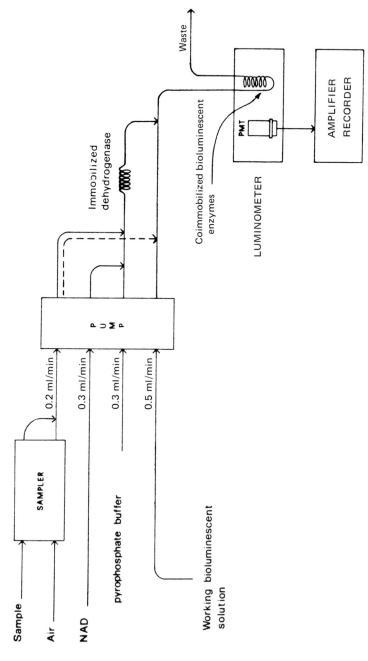

Fig. 1. Manifold for bioluminescent continuous-flow assay.

Solutions

NAD(P)H Assay. The working bioluminescent solution was 0.1 M potassium phosphate buffer, pH 7.0, containing 10 μM FMN, 27 μM decanal, and 0.5 mM dithiothreitol. The solution was prepared 20–30 min before analysis. Decanal was previously dissolved in isopropanol (0.05%, v/v) where it was stable for several weeks at 4°. The working bioluminescent solution showed no remarkable alteration after 8–10 hr at room temperature, in the dark. NAD(P)H standard solutions were 0.1–100 μM.

Bile Acids Assay. Bile acid standard solutions were 0.1–100 μM. The working bioluminescent solution was the same as for the NAD(P)H assay, but at pH 6.9. NAD solution (1 mM) in water was used for 3α- or 7α-hydroxy bile acids assay whereas NADP solution (1 mM) in water was used for 12α-hydroxy bile acids assay. Sodium pyrophosphate buffer (0.02 M), pH 9.0, was supplemented separately with a fourth stream (Fig. 1). Serum samples were heated at 70° for 15 min and diluted with 0.01 M sodium phosphate buffer, pH 7.4.

Ethanol Assay. Solutions were the same as for the bile acids assay. Blood samples were collected by venous puncture, before and after intake of alcoholic foodstuffs (mainly wine). The sera collected were stored at −20°. Before the analysis they were diluted 1 + 99 with 0.02 M sodium pyrophosphate buffer, pH 9.0, mixed by vortexing for 15 sec, and injected (10–25 μl) for the analysis. Saliva (1 ml) was collected before and 1–2 hr after intake of 15–33 ml of wine or beer and frozen at −20° just before analysis. Once thawed, it was mixed by vortexing and centrifuged at 1200 g for 10 min. The supernatant was carefully collected, diluted 1 + 99 with 0.02 M sodium pyrophosphate buffer, pH 9.0, and analyzed.

Glycerol Assay. Solutions were the same as for bile acids assay.

Aldehyde Assay. Solutions were the same as for bile acids assay.

Properties of Immobilized Enzymes

The activity and stability of immobilized enzymes are shown in Table I. Activity recoveries, that varied from 1.5 to 15% depending on the enzyme, were in the range of those found with other nylon tube-immobilized enzymes.[11] Variations of immobilization conditions including time of alkylation with triethyloxonium tetrafluoroborate (5–20 min), age of commercial glutaraldehyde (fresh or after 1 year storage at 4°), and use of bisimidate[8] (dimethyl pimelimidate) instead of glutaraldehyde scarcely influenced the activity recovery of dehydrogenases. Instead, the activity

[11] D. L. Morris, J. Campbell, and W. E. Hornby, *Biochem. J.* **147**, 593 (1975).

TABLE I
ACTIVITY AND STABILITY OF NYLON TUBE-IMMOBILIZED ENZYMES

Enzyme	Added enzyme (U/m nylon tube)	Immobilized enzyme (U/m nylon tube)	Activity recovery (%)	Stability[a] (half-life, days)
NAD(P)H:FMN oxidore-ductase	4.1[b]	0.06	1.5	20
7α-Hydroxysteroid dehydrogenase	8.4[b]	0.76	9.0	30
3α-Hydroxysteroid dehydrogenase	2.5[b]	0.38	15.2	30
12α-Hydroxysteroid dehydrogenase	2.2[b]	0.12	5.4	30
Alcohol dehydrogenase	120.0[c]	1.80	1.5	60
Glycerol dehydrogenase	14.0[d]	0.45	3.2	60
Aldehyde dehydrogenase	3.5[e]	0.32	9.1	30

[a] At room temperature 0.1 M potassium phosphate buffer, pH 7.0, 0.1 mM DTT, and 0.02% sodium azide.

[b] All of the enzyme present in the coupling solution (added enzyme) was immobilized onto the nylon tube.

[c] Seventy-five percent of the enzyme present in the coupling solution (added enzyme) was immobilized onto the nylon tube.

[d] Seventy-eight percent of the enzyme present in the coupling solution (added enzyme) was immobilized onto the nylon tube.

[e] Eighty-five percent of the enzyme present in the coupling solution (added enzyme) was immobilized onto the nylon tube.

recovery of NAD(P)H:FMN oxidoreductase was somewhat erratic and without a strict relationship to the immobilization conditions.

The K_m values for substrates and coenzymes of immobilized enzymes were higher than those of the free ones, except for NAD(P)H:FMN oxidoreductase where the values were similar (Table II). The increased K_m values should be due to diffusional limitations frequently present in immobilized enzyme systems.[12,13]

NAD(P)H Assay

The effect of several parameters, such as FMN and decanal concentration, flow rate, and sample volume on the performance of the luminescent reactor, have been optimized.[7] The best signal-to-noise ratio was obtained with 10 μM FMN and 27 μM decanal. The working bioluminescence-

[12] M. A. Mazid and K. J. Laidler, *Biochim. Biophys. Acta* **614**, 225 (1980).
[13] G. Carrea, R. Bovara, and P. Cremonesi, *Anal. Biochem.* **136**, 328 (1984).

TABLE II
MICHAELIS CONSTANTS OF FREE AND NYLON TUBE-IMMOBILIZED ENZYMES

Enzyme	Substrate or coenzyme	K_m of free enzyme[a] (M)	$K_{m,app}$ of immobilized enzyme[a] (M)
NAD(P)H : FMN oxidoreductase	FMN	4.2×10^{-5}	3.9×10^{-5}
	NADH	2.0×10^{-4}	4.0×10^{-4}
7α-Hydroxysteroid dehydrogenase	Cholic acid	3.2×10^{-4}	7.1×10^{-4}
	NAD	1.0×10^{-3}	1.4×10^{-3}
3α-Hydroxysteroid dehydrogenase	Cholic acid	2.5×10^{-6}	1.8×10^{-4}
	NAD	1.1×10^{-4}	3.7×10^{-4}
12α-Hydroxysteroid dehydrogenase	Cholic acid	1.1×10^{-4}	3.2×10^{-4}
	NADP	2.5×10^{-5}	1.0×10^{-4}
Alcohol dehydrogenase	Ethanol	1.3×10^{-2}	1.1×10^{-2}
	NAD	7.4×10^{-5}	1.0×10^{-3}
Glycerol dehydrogenase	Glycerol	3.9×10^{-2}	2.4×10^{-2}
	NAD	15.0×10^{-4}	7.5×10^{-4}
Aldehyde dehydrogenase	Acetaldehyde	9.0×10^{-6}	1.2×10^{-4}
	NAD	3.0×10^{-5}	5.1×10^{-4}

[a] The Michaelis constants were obtained from Lineweaver–Burk plots.

cent solution prepared by previously dissolving decanal in isopropanol was more stable and gave a better signal-to-noise ratio than those prepared using methanol or a suspension of decanal in water. The NAD(P)H assay was independent of flow rate in the range 0.36–1.50 ml/hr. Also sample volume and coil length did not influence the response provided the total volume resulting from the sample plus the working bioluminescent solution was smaller than the reactor volume.

The sensitivity of the NADH assay was very high since as little as 1 pmol of standard was detected (signal-to-noise ratio 3 : 1); the assay was linear from 1 to 2000 pmol. The NADPH assay was about three times less sensitive due to the lower activity of NAD(P)H : FMN oxidoreductase toward NADPH. Up to 30 samples per hour were analyzed. Washing between samples was not strictly necessary since in its absence no appreciable carryover was observed.

Two samples with low (2.5 pmol) and high (250 pmol) NADH levels were assayed to determine intra- and interassay variations. The relative standard deviation values were lower than 9 or 5% at low or high levels, respectively.

The residual activity of a reactor used for 2 months, analyzing more than 50 samples per day, was about 20%.

Bile Acids Assay

With separately immobilized 3α- or 7α-hydroxysteroid dehydrogenase the sensitivity was of 1 pmol of bile acid, because the bile acid oxidation was carried out at pH 9.0 where the transformation of substrate was pratically complete. High flow rates (0.5 ml/hr) decreased the assay sensitivity by decreasing bile acid transformation. Potential enzyme contamination with aldehyde dehydrogenase,[14] that would increase noise values in the presence of an aldehyde, could not affect the assay based on separately immobilized enzymes since decanal was not present in the buffer feeding immobilized hydroxysteroid dehydrogenase.

Serum samples (30) analyzed by the bioluminescent method gave results in good agreement with those obtained by radioimmunoassay, enzyme immunoassay, and high-performance liquid chromatography.[10] Two serum samples with low (2.5 μM) and high (22 μM) concentrations of bile acids were assayed to determine intra- and interassay variations. The relative standard deviation values were lower than 8 or 10% at low or high concentration, respectively. Up to 25 samples per hour were analyzed with no carryover.

The assay of 12α-hydroxy bile acids by means of NADP-dependent 12α-hydroxysteroid dehydrogenase was less sensitive due to the lower activity of NAD(P)H : FMN oxidoreductase toward NADPH, but gave similar results.

Ethanol Assay

The detection limit of the assay (mean ± 3 SD) was 1 μmol ethanol per liter of sample. The sensitivity was of 50 pmol, but such a low value normally was not necessary because ethanol content in saliva and serum samples was higher than 250 pmol/injection. With buffered ethanol solutions the response was linear between 50 and 2500 pmol and a similar linear range (250–2500 pmol) was achieved by serial dilution of saliva (1.38 g/liter of ethanol) and serum (0.87 g/liter of ethanol). The relative standard deviation values of intra- and interassay reproducibility study were always less than 10% both at high and low concentrations of ethanol in serum and saliva. No systematic difference was detected between the biological media and buffered ethanol solutions, and analytical recovery was between 94 and 103%.

Typical time courses of light peaks from the continuous-flow system when serum samples were injected are shown in Fig. 2. Due to the good sensitivity of the bioluminescent system it was possible to measure the

[14] J. Ford and M. DeLuca, *Anal Biochem.* **110,** 43 (1980).

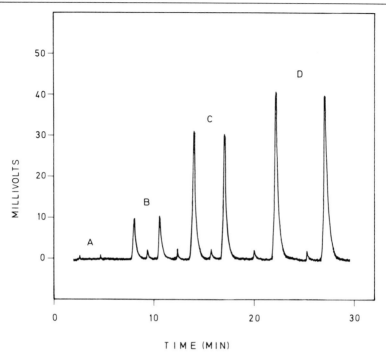

FiG. 2. Representative traces obtained by the bioluminescent continuous-flow assay of ethanol. (A) Serum sample before ethanol intake, (B) serum sample after ethanol intake, (C) standard ethanol (150 pmol), (D) serum sample (B) plus standard ethanol (C).

alcoholic content in saliva even 2 hr after injection. The alcohol concentration determined in saliva and serum (0.08–3.3 g/liter) was in good agreement with the previous observed values.[15] About 25–30 measurements could be carried out per hour. Washing between samples is advisable to avoid carryover.

Glycerol and Aldehyde Assay

With acetaldehyde standards the response was linear between 50 and 1000 pmol, and with glycerol standards between 50 and 500 pmol. In both cases satisfactory reproducibility (10%) was obtained.

Conclusion

Nylon tube-immobilized bioluminescent enzymes made possible specific assays of NAD(P)H, bile acids, ethanol, glycerol, and aldehyde at

[15] H. O. Beutler, *in* "Methods of Enzymatic Analysis" (H. U. Bergmeyer, ed.), 3rd ed., Vol. 6, p. 598. Verlag Chemie, Weinheim, 1984.

picomole levels. The precision of the method as well as its correlation with other methods such as radioimmunoassay, enzyme immunoassay, and high-performance liquid chromatography was satisfactory.[9,10]

The adopted continuous-flow system was simple, required only minor modifications of a commercial detector, and allowed analysis of about 20–30 samples per hour. Unlike a Sepharose column,[4,5] nylon reactors presented no problems with packing or disruption of gel matrix or bacterial growth which markedly enhances background light level.[5] This, together with its handiness, makes nylon tubes a very suitable enzyme support for continuous-flow analysis, in spite of the relatively low activity recovery of immobilized enzymes, which, however, is sufficient to achieve high sensitivity. Up to 700–900 samples were analyzed with use of only a few milligrams of enzymes, and therefore this bioluminescent method appears highly competitive with other methods such as radioimmunoassay, enzyme immunoassay, high-performance liquid chromatography, and fluorometry, where radioactive materials, separation steps, sample manipulation, and expensive equipment are needed.

The feasibility of continuous-flow bioluminescent systems has been recently verified with firefly luciferase, which was immobilized on nylon coils. This allowed the determination of ATP and ADP at picomole levels in human platelets in a simple and reliable way.[16]

[16] G. Carrea, R. Bovara, G. Massola, S. Girotti, A. Roda, and S. Ghini, *Anal. Chem.* **58**, 331 (1986).

[22] Active Center-Based Immunoassay Approach Using Bacterial Luciferase

By THOMAS O. BALDWIN, THOMAS F. HOLZMAN, and RITA B. HOLZMAN

Overview

Since the report in 1960 by Yalow and Berson[1] of the first radioimmunoassay (RIA) for insulin, similar assays have been reported for several hundred compounds of vastly different biological activity and chemical structure. RIA has been the principal assay strategy of clinical laboratories for many years, and the reasons are clear. The use of antibodies lends great specificity to the assay, and the use of radioactive isotopes of iodine as labels for either antigen or antibody makes the method very sensitive.

[1] R. S. Yalow and S. A. Berson, *J. Clin. Invest.* **39**, 1157 (1960).

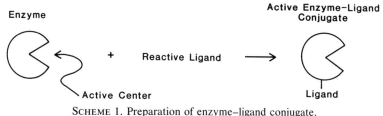

SCHEME 1. Preparation of enzyme–ligand conjugate.

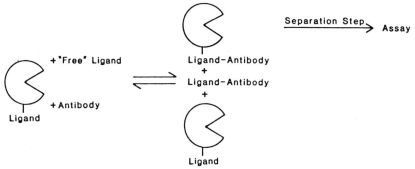

SCHEME 2. Use of enzyme–ligand conjugate in immunoassay.

The technique has been automated in many cases, and its popularity is not difficult to understand.

In recent years, however, growing concern over use of radioisotopes and the associated questions of disposal has led to interest in and increasing use of enzyme immunoassay methods. The heterogeneous enzyme immunoassays, which include the popular enzyme-linked immunosorbent assay (ELISA), are based on the same principles as RIA.[2,3] The basic difference between the ELISA and RIA procedures is the replacement of the radioisotope tag used in RIA with an enzyme whose activity can be assayed using standard methods. The ligand (antigen or antibody) of interest is attached to an enzyme (Scheme 1), whose activity permits quantitation of the amount of ligand bound by anti-ligand antibody in an immunoassay procedure (Scheme 2). The principle behind both the RIA and ELISA methods is the competitive binding of the enzyme-labeled antigen and unlabeled antigen with a specific antibody (Scheme 3). In the case of ELISA, the label is quantitated by enzyme activity assay while in RIA, the label is quantitated by measurement of radioactivity.

[2] E. Engvall and P. Perlmann, *Immunochemistry* **8**, 871 (1971).
[3] E. Engvall, K. Jonsson, and P. Perlmann, *Biochim. Biophys. Acta* **251**, 427 (1971).

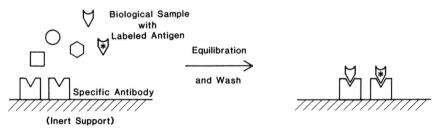

SCHEME 3. Competitive binding assay.

Although the ELISA methods are specific and simple, they suffer from several disadvantages. First, the conjugate between the enzyme and the antibody or antigen must be formed without serious reduction in enzyme activity or stability. Second, the conjugate in a direct competitive binding assay must be exposed to biological samples which quite often contain proteolytic enzymes or inhibitors of enzyme activity. This latter difficulty can generally be circumvented by noncompetitive ELISA techniques in which the incubation with test samples is carried out separately from the incubation with the enzyme-labeled antigen or antibody (Scheme 4).

Principle of the Active Center-Based Enzyme Immunoassay

Active center-based enzyme immunoassay is a new approach to immunoassay in general and ELISA in particular. During 1981 at Texas A&M University, while working toward development of a standard ELISA method using bacterial luciferase (an $\alpha\beta$ heterodimer; see Ziegler

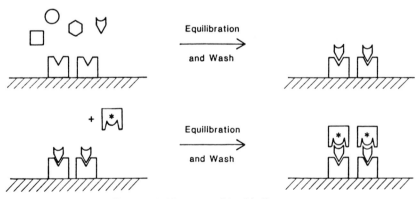

SCHEME 4. Noncompetitive binding assay.

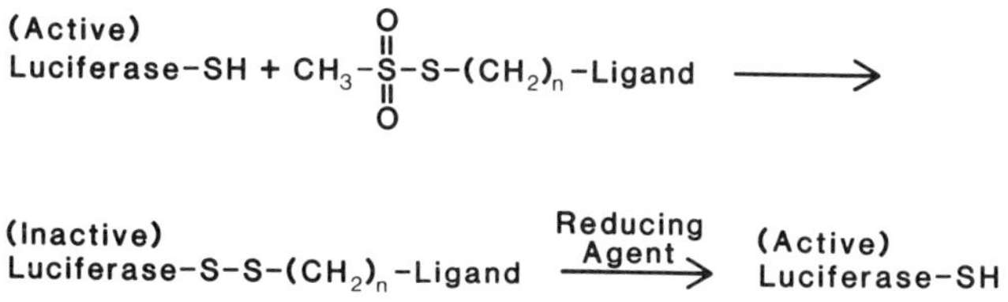

(Active)   O
                                  ‖
Luciferase-SH + CH$_3$-S-S-(CH$_2$)$_n$-Ligand   ⟶
                                  ‖
                                  O

(Inactive)   Reducing   (Active)
                                    Agent
Luciferase-S-S-(CH$_2$)$_n$-Ligand   ⟶   Luciferase-SH

SCHEME 5. Method of inactivating the luciferase.

and Baldwin[4] for a review of the system) as the reporter enzyme, we encountered the difficulties that have faced other investigators who have attempted to develop such assay methods.[5] The enzyme is rapidly inactivated by derivatization of either a highly reactive thiol in the active center or the α-amino group of either chain.[6,7] Without very careful attention to details of the coupling reaction to protect the residues in the active center, the enzyme was inactivated by the coupling chemistry and therefore of no use to the ELISA technology. We decided that instead of dealing with such difficult chemistry, we would try a new approach, inactivating the luciferase with a reagent that can be removed at a later time, thereby regaining full enzymatic activity. We were assisted by scientists at the Upjohn Company who synthesized several very useful methyl alkanethiolsulfonates (Table I). The method was presented in detail at the 1983 meeting of the American Chemical Society.[8] From a conceptual standpoint, the method is quite simple (Scheme 5), and the principle is based on a large body of literature on the reversible inactivation of enzymes by chemical modification (see ref. 9 for background).

In distinct contrast to ELISA and certain EMIT methods, where it is essential in the production of the conjugate to retain enzyme activity, in the active center-based method the reporter enzyme is purposely inactivated by allowing it to react with a chemical derivative of the antigen or antibody that carries a reactive group (Scheme 6). The derivative of the antigen or antibody is designed so that it can be removed from the reporter enzyme whenever desired, thereby releasing the active form of the enzyme into solution for quantitation in the final step of the immunoassay

---

[4] M. M. Ziegler and T. O. Baldwin, *Curr. Top. Bioenerg.* **12**, 65 (1981).

[5] J. Wannlund and M. DeLuca, in "Bioluminescence and Chemiluminescence" (M. DeLuca and W. D. McElroy, eds.), p. 693. Academic Press, New York, 1981.

[6] M. Ziegler-Nicoli, E. A. Meighen, and J. W. Hastings, *J. Biol. Chem.* **249**, 2385 (1974).

[7] W. R. Welches and T. O. Baldwin, *Biochemistry* **20**, 512 (1981).

[8] T. O. Baldwin, *Abstr. Pap., 188th Natl. Meet., Am. Chem. Soc.*, MBTD 0056 (1984).

[9] G. L. Kenyon and T. W. Bruice, this series, Vol. 47, p. 407.

TABLE I
The Thiolsulfonates: Structure and Nomenclature

C_2 Acid:
$$CH_3-\overset{\overset{O}{\|}}{\underset{\underset{O}{\|}}{S}}-S-CH_2-CH_2-COOH$$

S-(methanesulfonyl)-3-thiopropanoic acid

C_4 Acid:
$$CH_3-\overset{\overset{O}{\|}}{\underset{\underset{O}{\|}}{S}}-S-CH_2-CH_2-CH_2-CH_2-COOH$$

S-(methanesulfonyl)-5-thiopentanoic acid

C_5 Acid:
$$CH_3-\overset{\overset{O}{\|}}{\underset{\underset{O}{\|}}{S}}-S-CH_2-CH_2-CH_2CH_2-CH_2-COOH$$

S-(methanesulfonyl)-6-thiohexanoic acid

C_2 Amine:
$$CH_3-\overset{\overset{O}{\|}}{\underset{\underset{O}{\|}}{S}}-S-CH_2-CH_2-NH_2 \cdot HBr$$

S-(methanesulfonyl)-2-thioethylamine HBr

C_3 Amine:
$$CH_3-\overset{\overset{O}{\|}}{\underset{\underset{O}{\|}}{S}}-S-CH_2CH_2-CH_2-NH_2 \cdot HBr$$

S-(methanesulfonyl)-3-thiopropylamine HBr

C_5 Amine:
$$CH_3-\overset{\overset{O}{\|}}{\underset{\underset{O}{\|}}{S}}-S-CH_2-CH_2-CH_2-CH_2-CH_2-NH_2 \cdot HBr$$

S-(methanesulfonyl)-5-thiopentylamine HBr

Scheme 6. Preparation of active center-based enzyme–ligand conjugates.

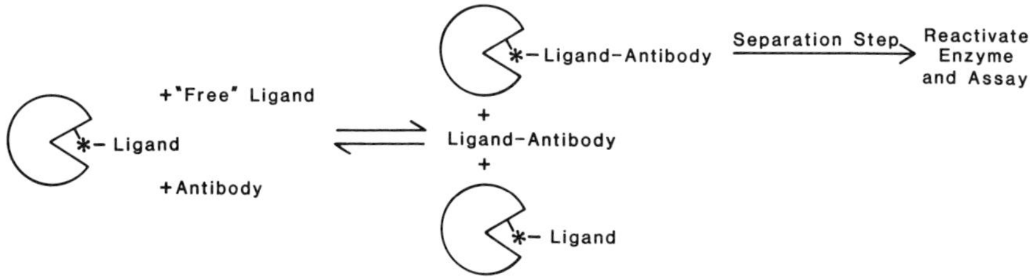

SCHEME 7. Use of active center-based enzyme–ligand conjugate in immunoassay.

(Scheme 7). The basis of this approach is the reversible modification of a single amino acid residue in or near the active center of the enzyme. For a given enzyme, the site of covalent modification is selected to be (1) unique, so that each enzyme molecule will be modified only once, and (2) easily accessible to a variety of coupling reagents. While numerous reagents and enzymes have been described that could potentially fulfill these requirements, we have focused on the enzyme bacterial luciferase and the thiol-directed reagents, the alkyl alkanethiolsulfonates[9] (see Scheme 5). Alkyl alkanethiolsulfonates inactivate luciferase by formation of a mixed disulfide with a reactive "essential" cysteinyl residue[10,11] (Cys-106 of the α subunit[6,12]), and the modified enzyme can be readily reactivated by addition of reducing agents. Alkyl alkanethiolsulfonates derivatized and conjugated with another ligand (antigen or antibody) permit reversible attachment of the ligand to luciferase.

Synthesis of the Alkyl Methanethiolsulfonate Reagents. The alkyl methanethiolsulfonates used in this study were synthesized by a modification of the procedures reported by Kenyon and Bruice.[9] All compounds were synthesized from a common precursor, potassium methanethiolsulfonate, by a simple S_{N^2} displacement reaction.

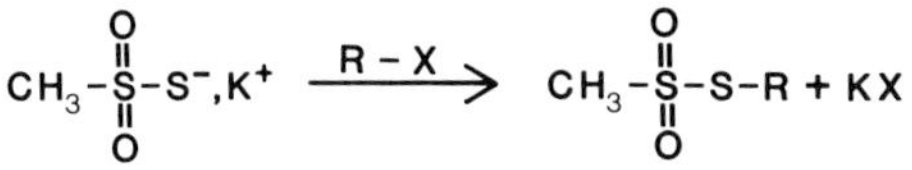

$$CH_3-\underset{\underset{O}{\|}}{\overset{\overset{O}{\|}}{S}}-S^-,K^+ \xrightarrow{R-X} CH_3-\underset{\underset{O}{\|}}{\overset{\overset{O}{\|}}{S}}-S-R + KX$$

Potassium Methanethiolsulfonate: Note of Caution. Many of the compounds used are either toxic or potent lachrymators or both, so all procedures should be carried out in a well-ventilated fume hood.

[10] M. M. Ziegler and T. O. Baldwin, *in* "Bioluminescence and Chemiluminescence: Basic Chemistry and Analytical Applications" (M. DeLuca and W. McElroy, eds.), p. 155. Academic Press, New York, 1981.

[11] S. K. Rausch, Ph.D. Thesis, Department of Biochemistry, University of Illinois, Urbana (1983).

[12] D. H. Cohn, A. J. Mileham, M. I. Simon, K. H. Nealson, S. K. Rausch, D. Bonam, and T. O. Baldwin, *J. Biol. Chem.* **260,** 6139 (1985).

Hydrogen sulfide gas was passed through a stirred solution of 10 g of KOH in 80 ml H_2O, cooled in an ice bath, until the solution was saturated with H_2S. While the solution was on ice, 5.8 ml of methanesulfonyl chloride (mesyl chloride) was added by syringe pump over a 1-hr period. Stirring was continued for another hour and the mixture was then filtered and evaporated to dryness under reduced pressure. The resultant residue was suspended in 25 ml dimethyl formamide and the mixture was warmed to 60° with stirring for 45 min under N_2. The mixture was then filtered, washed with DMF, dried under reduced pressure, and subsequently recrystallized from isopropanol to give the compound $CH_3-SO_2-S^-$, K^+. It is very important that the compound be stored dry; it is highly hygroscopic.

The potassium methanethiolsulfonate can be used to synthesize a wide variety of compounds. For the sake of illustration, we will discuss here the synthesis of the compound S-(methanesulfonyl)-6-thiohexanoic acid that we have used in the preparation of conjugates with insulin.

S-(Methanesulfonyl)-6-thiohexanoic Acid (C_5 Acid Thiolsulfonate). A mixture of 1 g of 6-bromohexanoic acid, 1 g of $CH_3-SO_2-S^-$, K^+, and 10 ml dimethyl formamide was stirred under N_2 at 60° for 3 hr. The mixture was then filtered through Celite, washed with DMF, and evaporated to an oil. The oil was combined with 7 ml of 1 : 1 acetonitrile–ether and stirred for 1 hr. The resulting precipitate was washed with ether, dried, dissolved in hot acetonitrile, filtered, and cooled. Upon cooling a precipitate formed which was recrystallized from methanol : ethyl acetate (1 : 1) to give the desired compound (C_5 acid thiolsulfonate).

N-Hydroxysuccinimide Ester of the S-(Methanesulfonyl)-6-thiohexanoic Acid. The N-hydroxysuccinimide esters are commonly used derivatives for coupling reactions. They have the advantage of ease of synthesis and stability, and can be used to derivatize virtually any compound containing a primary amine. The N-hydroxysuccinimide ester was prepared by carbodiimide coupling. A solution containing 67.8 mg C_5 acid thiolsulfonate, 34.5 mg N-hydroxysuccinimide, and 77.4 mg dicyclohexylcarbodiimide in 3 ml tetrahydrofuran was incubated overnight at 0°. The N-hydroxysuccinimide and the C_5 acid thiolsulfonate were both at 0.1 M concentration and the carbodiimide was 0.125 M. The reaction is shown below.

N–hydroxysuccinimide C_5 acid thiolsulfonate   NHS–C_5–thiolsulfonate

Preparation of C_5 Acid Thiolsulfonate-Derivatized Porcine Insulin.
Porcine insulin was dialyzed exhaustively against 50 mM phosphate, 10
mM EDTA, pH 7.0, at 4°. The protein (1 ml of 2 mg/ml solution) was then
mixed with 50 μl of the 0.1 M C_5 acid thiolsulfonate N-hydroxysuccinimide ester. The reaction was allowed to proceed at 4° with stirring for
24 hr. Excess reagent was removed by dialysis of the mixture against 50
mM phosphate, 10 mM EDTA, pH 7.0, at 4°, using 3000 molecular weight
cutoff dialysis tubing. After exhaustive dialysis, the concentration of the
insulin–C_5 thiolsulfonate conjugate was determined spectrophotometrically at 280 nm using an extinction coefficient of 1.04 (0.1%).

Preparation of Insulin–Luciferase Conjugate. Luciferase from *Vibrio
harveyi* was dialyzed exhaustively against 50 mM phosphate, 10 mM
EDTA, pH 7.0, to remove DTE used for enzyme storage. Typically 0.3
mg (3.9 nmol) luciferase was allowed to react with a 5-fold molar excess of
insulin—C_5 thiolsulfonate at room temperature until 5% of the initial luciferase activity remained, about 8 min. The reaction was terminated by
addition of bovine serum albumin to a concentration of 1%. The mixture
was immediately dialyzed against 50 mM phosphate, 10 mM EDTA, pH
7.0, at 4°, using 12,000–14,000 molecular weight cutoff dialysis tubing to
remove excess insulin–C_5 thiolsulfonate.

The inactive luciferase in the insulin-luciferase conjugate was readily
reactivated by addition of reducing agents such as 2-mercaptoethanol.

Synthesis of Estriol–Luciferase Conjugate. S-(Methanesulfonyl)-5-thiopentylamine (C_5 amine thiolsulfonate) was synthesized as described for the C_5 acid thiolsulfonate (above), but starting with 5-bromopentyl-amine HBr. Estriol-6-CMO (1,3,5-estratrien-3,16α,17β-triol-6-carboxy-methyl oxime; Steraloids, Inc.) dissolved in tetrahydrofuran (5 mg/0.2 ml) was mixed with 2.5 mg of carbonyldiimidazole and incubated for 30 min at 25°. Then 3.9 mg of C_5 amine thiolsulfonate was added and the pH carefully adjusted to ~8.0 with 1.0 M NaOH. The reaction was allowed to proceed for 18–19 hr at 25°. The product was isolated by thin-layer chromatography on silica gel G in 7:3 chloroform:methanol (R_f 0.76). The product was identified by NMR and quantitated by UV spectroscopy ($\varepsilon_{260\,nm}$ 1 × 10^4 M^{-1} cm^{-1}). The proposed structure of the thiolsulfonate derivative is presented below.

Bacterial luciferase was labeled with the estriol–thiolsulfonate derivative by allowing the two molecules to react in phosphate-buffered saline at 0°. The luciferase (1.7 μM) had been dialyzed exhaustively to remove all traces of reducing agent. The reaction was initiated by addition of the estriol–thiolsulfonate derivative (at ~11 μM); the reaction was terminated after ~2.5 hr by chromatography on BioGel P6, which resolved the derivatized enzyme from excess reactant and reaction product. The labeled luciferase was stored on ice prior to use.

Synthesis of Progesterone–Luciferase Conjugate. The progesterone–thiolsulfonate derivative was synthesized by carbodiimide coupling of the 11α-hemisuccinate with the C_5 amine thiolsulfonate. The reaction product was purified by reversed-phase HPLC. Coupling of the progesterone to the luciferase was accomplished by allowing luciferase (at 1.57 μM) to react with the progesterone thiolsulfonate (at 1.88 μM) in 0.05 M phosphate, pH 7.0, at 25°. The derivatized luciferase was resolved from excess reagent and reaction products by chromatography on BioGel P6.

C$_5$ Amine thiolsulfonate

Progesterone 11α-Hemisuccinate

Operational Considerations of the Active Center-Based Immunoassay Method

In order to use this method on a routine basis, we performed a series of background experiments to determine rates of reaction, stability of conjugates, and optimal conditions for reactivation of the luciferase. The first series of experiments was to determine the rates of reaction of the various parent thiolsulfonate compounds with the luciferase and the optimal conditions for reactivation of the derivatized enzyme. The second series of experiments was to determine the rates of reaction with the thiolsulfonate conjugates described above and the conditions required for reactivation. The third series of experiments was to demonstrate the validity of this strategy by using the derivatized luciferase in a competitive immunoassay format.

Chemical Modification of Bacterial Luciferase with the Methyl Alkanethiolsulfonates. The apparent second-order rate constants for inactivation of luciferase with the various linker compounds (see Table I) were determined in 50 mM phosphate, pH 7.0, at 22° both under pseudo-first-order conditions of reagent excess and under second-order conditions of equimolar luciferase and thiolsulfonate (see Table II). The apparent second-order rate constants determined by the two methods were the same within experimental error. It should be noted that luciferase inactivated under pseudo-first-order conditions in which the thiolsulfonate concentration was at least 10-fold higher than the luciferase concentration could not be reactivated by the addition of 2-mercaptoethanol. Reactivation experiments were always done under pseudo-first-order conditions

TABLE II
RATES OF INACTIVATION OF LUCIFERASE FROM *Vibrio harveyi* BY THE
METHYL ALKANETHIOLSULFONATES AND OF REACTIVATION OF THE
S-ALKYL LUCIFERASES[a]

Inactivation of luciferase at 25°, pH 7.0, $k_{2(obs)}$ (M^{-1} min^{-1})			
Chain length	Acids[b]	Acids[c]	Amines
C_2	3.75×10^4	1.03×10^4	5.22×10^5
C_3			9.34×10^6
C_4	2.77×10^5	3.00×10^4	
C_5	1.99×10^5	1.50×10^4	4.73×10^5

Reactivation of S-alkyl luciferases at 25°, pH 7.0, by 0.1 M 2-mercaptoethanol			
	Acids		Amines	
Chain length	k_1 (min^{-1})	$t_{1/2}$ (min)	k_1 (min^{-1})	$t_{1/2}$ (min)
C_2	4.20×10^{-2}	16.5	2.76	0.25
C_3			3.78×10^{-2}	18.3
C_4	0.60	1.2		
C_5	0.41	1.7	0.25	2.8

[a] The apparent second-order rate constants for inactivations were determined as described in the text. All reactions were at 25° at pH 7.0. With the exception of the single set of experiments (see footnote c), all reactions were in 50 mM phosphate.

[b] Reactions were in 50 mM phosphate, as described. The effect of phosphate on the rate of reaction of the acid thiolsulfonates with the reactive thiol of luciferase is apparent from a comparison of these data with those obtained as described in footnote c.

[c] Reactions were in 20 mM Bis–Tris, pH 7.0, 25°. All other conditions of the experiments were identical with those of footnote b.

with a vast excess of 2-mercaptoethanol (~0.1 M). The pseudo-first-order constants for reactivation of the S-alkyl luciferases are presented in Table II.

The rates of inactivation measured in 50 mM phosphate were exceedingly fast compared to the rate of inactivation of bacterial luciferase by N-ethylmaleimide under comparable conditions,[6] which showed an apparent second order rate constant of 1.67×10^3 M^{-1} min^{-1}. All were in the range of the rate observed for the parent thiolsulfonate, methylmethanethiolsulfonate (MMTS), 3.54×10^5 M^{-1} min^{-1}.[11] The rates of reactivation of the luciferase derivatized with the various thiolsulfonates were also quite fast, with the exception of the C_2 acid and the C_3 amine. We have no explanation at this time for the two slower reactivations.

The apparent second-order rate constants for inactivation of luciferase were also determined with the C_2, C_4, and C_5 acid thiolsulfonate compounds in 20 mM Bis–Tris, pH 7.0, 25° under conditions of reagent excess (Table II). From these data, it is clear that the rate of inactivation of the luciferase by the acid thiolsulfonates is significantly faster (2- to 10-fold, depending on chain length) in phosphate-containing buffers than in 20 mM Bis–Tris. With the C_5 acid thiolsulfonate, the rate of inactivation in 50 mM phosphate was more than 10 times the rate in Bis–Tris, but the rate of reactivation was the same in both buffers. It is presumed at this time that the phosphate dependence of the inactivation reaction is the result of the structural changes that occur in the enzyme in response to anion binding.[13,14]

Derivatization of Luciferase with the C_5 Amine Thiolsulfonate and with the Progesterone–Thiolsulfonate Conjugate. Samples of luciferase (1.57 μM) were incubated with either the C_5 amine thiolsulfonate or the C_5 amine progesterone–thiolsulfonate (at 1.88 μM) in 50 mM phosphate, pH 7.0, 25°. The inactivation of each enzyme sample was monitored as a function of time, and after the majority of enzyme in each sample had been inactivated, reactivation was initiated by addition of 2-mercaptoethanol (75 mM; Fig. 1). The inactivation of the luciferase by the C_5 amine progesterone–thiolsulfonate was biphasic, with about 80% of the conjugate reacting rapidly (second-order rate constant ~4.5 × 10^6 M^{-1} min^{-1}), and the remainder reacting with a second-order rate constant similar to that of the unmodified C_5 amine thiolsulfonate. The rapid rate of reaction of the C_5 amine progesterone–thiolsulfonate was not expected, but almost certainly is a reflection of the hydrophobic character of the environment of the reactive thiol of the luciferase.[15,16] The reactivation of the luciferase derivatized with the two reagents showed very little difference in rate, suggesting that there is very little difference in accessibility of the disulfide linkage in the two compounds (see Fig. 1).

The stability of the C_5 amine progesterone–luciferase conjugate was analyzed by incubating a sample of the conjugate prepared as described above for 48 hr at 4°. The sample was then warmed to 25° and 2-mercaptoethanol added to a final concentration of 0.1 M. Both the rate and extent of reactivation of the luciferase were indistinguishable from the sample that had been reactivated immediately following inactivation.

[13] T. O. Baldwin and P. L. Riley, *in* "Flavins and Flavoproteins" (K. Yagi and T. Yamano, eds.), p. 139. Japan Scientific Societies Press, Tokyo, 1980.

[14] T. F. Holzman and T. O. Baldwin, *Biochem. Biophys. Res. Commun.* **94,** 1199 (1980).

[15] M. Ziegler-Nicoli and J. W. Hastings, *J. Biol. Chem.* **249,** 2393 (1974).

[16] M. V. Merritt and T. O. Baldwin, *Arch. Biochem. Biophys.* **202,** 499 (1980).

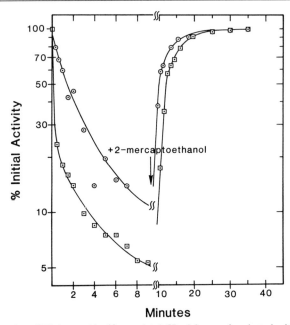

Fig. 1. Samples of *V. harveyi* luciferase (at 1.57 μM) were incubated with either the C_5 amine thiolsulfonate ($\odot$, 1.88 μM) or with C_5 amine progesterone ($\square$, 1.88 μM). Aliquots were removed from the reactions and diluted at the indicated times into luciferase assay buffer and the remaining luciferase activity determined by the reduced flavin injection method. At the time indicated by the arrow in the figure, 2-mercaptoethanol (75 mM final concentration) was added and the reactivation of the luciferase monitored by the standard luciferase assay.

The stability of the C_5 amine progesterone–luciferase conjugate to lyophilization was tested in a similar fashion. A sample of luciferase that had been inactivated as described above was diluted 1 : 2 with 0.2 M NaCl, 0.2 M phosphate, pH 7.0, frozen, and lyophilized. The dried sample was then redissolved in the original volume (before dilution) of water containing 0.1 M 2-mercaptoethanol. The rate of reactivation of the enzyme sample was about 2-fold slower than for the sample that was reactivated immediately following inactivation, but 100% of the original activity was regained, indicating that the material is fully stable to lyophilization.

Properties of the Insulin–C_5 Thiolsulfonate Derivative and of the Insulin–Luciferase Conjugate. The *N*-hydroxysuccinimide–C_5 thiolsulfonate was stable for at least 6 months stored at $-20°$ in tetrahydrofuran. The insulin–C_5 thiolsulfonate was unstable to prolonged freezing in water at $-20°$, and the insulin–luciferase conjugate stored at $-20°$ for several months could not be reactivated by 2-mercaptoethanol. The insulin–C_5

thiolsulfonate was found to be stable for at least 2 months when stored frozen in 10 mM EDTA, 1% bovine serum albumin. No conditions have been found to stabilize the insulin–luciferase conjugate to freezing, but the native enzyme and the insulin–C$_5$ thiolsulfonate are both stable to freezing. We have therefore adopted the practice of preparing fresh insulin–luciferase conjugate on a daily basis.

The Assay for Insulin. The insulin–luciferase conjugate was prepared as described above. Insulin–luciferase was used as reporter enzyme conjugate in an assay for insulin. Porcine insulin standards were prepared at 0.065, 0.131, 0.260, 0.520, 1.05, and 2.09 nM concentrations in immunoassay buffer (50 mM phosphate, 10 mM EDTA, pH 7.0, 1% BSA). Antibody titration curves determined that 1 : 2500 working titer (1 : 10,000 final titer) of guinea pig anti-porcine insulin was the optimal concentration, yielding 35–40% bound insulin–luciferase at a concentration of 0.5 nM. Addition of normal guinea pig serum to secondary antibody (goat anti-guinea pig IgG) resulted in nonspecific binding levels of 5–10% of the total insulin–luciferase added to the assay. The guinea pig anti-porcine insulin antibody, the guinea pig normal serum, and the goat anti-guinea pig IgG were obtained from Cambridge Medical Diagnostics, Inc.

The greatest sensitivity in the insulin standard curve was obtained by preincubation of primary antibody with the insulin standards for a period of 30 min at room temperature. The insulin–luciferase was then added and incubation continued for an additional 60 min at room temperature. The final incubation mixture (200 μl) contained the primary antibody, the insulin standards, and the insulin–luciferase (0.5 nM). The secondary antibody was then added as a 200 μl solution prepared from 50 μl goat anti-guinea pig IgG (stock concentration of 3.3 mg/ml), 10 μl undiluted normal guinea pig serum, and 140 μl of 7.3% PEG-8000. Precipitation was allowed to proceed for an additional 30 min. The samples were then centrifuged in a microfuge at 4° for 10 min. The supernatant was removed from each tube and 2-mercaptoethanol added to a final concentration of 60 mM. The reactivation reaction was allowed to proceed at room temperature for 60 min. The solutions were then placed on ice and the luciferase activity determined by the flavin injection assay.[17] The standard curve shown in Fig. 2 was determined with two separate preparations of insulin–luciferase. The flavin injection assays were performed in triplicate.

The Assay for Estriol. The estriol–luciferase conjugate was prepared as described above and stored on ice prior to use. The primary antibody incubations were performed in solutions containing 40 fmol of estriol–luciferase conjugate, rabbit anti-estriol antiserum (1 : 96,000 final titer),

---

[17] J. W. Hastings, T. O. Baldwin, and M. Ziegler-Nicoli, this series, Vol. 57, p. 135.

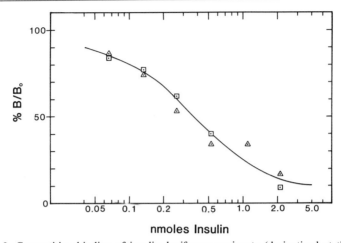

nmoles Insulin

FIG. 2. Competitive binding of insulin–luciferase conjugate (derivatized at the active center thiol) to antiporcine insulin antibody. The guinea pig antiporcine insulin antibody was incubated with free insulin standards prior to addition of insulin–luciferase conjugate. The ratio of the amount of luciferase bound (and reactivated) (B) to total luciferase bound (and reactivated) in the presence of zero-added insulin standard (B_0) is plotted as a function of the insulin standard added (in nanomoles per tube). Data are plotted from two experiments using different preparations of the insulin–luciferase conjugate.

and the estriol standard in a 300 μl total volume prepared with PBS and 0.2% BSA. After 1 hr incubation at room temperature, 100 μl of secondary antibody (goat anti-rabbit IgG) was added at a final dilution of 1 : 80 with 1.6% PEG-8000. After a second hour of incubation, the samples were centrifuged for 15 min in a microfuge and the supernatants removed from each pellet. The immunoprecipitates were resuspended in 0.5 ml of luciferase assay mix (0.2% BSA, 20 mM phosphate, pH 7.0) with dithiothreitol added to a final concentration of 100 mM. The luciferase was released as the active enzyme from the immune complex by the action of the reducing agent. After a 2 hr incubation, the samples were assayed for bioluminescence activity. Both flavin injection assays and the coupled assay with the flavin oxidoreductase were used and found to be acceptable.[17] The standard curve for this assay is shown in Fig. 3.

Summary and Future Directions

The basic methodology that is presented here was the product of the collaboration of the authors in the Department of Biochemistry and Biophysics at Texas A&M University. The technology has been transferred to the Upjohn Company where a variety of assays have been developed. There are several difficulties that remain to be solved. Most important is

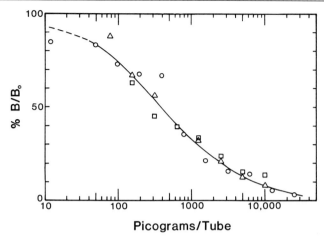

FIG. 3. Competitive binding of estriol–luciferase conjugate (derivatized at the active center thiol) to antiestriol antibody in the presence of free estriol. The ratio of bound (and reactivated) luciferase (B) to total bound (and reactivated) luciferase at zero-added estriol (B_0) is plotted as a function of total (underivatized) estriol added (as picograms per tube). Data are plotted from three separate experiments using three samples of the estriol–luciferase conjugate.

the problem of the stability of the luciferase conjugates. There are a variety of potential routes to the solution of the stability problem. We have found that conjugates with longer, more polar linkers form more stable conjugates with the luciferase. Compare, for example, the good stability of the luciferase–progesterone conjugate with the poor stability of the luciferase–estriol conjugate. The luciferase–progesterone conjugate was prepared from the 11α-hemisuccinate with the resulting additional carbonyl-containing linker. The chemistry of the linker compounds is an obvious parameter to be investigated in optimization of this methodology.

The complete covalent structure of the enzyme from *V. harveyi* is now known,[12] and with the cloned *lux* genes,[18–21] we are now doing mutagenesis experiments with the hope of designing a "better" luciferase. We have

[18] T. O. Baldwin, *38th Southwest & 6th Rocky Mt. Combined Reg. Meet. Am. Chem. Soc.* (1982).

[19] R. Belas, A. Mileham, D. Cohn, M. Hilmen, M. Simon, and M. Silverman, *Science* **218,** 791 (1982).

[20] T. O. Baldwin, T. Berends, T. A. Bunch, T. F. Holzman, S. K. Rausch, L. Shamansky, M. L. Treat, and M. M. Ziegler, *Biochemistry* **23,** 3663 (1984).

[21] D. H. Cohn, R. C. Ogden, J. N. Abelson, T. O. Baldwin, K. H. Nealson, M. I. Simon, and A. J. Mileham, *Proc. Natl. Acad. Sci. U.S.A.* **80,** 120 (1983).

also recently obtained quality crystals of the enzyme[22] and the structure is being solved in the laboratory of Prof. Brian Matthews (University of Oregon). In the next few years, we hope that our understanding of the structural chemistry of the luciferase will be sufficiently advanced to allow us to solve any difficulties which may arise. The technology of active center-based immunoassay using bacterial luciferase appears to have enormous potential for application to a variety of problems in the area of diagnostics.

Acknowledgments

This research was supported by grants from the National Science Foundation (PCM 8208589), the National Institutes of Health (AG 03697), the Robert A. Welch Foundation (A-865), the Upjohn Company, and a Hatch Grant (RI 6545).

We are grateful to numerous colleagues for their contributions. We are especially grateful to Dr. Miriam Ziegler, who was the first to demonstrate the reversible inactivation of luciferase by alkyl alkanethiolsulfonates on which this method is based, and who offered helpful criticisms of this manuscript. We are also indebted to Paul Satoh, Fred Yein, John Dougherty, Peter Riley, and others of the Upjohn Co.

[22] R. F. Swanson, L. H. Weaver, S. J. Remington, B. W. Matthews, and T. O. Baldwin, *J. Biol. Chem.* **260**, 1287 (1985).

[23] Bioluminescence Test for Genotoxic Agents

By S. ULITZUR

The Principle

The bioluminescence test (BLT) for genotoxic agents uses dark mutants of luminous bacteria and determines the ability of the tested agent to restore the luminescent state. Three groups of genotoxic agents have been shown to be active in the BLT[1-8]: (1) direct mutagens being either base-

[1] S. Ulitzur, I. Weiser, and Y. Yannai, *Mutat. Res.* **74**, 113 (1980).
[2] S. Ulitzur and I. Weiser, *Proc. Natl. Acad. Sci. U.S.A.* **78**, 3338 (1981).
[3] I. Weiser, S. Ulitzur, and S. Yannai, *Mutat. Res.* **91**, 443 (1981).
[4] S. Ulitzur, *Trends Anal. Chem.* **1**, 329 (1982).
[5] S. Ulitzur and M. Barak, *J. Bioluminescence and Chemiluminescence* (submitted for publication).
[6] B. Z. Levi, J. Kuhn, and S. Ulitzur, *Mutat. Res.* **173**, 233 (1986).
[7] Y. Ben-Issak, B. Z. Levi, H. Bassan, A. Lanir, R. Shor, and S. Ulitzur, *Mutat. Res.* **147**, 107 (1985).
[8] B. Z. Levi and S. Ulitzur, *Arch. Microbiol.* **134**, 281 (1983).

substitution or frame-shift agents, (2) DNA-damaging agents and DNA synthesis inhibitors, and (3) DNA-intercalating agents. The primary lesion of the dark mutant is still unknown; however, accumulating evidence suggests that the transduction of the luminescence operon is under continuous repression[4] probably by an intercistronic repressor. Restoring the luminescence of the repressed dark mutant can theoretically be achieved by three independent events: (1) blocking the formation of the repressor; altering its or the operator site's structure, (2) inactivating the repressor of the luminescence system, and (3) changing the physical configuration of the DNA, thus allowing unrepressed transcription of the luciferase operon. Blocking the formation of the repressor or altering its ability to bind to the operon site is expected from direct mutagens. As one can predict these mutagenic agents bring about the appearance of genetically stable luminous revertants; most of them are constitutive at different degrees with regard to the formation of the luminescence system.[1] The second event, namely the inactivation of the repressor of the luminescence system, was found to be associated with the activity of different DNA-damaging agents such as mitomycin C or UV irradiation and with the action of DNA synthesis inhibitors such as nalidixic acid, novobiocin, or methotrexate. Present evidence supports the possibility that all these agents act through their ability to trigger the "SOS functions" that involve the inactivation of the luminescence system's repressor.[3] DNA configurational changes due to interactions of chemicals between DNA bases seem to be the most potent and rapid way to restore the luminescence of the dark mutant. Different DNA-intercalating agents such as acridine dyes, caffeine, or norharman result in a prompt induction of the luminescence in the dark cells.[2,4,8]

The different end points and modes of action of the genotoxic agents are reflected also in the timing of the onset of the induced luminescence in the dark mutant culture. The intercalating agents act within 60 min and result in an almost full recovery of the wild-type luminescence. The SOS-inducing agents act after 2–4 hr, while the direct mutagens required 5–8 hr to increase the luminescence of the treated culture over that of the control level. Another major difference between these groups of genotoxic agents is that only the direct mutagens bring about the appearance of genetically stable revertants while the SOS-inducing agents and the DNA intercalators result in only phenotypic reversion of the luminescence.

Technically the BLT is very simple; sterility of the assay mixture is not essential, volatile agents could be tested, and the toxicity of the tested chemical is determined concomitant with its genotoxicity. The BLT can be run automatically and up to 50 chemicals can be tested in 1 day.

The Bioluminescence Test

Bacteria

SD-18 is a spontaneous dark variant of the luminous bacteria *Photobacterium leiognathi* strain BE8.[1] P.f-13 is a dark mutant of the luminous bacteria *Photobacterium fischeri* (NRRL-B111777). Both strains are available as lyophilized preparation from Microbics Corp., CA. The lyophilized cultures are stable for at least 1 year at 4°. Prior to the test the lyophilized culture is rehydrated by 3 ml of cold ASWRP medium. The rehydrated cultures should be kept cold where it is stable for up to 12 hr.

Assay Medium

The BLT is performed in the growth medium designated as ASWRP.[1] The medium contained (in g/liter) Biolife (Italy) peptone, 5; glycerol, 3; NaCl, 17.5; KCl, 0.75; $MgSO_4 \cdot 7H_2O$, 12.3; $CaCl_2 \cdot 2H_2O$, 1.45; $K_2HPO_4 \cdot 2H_2O$, 0.075; NH_4Cl, 1.0. The medium is buffered by 0.05 M morpolinopropanesulfonic acid (Sigma) to pH 7.0. When $MnSO_4$ is included in the assay medium the $MnSO_4$ (1 M) solution should be autoclaved separately and added to a cold and sterile solution of ASWRP.

S-9 Fraction

The S-9 fraction contains rat liver microsomes induced by a polychlorinated diphenyl preparation.[9] A ($\times 10$) concentrated S-9 mix contains glucose 6-phosphate, 5 mM; NADP, 4 mM; $MgCl_2$, 8 mM; KCl, 33 mM; MOPS buffer (Sigma), 100 mM, pH 7.5; S-9 microsomal fraction, 200 μl/2 ml of S-9 mix.

Luminescence Determination

The ideal photometer for the bioluminescence determination in the BLT is a scintillation counter. A thermostat-controlled electric heater fan should be attached to the outside wall of the counter. The hot air blows through holes that are made in the wall of the counting chamber (an old counter such as Packard model 2001 is suitable for this application). The counter is operated without coincidence at the ^{3}H setting. The high voltage should be set to a level that would result in a quantum efficiency of about 100 quanta/sec for 1 cpm. Alternatively, the sensitivity of the

---

[9] B. N. Ames, J. McCann, and E. Yamasaki, *Mutat. Res.* **31**, 347 (1975).

counter should be set to ensure that the maximal counting of the control cells (without added chemicals) will not exceed 50,000 cpm.

When a scintillation counter is not available, one can incubate the vials at 28–30° with slight shaking. The developed luminescence can be determined periodically; one determination after 8 hr and a second one after 18 hr of incubation are the minimal requirements. For light determination one can use all the commercial photometers that have been designed for ATP determination. A lowest light detected limit of 10^6 quanta/sec is desired.

Running the BLT in a scintillation counter one should be aware of the possibility of high counting resulting from an electrostatic current that may build up through the continuous movement of the glass vials. An antielectrostatic spray or high humidity avoids this phenomenon.

The Bioluminescence Test

The test requires new clean scintillation glass vials (26 mm diameter). The vials are equipped with cellulose stoppers and are sterilized in an autoclave for 15 min or in an oven (160°) for 1 hr. Testing an unknown chemical requires 13 vials; the first vial in the set contains 2 ml of ASWRP medium while all the rest contain 1 ml of the assay medium. A solution of the chemical in question is properly diluted in the first vial. Compounds that are not dissolved in water could be added from ethanolic, methanolic, or DMSO solutions; care should be taken not to exceed 0.5% of the organic solvent in the first vial. Since some DMSO preparations increase the luminescence of the dark mutants, controls with DMSO alone at the corresponding concentrations should be included in each test. The compound to be tested is serially double diluted in the assay medium nine times. It is important to include at least one toxic concentration of the tested chemical in the assay (a toxic concentration will inhibit a visible growth of the bacteria after 24 hr of incubation at 28°). Each test should contain at least 3 controls when the aqueous solution is being tested and 5 controls when the organic solvents are used. When the S-9 fraction is being used, 0.1 ml of the ×10 concentrated S-9 mix is added to each vial, including the controls. To start the test add 50 μl of the rehydrated dark mutant suspension to each vial. The final cell concentration at the beginning of the assay is assumed to be 10^5 cells/ml. The vials are then taken to a temperature-controlled (28°) scintillation counter, when the luminescence is determined repeatedly for 16–20 hr (one reading every 45–60 min is sufficient).

When the S-9 fraction is included, the vials are preincubated at 37° for 1 hr before being taken to the scintillation counter.

Evaluation of the Results

A typical kinetic curve of the light development with S.D-18 cells in the absence or presence of some representatives of genotoxic agents is shown in Fig. 1. It can be seen that the light of the control starts to increase (above the background noise) after about 4 hr of incubation. The increase in the luminescence of the control reflects mainly the increase in cell number in the growth medium. (The concentration of the bacteria increases about 1000-fold during 12 hr of incubation.) The luminescence of the control cells reaches its maximal value usually after 8–10 hr of incubation and then decays slowly.

In the presence of genotoxic agents, the onset of luminescence is typical to each group of agents, the earliest being the DNA intercalators and the latest with the direct mutagens. The rate of increase in luminescence and its maximal value depend very much on the nature of the chemical in question, and on its concentration. DNA intercalators and SOS-inducing agents result in a steeper increase in luminescence than the direct mutagens. Partially lethal concentrations of the tested agents result in a somewhat delayed increase in luminescence; however, these vials usually show maximal luminescence after longer incubation periods.

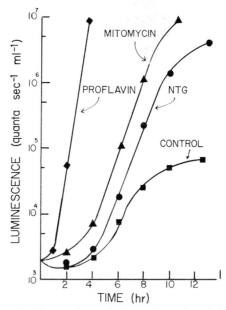

FIG. 1. The effect of proflavin, mitomycin C, and *N*-methyl-*N*-nitro-*N*-nitrosoguanidine on the kinetic of light development in the dark mutant SD-18 in ASWRP medium at 28°.

The best way to evaluate the activity of the chemical in question is to determine the maximal luminescence value that has been reached at any time along the assay. One can define a significant genotoxic activity when the maximal luminescence developed due to the chemical in question is 3–4 times higher than the maximal luminescence of the corresponding control.

Evaluation of the Mode of Action of the Active Genotoxic Agent

As mentioned above it is possible to predict the genotoxic end point of the active chemical by two criteria: the timing of the onset of the induced luminescence and the appearance of genetically stable revertants. In order to determine the appearance of stable revertants, the luminescent culture is properly diluted in 3% NaCl solution and 200–300 bacteria are spread with the aid of a glass rod over the surface of a petri dish with complex solid medium.[1] All the direct mutagens tested resulted in an appearance of 2–3 bright colonies per plate after incubation for 24 hr at 30°. Neither SD-18 nor P.f-13 reverts to luminescent colonies at a rate higher than 10^{-6} (less than one luminous colony out of a million dark ones).

Modification of the BLT

Determination of Volatile or Gaseous Samples

Highly volatile agents or gaseous samples require a slight modification of the standard BLT. To avoid escape of the volatile agent, the standard cellulose stopper is replaced by a rubber one. To ensure enough oxygen supply for the growth and for the luminescence expression, the volume of the assay medium is reduced to 0.5 ml (instead of 1 ml). Gaseous samples can be injected directly into the vial through the rubber stopper.

Determination of the Genotoxic Activity in Air Samples

Polluted air or other gaseous samples can be sucked through a filter. The filter is then placed at the bottom of the scintillation vial followed by the addition of ASWRP medium and the dark mutant suspension.

BLT on Solid Medium

This modification of the BLT requires complex solid medium described by Ulitzur et al.[1] The lyophilized bacteria are diluted in 3% NaCl to a final cell concentration of 5×10^3 cells/ml. Of this suspension 0.1 ml is spread, with the aid of a glass rod, over the solid medium, where 500 colonies are expected to grow. The chemical in question (at concentrations that would result in an inhibition zone of 1–2 cm) is adsorbed on a filter paper disk (e.g., Whatman AA) and the disk is placed in the center of the plate that is incubated at 30°. The results are recorded every 12–15 hr for a period of 3 days. Luminescence determination can be monitored by the aid of a cone-like apparatus with a proper light-sealed cover, which is hooked in front of the photomultiplier. Alternatively, the plates can be viewed in a dark room after 10–15 min of preadaptation to the dark.

Synergistic Factors

The activity of many genotoxic agents increases considerably under conditions that affect the fidelity of the genetic repair mechanism. Many metallic compounds including Zn^{2+}, Co^{2+}, and Mn^{2+} synergistically increase the genotoxic effect of different mutagenic agents in the BLT.[5] It is thus possible to further increase the sensitivity of the BLT by addition of $MnSO_4$ (15 mM) to the ASWRP medium.

Interfering Factors

Different detergents that effect the integrity of bacterial membrane (e.g., SDS) were found to increase the luminescence of the dark mutants. High concentrations of organic solvents including DMSO act similarly. It is therefore recommended that the concentration of organic solvents in the assay not exceed 0.5%. It is suggested that these membrane-damaging agents act through their effect on the DNA attachment site in the membrane that consequently results in the induction of the "SOS functions."

The presence of long-chain aliphatic aldehyde (C_6–C_{16}) results in a prompt but temporary increase in the luminescence of the dark mutants.[2] Since the aldehydes are used up in the process of luminescence, the increase in luminescence lasts only 10–20 min.

Reproducibility

Using new and clean vials and a standard procedure the BLT shows highly reproducible results with a standard deviation of ±15–20%. Smoking should not be permitted in the room (the mutagenic activity associated with 1/100 of one smoke puff can be determined in the BLT).

Results

Figure 2 shows the activity of different concentrations of several genotoxic agents with P.f-13 cells in the BLT. It can be seen that all the tested chemicals result, at a certain concentration range, in a linear increase in the luminescence of the dark mutant. Most of the genotoxic agents increase the luminescence of the dark mutant by a factor of 1000 times or more. The table sums up the results with about 90 chemicals that were tested with the dark strains SD-18 and P.f-13. The chemicals were grouped according to their structural or functional groups. The lowest detected concentration (in μg/ml) of each chemical is given together with the requirements for microsomal activation (asterisks). The comparative mutagenic activity determined in the Ames test is given as + or − for each compound for which such information was available. All the chemicals that are known to be active in the Ames test gave positive results in the BLT. The minimal detected concentrations were always lower in the BLT than in the Ames test. The dark mutant P.f-13 shows higher sensitivity than the spontaneous dark variant SD-18. The table (B) lists the chemicals that were preferentially or solely detected with the P.f-13 strain. Using this strain it is possible to detect different carcinogenic agents that were not detected by the Ames test, e.g., beryllium, nickel, lindane, safrole, *o*-tolidine, and *p*-tolylhydrazine. Not all the chemicals that increased the

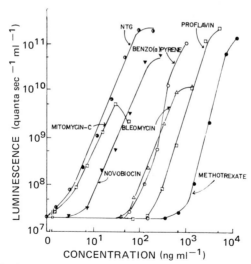

FIG. 2. Determination of some genotoxic agents in the BLT. The BLT was carried out with the dark mutant P.f-13 in ASWRP medium at 28°.

Chemical compound	BLT (MDC μg/ml)	Ames test
A. Chemical agents that were determined with SD-18		
Quinacrine	0.5	+
Ethidium bromide	2.5	+
Acriflavin	0.1	+
2-Aminobiphenyl	25	+
Proflavin	0.2	+
9-Aminoacridine	0.2	+
Acridine orange	0.2	+
N-Methyl-N-nitrosoguanidine	0.002	+
2-Nitrofluorene*	0.7	+
Mitomycin C	0.002	+
Hydrazine	0.07	+
Isoniazide	650	+
Phenylhydrazine	0.32	+
2-Nitrophenyl hydrazine	25	+
2,4-Nitrophenyl hydrazine	6	+
p-Tolyhydrazine	1	+
Benzylhydrazine	12	+
m-Hydroxybenzyl hydrazine	1	+
4-Nitro aniline	6	+
Caffeine	20	−
Theophylline	20	−
Methylmethane sulfonate	6	+
Ethylaminosulfonate	0.005	+
Ethyl isopropylaminosulfonate	150	+
Skatol	60	−
Norharman	3	−
Trp-p-2	3	−
Glu-p-1	12	−
Glu-p-2	3	−
Desulfiran	0.15	+
Hydroxyl amine	0.1	+
Ascorbic acid	300	+
Emodin	3	+
Coumermycin	2	+
Fistein*	30	+
Kampferol*	1.5	+
Quercetin*	1.5	+
B. Chemical agents preferentially detected with P.f-13 strain		
4-Nitroquinoline-N-oxide	0.007	+
1,3-Propansultone*	6	+
1,3-Propansulfonate*	6	+
Epichlorophydrin	50	
Nitrosobenzene*	6	
Niridazole	0.1	+
O-Tolidine	1	−
Thio-Tepa	50	+

Chemical compound	BLT (MDC μg/ml)	Ames test
9,10-Dimethylbenzo[a]anthracene*	25	+
Benzo[a]pyrene*	0.1	+
Adriamycin	50	+
2-Anthramine	0.5	+
Indol	2	
Nitrogen mustard	1.5	+
Methothrexate	2	–
Vincristine	1	–
Bleomycin	0.1	–
Vinyl chlorid (gas)	+	
6-Mercaptopurine	1	+
cis-Platinum-II-diaminodichloride	20	+
Phenol	20	–
Safrole	7	–
Shikimic acid	25	–
Lindane	50	–
Aflatoxin-B_1*	0.01	+
Novobiocin	0.02	–
Malonylaldehyde	0.5	
4-Amino phenol	6	–
Na_2WO_4	4000	–
$ZnCl_2$	30	–
KH_2ASO_4	100	+
$BeCl_2$	1.5	–
$K_2Cr_2O_7$	2	+
$MnCl_2$	20.000	–
Ni-acetate	10	–
$HgCl_2$	1	–
$CoCl_2$	40	–
$CdCl_2$	50	–

C. Chemical agents that were not active with either SD-18 or with strains P.f-13

Benzo[e]pyrene*	–	–
Chloramphenicol	–	–
Actinomycin D	–	–
Patulin	–	–
Penicillic acid	–	–
Acetamide	–	–
Flavone	–	–
Chrysen	–	–
Chloroform	–	–
$FeSO_4$	–	–
Na ASO_2^-	–	–
$RbCl_2$	–	–
$CuCl_2$	–	–
$SnCl_2$	–	–
$NaSeO_2$	–	–

luminescence of the dark mutants are carcinogens or mutagens; this group consists mainly of the DNA intercalators. The genotoxic effect of the DNA intercalators is controversial; many DNA intercalators act as co-mutagens that synergistically increase the activity of weak mutagens.

Application of the BLT

The BLT can be applied as an analytical test in different fields of toxicology, pharmacology, and chemotherapy as well as a powerful tool in research.

BLT as a Prescreening Test for Carcinogenic Agents

Being a general, sensitive, simple, and cheap assay, the BLT can be applied as a prescreening test for carcinogenic agents. Unlike the Ames test the BLT is not affected by the presence of amino acids or other nutrients; thus it is possible to assay complex organic matter such as foodstuffs and biological fluids.[7] Other advantages of the BLT are the ability to assay volatile and gaseous samples and to determine the toxicity of the chemical in question along with its genotoxicity. Most important is the fact that the BLT detects all the 46 tested chemicals that are known to be active in the Ames test. Moreover, quite a large number of carcinogenic agents that are not active in the Ames test are detected in the BLT (see the table).

The chemicals which respond positively in the BLT should then be tested by a battery of *in vivo* and *in vitro* tests which will confirm or refute their potential carcinogenicity.

Determination of Anticancer Agents

Many anticancer drugs act as DNA synthesis inhibitors or DNA damaging agents. The table shows that the most common anticancer agents are active in the BLT. It was also possible to detect these agents in the patient's urine and to determine the pharmacokinetics of the drug (unpublished). The BLT can also be used to screen potentially active anticancer agents on the one hand and to evaluate the genotoxicity of newly developed drugs on the other hand.

Research

The BLT is a potentially valuable tool for studying the activity of different physicochemical agents that interact with DNA. The ability to follow the kinetics of genetic events continuously and nondestructively is not offered by any other system.

[24] Determination of Antibiotic Activities with the Aid of Luminous Bacteria

By S. ULITZUR

Introduction

Rapid assays for antibiotics are desirable for the determination of their concentration in a system in which they are not constant. In medicinal practice such tests are essential for detection of antibiotics that are more likely to exert a toxic effect above a certain therapeutic concentration. In the dairy industry, such tests are required to detect the presence of antibiotics in milk. Similarly, in plants that produce antibiotics, it is often necessary to determine the momentary concentration of the antibiotics in fermenters.

The bacterial luminescence system can be used to determine the activity of different kinds of antibiotics. With regard to the specific end point of the antibiotics, the developed tests can be divided into three groups: (1) the lysis test, (2) the induced test, and (3) the bacteriophage coupled test.

The lysis test used a sensitive and highly luminescent bacteria and compares the level of the *in vivo* luminescence with and without the antibiotic in question. Antibiotics that affect the integrity of the cell wall or the cytoplasmic membrane result in a decrease of the *in vivo* luminescence of the treated culture. This principle has been applied to determine the activity of β-lactam antibiotics and polymyxins. Since the naturally occurring luminous gram-negative bacteria do not show high susceptibility to β-lactam antibiotics, it was necessary to clone the luminescence system into the gram-positive bacteria, *Bacillus subtilis*. The luminous *B. subtilis* cells are incubated in the milk containing the β-lactams for 45–60 min followed by the luminescence determination. This test allows detection of nanogram quantities of different β-lactam antibiotics.

The determination of polymyxin B in milk and other fluids is carried out with the aid of the luminous bacteria *Photobacterium leiognathi*. To increase the sensitivity of the test, crystal violet at sublethal concentration is added to the assay buffer. It was found that polymyxin B largely increases the susceptibility of the bacteria to the dye. This test allows a determination of 40 ng/ml of polymyxin B in a procedure lasting 15 min.

The induced test is applied to determine the activity of protein synthesis inhibitors. The test is based on the ability of the tested antibiotic to inhibit the luciferase synthesis in the treated cells. The test uses a dark mutant of luminous bacteria that undergoes prompt induction of the luminescence system in the presence of certain DNA-intercalating agents.[1,2] Compounds such as acridine dyes, norharman, or caffeine result in a prompt (10–12 min) induction of the luminescent system that increases more than 30-fold within 30 min. Protein synthesis inhibitor antibiotics that inhibit the *de novo* synthesis of proteins can be determined in this assay in a procedure lasting less than an hour.

The third test uses a biological coupled system in which highly luminescent bacteria are infected by a specific bacteriophage. In the absence of the antibiotic, the luminous bacteria are lysed within 45 min and their *in vivo* luminescence drops to almost zero. Antibiotics that inhibit DNA, RNA, or protein synthesis also inhibit the intracellular phage development and thus rescue the luminous bacteria.

The main advantages of the bioluminescence test for antibiotics are sensitivity and rapidity. The sensitivity for most of the antibiotics is far beyond the standard disk assay and the time required for the complete procedure is less than 1 hr. Another advantage of the bioluminescent test is the potential to detect any desired antibiotic in a mixture of unknown antibiotics. For this purpose an antibiotic-resistant strain of the relevant luminous bacteria is selected. All the luminescence tests can use lyophilized cultures of the luminous bacteria; such cultures are stable for at least 1 year at 4°.

The Lysis Test

The luminescence system of *Photobacterium fischeri* MJ1 was cloned into *Escherichia coli* according to the procedure of Engebrecht *et al.*[3]; Carmi and Kuhn[4] have succeeded in cloning the luciferase gene into the gram-positive bacteria *B. subtilis*. In the presence of external aldehyde the luminous *B. subtilis* cells emit about 1000 quanta/sec/cell.

The assay can use either a lyophilized preparation or a freshly growing culture. The luminous bacteria are grown on nutrient agar (Difco) containing chloramphenicol (3 μg/ml) at 30° for 24 hr. The cells are suspended in saline to a concentration of 0.5 OD (Klett units) K.U.-66. To determine the activity of β-lactams in milk, 25 μl of the luminescent *B. subtilis* cells is added to 1 ml of the suspected milk as well as to an antibiotic-free milk (control). The vials are incubated at 37° for 50 min followed by 10 min of incubation at room temperature. To determine the luminescence, 20 μl of 0.5% ethanolic solution of dodecyl aldehyde is added to each vial and the

[1] S. Ulitzur and I. Weiser, *Proc. Natl. Acad. Sci. U.S.A.* **78**, 3338 (1981).
[2] A. Naveh, I. Potasman, H. Basan, and S. Ulitzur, *J. Appl. Bacteriol.* **56**, 457 (1983).
[3] J. Engebrecht, K. Nealson, and M. Silverman, *Cell* **32**, 773 (1983).
[4] D. Carmi and J. Kuhn, Dept. of Biology, Technion, Haifa, Israel.

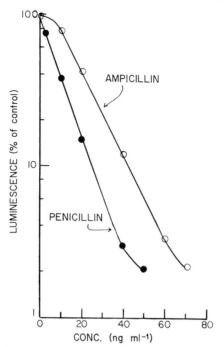

FIG. 1. Determination of β-lactam antibiotics in milk with the aid of luminescent *B. subtilis* bacteria (see text for details).

luminescence is immediately determined. Figure 1 shows the effect of different concentrations of penicillin G and ampicillin in milk on the luminescence of *B. subtilis*.

Bioluminescence Test for Polymyxin B in Milk[5]

The test uses a highly luminescent culture of *P. leiognathi* BE8 cells. Other species of luminous bacteria are also suitable for this test. The luminous bacteria are growing with shaking in liquid ASWRP medium[1] at 30° to a highly luminescent stage. The culture is washed twice with 3% NaCl in potassium phosphate buffer (0.02 M), pH 7.0. The cells are suspended in the same buffer to give a cell concentration of 0.2 OD K.U.-66. A lyophilized culture of luminous bacteria *P. phosphoreum* available from Microbics Corp., CA, is also suitable for this test. The lyophilized culture is rehydrated by 1 ml of cold distilled water and kept in ice until required.

One milliliter of the tested milk and 1 ml of a reference milk are placed in clean scintillation vials. To each vial 50 μl of the luminescent culture is

[5] S. Ulitzur, in preparation.

added and the vials are incubated for 10 min at room temperature. Aliquots (0.2 ml) from each vial are diluted into 2 ml of the NaCl phosphate buffer containing gentian violet (1 μg/ml). The luminescence is recorded after additional incubation for 5 min at room temperature.

Figure 2 shows the effect of different concentrations of polymyxin B in milk on the luminescence of the BE8 strain; it can be seen that as low as 40 ng/mg of PMB can be determined in milk. The test can be further synthesized by using spheroplasts of BE8 cells. A polymyxin B-resistant strain of BE8 can be used to ensure that the recorded activity is due to polymyxin B and not to other lytic agents.

The Induced Test

This test uses a dark mutant of luminous bacteria that undergoes a prompt induction of the luminescence system in the presence of certain DNA-intercalating agents.[1,2] Antibiotics that block protein synthesis inhibit the induction of the luminescence.

General Procedure[2]

The dark mutant *P. leiognathi* 8SD18 cells are inoculated on complex solid medium ASW-3 and the plate is incubated overnight at 30°. Some colonies are suspended in SMG buffer consisting of NaCl (2%), MOPS (Sigma) buffer (20 mM), pH 6–8, and glycerol (0.3%) to give a final cell concentration of 60 Klett Summerson units filter 66 (about 3×10^8 cells/ml). The bacterial suspension is kept at 4° where it is stable for up to 24 hr.

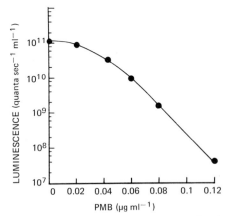

FIG. 2. Determination of polymyxin B in milk with the aid of *P. leiognathi* BE8 (see text for details).

Lyophilized cultures of 8SD18 cells are available from Microbics Corp., CA. Such cultures are rehydrated before use with 1 ml of cold distilled water.

The antibiotic solution is diluted in 1 ml of SMG buffer containing the DNA-intercalating agent proflavin hemisulfate. The final concentration of proflavin in the assay mixture is pH dependent being 35, 25, 10, 5, 2.5, and 1.5 μg/ml for the pH values of 5.6, 6, 6.8, 7, 7.5, and 7.9, respectively. The assay is started by the addition of 20 μl of the 8SD18 cell suspension to each vial. Control vials that do not contain the antibiotic should be included in each test. The vials are incubated with gentle shaking at 30° and the luminescence is determined during certain time intervals (usually between 40 and 60 min). The minimal detected antibiotic concentration in the bioluminescence test is defined as the concentration of antibiotic that inhibits 50% of the luminescence after a given period of incubation at 30°.

Results

The effect of different concentrations of gentamicin on the proflavin-induced luminescence of 8SD18 cells is shown in Fig. 3. The intensity of

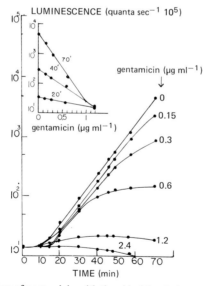

FIG. 3. Determination of gentamicin with the aid of the dark mutant *P. leiognathi* 8SD18. Gentamicin was serially double diluted in 1 ml of SMG buffer (pH 7.9) containing 1.5 μg/ml proflavin hemisulfate. The test was carried out as described in the text and the results were recorded after different times of incubation. The inset shows the correlation between the log of the *in vivo* luminescence and the gentamicin concentration after 20 and 70 min of incubation.

the developed luminescence is inversely proportional to the concentration of the antibiotic. Like many other biological systems that are subjected to inhibition by chemical agents, the relationship between antibiotic concentration and the luminescence is linear only within a certain range. This characteristic dose–response curve could be rendered linearly by expressing the luminescence level as the Γ (gamma) function:

$$\Gamma = \frac{L_{max} - L_{ob}}{L_{ob}}$$

L_{max} refers to the luminescence level of the antibiotic-free control while L_{ob} stands for the luminescence in the presence of given antibiotics concentrations.

Figure 4 shows the inhibition of luminescence (expressed as Γ) by different antibiotics. Statistical analysis of the results shows a linear correlation coefficient (r) of 0.87–0.97 for the different antibiotics given.

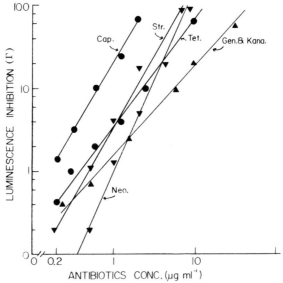

FIG. 4. Dose–response line for some protein synthesis inhibitors as determined by the induced test. The figure shows the correlation between Γ (gamma) function of the luminescence after 60 min of incubation and the concentration of different antibiotics. Tetracycline and chloramphenicol were tested in SMG buffer at pH 6.0; all the rest of the antibiotics were tested in SMG buffer at pH 7.9. Chloramphenicol (Cap.), tetracycline (Tet.), streptomycin (Str.), gentamicin (Gen.), neomycin (Neo.), and kanamycin (Kana.).

The Effect of Different Factors on the Assay

Cations. Being marine microorganisms the luminous bacteria require minimal concentration of salt in the assay buffer. This requirement can be provided by addition of 2% NaCl to the assay buffer. Care should be taken to minimize the concentration of divalent cations such as Ca^{2+} or Mg^{2+} that are known to inhibit the activity of aminoglycosides.

pH. The activity of certain antibiotics is pH dependent. Aminoglycosides should be assayed at pH 7.9 while tetracyclines show higher activity in the acidic range. The concentration of proflavin in the assay mixture should be adjusted to the pH of the assay (see above).

The dependence of the activity of some antibiotics on the pH of the medium can be used to discriminate between the activity of two antibiotics that coexist in the tested medium. For this purpose, the activity of the antibiotic in the tested solution (e.g., serum or milk) is determined at pH 7.9 as well as at pH 6. While aminoglycosides are at least 100 times more active at pH 7.9 than at pH 6.0 the reverse is true for tetracyclines.

The copresence of β-lactam antibiotics in the assay mixture does not affect the results; this can be attributed to both the short incubation time in the assay that does not allow cell division as well as to the presence of high salt concentration that stabilizes any spheroplasts that may be formed.

Determination of Antibiotics in Serum[2]

The procedure for determination of antibiotics in serum consists of the following steps.

1. The bactericidal activity of the serum is inactivated by heat at 56° for 30 min. To 0.8 ml of serum 0.2 ml of 10% NaCl containing 0.1 M MOPS buffer and 1.5% glycerol are added. The final pH value should be 7.9 for all antibiotics except chloramphenicol and tetracycline that are tested at pH 6.

2. 8SD18 cell suspension (200 μl) (3×10^8 cells/ml) is added to 0.8 ml of the serum as well as to 1 ml of antibiotic-free pooled serum.

3. After 10 min of preincubation at 30°, proflavin is added to give a final concentration of 1.5 μg/ml for pH 7.9 or 25 μg/ml for pH 6.0 and the vials are incubated with gentle shaking at 30°. The results are recorded after 40–60 min of incubation.

The table shows the concentration of antibiotics that was detected in serum with this procedure. It can be seen that most antibiotics tested are detected at subtherapeutic concentrations; thus, it is possible to dilute the serum 10- to 20-fold in SMG buffer and still have enough sensitivity to

DETERMINATION OF DIFFERENT ANTIBIOTICS IN SERUM
BY THE INDUCED TEST[a]

Antibiotic	pH	Range (μg/ml)	r	MDC
Chloramphenicol	6.0	0.08–2.5	0.98	0.10
Tetracycline	6.0	0.2–5.0	0.95	0.4
Gentamicin	7.9	0.1–5.0	0.90	0.2
Streptomycin	7.9	0.2–25.0	0.87	0.5
Tobramycin	7.9	0.3–50.0	0.90	0.5
Amikacin	7.9	7.9–15.0	0.95	0.6
Kanamycin	7.9	0.2–15.0	0.90	0.6
Neomycin	7.9	0.5–50.0	0.88	0.5

[a] The antibiotics were diluted in pooled antibiotic-free sera of healthy volunteers and the sera were heat inactivated (56° for 30 min). The induced test was carried out as described in the text in either pH 6.0 or pH 7.9; the results were recorded after 60 min. The table gives the range of antibiotic concentrations that affect the luminescence, the correlation coefficient for the linear regression (r), and the minimal detected concentration (MDC). The MDC has been defined as the minimal concentration of the antibiotic that inhibits 50% luminescence.

detect the desired concentration of the antibiotic in serum. (The activity of most antibiotics in 10% serum in SMG is higher than in 80% serum.) A correlation of 0.89 was obtained between the bioluminescence test and agar diffusion assay for gentamicin in serum.[2]

The new bioluminescence test for protein synthesis antibiotic inhibitors offers some obvious advantages over the currently used short-term tests. The bioluminescence test specifically determines the activity of the tested antibiotic as a *de novo* protein synthesis inhibitor. Antibiotics that affect DNA or cell wall synthesis are not detected in this test. The bioluminescence test is also more sensitive than most of the other bioassays available. This high sensitivity could be attributed to the higher susceptibility of newly synthesized proteins toward the inhibitory action of different antibiotics. Another advantage of the new bioluminescence test is that it may be run automatically, with the aid of a scintillation counter combined with a microprocessor.

The equipment required for light determination is already being manufactured. Most of the photometer–photomultipliers that are made for ATP determination using firefly luciferase can be employed. When a higher sensitivity of photometer is required, a higher concentration of SD18 cell can be used. However, when the bacterial density exceeds 2×10^7 cells/ml the culture should be aerated during the assay.

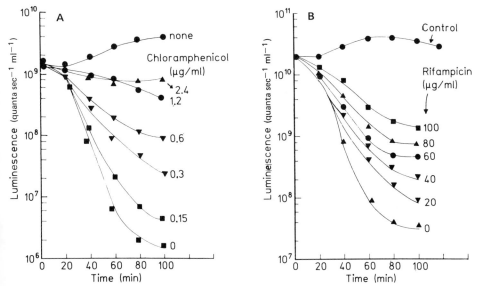

FIG. 5. Effect of chloramphenicol (A) and rifampicin (B) on the luminescence of a bacteriophage V_1-infected *V. harveyi* bacteria (see text for details).

In general, the bioluminescence test can be adapted to almost any set of conditions. The only basic requirements are to keep the assay buffer isotonically to the luminous bacteria and to use the proper combination of pH value and inducer concentration.

The Bacteriophage Coupled Test[6]

This test was designed to assay antibiotics that inhibit DNA or RNA synthesis, although protein synthesis inhibitor antibiotics are detected as well. The test uses the luminous bacteria *Vibrio harveyi* strain MAV, against which the bacteriophage V_1 was isolated from the sea. A highly luminescent culture of *V. harveyi* (2×10^8 cells/ml) is infected with bacteriophage V_1 with a multiplicity ratio of 100 phages per one bacterial cell. After 5 min of preincubation at 37°, to allow phage adsorption, the culture is cooled down to 4° where it is ready for use for a period of 3–4 hr. This culture can also be lyophilized; such lyophilized cells when rehydrated by cold water undergo 99% lysis within 60 min of incubation at 37°.

The assay is carried out in scintillation vials; the antibiotic in question is diluted in the growth medium ASWRP[1] and 0.1 ml of the phage-infected

[6] A. Naveh, Ph.D. Thesis, Technion, Haifa, Israel (1984).

culture (10^6 cells/ml) is added to each vial. The vials are incubated with shaking at 37° and the luminescence is determined after 45 min of incubation. Figure 5 shows the effect of different concentrations of rifampicin and chloramphenicol on the luminescence of the infected culture. It can be seen that the luminescence of the untreated culture has decreased about 99% after 60 min of incubation. Antibiotics that block the phage maturity save the bacteria and thus allow higher luminescence.

A similar test uses a cloned luminescent *E. coli*[3] and the bacteriophage T_5 has been successfully applied for determination of different RNA and DNA synthesis inhibitor antibiotics.

[25] Amplified Bioluminescence Assay Using Avidin–Biotin Technology

By G. Barnard, E. A. Bayer, M. Wilchek, Y. Amir-Zaltsman, and F. Kohen

The high affinity of avidin for biotin ($K_D = 10^{-15} M$) provides a powerful tool for studies in many areas such as (1) isolation of biotin-derivatized materials by affinity chromatography, (2) localization and visualization of various antigens, (3) drug delivery, (4) lymphocyte stimulation, and (5) immunoassays.[1,2] In immunoassay, advantage is taken of the four biotin-binding sites of avidin to amplify the sensitivity of the assay.

Here we will describe a specific example which illustrates this general approach.[3] A monoclonal or polyclonal antibody to a peptide hormone (e.g., hCG) is immobilized onto a solid matrix. The antigen is added, followed by a biotinylated preparation of an antibody directed against a second epitope on the antigen. After the immunological reaction, a secondary probe consisting of avidin and a biotinylated NAD^+-dependent enzyme [e.g., biotinylated glucose-6-phosphate dehydrogenase (G6PDH)] is added. The end point is determined by bioluminescence using glucose 6-phosphate and NAD^+ as substrates and bacterial luciferase/FMN/ decanal for initiation of light output. Figure 1 shows a flow sheet of the

[1] E. A. Bayer and M. Wilchek, *Methods Biochem. Anal.* **26,** 1 (1980).

[2] M. Wilchek and E. A. Bayer, *Immunol. Today* **5,** 39 (1984).

[3] F. Kohen, E. A. Bayer, M. Wilchek, G. Barnard, J. B. Kim, W. P. Collins, I. Beheshti, A. Richardson, and F. McCapra, *in* "Analytical Applications of Bioluminescence and Chemiluminescence" (L. Kricka and T. P. Whitehead, eds.), p. 149. Academic Press, New York, 1984.

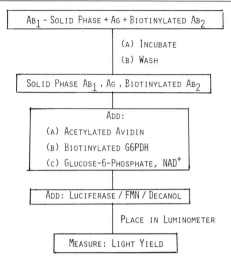

FIG. 1. A schematic outline of an immunobioluminometric (IBMA) assay for hCG mediated via avidin–biotin complexes.

method. The results of an immunobioluminometric assay (IBMA) for hCG will be reported here.

Procedures

Preparation of Biotinyl-G6PDH

Glucose-6-phosphate dehydrogenase (Sigma Type XXIV from *Leuconostoc mesenteroides* 1000 U, ~2.9 mg protein) is dissolved in 10 ml 0.1 M NaHCO$_3$ containing 0.15 M NaCl. A 20-fold excess of biotinyl-N-hydroxysuccinimide ester (BNHS, 110 μl of a 1.7 mg/ml dimethyl formamide solution) is added,[4] and the reaction is carried out for 2 hr at room temperature. The solution is dialyzed exhaustively first against phosphate-buffered saline, pH 7.2, and then against saline. The dialyzed enzyme is stored at 100 U/ml in aliquots at $-20°$.

Preparation of Biotinylated Monoclonal Anti-β-hCG

BNHS (250 μg in 10 μl of dimethyl formamide solution) is added to a solution containing monoclonal anti-β-hCG IgG at a concentration of 1 mg protein/ml buffer (20 mmol/liter phosphate, pH 7.4). The solution is

[4] E. A. Bayer and M. Wilchek, this series, Vol. 34, p. 265.

stirred for 6 hr at room temperature, and then dialyzed exhaustively against PBS. The biotinylated antibody (~1 mg/ml) is stored in aliquots at −20°.

Preparation of Acetyl N-Hydroxysuccinimide Ester

Glacial acetic acid (2.85 ml) is added to 150 ml dichloromethane. N-Hydroxysuccinimide (8 g) and dicyclohexylcarbodiimide (Fluka, 11.5 g) are dissolved in succession. The reaction is allowed to proceed for 16 hr at 0°. The precipitate (dicyclohexylurea, 11.9 g) is filtered and discarded. The filtrate is evaporated to dryness, the crude produce (8.8 g) is dissolved in a minimal amount of hot ethyl acetate, the precipitate is discarded, and the product recrystallized [yield, 6.9 g (90%)].

Preparation of Acetylated Avidin

Avidin (Belovo, Belgium; 35.2 mg, 0.52 μmol) is dissolved in an aqueous solution (7.9 ml) containing 0.05 M NaHCO$_3$. Acetyl N-hydroxysuccinimide ester (0.32 ml of a 5 mg/ml solution in dimethyl formamide) is added and the reaction is carried out for 2.5 hr at 25°. The solution is dialyzed exhaustively against phosphate-buffered saline and the acetylated protein is stored at −20° in aliquots at a concentration of 1.6 mg/ml.

Preparation of Antibody-Coated Microtiter Plates

Each well of a microtiter plate (Flow Lab, U.K.) is filled with 100 μl of polyclonal anti-rabbit hCG IgG antibodies in carbonate buffer, pH 9.6, at a concentration of 10 μg protein/ml buffer. The plate is incubated overnight at 4°. The plate is then rinsed once with the wash solution (PBS containing 0.05% Tween 20). Each well is then filled with 100 μl of 0.3% bovine serum albumin in saline. The plate is incubated for 2 hr at room temperature, and rinsed once with wash solution. The plate is covered with paraffin, and stored at −70° until use.

Immunoassay Procedure

1. Add in triplicate to each well 100 μl of hCG standards (0, 4, 10, 25, 50, 100 mIU/ml) prepared in PBS containing 0.1% BSA (assay buffer). To the remaining wells add in triplicate 100 μl of urine or plasma samples derived from pregnant or nonpregnant subjects.

2. Incubate overnight at 4°.

3. Wash plates five times with wash solution.

4. Add to each well 100 μl of biotinylated anti-β-hCG at a 1 : 1000 dilution in assay buffer and incubate at 4° for 2 hr.

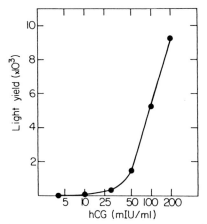

Fig. 2. hCG IBMA: dose–response curve. From Kohen et al.,[3] p. 156, by permission.

5. Wash the plate (at least three times), and add 100 μl of streptavidin or acetylated avidin[5] at a dilution of 1 : 200 in assay buffer. Incubate at room temperature for 15 min.

6. Wash the plate three times. Add 100 μl of biotinylated G6PDH at a dilution 1 : 50 in assay buffer to each well. Incubate at room temperature for 45 min.

7. Wash the plate three times. Add to each well the 80 μl of the substrates [glucose 6-phosphate/NAD (10 mg/mg/ml)] in phosphate buffer 0.1 M, pH 7.0, and 20 μl of the bacterial bioluminescence monitoring reagent (from LKB), containing luciferase/FMN/decanal. Incubate for 15 min at room temperature.

8. Transfer 80 μl of the reaction mixture from each well to a Luma-cuvette. Measure the light output for 10 sec.

Results

A typical dose–response curve for hCG utilizing the avidin–biotin system is shown in Fig. 2. A sensitivity of 15 mIU/ml of hCG is obtained. This sensitivity is satisfactory for detecting pregnancy in plasma or urine since the threshold value for pregnancy is 50 mIU hCG/ml.

---

[5] Avidin is a positively charged glycoprotein from egg white. In many cases either the charged properties or presence of carbohydrates cause high background levels in immunoassay. To counter this, bacterial streptavidin, which is a neutral protein possessing biotin-binding characteristics similar to that of avidin, can be substituted. We have found that by acetylating the avidin molecule as described here, the p*I* decreases to near neutrality, and the nonspecific background is substantially diminished.

Measurement of hCG in urine or plasma samples from pregnant and nonpregnant women was determined by immunobioluminometric assay and by radioimmunoassay. A good correlation was obtained ($n = 49$, $r = 0.94$).

Remarks

The work described here indicates that an immunobioluminometric assay for hCG can be used for the early detection of pregnancy.

The use of the avidin–biotin complex as probes in immunoassay systems is appealing for a variety of reasons: (1) biotinylation of antibodies or enzymes can be achieved under mild conditions and the immunological or enzymatic activity of the conjugates are only nominally affected; (2) biotinylated enzymes can be used as universal reagents; (3) in many cases, mediation via the avidin–biotin complex enables an amplified signal and increased sensitivity; and (4) avidin, streptavidin, and biotin, as well as conjugates, complexes, and derivatives thereof, are all available commercially.

In this procedure we described the stepwise addition of the various probes. Premade complexes, which comprise the appropriate probes (i.e., acetylated avidin or streptavidin, biotinyl antibody and/or biotinyl enzyme premixed in appropriate ratios) can also be used in order to minimize the number of steps necessary for the assay.

[26] Isolation and Expression of a cDNA Coding for Aequorin, the Ca²⁺-Activated Photoprotein from *Aequorea victoria*

By DOUGLAS C. PRASHER, RICHARD O. McCANN, and MILTON J. CORMIER

Aequorin is a bioluminescent protein isolated from the luminous hydromedusan *Aequorea victoria*[1] which emits blue light upon binding Ca²⁺ ($K_d \approx 1 \ \mu M$). This photoprotein was first described by Shimomura *et al.*[2]

---

[1] M. N. Arai and A. Brinckmann-Voss, *Can. Bull. Fish. Aquat. Sci.* **204,** 1 (1980).
[2] O. Shimomura, F. H. Johnson, and Y. Saiga, *J. Cell. Comp. Physiol.* **59,** 223 (1962).

METHODS IN ENZYMOLOGY, VOL. 133

It has been subsequently shown to be a single polypeptide whose approximate molecular weight is 20,000 and contains three high-affinity Ca^{2+}-binding sites.[3-6] In addition, aequorin contains coelenterate luciferin (1 mol/mol) whose structure is identical to that of *Renilla* luciferin.[7] Aequorin also contains some form of bound oxygen and, upon the binding of Ca^{2+}, catalyzes the oxidation of luciferin to oxyluciferin, CO_2, and light. The mechanism of this oxidation, as well as the nature of the products produced, is the same as that described for the luminescent oxidation of *Renilla* luciferin (reviewed in Cormier[8]). The resulting polypeptide, apoaequorin, can be separated from other reaction products and converted to aequorin by incubation in the presence of luciferin, dissolved oxygen, and 2-mercaptoethanol as illustrated in Fig. 1.[9-12]

Isolating the aequorin cDNA was desirable for a number of reasons. (1) It is a very laborious procedure to collect and process *Aequorea* so purification would be greatly simplified from a strain of *Escherichia coli* expressing the apoprotein. (2) Such a strain would yield unlimited quantities of apoaequorin not only for biochemical investigations but also for use as a nonisotopic label for immunodiagnostics. A bioluminescent label is ideal for such applications because commercially available instrumentation can easily detect attomol (10^{-18}) levels of aequorin (Fig. 2). (3) Site-directed mutagenesis could be performed on the gene to study structure–function relationships in the binding and oxidation of luciferin, the binding of Ca^{2+}, and the energy transfer to the *Aequorea* green fluorescent protein. (4) The translated product from a single aequorin cDNA would result in an aequorin preparation useful for X-ray crystallography. Native aequorin preparations are composed of numerous isoforms[13] which are believed to hinder crystallization of the protein. (5) Using the aequorin cDNA as a hybridization probe would not only permit us to isolate *Ae-*

[3] O. Shimomura, F. H. Johnson, and Y. Saiga, *J. Cell. Comp. Physiol.* **62,** 1 (1963).

[4] O. Shimomura and F. H. Johnson, *Nature (London)* **227,** 1356 (1970).

[5] J. R. Blinks, W. G. Wier, P. Hess, and F. G. Prendergast, *Prog. Biophys. Mol. Biol.* **40,** 1 (1982).

[6] D. G. Allen, J. R. Blinks, and F. G. Prendergast, *Science* **195,** 996 (1977).

[7] K. Hori, H. Charbonneau, R. C. Hart, and M. J. Cormier, *Proc. Natl. Acad. Sci. U.S.A.* **74,** 4285 (1977).

[8] M. J. Cormier, *in* "Bioluminescence in Action" (P. J. Herring, ed.), p. 75. Academic Press, London, 1978.

[9] O. Shimomura and F. H. Johnson, *Nature (London)* **256,** 236 (1975).

[10] O. Shimomura, F. H. Johnson, and H. Morise, *Biochemistry* **13,** 3278 (1974).

[11] O. Shimomura and F. H. Johnson, *Proc. Natl. Acad. Sci. U.S.A.* **75,** 2611 (1978).

[12] S. Inoue, H. Sugiura, H. Kakoi, K. Hasizuma, T. Goto, and H. Iio, *Chem. Lett.* p. 141 (1975).

[13] J. R. Blinks and G. C. Harrer, *Fed. Proc., Fed. Am. Soc. Exp. Biol.* **34,** 474 (1975).

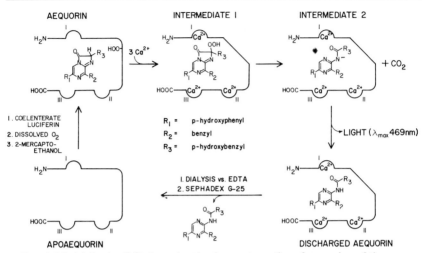

FIG. 1. Model for the Ca^{2+}-dependent luminescent reaction of aequorin and the conversion of apoaequorin to aequorin. Upon the addition of Ca^{2+}, the three high-affinity Ca^{2+} sites are filled causing a conformational change in the protein during which the oxygenated species is transferred to luciferin to form the putative intermediate, luciferin hydroperoxide. This intermediate decomposes to yield an apoaequorin–oxyluciferin complex (discharged aequorin) plus CO_2 and light.[8,9] The discharged aequorin can be dissociated by removal of bound Ca^{2+}, which is accomplished by dialysis against EDTA and gel filtration. The resulting apoaequorin can be converted to aequorin in the presence of synthetic coelenterate luciferin, dissolved oxygen, and 2-mercaptoethanol.[9–12]

quorea genomic clone(s) but also genes for similar photoproteins in other luminous hydrozoans and ctenophores.[14–18]

An Aequorea cDNA library was constructed using RNA isolated from the circumoral rings of the jellyfish. The poly(A)$^+$ RNA fraction was used to prepare double-stranded cDNA which was cloned into PstI-digested pBR322. E. coli transformants were screened for the presence of the aequorin cDNA using an oligonucleotide mixture whose sequence was based on the aequorin amino acid sequence.[19] Six transformants in the 6000 comprising the Aequorea cDNA bank hybridized to the oligonucleotide mixture. Aequorin activity was demonstrated in an extract containing one of the putative aequorin cDNAs.

[14] O. Shimomura, F. H. Johnson, and Y. Saiga, J. Cell. Comp. Physiol. 62, 9 (1963).
[15] J. G. Morin and J. W. Hastings, J. Cell. Physiol. 77, 305 (1971).
[16] A. K. Campbell, Biochem. J. 143, 411 (1974).
[17] W. W. Ward and H. H. Seliger, Biochemistry 13, 1500 (1974).
[18] W. W. Ward and H. H. Seliger, Biochemistry 13, 1491 (1974).
[19] H. Charbonneau, K. A. Walsh, R. O. McCann, F. G. Prendergast, M. J. Cormier, and T. C. Vanaman, Biochemistry 24, 6762 (1985).

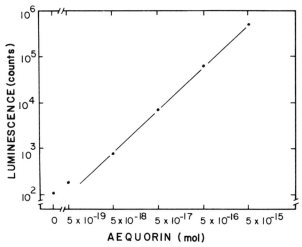

FIG. 2. Standard curve of the luminescence as a function of the aequorin concentration. Luminescence using native aequorin was determined as described in the text.

Aequorea cDNA Library Construction

Aequorea victoria are collected at the University of Washington's Friday Harbor Laboratories at Friday Harbor, Washington. The circumoral rings are cut from the jellyfish,[20] frozen in a dry ice/methanol bath, and stored at $-70°$ until needed. RNA is isolated and poly(A)$^+$ RNA is prepared as previously described.[21,22] We obtain 1.6 μg of poly(A)$^+$ RNA per gram of frozen tissue. Double-stranded cDNA is synthesized from the *Aequorea* poly(A)$^+$ RNA as described by Wickens *et al.*[23] We average 670 ng single-stranded cDNA synthesized from 5 μg of poly(A)$^+$ RNA, i.e., approximately 13% of the poly(A)$^+$ RNA is utilized as a template during the first strand synthesis. Upon treatment of the double-stranded cDNA with nuclease S1 the cDNAs are size fractionated by gel filtration.[24,25] After addition of homopolymeric dC tails, the double-stranded cDNA is

[20] J. R. Blinks, P. H. Mattingly, B. R. Jewell, M. van Leenwen, G. C. Harrer, and D. G. Allen, this series, Vol. 57, p. 292.

[21] Y.-J. Kim, J. Shuman, M. Sette, and A. Przybyla, *J. Cell Biol.* **96,** 393 (1983).

[22] H. Aviv and P. Leder, *Proc. Natl. Acad. Sci. U.S.A.* **69,** 1408 (1972).

[23] M. P. Wickens, G. N. Buell, and R. T. Schimke, *J. Biol. Chem.* **253,** 2483 (1978).

[24] H. M. Goodman and R. J. MacDonald, this series, Vol. 68, p. 75.

[25] H. Land, M. Grez, H. Hauser, W. Lindenmaier, and G. Schutz, this series, Vol. 100, Part B, p. 285.

annealed to dG-tailed *Pst*I-digested pBR322[26] and used to transform *E. coli* strain SK1592.[27]

We find the tailing reactions to be a very important factor in the ability to obtain *E. coli* transformants. Peacock *et al.*[28] have shown the lengths of the tails to affect the transformation efficiency to a large extent. We find that terminal transferase (Pl Biochemicals) requires the use of polymerization conditions different from other reports such that the reaction rate is slowed. We recommend that pilot tailing reactions be performed using *Pst*I-digested pBR322 in the presence of $[\alpha\text{-}^{32}P]$dGTP and $[\alpha\text{-}^{32}P]$dCTP separately to determine the polymerization rate of any lot of terminal transferase.

Tailing of PstI-Digested pBR322. The cloning vector is tailed for 10 min at 15° in 20 μl total volume in the following reaction mixture:

1 μg *Pst*I-digested pBR322 (0.36 pmol)
0.2 *M* potassium cacodylate, pH 7.5
2 m*M* MnCl$_2$
0.1 m*M* DTT
50 μg/ml BSA
50 μ*M* dGTP
10 U terminal transferase

The reaction is terminated by adding 1/10 volume of 0.2 *M* EDTA and the enzyme is inactivated at 70° for 10 min.

Tailing of Aequorea cDNA. The polymerization of dC residues by terminal transferase is much more difficult to control than that of dG residues so we recommend that this tailing reaction be performed in a variety of conditions. We find the reaction time to be the simplest variable to control. Deoxycytidine residues are added to the *Aequorea* cDNA in 30 μl total volume under the following conditions:

approximately 1 pmol cDNA
0.2 *M* potassium cacodylate, pH 7.5
1 m*M* CoCl$_2$
0.1 m*M* DTT
50 μg/ml BSA
1 μ*M* dCTP
19 U terminal transferase

[26] A. Otsuka, *Gene* **13,** 339 (1981).
[27] S. R. Kushner, *in* "Genetic Engineering" (H. W. Boyer and S. Nicosia, eds.), p. 17. Elsevier/North-Holland Biomedical Press, Amsterdam, 1978.
[28] S. L. Peacock, C. M. McIver, and J. J. Monahan, *Biochim. Biophys. Acta* **655,** 243 (1981).

The reaction is incubated at 15° and 10-μl aliquots are removed 2, 5, and 10 min after initiating the reaction with terminal transferase. The reaction is terminated as noted previously.

Annealing Reactions. The aliquots removed in the previous reaction contain cDNAs tailed to different extents and there is no efficient method to determine which aliquot contains cDNA with the proper tail lengths. Hence each cDNA aliquot is annealed to the cloning vector and the recombinant frequency of the transformants is used as an indication of which tailed cDNA is most useful. The annealing reactions are performed in 25 μl total volume under the following conditions:

10 mM Tris, pH 8
1 mM EDTA
100 mM NaCl
0.01 pmol dG-tailed *Pst*I-digested pBR322
0.08 pmol dC-tailed *Aequorea* cDNA

The reaction is incubated at 80° for 5 min in a beaker of water (75 ml) and then the beaker and its contents are transferred to a 45° water bath. The DNA is used to transform *E. coli* after the temperature has equilibrated. Tetracycline-resistant, ampicillin-sensitive colonies are transferred to microtiter dishes and frozen at −70° enclosed in an ammunition box to prevent drying of the cultures.

Identification of the Aequorin cDNAs. The cDNA library is screened using the following synthetic oligonucleotide mixture:

$$\begin{matrix} & & A & & \\ 3'-\text{ACCATATGGTACCTAGG}-5' \\ & & G & \quad T & \quad G \\ & & & C & \end{matrix}$$

These oligonucleotides code for all of the possible codons which represent the amino acids of the peptide, Trp[173]-Tyr-Thr-Met-Asp-Pro[178], located in the carboxy-terminal region of the aequorin polypeptide.[19] The 17-mers are radioactively labeled with polynucleotide kinase and [α-^{32}P]dATP[29] and the unincorporated ^{32}P is removed by its purification on DEAE cellulose. We routinely obtain a specific activity of 10^8 cpm/μg when we label 40 pmol of oligomers in the presence of 200 μCi[α-^{32}P]dATP.

The *E. coli* recombinants stored in the microtiter dishes are transferred to nitrocellulose filters (7 × 11 cm) placed on Luria agar plates containing tetracycline (20 μg/ml). The cultures are transferred from the 96-well microtiter plates simultaneously to the nitrocellulose filters using

---

[29] A. M. Maxam and W. Gilbert, this series, Vol. 65, p. 499.

a metal 96-prong device. The colonies are grown for 12 hr at 37° and the filters are treated as Taub and Thompson[30] described for using Whatman 541 paper instead of nitrocellulose. We find their treatment with proteinase K significantly lowers nonspecific hybridization. The filters are baked under vacuum for 2 hr once they have air dried.

The filters are wetted in 1× SSC and incubated in a heat-sealable plastic bag at 55° for 12–20 hr in 3 ml per filter of a prehybridization solution (10× NET,[31] 0.1% SDS, 3× Denhardt's). The hybridization bag is cut open and the prehybridization solution is replaced with 1 ml per filter of the hybridization solution (10× NET, 0.1% SDS, 3× Denhardt's, 1 × 10^6 cpm ^{32}P-labeled 17-mers). The resealed bag is incubated with shaking in a water bath for 24 hr at 37°. The filters are washed four times in 10× SSC at 4° for 10 min and then wrapped in plastic film after they are air dried. Kodak XAR-5 film is exposed overnight to the wrapped filters at −70° using a DuPont Cronex intensifying screen.

Expression of Apoaequorin in *E. coli* Extracts

Using this procedure we found six *E. coli* strains in our *Aequorea* cDNA library which contained plasmid that hybridized to the oligonucleotide mixture. A simple method to prove that one of the cDNAs contains the aequorin sequence is to demonstrate photoprotein activity in one of the *E. coli* extracts. This is possible because *E. coli* does not contain any contaminating activities and the aequorin assay is very sensitive ($<10^{-18}$ mol).

The *E. coli* extracts are prepared as follows. Grow 25 ml cultures in Luria broth overnight at 37°. Pellet the cells by centrifugation and resuspend them in 5 ml of 10% sucrose, 50 mM Tris, pH 8, 12.5 mM EDTA, 2 mg/ml lysozyme, and 20 μg/ml RNase A. Incubate the mixtures on ice for 45 min and then centrifuge at 43,500 g for 1 hr. The supernatants are assayed for photoprotein activity. To 0.5 ml of each extract add 2-mercaptoethanol and synthetic coelenterate luciferin[7,32] to final concentrations of 2 and 0.1 mM, respectively. Incubate overnight at 4° to charge the apoaequorin. Measure the peak light intensity and total photons using a photometer[33] by injecting 5 μl of the charged extract into 0.5 ml of 0.1 M CaCl$_2$, 0.1 M Tris, pH 8.

[30] F. Taub and E. B. Thompson, *Anal. Biochem.* **126**, 222 (1982).
[31] 1× NET = 0.15 M NaCl, 15 mM Tris, pH 7.5, 1 mM EDTA; 1× Denhardt's = 0.02% BSA, 0.02% Ficoll, 0.02% poly(vinylpyrrolidone); 1× SSC = 0.15 M NaCl, 15 mM sodium citrate.
[32] K. Hori, J. M. Anderson, W. W. Ward, and M. J. Cormier, *Biochemistry* **14**, 2371 (1975).
[33] J. M. Anderson, G. J. Faini, and J. E. Wampler, this series, Vol. 57, p. 529.

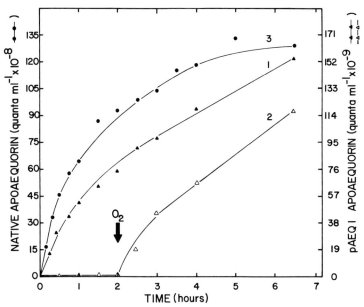

FIG. 3. Time- and oxygen-dependent formation of Ca^{2+}-dependent luminescence in AEQ1 extracts and native apoaequorin. For this experiment, the apoaequorin activity in AEQ1 extracts was partially purified as described.[34] Conditions used: in curves 1 and 2, 0.5-ml aliquots of the active fractions charged as described in the text. At appropriate times, 5-μl aliquots were removed and assayed for photoprotein activity. In curve 2, dissolved O_2 levels were reduced by bubbling with argon and the mixture exposed to oxygen at the time indicated. In curve 3 native apoaequorin was used in the incubation mixture instead of the AEQ1 extract.

Of the six recombinants we isolated from our *Aequorea* cDNA library one contained apoaequorin which could be charged with luciferin. Light emission from the extract of this recombinant was dependent upon the addition of Ca^{2+}. This strain (AEQ1) contained a plasmid designated pAEQ1.[34]

The *E. coli*-expressed photoprotein has properties similar to native aequorin.[34] The kinetics of formation (Fig. 3) and the requirements for the formation of this photoprotein activity were identical to that observed when authentic apoaequorin was used. Figure 3 shows dissolved O_2 was required. Furthermore, the elimination of either 2-mercaptoethanol or coelenterate luciferin from the reaction mixture resulted in no production of Ca^{2+}-dependent photoprotein activity. Injection of the active compo-

[34] D. Prasher, R. O. McCann, and M. J. Cormier, *Biochem. Biophys. Res. Commun.* **126,** 1259 (1985).

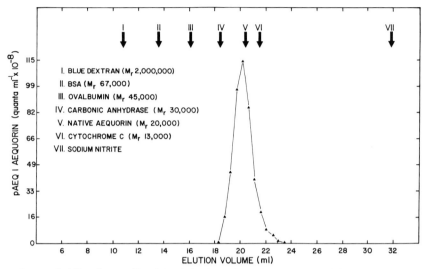

FIG. 4. Gel filtration profile of the Ca^{2+}-dependent photoprotein activity generated from AEQ1 extracts. The same preparation used in Fig. 3 was chromatographed on a G-75 column (0.8×32 cm) equilibrated with 10 mM EDTA, 15 mM Tris, pH 7.5, and 100 mM KCl. Photoprotein activity was assayed as described in the text. The elution volumes of various molecular weight markers are indicated.

nent into Ca^{2+}-free buffers produced no luminescence. The subsequent addition of Ca^{2+} resulted in a luminescence flash.

The extract was further characterized by ion-exchange chromatography and gel filtration. The apoaequorin activity eluted from a DEAE-cellulose column at approximately 0.3 M NaCl, which is similar to that observed for authentic apoaequorin. After the partially purified component in AEQ1 extracts was charged with luciferin, it was subjected to gel filtration, and it eluted from the column slightly ahead of aequorin (Fig. 4). We calculated an M_r of 20,600 for the AEQ1 photoprotein and 19,600 for native aequorin. Similar results were observed during *in vitro* translation experiments.[34] We also concluded from the data of Fig. 4 that the luciferin becomes tightly associated with the active component in AEQ1 extracts under the charging conditions used. The Ca^{2+}-dependent luminescence of the pooled photoprotein (Fig. 4) had kinetics and a wavelength distribution similar to that observed for native aequorin. Hence we concluded pAEQ1 codes for apoaequorin.[34]

Increasing the Expression Level of Apoaequorin in E. coli. A low level of apoaequorin was expressed in an *E. coli* strain containing pAEQ1.[34] The expression level was increased by subcloning the apoaequorin gene from pAEQ1 into another plasmid (pUC9) in such a manner that tran-

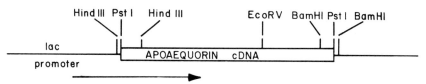

FIG. 5. Schematic representation of the aequorin cDNA in pAEQ8 which contains the subcloned *Pst*I fragment from pAEQ1. The apoaequorin coding sequence reads from left to right as drawn. The arrow indicates the direction of transcription from the lac promoter of pUC9. The *Pst*I fragment has the reverse orientation in pAEQ7.

scription of the apoaequorin gene was increased due to its position relative to the inducible lac promoter.

Two different plasmids, pAEQ7 and pAEQ8, were constructed by subcloning the *Pst*I insert of pAEQ1 into pUC9. pAEQ8 contains the *Pst*I insert in the desired orientation shown in Fig. 5, whereas pAEQ7 contains the fragment inserted in the opposite orientation.

Individual *E. coli* strains (host: JM105) containing pAEQ7 and pAEQ8 were grown at 37° in 20 ml Luria broth containing ampicillin (50 μg/ml). When the OD$_{550}$ reached 0.6, 0.2 ml of a 100 mM IPTG (isopropyl-β-D-thiogalactopyranoside) solution was added. After 0, 1, and, 2 hr, 5-ml aliquots were removed, and the cell pellets frozen.

APOAEQUORIN EXPRESSION LEVELS IN *E. coli* STRAINS CONTAINING
pAEQ1, pAEQ7, AND pAEQ8

Plasmid	Hours after induction with IPTG	Aequorin specific activity (photons mg^{-1})	Soluble protein represented by aequorin (%)	
			Assuming 1% charging efficiency	Assuming 20% charging efficiency
pAEQ7	0	≤0.02	—	0
	1	≤0.04	—	0
	2	≤0.02	—	0
pAEQ8	0	0.668	0.042 (11)[a]	0.002 (10)
	1	17.2	1.07 (268)	0.05 (263)
	2	38.4	2.40 (600)	0.12 (600)
pAEQ1	—	0.061	0.004	0.0002
Strain				
SK1592	—	0	—	0

[a] The numbers in parentheses represent the fold increase over the activity observed in extracts containing pAEQ1.

Extracts of the cells were prepared and photoprotein activity was determined as described previously. The table contains the results obtained when we grew strains containing pAEQ1, pAEQ7, pAEQ8, or no plasmid. The aequorin activity in these extracts was dependent upon the charging efficiency with luciferin. It has not yet been possible to quantitate the charging efficiency in the crude *E. coli* extracts so calculations based on 1 and 20% charging efficiency are included since we have observed no higher than a 20% charging efficiency using apoaequorin isolated from *Aequorea*. The expression from pAEQ8 was significantly higher (600-fold) 2 hr postinduction when compared to expression from pAEQ1. No induction of the aequorin gene was observed from pAEQ7 which contains the gene in the opposite orientation as in pAEQ8. These data demonstrate that the apoaequorin expression level can be significantly increased.

Acknowledgments

We are indebted to Dr. Alan Przybyla and Jon Shuman from the University of Georgia for their many helpful discussions and for providing the laboratory space where this work was performed. This work would not have been possible without the cooperation of the personnel at The Friday Harbor Laboratories under the directorship of Dr. A.O.D. Willows. We would also like to thank Dr. Harry Charbonneau for providing us with amino acid sequence information on aequorin.

[27] Molecular Cloning of Apoaequorin cDNA

By MASATO NOGUCHI, FREDERICK I. TSUJI, and YOSHIYUKI SAKAKI

Introduction

Aequorea victoria has photogenic organs on each side of the base of tentacles present along the margin of the umbrella. A photoprotein aequorin (MW 21,000) has been isolated from the photogenic organ.[1,2] The protein emits a blue light in the presence of a trace amount of Ca^{2+} using coelenterazine as a functional chromophore. The chemical structure of the chromophore has been elucidated,[3] but little is known about the

[1] O. Shimomura, F. H. Johnson, and Y. Saiga, *J. Cell. Comp. Physiol.* **59**, 223 (1962).
[2] J. R. Blinks, P. H. Mattingly, B. R. Jewell, M. van Leeuwen, G. C. Harrer, and D. G. Allen, this series, Vol. 57, p. 292.
[3] F. H. Johnson and O. Shimomura, this series, Vol. 57, p. 271.

molecular characteristics of the protein moiety which also plays an essential role in the luminescent reaction. Molecular cloning studies of apoaequorin cDNA may be expected to provide important information on the structure of the protein moiety, the mechanism of the reaction, and also open ways to produce large amounts of aequorin by recombinant DNA techniques.[4] This chapter describes a method for cloning apoaequorin cDNA from *Aequorea* mRNA. The method would also be applicable for the cloning of DNAs of jellyfish proteins other than apoaequorin.

Preparation of Total Cellular RNA

The jellyfish, *A. victoria,* about 3 in. in diameter, can be obtained in large numbers at the Friday Harbor Laboratories, Friday Harbor, Washington, from July through September. Preparation should be started with at least 400 jellyfish. Total cellular RNA can be extracted from the outer margin of the umbrella by the phenol extraction technique. Immediately after cutting off thin rings of tissue containing photogenic organs of *A. victoria,* the tissues (350 g wet weight) are mixed with 1/10 (v/w) of buffer A [10 mM sodium acetate, pH 5.5, 10 mM Na₃ EDTA and 5 μg of poly(vinyl sulfate) per ml] and 1/100 (v/w) of 10% SDS (sodium dodecyl sulfate) and homogenized with 350 ml of buffer A-saturated phenol for 7 min in a Waring blender at top speed. The homogenate is separated into two phases by centrifugation at 9000 rpm for 15 min at 4° in a Sorvall SS34 rotor. The upper aqueous phase is treated again with an equal volume of the buffer A-saturated phenol and centrifuged. The organic phase from the first centrifugation is also subjected to reextraction with 350 ml of buffer A and centrifuged. After the centrifugations, the two aqueous phases are combined and mixed with 2 volumes of ethanol. The solution is kept at −80° for 1 hr and the resulting precipitate is harvested by centrifugation at 9000 rpm for 15 min at 4° in a Sorvall SS34 rotor.

The pellet containing RNA is dissolved in the minimum volume of 20 mM Na₃ EDTA, pH 7.0, and any undissolved white material is removed by centrifugation at 6000 rpm for 20 min at 0°. The supernatant is mixed with an equal volume of 6 *M* sodium acetate, pH 5.5, and kept at −20° for 30 min. The precipitate containing high-molecular-weight RNA is harvested by centrifugation at 6000 rpm for 15 min at −10° in a Sorvall SS34 rotor. The dissolution of RNA with 20 mM Na₃ EDTA and precipitation with 3 *M* sodium acetate should be repeated at least 3 times to remove

[4] S. Inouye, M. Noguchi, Y. Sakaki, Y. Takagi, T. Miyata, S. Iwanaga, T. Miyata, and F. I. Tsuji, *Proc. Natl. Acad. Sci. U.S.A.* **82,** 3154 (1985).

contaminants such as proteins, polysaccharides, and low-molecular-weight RNAs. The pellet is dissolved in a minimum volume of 10 mM Tris–HCl, pH 7.5/1 mM EDTA. The RNA is reprecipitated by adding 2 volumes of ethanol in the presence of 0.15 M sodium acetate. The final pellet is rinsed with 66% (v/v) ethanol containing 0.1 M sodium acetate, pH 5.5, and the RNA is dried in a vacuum desiccator and stored at $-80°$. The RNA at this stage can be stored at $-80°$ for at least 2 months for subsequent manipulation. In our experience, the extraction methods using guanidinium–HCl[5] or guanidinium isothiocyanate,[6] followed by centrifugation with a CsCl cushion, were not effective for jellyfish RNA. Jellyfish appears to contain large amounts of material which cosediments with RNA through a CsCl cushion and interferes with the subsequent purification steps.

Oligo(dT)-Cellulose Column Chromatography

RNA is dissolved in TE buffer (10 mM Tris–HCl, pH 7.5/1 mM Na$_2$ EDTA) to give an absorbance of about 25 at 260 nm. After heat denaturation at 65° for 5 min, the RNA solution is chilled immediately in an ice both and 1/10 volume of 5 M NaCl is added to the solution. The solution is applied to an oligo(dT)-cellulose column [50 mg of type 7 oligo(dT)-cellulose, P-L Biochemicals, Milwaukee, WI] at room temperature, and the column is washed with 10 bed volumes (10 ml) of 10 mM Tris–HCl, pH 7.5/1 mM Na$_3$ EDTA/0.5 M NaCl to remove unbound RNA. Poly(A)$^+$ RNA is eluted with 1 ml of TE buffer. About 5 μg of poly(A)$^+$ RNA can be obtained from 5 mg of total cellular RNA. In a typical experiment, the ratios of A_{280}/A_{260} and A_{230}/A_{260} were 0.52 and 0.66, respectively. Relatively high values of A_{230}/A_{260} may indicate that the RNA preparation still contains some contaminants, but the RNA preparation is pure enough for the following experiments.

Translation of RNA in a Rabbit Reticulocyte Cell-Free Translation System

The activity of the RNA for translation should be checked. The Mg^{2+} and K$^+$ concentrations are critical for the rabbit reticulocyte translation system. Results of a typical experiment are presented in Fig. 1. The

[5] R. G. Deeley, J. I. Gordon, A. T. H. Burns, K. P. Mullinix, M. Bina-Stein, and R. F. Goldberger, *J. Biol. Chem.* **252,** 8310 (1977).
[6] A. Ullrich, J. Shine, J. Chirgwin, R. Pictet, E. Tischer, W. J. Rutter, and H. M. Goodman, *Science* **196,** 1313 (1977).

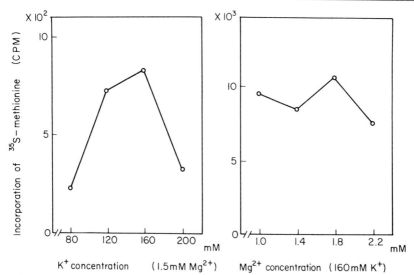

FIG. 1. Effects of K^+ and Mg^{2+} concentration on translation activity of *Aequorea* RNA in rabbit reticulocyte lysate. Ten micrograms of total RNA from *Aequorea* was subjected to *in vitro* translation using rabbit reticulocyte lysate (New England Nuclear, Boston, MA) in the presence of K^+ and Mg^{2+} at the concentrations indicated. Incorporation of [^{35}S]methionine into acid-insoluble fraction was measured.

optimum concentrations for K^+ and Mg^{2+} were 120–160 and 1–2 mM, respectively. Reaction products at the optimum concentration of K^+ and Mg^{2+} should be analyzed by SDS-polyacrylamide gel electrophoresis according to Laemmli.[7] Fig. 2 shows an example of such an analysis. The electrophoretic profiles of the translated products by total RNA and poly(A)$^+$ RNA appear almost identical, suggesting that most mRNAs of jellyfish have poly(A) tails.

Construction of cDNA Library

Three micrograms of poly(A)$^+$ RNA prepared as above are converted to cDNA with avian myeloblastosis virus reverse transcriptase (Life Sciences, St. Petersburg, FL) using 0.07 μg of primer DNA having a poly(T)$_{50-70}$ tail as described by Okayama and Berg.[8] The average length of cDNA synthesized (single-stranded) can be estimated from the incorporation of (α-^{32}P]dCTP into a TCA-insoluble fraction. In the present work it

[7] U. K. Laemmli, *Nature* (*London*) **227**, 680 (1970).
[8] H. Okayama and P. Berg, *Mol. Cell. Biol.* **2**, 161 (1982).

total RNA poly(A⁺)RNA

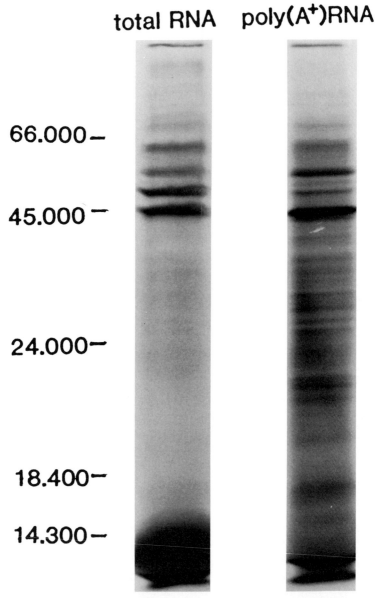

FIG. 2. Gel electrophoretic pattern of *in vitro* translation products of *Aequorea* RNA Total and poly(A)⁺ RNAs from *Aequorea* were subjected to *in vitro* translation in rabbit reticulocyte lysate in the presence of [³⁵S]methionine. Products were analyzed by SDS–polyacrylamide gel electrophoresis followed by autoradiography.

was about 230 base pairs. The cDNA is then converted to double-stranded DNA and cloned into *Escherichia coli* DH1 according to the method of Okayama and Berg.[8] Using this procedure about 1.5×10^4 independent clones were obtained. Clones are grown in L-broth containing 100 μg/ml of ampicillin and can be stored at $-80°$ in the presence of 14% glycerol.

Partial Amino Acid Sequences and Oligonucleotide Probes

For selecting the clones carrying aequorin cDNA, two methods can be used: colony hybridization using oligonucleotide probes and screening by specific antibody. Herein colony hybridization is employed. For preparing probes, the partial amino acid sequences are at first required. The sequences can be determined by the method of Edman degradation.[9] In the work described here the purified aequorin was cleaved by BrCN and the fragments were isolated by high-performance liquid chromatography. Each fragment was analyzed with a Beckman 890D Amino Acid Sequencer and an Applied Biosystems Gas-Phase Sequenator.[4] A few portions of the sequences having minimum codon ambiguity are selected to synthesize the oligonucleotide probes (Fig. 3). Oligonucleotides can be chemically synthesized by the phosphoramidate method using an Applied Biosystems DNA Synthesizer. Probes are labeled with ^{32}P using T$_4$ polynucleotide kinase and [α-^{32}P]ATP as described by Maniatis *et al.*[10]

Colony Hybridization

Bacteria are grown on nitrocellulose filter (Schleicher and Schuell, BA 85) and plasmid DNA is fixed on the filter by the standard procedure as described by Maniatis *et al.*[10] For the first screening, hybridization is carried out in 1.1 M NaCl/60 mM NaH$_2$PO$_4$, pH 7.4/6 mM EDTA/0.2% Ficoll/0.2% poly(vinylpyrrolidone)/0.2% bovine serum albumin/0.5% SDS/250 μg of yeast tRNA per ml/0.1 ng of ^{32}P-labeled probe DNA (AQ) per ml at 30° for 12 hr. The filter is washed three times with 6× SSC (1× SSC = 0.15 M NaCl/0.015 M sodium citrate, pH 7.0) at room temperature, then twice with the same solution at 37°, and subjected to autoradiography. More than 100 colonies were found to give positive signals in the first screening. These positive clones are subjected to hybridization with probes AQ1, AQ2, AQ3, and AQ4 under the same conditions as described above. Only AQ1 can hybridize with these clones. In the second screen-

---

[9] T. Miyata, K. Usui, and S. Iwanaga, *J. Biochem. (Tokyo)* **95,** 1793 (1984).
[10] T. Maniatis, E. F. Fritsch, and J. Sambrook, "Molecular Cloning, A Laboratory Manual." Cold Spring Harbor Lab., Cold Spring Harbor, New York, 1982.

```
Val-Lys-Leu-(Asp)-(Phe)-Asp-Phe-Asp-Asn-Pro-----
Ser
Gly
Leu


Val-Lys-Leu-Asn-Thr-Asp-Phe-Asp-Asn-Pro-----

 Probe AQ        5'      GAC TTC GAC AAC CC        3'
                          T   T   T   T


Met-(Phe or Thr)-Asn-(Phe)-Leu-Asp-Val-Asn-----

                        Met-Asp-Pro-Ala-Cys-Glu-Lys-Leu-Tyr-Gly-

 Probe AQ 1     5'      ATG GAT CCT GCT TG        3'
                                 C   C
                                     A
                                     G


 Probe AQ 2     5'      ATG GAT CCA GCT TG        3'
                                 G   C
                                     A
                                     G


 Probe AQ 3     5'      ATG GAT CCT GCT TG        3'
                                 C   C
                                     A
                                     G


 Probe AQ 4     5'      ATG GAC CCA GCT TG        3'
                                 G   C
                                     A
                                     G
```

FIG. 3. Probe sequences for apoaequorin used in cDNA screening. Amino acid sequences of peptides from aequorin were determined as described.[4] Portions having minimal codon ambiguity were chosen for chemical synthesis of oligonucleotide probes.[4]

ing, hybridization conditions are varied. The melting temperature (T_m) for AQ and AQ1 probes can be calculated to be 38–46 and 44–48°, respectively, assuming that the AT pair and GC pair make a contribution of 2 and 4°, respectively, to T_m of double-stranded DNA. Hybridization is carried out in the same solution for the first screening at 33° for probe AQ, or at 39° for AQ1. In addition, hybridization is also carried out at 35° for AQ and 41° for AQ1. After hybridization for 12 hr, the filters are each washed four times with 6× SSC at 0° for 10 min and twice with 6× SSC at 37° for 10 min. By these tests, several highly positive and moderately positive clones can be isolated. The restriction maps of the inserted DNA should be constructed. The size of apoaequorin cDNA will be about 700 base pairs. The maps of the inserted DNA may be very similar but not identical with each other. This may indicate that aequorin has some isoproteins.

```
                                                                    (G)₃₃AA  -80

TGCAATTCATCTTTGCATCAAAGAATTACATCAAATCTCTAGTTGATCAACTAAATTGTCTCGACAACAACAAGCAAAC   -1
                                  -1   1
Met Thr Ser Lys Gln Tyr Ser Val Lys Leu Thr Ser Asp Phe Asp Asn Pro Arg Trp Ile
ATG ACA AGC AAA CAA TAC TCA GTC AAG CTT ACA TCA GAC TTC GAC AAC CCA AGA TGG ATT   60
                                                                        ─────────→

                           20
Gly Arg His Lys His Met Phe Asn Phe Leu Asp Val Asn His Asn Gly Lys Ile Ser Leu
GGA CGA CAC AAG CAT ATG TTC AAT TTC CTT GAT GTC AAC CAC AAT GGA AAA ATC TCT CTT   120

              40
Asp Glu Met Val Tyr Lys Ala Ser Asp Ile Val Ile Asn Asn Leu Gly Ala Thr Pro Glu
GAC GAG ATG GTC TAC AAG GCA TCT GAT ATT GTC ATC AAT AAC CTT GGA GCA ACA CCT GAG   180
              ────→

                           60
Gln Ala Lys Arg His Lys Asp Ala Val Glu Ala Phe Phe Gly Gly Ala Gly Met Lys Tyr
CAA GCC AAA CGA CAC AAA GAT GCT GTA GAA GCC TTC TTC GGA GGA GCT GGA ATG AAA TAT   240
              ────→

                           80
Gly Val Glu Thr Asp Trp Pro Ala Tyr Ile Glu Gly Trp Lys Lys Leu Ala Thr Asp Glu
GGT GTG GAA ACT GAT TGG CCT GCA TAT ATT GAA GGA TGG AAA AAA TTG GCT ACT GAT GAA   300
                                                            ←────

                           100
Leu Glu Lys Tyr Ala Lys Asn Glu Pro Thr Leu Ile Arg Ile Trp Gly Asp Ala Leu Phe
TTG GAG AAA TAC GCC AAA AAC GAA CCA ACG CTC ATC CGT ATA TGG GGT GAT GCT TTG TTT   360
              ────→           ────

                           120
Asp Ile Val Asp Lys Asp Gln Asn Gly Ala Ile Thr Leu Asp Glu Trp Lys Ala Tyr Thr
GAT ATC GTT GAC AAA GAT CAA AAT GGA GCC ATT ACA CTG GAT GAA TGG AAA GCA TAC ACC   420
─────────────             ←────                              ────→

                           140
Lys Ala Ala Gly Ile Ile Gln Ser Ser Glu Asp Cys Glu Glu Thr Phe Arg Val Cys Asp
AAA GCT GCT GGT ATC ATC CAA TCA TCA GAA GAT TGC GAG GAA ACA TTC AGA GTG TGC GAT   480
────→

                           160
Ile Asp Glu Ser Gly Gln Leu Asp Val Asp Glu Met Thr Arg Gln His Leu Gly Phe Trp
ATT GAT GAA AGT GGA CAA CTC GAT GTT GAT GAG ATG ACA AGA CAA CAT TTA GGA TTT TGG   540

                           180
Tyr Thr Met Asp Pro Ala Cys Glu Lys Leu Tyr Gly Gly Ala Val Pro ***
TAC ACC ATG GAT CCT GCT TGC GAA AAG CTC TAC GGT GGA GCT GTC CCC TAA GAAGCTCTACG   602
                                          ←────

GTGGTGATGCACCCTGGGAAGATGATGTGATTTTGAATAAAACACTGATGAATTCAATCAAAATTTCCAAATTTTTGA   681

ACGATTTCAATCGTTTGTGTTGATTTTTGTAATTAGGAACAGATTAAATCGAATGATTAGTTGTTTTTTTAATCAACAG   760

AACTTACAAATCGAAAAAGT(A)₆₄
```

FIG. 4. Nucleotide sequence of an aequorin cDNA clone and the predicted amino acid sequence.[4] The amino acid sequences which coincided with those obtained by Edman degradation are shaded. The vertical arrow shows the presumed NH₂-terminus of aequorin. The horizontal arrows show the regions where the amino acid compositions coincided with those obtained from lysyl endopeptidase digestion. The numbers at the right and those on the amino acid residues show the nucleotide positions and the amino acid positions, respectively. (G)ₙ and (A)ₙ are oligo(G) and oligo(A) residues. The putative poly(A) addition signals AATAAA, ATTAAA, and AATCAA are underlined.

Identification of Apoaequorin cDNA

The clones obtained above are subjected to final tests as to whether they are actually apoaequorin clones. It is possible to identify the apoaequorin clones using the following criteria: (1) the cDNA sequences should encode the amino acid sequences obtained by protein analysis, (2) the amino acid sequence deduced from the cDNA sequence should be compatible with the properties of aequorin, and (3) the protein produced by the cDNA sequence should possess aequorin activity in the presence of coelenterazine and Ca^{2+}.

Sequence analysis of cDNA should provide information regarding criteria (1) and (2). DNA sequencing may be carried out by the method of Maxam and Gilbert.[11] Apoaequorin may be detected by regenerating aequorin in the presence of coelenterazine and 2-mercaptoethanol.[12] The clone which we analyzed fulfilled the above three criteria. (1) The cDNA contained sequences encoding amino acid sequences obtained by protein analysis (Fig. 4). (2) The amino acid composition deduced from the cDNA sequence is consistent with that obtained by protein analysis, the molecular weight of protein deduced from the cDNA sequence is compatible with that determined by SDS–gel electrophoresis.[4] (3) The amino acid sequence deduced from the cDNA sequence contained three E-F hand structures which are characteristic for Ca^{2+}-binding sites of proteins. (4) Extracts obtained from cultures of the clone gave aequorin activity after incubation with coelenterazine and 2-mercaptoethanol. These results provide conclusive evidence that the clone obtained by the above procedure is an apoaequorin cDNA clone. A paper reporting the expression of apoaequorin cDNA has appeared elsewhere[13] and the cloning of the cDNA by the same group is described in this volume.[14]

Acknowledgments

The research was supported in part by grants from the Japan Society for the Promotion of Science and the National Science Foundation (INT 82-10587, PCM 82-15773).

[11] A. M. Maxam and W. Gilbert, this series, Vol. 65, p. 499.
[12] O. Shimomura and F. H. Johnson, *Nature (London)* **256,** 236 (1975).
[13] D. Prasher, R. O. McCann, and M. J. Cormier, *Biochem. Biophys. Res. Commun.* **126,** 1259 (1985).
[14] D. C. Prasher, R. O. McCann, and M. J. Cormier, this volume [26].

[28] Cell-Free Components in Dinoflagellate Bioluminescence. The Particulate Activity: Scintillons; The Soluble Components: Luciferase, Luciferin, and Luciferin-Binding Protein

By J. Woodland Hastings and Jay C. Dunlap

Introduction

The bioluminescent marine dinoflagellates emit light as rapid (0.1 sec) flashes following stimulation,[1,2] and also spontaneously.[3] In the *Gonyaulax polyedra* cell, the light is emitted from many organelles, recently identified as dense bodies resembling microbodies, or peroxisomes.[4,5] Comparative studies indicate that the cellular and the biochemical components are the same or similar in different species.[6]

In extracts of *G. polyedra* made at pH 8, particulate bodies that possess the potential for bioluminescence can be isolated: after centrifugation and resuspension of the particles, a (rapid) decrease in the pH to 5.7 results in the emission of a flash which closely mimics the flash of the living cell.[7] These particles, referred to as scintillons, and presumably derived from the dense bodies mentioned above, possess the several molecular elements involved in the bioluminescent system[8]: the enzyme luciferase (Lase),[8a] the substrate luciferin (LFN), and a LFN-binding protein (LBP). These biochemical components, which are also found in varying amounts in the initial soluble fraction of extracts of *G. polyedra* and other species,[6] are believed to be functionally assembled in scintillons so

---

[1] J. W. Hastings and B. M. Sweeney, *J. Cell. Comp. Physiol.* **49**, 209 (1957).

[2] R. Eckert, *Science* **151**, 349 (1966).

[3] R. Krasnow, J. Dunlap, W. Taylor, J. W. Hastings, W. Vetterling, and V. D. Gooch, *J. Comp. Physiol.* **138**, (1980).

[4] C. H. Johnson, S. Inoué, A. Flint, and J. W. Hastings, *J. Cell Biol.* **100**, 1435 (1985).

[5] M. T. Nicolas, C. H. Johnson, J. M. Bassot, and J. W. Hastings, *Cell Biol. Int. Rep.* **9**, 797 (1985).

[6] R. Schmitter, D. Njus, F. M. Sulzman, V. Gooch, and J. W. Hastings, *J. Cell. Physiol.* **87**, 123 (1976).

[7] M. Fogel, R. Schmitter, and J. W. Hastings, *J. Cell Sci.* **11**, 305 (1972).

[8] J. P. Henry and J. W. Hastings, *Mol. Cell. Biochem.* **3**, 81 (1974).

[8a] Abbreviations: Lase, luciferase; LFN, luciferin; LBP, luciferin-binding protein; EDTA, ethylenediaminetetraacetic acid; SDS, sodium dodecyl sulfate, DTT, dithiothreitol; 2ME, 2-mercaptoethanol; ϕ, quantum yield; q, quanta; Tris, tris(hydroxymethyl)aminoethane.

as to facilitate the production of flashes. In a proposed model,[9] the luciferin-binding protein is postulated to act by keeping the luciferin sequestered from the luciferase until a pH change releases it, thus triggering the flash. We describe here the isolation and purification of the scintillons and several soluble components (Lase, LBP) from *G. polyedra* and LFN from *Pyrocyctis lunula*.

Buffers and Medium

Buffers

A: 0.05 M Tris–HCl, 0.01 M EDTA, 1 mM DTT, pH 8
B: 0.1 M Tris–HCl, 0.01 M EDTA, 3 mM DTT, pH 8.5
C: 1 mM Tris–HCl, 0.1 mM EDTA, 3 mM DTT, pH 8.5

Supplemented Sea Water Culture Medium (Modified f/2)[10]

Metal mix, 1.0 ml (see below for final concentration)
Vitamin mix, 0.1 ml
$NaH_2PO_4 \cdot H_2O$ (5 g/liter), 1.0 ml (36 μM final concentration)
$NaNO_3$ (75 g/liter), 1.0 ml (880 μM final concentration)
Soil extract, 5.0 ml
Glass distilled water, 193 ml
Filtered seawater, 800 ml

Stock Solutions (in Glass Distilled Water)

Metal Mix

Primary stocks, each in 100 ml (final concentrations in f/2 given in parentheses)

$CuSO_4 \cdot 5H_2O$	0.98 g (0.04 μM)
$ZnSO_4 \cdot 7H_2O$	2.2 g (0.08 μM)
$CoCl_2 \cdot 6H_2O$	1.05 g (0.05 μM)
$MnCl_2 \cdot 4H_2O$	18 g (0.9 μM)
$NaMoO_4 \cdot 2H_2O$	0.63 g (0.03 μM)

For metal mix stock dissolve 3.15 g $FeCl \cdot 6H_2O$ (11.7 μM final) and 4.36 g Na_2EDTA (11.7 μM final) in ~900 ml distilled water; add 1 ml of each trace metal primary stock, and bring to 1 liter. Use 1 ml/liter to make medium f/2.

---

[9] J. W. Hastings, *in* "Luminescence in Plants" (Govindjee, Fork, and Amesz, eds.). Academic Press, Orlando (in press).
[10] R. R. L. Guillard and J. H. Ryther, *Can. J. Microbiol.* **8**, 229 (1962).

Vitamin Mix

Stock
  Thiamine–HCl, 100 mg
  Biotin, 0.5 mg
  Vitamin B_{12}, 0.5 mg
  Distilled water, 100 ml

The vitamin mix stock is dispensed in 1 or 2 ml lots in ampoules, autoclaved, and stored in the ice box. Use 0.1 ml/liter of f/2 medium.

Cell Cultures, Harvesting, and Extraction

Unialgal (not necessarily axenic) cultures of the photosynthetic marine dinoflagellate *G. polyedra* are grown at 20 ± 5° in 2.8-liter Fernbach flasks containing 1 liter of the supplemented sea water medium,[10] and maintained on a 12 hr light:12 hr dark cycle, with light provided by cool white fluorescent bulbs. Cultures are inoculated at densities of 100 to 500 cells/ml, as measured with a Coulter model ZBI cell counter, with a 140 μM aperture (Coulter Electronics, Hialeah, FL). With a high light intensity, doubling time is 2 to 3 days; after 2 to 4 weeks, at cell densities of 7,000–15,000 cells/ml, cells are harvested by filtration with Whatman #541 paper; the yield should be 0.3 to 0.7 g wet weight per liter of culture.

Pyrocystis lunula is grown in the same medium under similar conditions; continuous illumination may be used instead of light–dark cycles. Growth is somewhat slower (doubling time, about 4 days), and harvesting may be carried out at somewhat higher cell densities (15,000–20,000 cells/ml).

For the extraction of the macromolecular components of the luminescent system, operations should be performed in the cold (4°) and as rapidly as possible. Cells are scraped from the filter paper and resuspended in extraction buffer. The choice of extraction method may be determined in part by the volume of extract to be processed. Suitable extraction methods include a single passage through a French press (greater than 4000 psi), a hand-held Kirkland emulsifier (Brinkman), a Ten Broeck glass homogenizer, a Fisher tissue homogenizer, or vortexing in the presence of glass beads. Each of these has been found to result in optimum or satisfactory yields of one component or another; different methods should be evaluated for achieving an optimal yield of the component of interest. The cell walls and large particles are removed by a low-speed centrifugation (5000 g, 5 min); the first supernatant is then subjected to high-speed centrifugation (20 min at 27,000 g). The supernatant contains the several active soluble components: LFN, Lase, and LBP. The pellet contains the activity of the particulate or scintillon fraction, a single component whose activity is triggered by a shift of the pH from alkaline (pH 8) to acid (pH 5.7). If proteolysis is to be avoided during extraction, care must be exercised to maintain the pH of the buffer above 8; cell extracts themselves are acidic. Tris buffers are preferred over phosphate.

The Particulate Activity: Scintillons

Assays

The Particulate (Scintillon) Assay. The activity is assayed by measuring the light emission following the rapid injection of acid (by syringe) or mixing (by stopped flow) of the sample at pH 8 with acid to bring the final pH to about 5.7.[11] A small volume (5–50 μl) of the scintillon sample is added to 1 ml of buffer A (pH 8) and mixed with 1 ml of 0.2 M sodium citrate, pH 5.2 (or 0.03 M acetic acid) directly in front of the phototube. The instrumentation must be capable of registering rapid (1 msec) changes. Either the peak light intensity or the total (integrated) light can be used as the measure of activity.

Supercharging Assay. If scintillons are incubated in advance for 5–50 min with luciferin (LFN) at pH 8 and 20°, more light will be emitted in the flash. The LFN is bound to the scintillons; excess soluble free LFN may be removed by pelleting and resuspending scintillons in fresh buffer.

Recharging Assay. The activity of scintillons that have been discharged by acidification in the normal assay may be restored by a similar incubation, but only after readjusting the pH to 8 subsequent to discharge.

Isolation and Purification

Scintillons are extracted by homogenization of cells in buffer B at pH 8.5, recovered in the pellet after high-speed centrifugation, and resuspended in 2 ml of the same buffer per 10^7 cells. Such preparations gradually lose activity at 1°, about 50% after 48 hr at 1° in the dark. Activity may be preserved by storage at −80°; the addition of 20% glycerol helps prevent losses due to freezing and thawing.

A sucrose gradient is prepared by layering 8 ml each of six sucrose solutions having densities ranging from 1.29 to 1.14 g/cm³, allowed to

---

[11] M. Fogel and J. W. Hastings, *Proc. Natl. Acad. Sci. U.S.A.* **69,** 1073 (1972).

"smooth" for about 36 hr at 4° so that there are no sharp density boundaries between the layers where particles could accumulate.

The crude scintillon preparation is layered on the preformed gradient and centrifuged in a swinging bucket rotor for 12 hr at 24,000 rpm. Fractions are collected for measurements of absorbance, fluorescence, and bioluminescence (scintillon) activity. Mitochondria and chloroplasts band at lower densities (1.18 g/cm³), but the tails may overlap the scintillon peak. Absorbance at 670 nm serves to track chloroplasts; fluorescence emission at 470 nm (excited at 390 nm) identifies the LFN, which coincides with bioluminescence as determined by the scintillon assay. Further purification may be achieved by the use of a second sucrose-density gradient: its shape may be varied or a different material (e.g., Percoll) may be employed. Spontaneous losses in activity during purification due to oxidation of LFN may be recovered in part by incubation of the scintillon preparation with free LFN (supercharging, see above). Protection against losses is potentiated by high concentrations (10 mM) of DTT or 2ME.

Properties

Scintillons are heterogeneous but band with a peak density of 1.23 g/ml. From velocity sedimentation measurements the size has been estimated to correspond to an M_r of 10^9, thus a particle about 0.5 μm in diameter. This matches the size of dense bodies seen in the electron microscope which are labeled with antibody against luciferase.[5] The number of fluorescent particles in different fractions after density gradient sedimentation is proportional to the scintillon activity.[4] Both the flash kinetics and the photon yield of scintillons are concentration independent over a wide range, indicating that the functional molecular components are not significantly dissociated upon dilution (within the time period required for assay), and that no factors exogenous to the scintillon particle are required for luminescence.

Although living cells can flash repeatedly, scintillons *in vitro* emit only a single flash following a pH jump. However, as mentioned above, discharged scintillons may be "recharged" *in vitro* simply by returning the pH to 8 and incubating with fresh luciferin. This procedure also restores the fluorescence of discharged scintillons, as seen microscopically.[4]

Although scintillon activity has been found in all dinoflagellate species examined, there are some differences, and the amount of scintillon activity may be very low in some species.[6] It is possible that more than one organelle or subcellular compartment may be involved in scintillon activity, and that these may differ in different species. This is suggested by the fact that in *G. polyedra* and *G. tamarensis* the major part of the activity bands at a density peaking between 1.20 and 1.23 g/ml, while in *Pyrocys-*

tis (Dissodinium) lunula and *Pyrocystis noctiluca* the values are centered around 1.15 to 1.16 g/ml.[6] Especially intriguing in this connection is that in extracts of *G. polyedra,* both of the above density classes of scintillon occur.[12]

Luciferase

The active luciferase in the soluble fraction of extracts of *G. polyedra* may occur in two different molecular weight forms. The larger is a 135,000 MW molecule, the native single chain, which occurs in extracts made at pH 8. If, instead, the extract is made with a buffer at pH 6, all of the activity is found in the 35,000 MW range, this due to the action of a protease.[13-15] These two forms are referred to as A135 and B35 luciferases. They possess distinctly different pH–activity profiles (Fig. 1): both are active pH 6, but A135 is inactive at pH 8, where A35 retains considerable activity. Larger aggregates or multimers of the A135 form (e.g., A420) also occur in extracts.

Assays

Since the two forms of luciferase have different pH profiles, assays at different pHs have diagnostic value. Assays (at 20°) may be carried out either at pH 6.3 with 0.2 M sodium citrate or at pH 8 with 0.2 M sodium phosphate; bovine serum albumin (0.1 mg/ml) may be included.

To 2 ml of buffer are added sequentially small volumes (5–50 μl) of Lase and LFN. The vial is then quickly placed in the photometer compartment and the shutter opened, this requiring less than 1 sec; alternatively, the last component is injected with the shutter open.

The initial maximum intensity, which is reached within a few seconds or less, is used as a measure of the relative quantity or activity of either LFN or Lase, holding constant the amount of the second component. For comparisons over longer time periods (weeks to years) many small aliquots of both LFN (stored under argon or reducing agents) and Lase should be dispensed and kept frozen at −85° where both are quite stable. That neither has lost activity can be verified by reacting together thawed samples of each, and each can then serve to construct anew a concentration–activity standard curve.

[12] R. Schmitter, Ph.D. Thesis, Harvard University, Cambridge, Massachusetts (1973).
[13] N. Krieger and J. W. Hastings, *Science* **161,** 586 (1968).
[14] M. Fogel and J. W. Hastings, *Arch. Biochem. Biophys.* **142,** 310 (1971).
[15] N. Krieger, D. Njus, and J. W. Hastings, *Biochemistry* **13,** 2871 (1974).

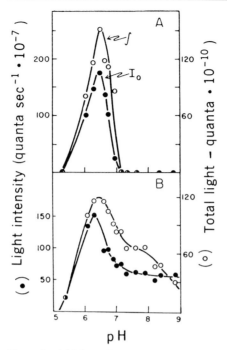

FIG. 1. Effect of pH on the initial maximum rate (I_0) and total light ($\int$) emitted in reactions catalyzed by high- (A135) (A) and low- (B35) (B) molecular-weight forms of *Gonyaulax* luciferase.[13]

Purification

Gonyaulax luciferase is present in extracts from both the day and night phases of cells grown in light–dark cycles, but there is much less actvity extractable from day phase cells.[16,17] The basis of this difference has been the object of intense study and it is now virtually certain that the 135-kDa enzyme, whether isolated from day or night phase cells, is the same molecule, and that night phase cells simply contain more of the protein.[18,19] Based on extracted activities, the smaller proteolytic fragment (B35 Lase) does not appear to occur normally in the cell at any detectable levels at any time, even though the disappearance of luciferase during the day phase implies that the A135 Lase is proteolysed. B35 Lase occurs only in extracts made at pH 6.

[16] J. W. Hastings and V. C. Bode, *Ann. N.Y. Acad. Sci.* **98**, 876 (1962).
[17] L. McMurry and J. W. Hastings, *Biol. Bull. (Woods Hole, Mass.)* **143**, 196 (1972).
[18] J. Dunlap and J. W. Hastings, *J. Biol. Chem.* **256**, 10509 (1981).
[19] C. H. Johnson, J. F. Roeber, J. W. Hastings, *Science* **223**, 1428 (1984).

TABLE I
PURIFICATION STEPS FOR *Gonyaulax* LUCIFERASE

Step	Volume (ml)	A_{280} (1 cm)	Activity (q)	Specific activity	Yield (%)
1. Crude extract (10^8 cells)	300	100	30×10^{12}	0.3×10^{12}	100
2. Supernatant	260	30	30	1	85
3. Ammonium sulfate precipitate	24	44	270	6	70
4. Affi-Gel Blue eluate	5	11	870	80	47
5. Gel filtration	5	2	5600	280	30
6. DEAE-BioGel A	5	0.42	3600	860	20

B35 (Proteolyzed) Luciferase[15]

Cells are harvested and extracted in 0.05 M Na$^+$/K$^+$ phosphate, pH 6, with 2×10^{-4} M DTT. Solid $(NH_4)_2SO_4$ is added to precipitate the active B35 Lase; the material precipitating between 35 and 65% saturation of $(NH_4)_2SO_4$ is redissolved in buffer A. This preparation, which contains no LFN activity, is then applied to a Sephadex G-100 column equilibrated and eluted with the same buffer (Fig. 2B). The fractions containing Lase activity elute at volumes corresponding to about 35 kDa; activity occurs in assays at both pH 6.3 and 8. An "inhibitor" elutes at volumes corresponding to about 120 kDa; it inhibits the reaction in assays at pH 8, but not pH 6.3. It is due to a luciferin-binding protein (LBP) which binds the LFN at pH 8 (but not at pH 6.3).

A135 Luciferase[18,20]

A summary of the steps and recovery in the purification scheme is given in Table I. Cells are extracted in buffer B (step 1); the particulate matter (including scintillons) is removed by centrifugation for 20 min at 27,000 g (step 2). The active Lase in the supernatant is precipitated at 4° by the addition of a solution of saturated $(NH_4)_2SO_4$, pH 8.5; material precipitating between 30 and 55% saturation is pelleted and redissolved in buffer C (step 3). Upon gel filtration (Sephadex G-100), the Lase activity, as determined by the assay at pH 6.3 with added LFN, is entirely in the higher molecular weight form (Fig. 2A). But there is also luminescence activity without added LFN, initiated simply by the change in pH. How-

[20] J. Dunlap, Ph.D. Thesis, Harvard University, Cambridge, Massachusetts (1979).

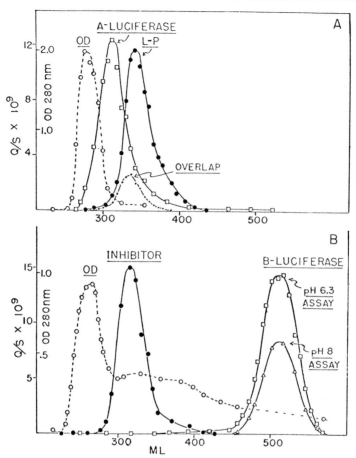

FIG. 2. Resolution by gel filtration of components involved in the soluble system from *Gonyaulax* showing differences dependent upon the pH during extraction. (A) After extraction at pH 8, with assays at pH 6.3, the higher molecular weight A-luciferase occurs along with the binding protein (L-P) for luciferin; "overlap" is light emission of the fractions at pH 6.3 with no additions. (B) After extraction at pH 6, a lower molecular weight form of luciferase occurs with activity at both pH 8 and 6.3 (Fig. 1). The "inhibitor" is material active in the assay at pH 8, as illustrated in Fig. 7.[14]

ever, its peak in the elution profile is in a position different from that of the Lase. This is attributable to the presence in these fractions of LBP carrying bound LFN, a consequence of the fact that the extracts have been kept at pH 8 or higher, where LFN remains bound. The profile of the binding protein itself is determined by assaying at pH 6.3 with an excess of added Lase; this establishes the location of the LFN (and thus the LBP to which it had been bound).

The $(NH_4)_2SO_4$ precipitate is dissolved in buffer C and treated to remove excess salt by hollow fiber dialysis against buffer C. The protein is then applied with a flow rate of about 1.5 ml/min onto a 2.0×15-cm column of 50–100 mesh Affi-Gel Blue (Bio-Rad Laboratories, Richmond, CA) previously equilibrated with buffer C (step 4). The column is washed with about 100 ml of buffer C; the luciferase, now free of proteases and most of the cellular pigments, is eluted stepwise with 200 ml of 50 mM Tris, 5 mM EDTA, 3 mM DTT, 500 mM NaCl, pH 8.5. The fractions exhibiting activity more than 10% of that of the peak fraction are pooled and concentrated by precipitation in 70% saturated ammonium sulfate. In recent work we have encountered difficulties with this Affi-Gel Blue step; poor binding and low recoveries have been experienced. It may be preferable to omit it.

Step 5 involves gel filtration on Sephacryl S-300. The precipitate is dissolved in 5 ml buffer (1.0 mM Tris, 0.1 mM EDTA, 100 mM NaCl, 3 mM DTT, pH 8) and loaded on a 2.6×97-cm Sephacryl S-300 column (Pharmacia, Piscataway, NJ) previously equilibrated in the same buffer. Lase is eluted overnight with the same buffer at a flow rate of 25 ml/hr. Activity appears in two peaks at M_r values of about 420,000 and 130,000 occurring in a ratio of about 8:1; these are pooled separately and the larger one, constituting 80–90% of the total, is concentrated either by ultrafiltration through an Amicon PM 30 membrane or by precipitation in 90% ammonium sulfate and resuspension in 10 ml of buffer C.

In step 6, the concentrated protein from gel filtration is dialyzed overnight against 4 liters of buffer C and the conductivity checked to assure that it is less than 150 μmho. The solution is then loaded onto a 1×25-cm column of DEAE-BioGel A (Bio-Rad) equilibrated in the same buffer, washed at a flow rate of 15 ml/hr with 40 ml of buffer C, and then eluted with a 150 ml continuous linear gradient of 0 to 100 mM NaCl in buffer C. The purified Lase elutes at a conductivity of about 1.4 mmho (27 mM NaCl). Assayed with 12.5 nM stock LFN, the purified Lase has a specific activity (based on initial maximum intensity) of approximately 8.5×10^{14} quanta sec^{-1} A_{280}^{-1}.

Properties: Stability and Quantities

Lase is not particularly stable during purification. In earlier stages this is probably due in part to proteolytic inactivation, but other factors are involved at later stages. Instability is aggravated in dilute solutions of low ionic strength such as those preceding the final DEAE step; this problem may make further purification of Lase difficult since large quantities of cells are not easily obtainable. With our facilities, the maximum rate of production of cells (about 500 liters of culture per month) permitted the purification of only about 1 mg of Lase in that time period. The loss of Lase activity was shown by Njus[21] to be about 12% per freeze thaw cycle (exclusive of the slower spontaneous loss in the frozen state). Although losses can be reduced by the inclusion of 20% glycerol, the instability of preparations also means that it is difficult to recover material falling in side fractions during the purification, so that the yield of 15 to 20% represents the yield obtained.

Protease activity in crude extracts may result in the appearance of Lase activity at pH 8. The protease activity itself has a broad pH profile with a peak around pH 5 to 5.5, with 10% of the activity remaining at pH 7. It is not inhibited (i.e., <10%) by EDTA (up to 0.1 M) or phenylmethylsulfonyl fluoride[21] 1975) but is inhibited by sulfhydryl reagents[20]; unfortunately so is luciferase. Thus, the simplest procedure to avoid proteolysis is to keep the pH high and work quickly in the cold until the protease is eliminated.

Numerous treatments have been attempted in order to completely solubilize the particulate luciferase activity.[8] Seventeen detergents spanning the range of hydrophobicity were tried at $\frac{1}{5}$, 1, and $5\times$ their critical micelle concentrations in hopes that luciferase could be lifted off its putative membrane without complete solubilization of the membrane itself.[20] Other treatments with various salts and chaotropic agents were also tried. Some detergents (notably Triton X-45, X-100, and X-102) enhanced activity in assays but none was found that would greatly enhance the specific activity of the soluble fraction at the expense of the pelletable fraction.

A135 Lase displays a highly asymmetric pH–activity profile; the enzyme has maximum activity at about pH 6.3 and virtually none above pH 7 (Fig. 1). The biological significance of this finding may relate to the activity of the LBP (see below).

During purification, at least under the conditions described, most of the Lase activity displays an apparent molecular weight of about 420,000. However, this Lase has a single chain molecular weight of 135,000 ± 5,000 on SDS–polyacrylamide gels and in gel filtration on Sepharose 6B-Cl in 6 M guanidine–HCl.[18,19] Proteolysis of Lase (see above) lowers its apparent size under both denaturing and nondenaturing conditions to about 35 kDa, reduces the quantum yield of the reaction, and alters the pH–activity profile.[20]

---

[21] D. Njus, Ph.D. Thesis, Harvard University, Cambridge, Massachusetts (1975).

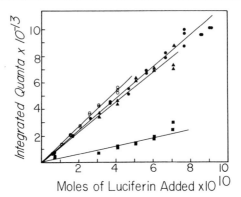

FIG. 3. Determinations of the quantum yield of luciferin. Determinations were with both night (circles) and day (triangles) unproteolyzed luciferases (purified through step 5) and the proteolytic fragment (squares). Least-squares regression lines were fit to each set of data points. The slope of the line gives the number of quanta produced per mole of luciferin oxidized; the quantum yield can be calculated by dividing the slope by Avogadro's number. For the two determinations with night luciferase, quantum yields of 21.7 and 22.4% were obtained, while day luciferase gave a value of 19%; these values are not statistically different at the 95% confidence level. The proteolytic fragment, however, had a significantly lower quantum yield of 6%.[18]

As shown in Fig. 3, the quantum yield of A135 Lase is 0.22 and that of the proteolytic fragment B35 about 0.06 based on the luminol standard.[22] The native enzyme displays a K_m for luciferin of about 2.5×10^{-7} M. The absorption spectrum measured at a concentration of 125 μg ml^{-1} shows no unusual chromophores or other substances absorbing light between 250 and 800 nm. The partial amino acid composition of Lase[18,20] is not remarkable in any way. Antibody to Lase has been raised in rabbits and used, along with comparisons of the physicochemical characteristics described above, to examine differences in Lases in extracts from day and night phase cells. Lases isolated from either time of day are indistinguishable with regard to any of the physicochemical or immunological criteria used.[19,20]

Luciferin

Assays

Luciferin (LFN) is assayed by one of three methods. Routine determinations are made by adding a small volume (less than 100 μl) of LFN to a

[22] J. Dunlap and J. W. Hastings, *Biochemistry* **20**, 983 (1981).

fixed given amount (usually 5 or 10 μl) of Lase (either A135 or B35) in 2 ml of assay buffer (pH 6.3) at 20°. With Lase B35 the assay may, alternatively, be carried out at pH 8. The components are rapidly mixed and placed in a photometer and the peak light intensity recorded.[14] Such an assay is linear with the amount of LFN added over several orders of magnitude[21] and is routinely used during extractions to determine relative amounts and yields through the various purification steps. However, care must be taken to use the same activity of Lase in different assays. A convenient method is to prepare a quantity of a suitable activity, freeze and store many small aliquots at $-80°$, one of which may be thawed for use at any time.

A second method is to add the LFN to the assay buffer (pH 6.3) in front of a photomultiplier and to inject sufficient Lase to exhaust light emission within about 3 min.[20,22] The total photon yield, determined by integration of the signal, is linear with the amount of LFN added over several orders of magnitude, but the assay consumes more Lase. The absolute concentration of LFN is calculated by dividing the photon yield by the quantum yield for Lase (0.22 for A135).

A third procedure takes advantage of the fact that LFN is stoichiometrically oxidized by potassium ferrocyanide. This allows a rigorous quantitation of LFN by titration with a known standard ferrocyanide solution.[22] A sample of LFN is treated with increasing amounts of ferrocyanide; after each addition a small aliquot of the LFN is assayed by integrating the total light produced upon addition of Lase. The plot of light produced per aliquot versus total ferrocyanide added yields an intercept where all of the LFN in the sample has been oxidized prior to the addition of the Lase. Assuming a two electron transfer oxidation, as is typical for the potassium ferrocyanide oxidation of organic compounds (however, see Johnson et al.[23]), the number of moles of LFN in the sample is equal to half the number of moles of ferrocyanide at the intercept.

Purification

The steps and yields for LFN purification are summarized in Table II. *P. lunula* cells grown in constant light are harvested on Whatman 541 filter paper, scraped off, dispersed into boiling extraction buffer (2 mM $K_xH_yPO_4$ 5 mM 2ME, pH 8.5, 5 ml/liter of culture), heated for 10 sec, and then immediately chilled in ice water (step 1). While cooling, the solution is saturated with argon. All subsequent operations are performed in the cold and under an argon atmosphere whenever possible. Solutions and

---

[23] F. H. Johnson, O. Shimomura, Y. Saigo, L. C. Gershman, G. T. Reynolds, and J. R. Waters, *J. Cell. Comp. Physiol.* **60**, (1962).

TABLE II
Purification Steps for *Gonyaulax* Luciferin

Step	Volume (ml)	Total activity (q)	Yield (%)
1. Crude boiled extract after centrifugation (6×10^7 cells)	450	5×10^{17}	100
2. DEAE-cellulose (coarse)	170	3.3	66
3. Butanol extraction	100	2.5	50
4. Alumina column eluate (after concentration)	14	2.4	48
5. DEAE-cellulose (fine)	90	1.9	36

reagents should be saturated with argon and all chromatographic steps performed under a positive argon pressure.

The pH of the boiled cell suspension is readjusted to pH 8.5 with 0.2 M NaOH, resulting in the formation of a white flocculent precipitate and a slight bathychromic shift from green to reddish-yellow. This slurry is centrifuged for 20 min at 27,000 g at 0°. The reddish yellow supernatant contains the LFN; its conductivity is reduced to less than 5 mmho by adding 1 to 2 volumes of ice cold 95% ethanol. It is then applied (step 2) to a 5×25-cm column of DEAE cellulose equilibrated with 2 mM $K_xH_yPO_4$, 5 mM 2ME in 50% ethanol, pH 8.0. The column must be developed quickly to avoid LFN oxidation, which will occur with even trace quantities of oxygen. Thus, a coarse grade of cellulose is required. Selectacel (Brown & Co.) or Sigma coarse grade are acceptable, but DEAE-Sephadex (Pharmacia), DEAE-BioGel (Bio-Rad), and DE-32 (Whatman) all result in lower yields, probably due to their lower maximum flow rates. After loading, the column is washed with 200 ml of equilibration buffer and eluted with a gradient of 400 ml each of equilibration buffer and 0.2 M $K_xH_yPO_4$, 5 mM 2ME, pH 8. A flow rate of at least 3 to 4 ml/min should be maintained; the LFN elutes at a conductivity of about 7 mmho. Unless further purification or immediate experiments are planned, the LFN preparation can be stored under argon at $-80°$; under these conditions a loss of approximately 10% per month may occur.

The further purification steps are modified from those used by Shimomura[24] for the euphausid (*Meganyctiphanes*) fluorescent substance. Columns are loaded and eluted with a positive argon pressure of 5–10 psi.

[24] O. Shimomura and F. H. Johnson, *Biochemistry* **6**, 2293 (1967).

In preparation for step 3, the DEAE eluate (about 150 ml) is concentrated *in vacuo* (with slight warming) to approximately 10–20 ml and then diluted with 4 volumes of cold *n*-butanol. Alternatively, concentration can be achieved by lowering the pH to 3.8 with 1 M citrate and extracting the LFN into ethyl acetate and subsequently back into about $1/10$ the volume of water at pH 9 (0.1 M Tris). By this method the LFN in an aqueous solution will partition greater than 90% into small volumes of ethyl acetate or dichloromethane, about 40% into equal volumes of benzene or 3:1 chloroform/methanol, but not at all into anisole or carbon tetrachloride, as measured by assaying for LFN in the aqueous and organic phases and correcting for solvent inhibition of the Lase. Greater than 50 mM salt in the aqueous phase is necessary to achieve this separation; the pH dependence displays a pK of around 4.3. Generally, LFN is more stable in ethyl acetate than dichloromethane, so the use of the former is preferred.

In either case, a powdery precipitate is formed and is removed by centrifugation (5000 g, 10 min); the greenish-yellow fluorescent (excitation, 340–420 mm; emission, 450–550 nm) supernatant is loaded on a 2.5 × 25-cm column of basic alumina in ice cold *n*-butanol (step 4). The LFN appears as a highly blue-fluorescent band in the top 4 cm of the column. After loading, the column is washed with 100 ml of 50% ethanol containing 2.5% ammonium hydroxide. The progress of elution in this and subsequent chromatographic steps can be monitored on the column and in the tubes by the location of the fluorescence of the material; blue fluorescence is correlated with LFN activity.

The active fractions from the alumina column (about 50 ml) are concentrated *in vacuo* to 5–10 ml, diluted with 1 volume of DEAE equilibration buffer, and (step 5) loaded onto a 2.5 × 8-cm of DEAE cellulose (Whatman DE-32 or equivalent grade, chosen for high binding capacity rather than fast elution) equilibrated in 10 mM $K_xH_yPO_4$, pH 8, in 50% ethanol. The column is washed with 100 ml of equilibration buffer and eluted with 300 ml of equilibration buffer containing 0.1 M NaCl. In collecting the active fractions (in 90 ml in Table II), care must be taken to exclude any oxidized LFN which is blue and nonfluorescent, and elutes slightly before the yellowish and highly fluorescent LFN.

Properties

Highly purified LFN is yellow in aqueous solution with maxima at 245 and 390 nm in 50% ethanol (Fig. 4) and an extinction coefficient of about 2.8×10^4 at 390 nm.[25] Increasing concentrations of ethanol shift the

---

[25] J. Dunlap, J. W. Hastings, and O. Shimomura, *FEBS Lett.* **135**, 273 (1981).

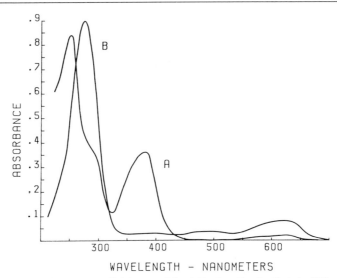

FIG. 4. Absorption spectra slightly autoxidized dinoflagellate luciferin in 50% ethanol (A) and blue oxidation product of luciferin in methanol (B).[25]

390 nm absorption peak slightly (and ethyl acetate even more; up to 30 nm) toward shorter wavelengths. The product of LFN oxidation (Fig. 4) is blue with an absorption maximum at 635 nm (in methanol). Oxidation is more rapid at extremes of pH and ionic strength; this becomes particularly important in steps where concentration is achieved either by *in vacuo* evaporation of an aqueous solution containing salt or by lowering the pH and extracting into an organic solvent. Acidity (below about pH 3.8) virtually eliminates the 390 nm peak, although it is recoverable if the pH is returned to greater than 5 before degradation has taken place. Prolonged exposure to low pH results in the formation of the same biologically inactive bluish compound that results from air oxidation.

The corrected fluorescence excitation and emission spectra for the purified compound are displayed in Fig. 5; the emission peaking at 474 nm is coincident with the *in vivo* bioluminescence emission spectrum for this and other species of dinoflagellates. No other fluorescent compounds are present in the purified material. The excitation peak for this compound corresponds to the maximum in the UV absorption spectrum. The pH dependence of fluorescence reveals a pK of around 4.7 for the fluorescer. This is near the pK already noted for extraction of the active compound into organic solvents, and is also near to the drop off point in the stability versus pH plot.

Gel filtration (BioGel P2) suggests an upper limit for the molecular

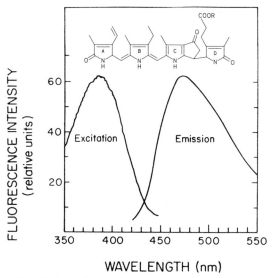

Fig. 5. Corrected excitation and emission spectra of partially purified *Dissodinium* lucif-erin, determined with a Farrand MK-1 spectrofluorometer. The proposed partial structure for the molecule is shown at the top.[21]

weight of LFN of about 600; a variety of physical and chemical character-istics support the identification of this compound as a substituted polypyr-role,[25] a bile pigment-type compound and not a porphyrin. It displays a positive Ehrlich reaction, the Schlesinger reaction, and yields substituted maleimides upon acid chromate oxidation.[22] Chromate oxidation has also been carried out in the presence of $^{18}O_2$ and $H_2^{18}O$ followed by mass spectroscopic analysis in order to carry out preliminary sequence analysis of the four pyrroles in the chain.

Interestingly, dinoflagellate LFN displays a limited cross-reaction with the fluorescence substance F involved with bioluminescent reaction of the marine euphausid *Meganyctiphanes norvegica*.[26] Chemical and mass spectroscopic analysis of F indicates that it is also a polypyrrole of the bile pigment type, but in this case the sequence of the four pyrrole rings is not the same. Both compounds could be derived from modified chlorophylls, but LFN would derive from pyrochlorophyll *a* or *c* by ox-idative cleavage between rings A and D, whereas F would derive from chlorophyll *b* by cleavage between rings A and B.

[26] J. Dunlap, O. Shimomura, and J. W. Hastings, *Proc. Natl. Acad. Sci. U.S.A.* **77**, 1394 (1980).

Luciferin-Binding Protein

Assays

Luciferin-binding protein (LBP) binds luciferin (LFN) at pH 8 but not at pH 6.[14] The apoprotein is assayed by measuring its capacity for binding LFN in either of two ways. In the first, LFN is mixed with LBP and the amount of LFN the LBP carries with it in the void volume upon gel filtration is measured. Alternatively, it can be measured by its ability to sequester LFN from (and thus inhibit) B35 Lase in the reaction at pH 8.

In the gel filtration procedure a 1 ml G-25 Sephadex column is made in a Pasteur pipet plugged with glass wool.[6,27] The LBP (5–50 μl) and sufficient LFN to saturate the LBP (25–50 μl) are mixed at pH 8, applied to the column, and eluted with 0.05 M Tris–HCl or Tris-maleate buffer with 5 mM 2ME, pH 7.8, all at room temperature. The LFN–LBP complex elutes first (Fig. 6A) followed by free LFN, if any. In the same procedure at pH 5.7 none of the luciferin is bound (Fig. 6B). Aliquots of each fraction are assayed by the standard LFN assay with A135 Lase in 2 ml assay buffer at pH 6.3.

The second method requires that Lase be absent from the LBP preparation, readily possible if the cell extract for isolating the LBP is made at pH 6, where only the lower MW form of Lase (B35) occurs and is readily separated from the LBP. For the assay (Fig. 7), 10–50 μl of the LBP preparation (adjusted to pH 8, if necessary) is added to a reaction mixture in progress containing B35 Lase and free LFN at pH 8.[13,15] Since the rate of this reaction is proportional to the concentration of free LFN, the difference between the light intensity before and after the LBP is added is a measure of the LFN sequestered.

Isolation and Partial Purification

Cells are extracted in buffer at pH 6 (see B35 Lase, above). Because of a protease active at this pH, all of the Lase is converted during extraction to a mixture of species with sizes ranging from 30 to 40 kDa. These can be separated from LBP (M_r 120,000) by gel filtration.

To the crude extract, $(NH_4)_2SO_4$ in phosphate buffer pH 6 is added to 75% saturation. The precipitate is dissolved in 5–6 ml 10 mM Tris–HCl buffer, pH 8, with 5 mM 2ME. A small amount (for tracking purposes) of stock-free LFN is then added and the material is loaded on a Sephadex G-200 column. The fractions containing LFN (standard Lase assay) are

[27] F. N. Sulzman, N. R. Krieger, V. D. Gooch, and J. W. Hastings, *J. Comp. Physiol.* **128,** 251 (1978).

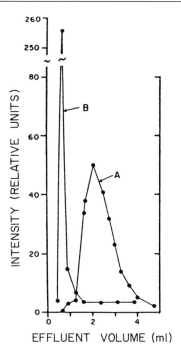

FIG. 6. Sephadex chromatography of luciferin binding protein–luciferin complex at pH 7.8 (A) and at pH 5.7 (B). When bound (at pH 7.8), luciferin elutes along with protein in the void volume; at pH 5.7 it is free and emerges in later tubes.[27]

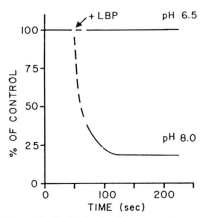

FIG. 7. Effect of LBP on the luciferin–luciferase reaction at pH 8.0 and 6.5. Arrow indicates time at which 10 μl of LBP solution was added to the reaction mixtures in two separate experiments. At pH 6.5 there was no inhibition; at pH 8.0 (lower curve) the light dropped rapidly.[27]

identified and pooled. The protein is precipitated as before with $(NH_4)_2SO_4$ and redissolved in 1–2 ml of 0.2 M phosphate buffer, pH 8, with 5 mM 2ME.

Properties

Molecular Weight. The molecular weight of LBP, about 120,000 ± 10,000, is not very different from the A135 Lase, a fact that makes it difficult to separate the two. It could be that the LBP is derived from Lase by the loss of a peptide carrying the active center but not the LFN binding site. This seems unlikely, since the A135 Lase does not bind LFN at pH 8.

pH Characteristics. LBP binds LFN at pH 8 but not at pH 6. The titration for this, as determined either by the binding assay or inhibition of Lase, is quite steep, with a half maximum at about pH 7. The ionizable group(s) responsible has not been pinpointed.

Binding Constant. Although the absolute concentrations of stock LBP solutions are uncertain, the amount of unbound LFN at equilibrium in any given case can be experimentally determined, since it is directly related to the initial light intensity when B35 Lase is added to the mixture at pH 8.[27] By determining the concentrations of unbound LH_2 (in q/liter/ϕ) for various mixtures of LFN and LBP that were brought to equilibrium, a value of K_d 3.5 × 10^{-9} M has been obtained, assuming $\phi = 0.2$.

Components from Other Dinoflagellate Species

While there is a basic similarity in the biochemistry of bioluminescence in different dinoflagellates, there are some significant intraspecies differences. One is that, whereas *soluble* LBP activity is readily isolated from *Gonyaulax* species, it is not detectable in *P. noctiluca* and *P. (Dissodinium) lunula.* However, some scintillon activity can be isolated from both these species, and soluble LFN is bound by the scintillons and carried by them in a density gradient centrifugation.[6] As mentioned above, this is true also for scintillons isolated from the *Gonyaulax* species.[11] Thus LFN binding activity *does* exist in each scintillon system, and the absence of soluble binding protein is still compatible with a model of the flash in which a rapid shift in the pH triggers the release of LFN from a particulate binding site.[9]

There are several other differences between the different luminescent systems.[6] The sizes of the single active native polypeptide chain are 140 kDa in the *Gonyaulax* species and 60 kDa in the other two; curiously, the active proteolyzed chains are slightly larger in the latter species. The buoyant densities of scintillons extracted from the *Gonyaulax* species are in the range of 1.20 to 1.23 g/cm^3, whereas those from the other species are 1.15 to 1.16 g/cm^3. However, it has been reported that *G. polyedra* possesses both denisty classes.[12]

Despite these differences there are fundamental similarities in dinoflagellate luminescence. Luciferases and luciferins cross-react in all combinations. Scintillons can be extracted from all four species and recharged (or supercharged) with *P. lunula* LFN. The pH characteristics of the four systems are strikingly alike; the intact luciferases have similar pH–activity profiles with a precipitous drop above pH 7. Scintillons from all species share the property of flashing when the pH is suddenly dropped from 8 to 5.7. This underscores the postulate that a pH drop is the *in vivo* event responsible for triggering the bioluminescent flash.[9]

Section II

Chemiluminescence

[29] Enhanced Chemiluminescent Reactions Catalyzed by Horseradish Peroxidase

By GARY H. G. THORPE and LARRY J. KRICKA

Chemiluminescent and Bioluminescent Assays for
Horseradish Peroxidase

Rapid and sensitive quantitation of the widely used enzyme horseradish peroxidase (EC 1.11.1.7) is possible with several chemiluminescent and bioluminescent systems. Horseradish peroxidase catalyzes the luminescent oxidation of a range of substrates including cyclic hydrazides,[1-4] phenol derivatives,[5,6] and components of bioluminescent systems.[7-9] Suitable substrates include luminol and related molecules, pyrogallol, purpurogallin or luciferins isolated from *Pholas dactylus*,[7] the firefly *Photinus pyralis*,[8] or *Cypridina*.[9] Such reactions differ widely in their detection limits, specificity, reagent availability, and the magnitude and kinetics of light emission and this restricts their applicability. The limited availability of reagents such as *P. dactylus* luciferin, for example, at present precludes the routine clinical use of this bioluminescent system. Also many of the chemiluminescent assays for peroxidase suffer from limitations such as rapid decay of light emission and require initiation of individual reactions in front of photodetectors or precise timing of measurements. Further limitations include poor signal to background ratio, poor specificity, and low intensity light emission under conditions in which interference is minimized. The latter limitation necessitates the use of sophisticated detectors. Many of these problems are overcome by the use of enhanced chemiluminescent assays and these are considered in detail in subsequent sections.

[1] L. Ewetz and A. Thore, *Anal. Biochem.* **71,** 564 (1976).
[2] H. Arakawa, M. Maeda, and A. Tsuji, *Anal. Biochem.* **97,** 248 (1979).
[3] A. D. Pronovost and A. Baumgarten, *Experientia* **38,** 304 (1982).
[4] K.-D. Gundermann, K. Wulff, R. Linke, and F. Stahler, *in* "Luminescent Assays: Perspectives in Endocrinology and Clinical Chemistry" (M. Serio and M. Pazzagli, eds.), p. 157. Raven Press, New York, 1982.
[5] B. Velan and M. Halmann, *Immunochemistry* **15,** 331 (1978).
[6] M. Halmann, B. Velan, T. Sery, and H. Schupper, *Photochem. Photobiol.* **30,** 165 (1979).
[7] K. Puget, A. M. Michelson, and S. Avrameas, *Anal. Biochem.* **79,** 447 (1977).
[8] D. Slawinska, J. Slawinski, W. Pukacki, and K. Polewski, *in* "Analytical Applications of Bioluminescence and Chemiluminescence" (E. Schram and P. Stanley, eds.), p. 239. State Publishing and Printing, Westlake Village, California, 1979.
[9] T. Kobayashi, K. Saga, S. Shimizu, and T. Goto, *Agric. Biol. Chem.* **45,** 1403 (1981).

General Mechanisms and Efficiency

Although the exact mechanistic details remain unknown and the precise nature of individual emitters remain unidentified, similar general reaction mechanisms have been proposed for the luminescent oxidation of several peroxidase substrates.[10] The peroxidase-catalyzed chemiluminescent oxidation of luminol has been subject to numerous mechanistic studies.[11] A generally accepted reaction scheme[12] involves the oxidation of luminol by a complex between oxidant and peroxidase to produce a luminol radical. Luminol radicals then undergo further reactions resulting in the formation of an endoperoxide which then decomposes to yield an electronically excited 3-aminophthalate dianion emitting light on return to its ground state. A simplified reaction scheme is shown in Fig. 1.

The intensity of chemiluminescence at any time depends on the chemiluminescence efficiency and the number of molecules reacting per unit time. The chemiluminescence efficiency is dependent on the efficiency of excited state production, the proportion of molecules in an excited state, the chemical yield of the particular reaction product, and the luminescence efficiency (the ratio of photons emitted to the number of electronically excited molecules formed). The luminescence efficiency also includes the efficiency of any transfer process. In theory a single molecule of chemiluminescent reactant could produce one electronically excited molecule which could then decay to emit one photon of light. Chemiluminescent reactions are however not this efficient and although a quantum yield of 1 einstein/mol is theoretically possible, the ratio of the number of emitted photons to the number of reacting molecules is seldom greater than 1% of this theoretical value. The reported quantum efficiency for typical peroxidase-catalyzed chemiluminescent reactions are luminol/ H_2O_2, 0.01,[11] and purpurogallin/H_2O_2, 1.5×10^{-8}.[8]

Enhancement of Light Emission from Luminescent Reactions Involving Horseradish Peroxidase

The multistep nature of peroxidase-catalyzed luminescent reactions, with rate limiting steps and competing side reactions, creates several areas in which improvements in efficiency can produce increases in light emission. Several examples of enhancement of light emission in peroxidase-catalyzed reactions are collected in Table I.[12-19]

[10] B. Arrio, B. Lecuyer, A. Dupaix, P. Volfin, M. Jousset, and A. Carrette, *Biochimie* **62**, 445 (1980).

[11] M. J. Cormier and P. M. Prichard, *J. Biol. Chem.* **243**, 4706 (1968).

[12] H. P. Misra and P. M. Squatrito, *Arch. Biochem. Biophys.* **215**, 59 (1982).

[13] J. K. Wong and M. L. Salin, *Photochem. Photobiol.* **33**, 737 (1981).

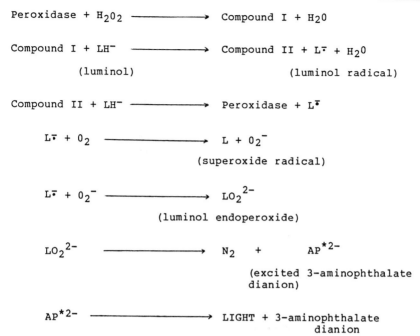

FIG. 1. Proposed reactions sequence in the peroxidase-catalyzed chemiluminescent oxidation of luminol.

Pretreatment of peroxidase with sodium hydroxide prior to quantitation using the luminol–perborate reaction at high pH enhances light emission.[1] The increased ability to induce light emission presumably reflects the degree of dissociation of protoporphyrin from the protein, but specificity is poor because similar reactions occur with other hemoproteins. Above certain peroxidase concentrations azide addition results in enhancement of light emission from the peroxidase–purpurogallin–H_2O_2 system.[14] Dimerization of peroxidase can occur and the active sites are

[14] M. Halmann, G. Gabor, and B. Velan, *Photochem. Photobiol.* **34,** 95 (1981).
[15] G. Ahnström and R. Nilsson, *Acta Chem. Scand.* **19,** 313 (1965).
[16] G. H. G. Thorpe, L. J. Kricka, E. Gillespie, S. Moseley, R. Amess, N. Baggett, and T. P. Whitehead, *Anal. Biochem.* **145,** 96 (1985).
[17] T. J. N. Carter, C. J. Groucutt, R. A. W. Stott, G. H. G. Thorpe, and T. P. Whitehead, European Patent 87959 (1982).
[18] G. H. G. Thorpe, L. J. Kricka, S. B. Moseley, and T. P. Whitehead, *Clin. Chem. (Winston-Salem, N.C.)* **31,** 1335 (1985).
[19] L. J. Kricka, G. H. G. Thorpe, and T. P. Whitehead, European Patent 116454 (1983).

TABLE I
ENHANCEMENT OF LIGHT EMISSION FROM LUMINESCENT REACTIONS INVOLVING
HORSERADISH PEROXIDASE

Luminescent system	Enhancers	Magnitude of enhancement (fold)	References
Luminol–perborate–peroxidase	Sodium hydroxide	0.5	1
Luminol–H_2O_2–peroxidase	Hydroxyl radical scavengers such as mannitol, benzoate, and histidine	3	12
Luminol–H_2O_2–peroxidase	Deuterated water	0.35	13
Purpurogallin–H_2O_2–peroxidase	Azide	7	14
Pyrogallol–H_2O_2–peroxidase	o-Phenylenediamine	—	15
Purpurogallin–H_2O_2–peroxidase	p-Phenylenediamine	8	6
Luminol–H_2O_2–peroxidase	6-Hydroxybenzothiazole derivatives, or para-substituted phenols	~1000	16–19

probably less accessible to bulky chemiluminescent substrates such as purpurogallin. Purpurogallin has a higher affinity for the monomeric form of the enzyme, and by forming a complex with peroxidase, azide inhibits the dimerization process enabling the monomeric form to survive at higher peroxidase concentrations.

When performed in a deuterated solvent the relative light emission of the peroxidase–luminol–H_2O_2 reaction is increased, possibly reflecting the greater radiative lifetime of the excited aminophthalate molecule in the solvent.[13] The addition of hydroxyl radical scavengers such as mannitol to the peroxidase–luminol–H_2O_2 reaction also enhances light emission by removing hydroxyl radicals and preventing their interference in the light producing steps.[12]

The peroxidase-catalyzed chemiluminescent oxidation of pyrogallol by peroxide is activated by addition of o-phenylenediamine.[15] The velocity of the luminescent process is accelerated and, although total light emission is unchanged, peak intensity is increased. Activation is proposed to arise from the coupled oxidation of pyrogallol (or a reaction product) by reaction products initially derived from the enhancer.[15] Incorporation of p-phenylenediamine into the peroxidase–purpurogallin–H_2O_2 system also results in an 8-fold enhancement of light emission.[6]

Particularly efficient enhancement of light emission from the peroxidase-catalyzed oxidation of cyclic diacyl hydrazides at approximately pH 8.5 is achieved by addition of molecules such as 6-hydroxybenzothiazole derivatives[16,17] or substituted phenols.[18,19] These reactions are considered in greater detail in the following sections.

Advantages of 6-Hydroxybenzothiazole or Phenol Derivative Enhanced Chemiluminescent Assays for Peroxidase

The major advantages of enhanced assays are that the intensity of light emission may be greater than 1000-fold that of the unenhanced reaction and conditions can be employed under which light emission is prolonged and decays slowly. Light emission from enhanced reactions is easily measured and peroxidase can be assayed sensitively in seconds without any need to initiate individual reactions directly in front of the photodetector. Although prolonged light emission can be produced with cyclic diacyl hydrazides alone, the ability to combine such kinetics with high intensity emission and improved discrimination between peroxidase concentrations is achieved only by addition of enhancer. Additional advantages of enhanced assays are that they are relatively specific with reduced background emission, require no lengthy preincubation,[20] and can be performed simply, under mild conditions, using a premixed cocktail of readily available, inexpensive reagents.

Characteristics of Enhanced Luminescent Reactions

Light Measurements

Light measurements were performed in a luminometer based on a side-window photomultiplier tube (EMI Type 9781A, 94 μA/lumen) using photocurrent measurement. Disposable, square plastic cuvettes (1 × 1 × 4.5 cm) (W. Sarstedt UK Ltd, UK) were employed as reaction vessels. Luminescence emission spectra were determined by use of a Model SRF 100 recording spectrofluorometer, Baird-Atomic Ltd, UK. Scans were determined between 350 and 600 nm at 4 nm/sec.

Reagents

Luminol and isoluminol were obtained from Sigma Chemical Company, 7-dimethylaminonaphthalene-1,2-dicarboxylic acid hydrazide (Boehringer Mannheim GmbH, Germany) and N-(6-aminobutyl)-N-ethyl-isoluminol from LKB Wallac, Finland. Luminol was purified by recrys-

[20] W. P. Collins, G. J. Barnard, J. B. Kim, D. A. Weerasekera, F. Kohen, Z. Eshhar, and H. R. Lindner, in "Immunoassays for Clinical Chemistry" (W. M. Hunter and J. E. T. Corrie, eds.), p. 373. Churchill-Livingstone, Edinburgh and London, 1983.

tallization from sodium hydroxide and decolorization through charcoal. Synthetic firefly D-luciferin and its 6-ethyl ether were purchased from Sigma. Dehydroluciferin was kindly supplied by Professor M. DeLuca (Chemistry Department, University of California, La Jolla, CA): all other benzothiazole derivatives were kindly supplied by Dr. N Baggett (Chemistry Department, University of Birmingham, UK).

p-Iodophenol and other halophenols, p-phenylphenol, p-hydroxycinnamic acid, naphthols, and 2,3-dichloro-5,6-dicyano-1,4-benzoquinone were purchased from Aldrich Chemical Company. Rabbit anti-human α-fetoprotein–horseradish peroxidase conjugate (2.6 mg protein/ml; 24 μM horseradish peroxidase) was purchased from Dakopatts a/s, Denmark. Horseradish peroxidase Type VI, horseradish peroxidase isoenzyme preparations (Types VII, VIII, and IX), hemoglobin (human, Type IV), myoglobin, and cytochrome c were obtained from Sigma. All other reagents were from Sigma or BDH Chemicals Ltd, UK.

Identification of Enhancers

Compounds were screened for their ability to act as enhancers by determining their influence on light emission from the peroxidase conjugate-catalyzed oxidation of luminol at pH 8.5. Antibody–peroxidase conjugate (10 μl 1:1000 dilution) and the compound under investigation (5, 10, or 90 μl; 4.54 mM in dimethyl sulfoxide or Tris buffer, 0.1 M, pH 8.5) were placed in separate corners of a cuvette. The chemiluminescent reaction was initiated by manual addition of 900 μl of a luminol (1.25 mM), hydrogen peroxide (2.7 mM) mixture prepared in Tris buffer (0.1 M, pH 8.5). Light emission was recorded and levels of light emitted 30 sec after initiation were compared with controls in which dimethyl sulfoxide or buffer replaced the candidate enhancer. The influence of potential enhancers on background light emission from the luminol, hydrogen peroxide reaction was determined, with antibody–peroxidase conjugate replaced with distilled water. Enhancement was deemed to have occurred when the compound under investigation increased the peroxidase-catalyzed light emission and had an insignificant effect on (or reduced) background light emission.

Many compounds, including 6-hydroxybenzothiazole derivatives, substituted phenols, and naphthols can act as enhancers (Table II). The degree of enhancement and improvement in signal to background ratio depends on the particular enhancer employed. Degree of enhancement was also dependent on the enhancer concentration and increases in light emission occurred if enhancer was added either during, or at the start of luminescent reactions.

TABLE II

COMPOUNDS PRODUCING ENHANCEMENT OF LIGHT EMISSION FROM THE HORSERADISH
PEROXIDASE CONJUGATE-CATALYZED PEROXIDATION OF LUMINOL

6-Hydroxybenzothiazole derivatives	Substituted phenols and naphthols
Firefly D-luciferin, firefly L-luciferin, dehydroluciferin, 6-hydroxyben-zothiazole, 2-cyano-6-hydroxyben-zothiazole	p-Iodophenol, p-bromophenol, p-chlorophenol, 2,4-dichlorophenol, p-hydroxycinnamic acid, p-phenyl-phenol, 1,6-dibromonaphth-2-ol, 1-bromonaphth-2-ol, 2-naphthol, 6-bromonaphth-2-ol

Under the conditions employed for enhanced reactions several compounds previously reported to increase light emission from luminescent systems (e.g., potassium bromide, potassium chloride, sodium azide, mannitol, histidine, thiourea, and sodium benzoate) produced no increase in emission from the peroxidase conjugate-catalyzed peroxidation of luminol, while others (e.g., pyrogallol, ascorbic acid, and p-phenylene-diamine) produced pronounced decreases (>99%) in light emission.

6-Hydroxybenzothiazole Derivatives. A series of 6-hydroxyben-zothiazole derivatives (Fig. 2) were examined for enhancer activity. Five of the benzothiazole derivatives tested, 6-hydroxybenzothiazole, 2-cyano-6-hydroxybenzothiazole, 2-(6-hydroxy-2-benzothiazolyl)thiazole-4-carboxylic acid (dehydroluciferin), and firefly D- or L-luciferin produced enhanced light emission from the peroxidase conjugate-catalyzed oxida-

I: $R_1 = H, R_2 = H$,   II: $R_1 = H, R_2 = OH$,   III: $R_1 = H, R_2 = OCH_3$,

IV: $R_1 = NH_2, R_2 = OH$,   V: $R_1 = NH_2, R_2 = OCH_3$,   VI: $R_1 = OH, R_2 = OH$,

VII: $R_1 = CN, R_2 = OH$,   VIII: $R_1 = CN, R_2 = OCH_2CH_3$,   IX: $R_1 = $ ⟨N⟩COOH, $R_2 = OH$,

X: $R_1 = $ ⟨N⟩COOH, $R_2 = OH$,   XI: $R_1 = $ ⟨N⟩COOH, $R_2 = OCH_2CH_3$.

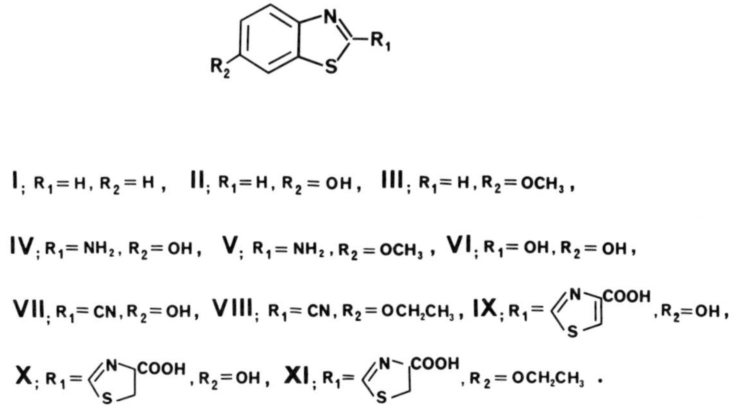

FIG. 2. Benzothiazole derivatives screened for the ability to enhance light emission from the peroxidase conjugate-catalyzed oxidation of luminol. From Thorpe et al.[16]

tion of luminol. The influence of benzothiazole derivatives on light emission from the reaction in the presence and absence of peroxidase conjugate is shown in Table III. 6-Hydroxybenzothiazole and dehydroluciferin produced greater enhancement than firefly luciferin and several hundred-fold increases of light emission over the unenhanced reaction were achieved. The enhancers also depressed the background light emission from the luminol–oxidant mixture by over 90%. This produced a further improvement in the signal to background ratio, reduced background light emission was however also produced by benzothiazole derivatives which did not enhance light emission.

Stringent structural requirements, notably a hydroxyl group at the 6-position, were required for a benzothiazole to act as an enhancer. 6-

TABLE III
INFLUENCE OF BENZOTHIAZOLE DERIVATIVES ON LIGHT EMISSION FROM THE LUMINOL,
HYDROGEN PEROXIDE SYSTEM IN THE PRESENCE AND ABSENCE OF
HORSERADISH PEROXIDASE CONJUGATE[a]

Benzothiazole derivative[b]	Relative light emission (mV)		Magnitude of enhancement of light emission		Percentage reduction in background light emission
	30 sec	60 sec	30 sec	60 sec	(5–180 sec)
DMSO (solvent control)	16	11	—	—	0
Benzothiazole (I)	10	8	—	—	
6-Hydroxybenzothiazole (II)	5950	7340	372	667	98.9
6-Methoxybenzothiazole (III)	13	10	—	—	94.5
2-Amino-6-hydroxybenzothiazole (IV)	8	8	—	—	99.9
2-Amino-6-methoxybenzothiazole (V)	9	6	—	—	95.0
2,6-Dihydroxybenzothiazole (VI)	5	5	—	—	99.5
2-Cyano-6-hydroxybenzothiazole (VII)	600	750	38	68	95.9
2-Cyano-6-ethoxybenzothiazole (VIII)	3	2	—	—	96.3
Dehydroluciferin (IX)	7240	6990	453	636	96.9
Firefly luciferin (X)	2480	2460	155	244	96.7
6-Ethoxyluciferin (XI)	5	4	—	—	99.7

[a] Antibody–peroxidase conjugate (10 μl of a 1:1000 dilution in 0.1 M Tris buffer, pH 8.0) and benzothiazole derivative (10 μl of a 0.5 mg/ml solution in dimethyl sulfoxide) were placed in separate corners of a cuvette. Luminol (50 μl, 1.2 mM) and hydrogen peroxide (50 μl, 40 mM) in Tris buffer (0.1 M, pH 8.0) were placed in the remaining corners and luminescent reactions initiated by manual addition of Tris buffer (900 μl, 0.1 M, pH 8.0). Light emission against time was recorded and intensity at different times after initiation determined. Control experiments in which antibody–peroxidase conjugate or benzothiazole derivatives were replaced by solvent were also performed. From Thorpe et al.[16]
[b] Structures are shown in Fig. 2.

Hydroxybenzothiazoles with a 2-amino or 2-hydroxy group did not act as enhancers and substitution of the 6-hydroxy group (e.g., 6-methyl ether or 6-ethyl ether derivatives) or its removal (e.g., benzothiazole) eliminated the enhancement properties of the derivatives.

Substituted Phenols and Naphthols. Several substituted phenols and naphthols have been identified which produce a greater degree of enhancement than achieved with 6-hydroxybenzothiazoles (Table IV). Although a large number of phenols produce no enhancement of light emission when incorporated into the peroxidase conjugate-catalyzed luminol–hydrogen peroxide reaction, compounds including p-iodophenol, p-phenylphenol, 1,6-dibromonaphth-2-ol, and p-hydroxycinnamic acid all produced pronounced enhancement of light emission. Several para-substituted halophenols, of which p-iodophenol was the most efficient, acted as effective enhancers and the increases in light intensity produced by such compounds are shown in Table V. The magnitude of enhancement was dependent on the concentration of enhancer and the halogen substituent (p-iodophenol > p-bromophenol > p-chlorophenol). In common with 6-hydroxybenzothiazole derivatives substitution of the hydroxyl group, as in p-chloromethoxyphenol, abolished the enhancement. Background light emission from the luminol–hydrogen peroxide reaction in the absence of peroxidase conjugate was reduced by variable degrees on addition of substituted phenols or naphthols (e.g., dependent upon conditions, reductions of 40–90% were produced with p-iodophenol).

TABLE IV

LIGHT EMISSION FROM THE HORSERADISH PEROXIDASE
CONJUGATE-CATALYZED OXIDATION OF LUMINOL ON ADDITION
OF ENHANCERS[a]

Enhancer	Quantity of enhancer added (nmol)	Light emission after 30 sec (mV)
No enhancer	—	11
p-Phenylphenol	22	5830
1,6-Dibromonaphth-2-ol	22	2401
p-Hydroxycinnamic acid	45	2317
p-Iodophenol	90	1958
1-Bromonaphth-2-ol	22	651
6-Hydroxybenzothiazole	45	297

[a] Enhancer and antibody–peroxidase conjugate (10 μl of a 1:1000 dilution) were placed in separate corners of a cuvette and luminescent reactions initiated by addition of 900 μl of a luminol (1.25 mM), hydrogen peroxide (2.7 mM) mixture prepared in Tris buffer (0.1 M, pH 8.5).

TABLE V

ENHANCEMENT OF LIGHT EMISSION FROM THE HORSERADISH PEROXIDASE
CONJUGATE-CATALYZED OXIDATION OF LUMINOL BY ADDITION OF HALOPHENOLS[a]

| | Quantity added | | | |
| | Light intensity after 30 sec (mV) | | Reduction in background light intensity after 30 sec (%) | |
Enhancer	45 nmol	408 nmol	45 nmol	408 nmol
Solvent control	9	15	—	—
p-Iodophenol	903	3056	43	89
p-Bromophenol	261	1379	24	80
p-Chlorophenol	190	1190	13	80
2-Chloro-4-bromophenol	80	555	7	69
2,4-Dichlorophenol	29	165	50	96
3,4-Dichlorophenol	20	82	49	88

[a] Enhancer (10 or 90 μl of a 4.54 mM solution in dimethyl sulfoxide) and antibody–peroxidase conjugate (10 μl of a 1:1000 dilution) were placed in separate corners of a cuvette and luminescent reactions initiated by addition of 900 μl of a luminol (1.25 mM), hydrogen peroxide (2.7 mM) mixture prepared in Tris buffer (0.1 M, pH 8.5). Reactions were also performed with enhancer or conjugate replaced by the appropriate solvent.

The Synergistic Action of Enhancers

Light emission from components of enhanced and unenhanced reactions involving luminol, and the synergistic nature of enhancement are shown in Table VI. Similar results were obtained using 6-hydroxybenzothiazole, firefly luciferin,[21] dehydroluciferin, p-iodophenol, or p-phenylphenol. Light emission from the luminol–H_2O_2 system catalyzed by peroxidase was increased by enhancer addition, and background light emission in the absence of peroxidase was decreased. The increased light emission could not be explained as an additive effect. Although some light was emitted with enhancers in the presence of H_2O_2 and peroxidase conjugate, the dramatic increase in light emission on addition of enhancer to the luminol–H_2O_2–peroxidase conjugate system significantly exceeded

[21] T. P. Whitehead, G. H. G. Thorpe, T. J. N. Carter, C. Groucutt, and L. J. Kricka, *Nature (London)* **305,** 158 (1983).

TABLE VI
LIGHT EMISSION FROM COMPONENTS OF ENHANCED CHEMILUMINESCENT REACTIONS[a]

	Enhancer			
	Light emission after 60 sec (mV)[b]		Light emission after 30 sec (mV)[c]	
Reaction components	6-Hydroxy-benzothiazole	Dehydro-luciferin	p-Iodophenol	2,3-Dichloro-5,6-dicyano-1,4-benzoquinone
Luminol, H_2O_2, peroxidase conjugate	22	22	119	153
Luminol, H_2O_2, peroxidase conjugate, enhancer	11,410	18,350	195,500	110,160
Luminol, H_2O_2	10	10	8	8
H_2O_2, enhancer	<1	<1	<8	8
H_2O_2, peroxidase conjugate, enhancer	8	5	85	25
Luminol enhancer	<1	<1	<8	<8
Luminol, peroxidase conjugate, enhancer	<1	<1	<8	<8
Luminol, H_2O_2, enhancer	<1	<1	<8	109,480

[a] Reactions were performed essentially as described in the text with combinations of components replaced by the appropriate solvent.
[b] Final reaction conditions: pH 8.0, luminol 59 μM, H_2O_2 1.9 mM, 6-hydroxybenzothiazole 32.3 μM, dehydroluciferin 17.6 μM.
[c] pH 8.0, luminol 290 μM, H_2O_2 1.95 mM, p-iodophenol 45 μM, 2,3-dichloro-5,6-dicyano-1,4-benzoquinone 43 μM.

that from systems with either luminol or enhancer alone. Very low levels of light emission were obtained in the absence of oxidant. 2,3-Dichloro-5,6-dicyano-1,4-benzoquinone, which has structural similarities to other chemiluminescent enhancers,[22] also produced enhancement of light emission from the luminol–H_2O_2–peroxidase conjugate reaction; however, in sharp contrast to 6-hydroxybenzothiazole and phenol derivative enhancers this increase also occurred in the absence of conjugate.

Specificity of Enhancement

Diacyl Hydrazide. Increases in light emission from the oxidation of a range of cyclic diacyl hydrazides, including luminol, isoluminol, N-(6-

[22] H. W. Yurow and S. Sass, *Anal. Chim. Acta* **88,** 389 (1977).

aminobutyl)-N-ethylisoluminol, and 7-dimethylaminonaphthalene-1,2-dicarboxylic acid hydrazide, are achieved by addition of 6-hydroxybenzothiazole derivatives such as 6-hydroxybenzothiazole, dehydroluciferin, or firefly luciferin.[16] Enhanced light emission with these diacyl hydrazides is also achieved by addition of p-iodophenol or p-phenylphenol.[18] At pH 8.5, in reactions employing cyclic diacyl hydrazides (5.6 μM), H_2O_2 (0.19 mM), and p-iodophenol (34 μM), 2492-, 120-, 88-, and 107-fold increases in light intensity were obtained with luminol, isoluminol, N- (6-aminobutyl)-N-ethylisoluminol, and 7-dimethylaminonaphthalene-1,2-dicarboxylic acid hydrazide, respectively. Similar reactions with p-phenylphenol (1.1 μM) produced 814-, 37-, 32-, and 70-fold increases in light intensity over the unenhanced reactions. Enhancement of the peroxidase conjugate-catalyzed reactions was also achieved at pH 8.0 and with various concentrations of cyclic diacyl hydrazide, and enhancer. Maximum increases in light emission, up to 2500-fold that of the unenhanced reaction, were obtained with luminol; this however had been carefully purified whereas the other compounds were used as supplied. Improvements in the magnitude of enhancement from other cyclic diacyl hydrazides may be possible on further purification.

Oxidant. Several oxidants including hydrogen peroxide and sodium perborate are capable of participating in enhanced luminescent reactions. Table VII shows the broadly similar increases in light emission obtained in enhanced reactions involving these oxidants.

Hemoprotein Catalyst. Although basic and acidic isoenzymes cata-

TABLE VII
INFLUENCE OF OXIDANT ON ENHANCEMENT OF LIGHT EMISSION FROM THE
LUMINOL–OXIDANT–PEROXIDASE CONJUGATE REACTION BY
6-HYDROXYBENZOTHIAZOLE DERIVATIVES OR p-IODOPHENOL[a]

Enhancer	Increase in light emission (fold) with oxidant	
	Hydrogen peroxide	Perborate
6-Hydroxybenzothiazole (pH 8.0)	372	406
Dehydroluciferin (pH 8.0)	453	451
Firefly luciferin (pH 8.0)	155	71
p-Iodophenol (pH 8.5)	553	544

[a] Light intensity was recorded after 30 sec. Reactions were performed at pH 8.0 or 8.5. Final reagent concentrations: oxidant, 1.96 mM; luminol, 59 μM; 6-hydroxybenzothiazole derivatives, 5 mg/liter; p-iodophenol, 350 μM; horseradish peroxidase conjugate, 240 pM.

TABLE VIII

LIGHT EMISSION FROM THE PEROXIDATION OF LUMINOL CATALYZED BY
HORSERADISH PEROXIDASE ISOENZYMES IN THE ABSENCE OR PRESENCE
OF p-IODOPHENOL[a]

| | Light emission (V) | |
HRP isoenzyme preparation	Unenhanced	p-Iodophenol enhanced
Two basic isoenzymes (Type VI)	0.04	96.56
One acidic isoenzyme (Type VII)	1.36	1.96
One acidic isoenzyme (Type VIII)	0.75	1.29
One basic isoenzyme (Type IX)	0.02	67.32

[a] Horseradish peroxidase preparation (10 μl, 1 μg/ml) and p-iodophenol
(20 μl, 18.2 mM) or solvent were placed in separate corners of a
cuvette. Luminescent reactions were initiated by addition of 1 ml of a
luminol (60 μM), hydrogen peroxide (2 mM) mixture in 0.1 M Tris
buffer, pH 8.5. Light intensity after 60 sec was recorded.

lyze the chemiluminescent oxidation of luminol the degree of enhancement produced on addition of enhancers differs considerably. The influence of p-iodophenol on light emission from the luminol–hydrogen peroxide reactions with different peroxidase isoenzymes is shown in Table VIII. More than 10,000% increases in light emission were achieved with basic isoenzymes in contrast to increases of less than 100% with the acidic isoenzymes. Enhancement was relatively specific for peroxidase; under the conditions employed light emission from luminol with other hemoprotein catalysts such as hemoglobin, myoglobin, or cytochrome c was reduced up to 70% by the addition of p-iodophenol. This improves the sensitivity and specificity with which peroxidase conjugates can be quantitated when interference from compounds such as hemoglobin is encountered.

Influence of pH

Enhancement of light emission from the peroxidase conjugate-catalyzed luminol–hydrogen peroxide reaction exhibited a marked pH dependence. Addition of p-iodophenol[18] or firefly luciferin[21] produced pronounced increases in light emission between pH 7 and 9.5 with maximal intensity at approximately pH 8.6 (Fig. 3). Similar effects were seen with p-chlorophenol, p-hydroxycinnamic acid, and 1,6-dibromonaphth-2-ol.

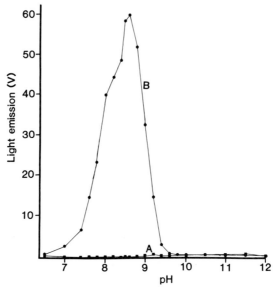

FIG. 3. Variation with pH of light emission from peroxidase conjugate-catalyzed luminescent reactions with luminol and hydrogen peroxide. A, Unenhanced; B, enhanced. Reactions were performed using Tris buffer (0.1 M, pH 6.5–9.0) or glycine/NaOH buffer (0.1 M,
pH 9.2–12.0) and light intensity after 30 sec was recorded. Final concentrations: luminol, 59
μM, hydrogen peroxide, 1.87 mM, and p-iodophenol, 336 μM. From Thorpe *et al.*[18]

Enhancement by 2,4-dichlorophenol, however, showed a slightly shifted
pH profile with maximal light emission at approximately pH 8.0.

Influence of Enhancer Concentration

The degree of enhancement was dependent on the enhancer concentration. Figures 4 and 5 illustrate the influence of various concentrations
of p-iodophenol or 6-hydroxybenzothiazole on light emission from the
peroxidase conjugate-catalyzed luminol–H_2O_2 reaction. Addition of enhancer increased light emission up to an optimal concentration above
which further addition of enhancer decreased light intensity and the degree of enhancement. The optimal enhancer concentrations depend on the
particular enhancer and the concentration of the luminol–peroxide mixture employed. In reactions employing luminol 56 μM, and hydrogen
peroxide 1.87 mM, optimal final concentrations for p-iodophenol and p-
phenylphenol were approximately 340 and 11 μM, respectively. Under
similar conditions, optimal concentrations for firefly luciferin and 6-hy-
droxybenzothiazole were approximately 55 and 24 μM.

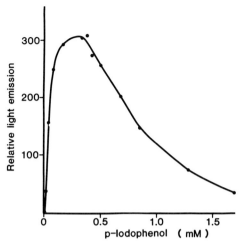

FIG. 4. Influence of p-iodophenol concentration on light emission from the peroxidase conjugate-catalyzed oxidation of luminol. Reactions were performed at pH 8.5 with luminol 56 μM and H_2O_2 1.87 mM. From Thorpe et al.[18]

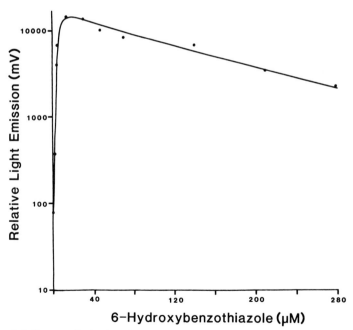

FIG. 5. Influence of 6-hydroxybenzothiazole concentration on light emission from the peroxidase conjugate-catalyzed oxidation of luminol. From Thorpe et al.[16]

Chemiluminescent Emission Spectra of Enhanced Reactions

Light emission from the unenhanced luminol–H_2O_2–peroxidase reaction exhibited maximal emission around 425 nm. On addition of enhancers such as 6-hydroxybenzothiazole derivatives or halophenols, the emission specta were found to be remarkably similar and essentially the same as in the unenhanced reaction (Figs. 6 and 7). Similar emission spectra were

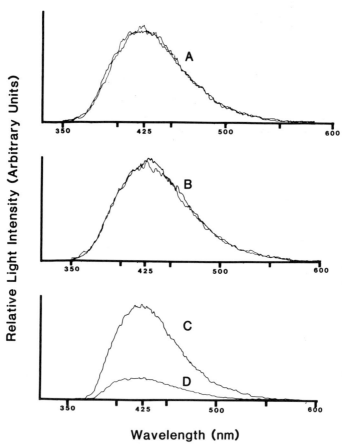

Wavelength (nm)

FIG. 6. Spectral energy distribution of the enhanced and unenhanced peroxidase conjugate-catalyzed oxidation of luminol. (A) 6-Hydroxybenzothiazole enhanced; (B) dehydroluciferin enhanced; (C) firefly luciferin enhanced; (D) unenhanced. Reactions A and B were performed at pH 8.5 with luminol 58.5 μM, H_2O_2 1.96 mM, and either 6-hydroxybenzothiazole 32 μM or dehydroluciferin 17.6 μM. Reactions C and D were performed at pH 8.0 with luminol 1.22 mM, H_2O_2 2.64 mM in the presence or absence of firefly luciferin 39 μM. From Thorpe et al.[16]

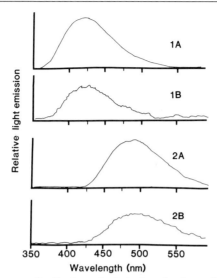

FIG. 7. Spectral energy distribution of p-iodophenol enhanced and unenhanced peroxidase conjugate-catalyzed oxidation of cyclic diacyl hydrazides. (1) Luminol, (2) 7-dimethylaminonaphthalene-1,2-dicarboxylic acid hydrazide; (A) enhanced, (B) unenhanced. Reactions were performed at pH 9.0, with luminol 139 μM or 7-dimethylaminonaphthalene-1,2-dicarboxylic acid hydrazide 103 μM and H_2O_2 3.7 mM, in the absence or presence of p-iodophenol 41.7 μM. From Thorpe et al.[18]

obtained with firefly luciferin, 6-hydroxybenzothiazole, dehydroluciferin, p-iodophenol, p-chlorophenol, or 2,4-dichlorophenol as enhancer.

The spectrum of light emitted from the peroxidase conjugate-catalyzed oxidation of 7-dimethylaminonaphthalene-1,2-dicarboxylic acid hydrazide is at a longer wavelength than that of luminol, with peak emission near 495 nm. Figure 7 illustrates how this difference is maintained in the p-iodophenol enhanced reactions. Although it was not possible to determine the emission spectra of the low-intensity light emitted with enhancer alone, the similarity of spectra in enhanced and unenhanced reactions and the apparent retention of the emission spectra of the acyl hydrazide employed suggest the aminophthalate products and not the enhancers are the emitters.

Enhancement of Luminescent Reactions Not Involving
   Horseradish Peroxidase

Enhancement of light emission is a phenomenon which has been observed in a wide range of chemiluminescent and bioluminescent reactions. Typical examples are collected in Tables IX and X.

TABLE IX
ENHANCEMENT OF LIGHT EMISSION FROM CHEMILUMINESCENT SYSTEMS

Luminescent system	Enhancers	Magnitude of enhancement (fold)	References
Luminol–H_2O_2–chromium(III)	Chloride or bromide ions	8	23
Luminol–H_2O_2	Organic compounds such as p-chlorobenzenediazonium fluoroborate, tribromonitromethane, 1-chloro-2,4-dinitrobenzene	5–250,000	22
Antibody bound isoluminol–steroid conjugates–H_2O_2–microperoxidase	Sodium hydroxide	—	20
Isoluminol–steroid–H_2O_2–microperoxidase	Specific antibody	6	25
Isoluminol–biotin–H_2O_2–lactoperoxidase	Avidin (biotin-specific binding protein)	10	26
Lucigenin–H_2O_2–NaOH	Micelle-producing surfactants such as cetyltrimethylammonium bromide	4	27
Short chain N-alkylamino-luminol–H_2O_2–persulfate	Micelle-producing surfactants such as cetyltrimethylammonium bromide	0.7	28
Lucigenin–H_2O_2–NaOH	Cyclodextrin	20	29

Addition of chloride or bromide ions to the chromium(III)-catalyzed luminol–H_2O_2 reaction increases light emission 8-fold.[23] The proposed mechanism of enhancement involves the formation of a chromium–bromide–peroxide complex in which the bromide facilitates electron transfer in the reaction with luminol. Enhancement of light emission from the luminol–H_2O_2 reaction at alkaline pH has been achieved by the addition of various organic compounds generally containing a diazonium or oxonium group, or various combinations of carbonyl, cyano, nitro, or halo groups.[22] It is proposed that the induction of light emission may involve the nucleophilic attack of hydroperoxide anion on the electrophilic organic compounds followed by oxidation of luminol by intermediate hydroperoxides of varying oxidizing power.

Two- to fivefold enhancement of light emission, with virtually no increase in background emission, has been achieved by combining pseudoperoxidase, sodium hydroxide, and hydrogen peroxide reagents prior to initiation of chemiluminescence from 6-[N-(4-aminobutyl-N-ethyl)iso-

[23] D. E. Bause and H. H. Patterson, *Anal. Chem.* **51**, 2288 (1979).

TABLE X
ENHANCEMENT OF LIGHT EMISSION FROM BIOLUMINESCENT SYSTEMS

Luminescent system	Enhancers	Magnitude of enhancement (fold)	References
Renilla luciferase–luciferin	Green-fluorescent protein	3	30
Pholas dactylus luciferase–luciferin–O_2	Reducing agents such as pyrogallol, ascorbic acid, or 1,2-dihydroxybenzene	3–6	32
Firefly luciferase–luciferin–ATP–O_2	Nonionic surfactants such as Triton X-100, poly(vinylpyrrolidone), or polyethylene glycol	3–6	33

luminol] hemisuccinamide labels in solid phase immunoassays.[24] Generation of OOH^- ions which optimize the light producing reaction via intermediate dioxetane production may be responsible for increased light production.

Preincubation of antibody-bound steroid isoluminol conjugates with sodium hydroxide in heterogeneous immunoassays increases light emission.[20] This may arise from either dissociation of the antibody complex or physicochemical changes in the label. Increased light emission is also produced when certain isoluminol-labeled antigens bind to specific antibodies[25] or binding proteins.[26] This forms the basis of several nonseparation immunoassays and it is suggested that enhancement occurs because a more hydrophobic environment is established for the emitting species.

Organized media such as micellar solutions have been shown to increase the quantum yields of chemiluminescent reactions of lucigenin with hydrogen peroxide.[27,29] Solutions containing lamellar aggregates of

[24] W. G. Wood, U. Hantke, and A. J. Gross, *J. Clin. Chem. Clin. Biochem.* **23**, 47 (1985).

[25] F. Kohen, J. B. Kim, G. Barnard, and H. R. Lindner, *Steroids* **36**, 405 (1980).

[26] H. R. Schroeder, P. O. Vogelhut, R. J. Carrico, R. C. Boguslaski, and R. T. Buckler, *Anal. Chem.* **48**, 1933 (1976).

[27] C. M. Paleos, G. Vassilopoulos, and J. Nikokavouras, *in* "Bioluminescence and Chemiluminescence. Basic Chemistry and Analytical Applications" (M. A. DeLuca and W. D. McElroy, eds.), p. 729. Academic Press, New York, 1981.

[28] K.-D. Gundermann, *in* "Bioluminescence and Chemiluminescence. Basic Chemistry and Analytical Applications" (M. A. DeLuca and W. D. McElroy, eds.), p. 17. Academic Press, New York, 1981.

[29] M. L. Grayeski and E. Woolf, *in* "Analytical Applications of Bioluminescence and Chemiluminescence" (L. J. Kricka, P. E. Stanley, G. H. G. Thorpe, and T. P. Whitehead, eds.), p. 565. Academic Press, Orlando, 1984.

didodecyl dimethyl ammonium bromide increase the efficiency of the lucigenin reaction by a factor of 12 and increased emission is thought to arise from an increase in the efficiency of the excitation step of the reaction caused by solubilization of an intermediate in the less polar environment of the organized media. β-Cyclodextrin, a rigid conical carbohydrate molecule with a hollow hydrophobic interior, enhances light emission from the lucigenin reaction.[29] Enhancement here is attributed to an increase in excitation efficiency and rate of the reaction through inclusion of a reaction intermediate in the cyclodextrin cavity.

Radiationless or resonance energy transfer has been proposed to account for the 3-fold enhancement of light emission when *Renilla* green fluorescent protein is added to the *Renilla* luciferase–luciferin system.[30] In this process an excited donor molecule transfers its excitation energy directly to an acceptor molecule with which it is in resonance. As the efficiency of excited state production may be greater than the fluorescence quantum yield of the donor molecule, it is possible to increase the overall fluorescence quantum yield of a resonance energy transfer system by the selection of a suitable acceptor molecule. Energy from excited sulfur dioxide can also be transferred to fluorescent molecules and very weak chemiluminescence from the permanganate-sulfite system can be enhanced by a factor of 330 on addition of fluorescent molecules such as riboflavin phosphate.[31]

Light emission from the *P. dactylus* luciferase-catalyzed oxidation of its luciferin by oxygen can be enhanced by reducing agents such as pyrogallol or 1,2-dihydroxybenzene.[32] The mechanism proposed involves either an induction of free-radical chain reactions or modification of enzyme bound product to facilitate its dissociation. Stimulation of light emission from the firefly luciferase reaction by solvents such as nonionic surfactants[33] may involve a similar mechanism. It is proposed that the polymers bind to the enzyme in such a manner that the product molecule, which is inhibitory and normally tightly bound, becomes more readily dissociated.

Mechanism of Enhancement

The full mechanism by which 6-hydroxybenzothiazoles and phenols enhance the peroxidase-catalyzed oxidation of cyclic hydrazides is not known. Many compounds reported to increase light emission from chemiluminescent and bioluminescent systems (Tables IX and X) do not en-

[30] W. W. Ward and M. J. Cormier, this series, Vol. 57, p. 257.
[31] M. Yamada, T. Nakada, and S. Suzuki, *Anal. Chim. Acta* **147**, 401 (1983).
[32] A. M. Michelson and M. F. Isambert, *Biochimie* **55**, 619 (1973).
[33] L. J. Kricka and M. DeLuca, *Arch. Biochem. Biophys.* **217**, 674 (1982).

hance the peroxidase-catalyzed system under conditions where 6-hy-droxybenzothiazole and phenol derivatives produce dramatic increases in light intensity suggesting they operate by a different mechanism. Increased light emission is unlikely to arise from solvent effects or the production of organized media, and the similar emission spectra of enhanced and unenhanced reactions would suggest that the enhancers themselves do not act as emitters nor that more efficient chemiluminescent compounds are formed between the acyl hydrazides and the enhancers. Most probably the enhancers directly or indirectly, accelerate one or more steps in the complex peroxidase-catalyzed reaction prior to the emission reaction.

Applications of Enhanced Luminescent Quantitation of Horseradish Peroxidase Conjugates

The use of an enhancer facilitates rapid and sensitive chemiluminescent quantification of peroxidase conjugates and should aid the introduction of chemiluminescently monitored immunoassays into routine use. Enhanced chemiluminescent end points have been incorporated into a range of immunoassays, both immunoextraction and competitive, in conjunction with solid supports such as beads, tubes, and microtiter plates (Table XI).[34–40] Enhanced end points are easily performed and relatively simple instrumentation based on photodetectors such as photomultiplier tubes or photographic film can be employed. The prolonged, relatively constant and intense light emission achieved with enhanced reactions simplifies the design of manual or automatic luminometers required for routine applications. Rapid and reproducible initiation of individual reactions in front of photodetectors is unnecessary and therefore expensive

[34] G. H. G. Thorpe, R. Haggart, L. J. Kricka, and T. P. Whitehead, *Biochem. Biophys. Res. Commun.* **119**, 481 (1984).

[35] G. H. G. Thorpe, L. J. Kricka, and T. P. Whitehead, *Clin. Chem. (Winston-Salem, N.C.)* **31**, 913 (1985).

[36] G. H. G. Thorpe, L. A. Williams, L. J. Kricka, T. P. Whitehead, H. Evans, and D. R. Stanworth, *J. Immunol. Methods* **79**, 57 (1985).

[37] G. H. G. Thorpe, E. Gillespie, R. Haggart, L. J. Kricka, and T. P. Whitehead, *in* "Analytical Applications of Bioluminescence and Chemiluminescence" (L. J. Kricka, P. E. Stanley, G. H. G. Thorpe, and T. P. Whitehead, eds.), p. 243. Academic Press, Orlando, 1984.

[38] G. H. G. Thorpe, T. P. Whitehead, R. Penn, and L. J. Kricka, *Clin. Chem. (Winston-Salem, N.C.)* **30**, 806 (1984).

[39] G. H. G. Thorpe, S. B. Moseley, L. J. Kricka, R. A. Stott, and T. P. Whitehead, *Anal. Chim. Acta* **170**, 101 (1985).

[40] H. X. Wang, J. George, G. H. G. Thorpe, R. A. Stott, L. J. Kricka, and T. P. Whitehead, *J. Clin. Pathol.* **38**, 317 (1985).

TABLE XI

ENHANCED CHEMILUMINESCENT ENZYME IMMUNOASSAYS BASED ON THE HORSERADISH
PEROXIDASE-CATALYZED LUMINOL–HYDROGEN PEROXIDE REACTION

Analyte	Enhancer	Solid support	References
α-Fetoprotein	Firefly luciferin	Beads	21
Rubella virus-specific IgG	Firefly luciferin	Beads	34
Choriogonadotropin	p-Iodophenol	Beads	18
Carcinoembryonic antigen	p-Iodophenol	Beads	35
IgE	Firefly luciferin	Tubes	34
IgE	p-Iodophenol	Tubes	36
Digoxin	Firefly luciferin	Tubes	34
Digoxin	p-Iodophenol	Tubes	18
Thyroxine	Firefly luciferin	Tubes	37
Ferritin	Firefly luciferin	Microtiter plate	38
Cytomegalovirus-specific IgG	6-Hydroxyben-zothiazole, dehydroluciferin	Microtiter plate	39
Factor VIII-related antigen	p-Iodophenol	Microtiter plate	40

injection systems or sophisticated detectors are not required. The enhanced end point is particularly attractive for use in luminescent immunoassays performed on microtiter plates, which are already widely used with colorimetric monitored immunoassays employing peroxidase labels. Previous instrumentation for monitoring chemiluminescent immunoassays on microtiter plates was complex[41] but with enhanced chemiluminescence multiple reactions can be initiated manually outside the luminometer and after several minutes groups of measurements can be conveniently made on batches of samples, with a single detector. Carryover of light from neighboring wells is eliminated by the use of opaque microtiter plates and rapid sequential measurements from a series of wells is possible. Enhanced chemiluminescent end points are employed in assays in the commercially available "Amerlite" system (Amersham International plc, Buckinghamshire HP7 9NA, United Kingdom).[42] Assays are performed in opaque wells which can be assembled in a 96-place holder. Enhanced chemiluminescent reactions in all wells are initiated manually by addition of a single reagent mixture. Light emission stabilizes within 2 min and is emitted for a minimum of 20 min. Light intensity can be measured at any time during this period using an automated luminescence analyzer which

[41] H. R. Schroeder, C. M. Hines, D. D. Osborn, R. P. Moore, R. L. Hurtle, F. F. Wogoman, R. W. Rogers, and P. O. Vogelhut, *Clin. Chem. (Winston-Salem, N.C.)* **27**, 1378 (1981).
[42] J. Edwards, *Med. Lab. World,* December, 35 (1985).

will take measurements from 96 wells, automatically fit a dose–response curve to standards, interpolate unknowns, and print out the assay results within 2 min.[42]

Instant photographic film has been employed to detect light emission in a range of chemiluminescent and bioluminescent reactions[43] and the intense and prolonged nature of chemiluminescent reactions enhanced by phenol or 6-hydroxybenzothiazole derivatives extends the applicability of this simple and convenient approach to analysis into immunoassays. Multiple samples can, for instance, be assayed rapidly and simultaneously with a camera luminometer which is inexpensive, small, portable, and requires no power source. Reactions can be initiated manually prior to exposure to the film and a permanent visual record of the semiquantitative results is obtained. Although no numerical results are obtained, the response is acceptable for many clinical situations where it is necessary only to know whether the analyte concentration lies outside a certain range. Suitable screening applications include the detection of infection-related antibodies and allergy testing.

Summary

Enhancement of light emission from the horseradish peroxidase-catalyzed oxidation of diacyl hydrazides on addition of 6-hydroxybenzothiazole or phenol derivatives forms the basis of rapid, specific, and sensitive chemiluminescent assays for peroxidase. The advantages and wide applicability of the technique have been demonstrated in a range of ligand-binding assays. Careful selection of chemiluminescent reagents, enhancer, their relative proportions and reaction conditions, and more detailed knowledge of the mechanism of enhancement should enable further improvements in sensitivity and the intensity or constancy of light emission.

Acknowledgments

The financial support of the Department of Health and Social Security is gratefully acknowledged. Aspects of this work are the subject of patent applications which have been assigned to the British Technology Group (NRDC).

---

[43] L. J. Kricka and G. H. G. Thorpe, this volume [33].

[30] Luminescence Immunoassays for Haptens and Proteins

By W. Graham Wood, Jörg Braun, and Uwe Hantke

Introduction

The aim of this chapter is to acquaint the reader with the problems and possible solutions for the assay of haptens and proteins in body fluids using luminescent labels based on aryl hydrazides. One question which arises and which must be answered here is why sensitive assays should be developed for components which can be measured using less sensitive methods, such as nephelometry. The answer lies in the field in which these assays are to be used. In the area of neonatal surveillance, for example, in cases of sepsis, the sample volume obtainable is minimal, and because of the regular controls needed, sensitive luminescence immunoassays have their place in the routine laboratory.

In the assays described here for serum proteins, 10 μl serum or plasma is usually sufficient for duplicate determinations of several relevant parameters. The assays are suitable for serum, plasma, sputum, and saliva, samples having to be diluted in many cases before assay.

The optimization of the assay system is as important as the ability to determine different parameters, and choices of systems are compared, using results gathered in routine and research over the past 3 years, during which time over 70 assays using a luminescent label have been developed in this laboratory.

All labels described here are based on derivatives of isoluminol or naphthalene-1,2-dicarboxylic acid hydrazide. Different heme-containing molecules have been compared with respect to light output and light kinetics, both in liquid and solid-phase systems.

Because of the large numbers of "optimized systems" published, assay design, oxidation system, and luminogen used play a major role in this chapter, and examples of assays in routine use are given to aid the reader in setting up assays.

Materials

Polystyrene balls (6.4 mm/0.25 in. diameter) were obtained from either Spherotech Kugeln, Fulda, D, Euromatic, Brentford, Middlesex, UK, or Precision Plastic Ball Co., Chicago, IL

Poly(phenylalanine–lysine) (1 : 1 copolymer) was purchased from either Miles-Yeda (Bayer), Munich, D, or Sigma, Munich, D, and had a relative molecular mass of 30–40 kDa

Microperoxidase MP-11, hemin, cytochrome c, horseradish peroxidase, and hematin were obtained from Sigma, hematoprotoporphyrin IX from Aldrich, Steinheim, D
6-[N-(4-Aminobutyl-N-ethyl)-2,3-dihydrophthalazine-1,4-dione]
hemisuccinamide (ABEI-H) was purchased from LKB-Wallac,
Turku, SF, 7-N-(4-aminobutyl-N-ethyl)naphthalene-1,2-dicarbox-
ylic acid hydrazide (ABEN) being a gift from Dr. Hartmut
Schroeder, Miles Laboratories, Elkhart, IN
Buffer substances were from Merck, Darmstadt, D, and Tween 20
from Sigma
All samples were measured in a 300-sample semiautomatic luminometer (Laboratorium Prof. Dr. Berthold, Wildbad, D), data processing being performed off-line on a Commodore 8296-D desk-top
computer (CBM business machines, Neu-Isenburg, D)

The following apparatus was needed to perform the assays here described under optimal conditions.

1. Luminometer, e.g., Berthold LB-950 (300 samples) or LKB 1251
   (25 samples), fitted with one dispenser for 300 μl (Berthold) or 2
   dispensers for parallel dispensing (LKB)
2. Reaction trays (20/60 wells), e.g., Abbott
3. Multiwash unit, e.g., Abbott Pentawash or Proquantum
4. Horizontal shaker, e.g., Heidolph, 170–200 rpm
5. Ball dispenser, e.g., Abbott RAB (single ball) or 60-ball dispenser

Assay Systems

The assay systems described here have been published in detail elsewhere[1-5] and are given here in brief for those not acquainted with them.

1. SPALT: solid phase antigen luminescent technique. This assay
principle is analogous to the RAST for allergens and uses an immobilized
antigen (protein or protein–hapten conjugate) and a luminescent-labeled
second (species-IgG directed) antibody. As no labeled antigen is used to
"dilute" the system, SPALT assays have a lower detection limit on the

[1] W. G. Wood, *J. Clin. Chem. Clin. Biochem.* **22,** 905 (1984).
[2] W. G. Wood, H. Fricke, J. Haritz, H. S. Krausz, B. Tode, C. J. Strasburger, and P. C. Scriba, *J. Clin. Chem. Clin. Biochem.* **22,** 349 (1984).
[3] C. J. Strasburger, H. Fricke, A. Gadow, W. Klingler, and W. G. Wood, *Aerztl. Lab.* **29,** 75 (1983).
[4] W. G. Wood, *Aerztl. Lab.* **30,** 336 (1984).
[5] W. G. Wood, *in* "Reviews on Immunoassay Technology" (S. B. Pal, ed.), p. 105. de Gruyter, Berlin, 1985.

whole than "conventional" immunoassays of the CELIA (chemiluminescent immunoassay) type (see 4 below).

2. ILMA: immunoluminometric assay. This system is analogous to the
immunoradiometric assay (IRMA).

3. ILSA: immunoluminometric-labeled second antibody assay. The
ILSA uses the same principle as the ILMA, but has an additional step in
which a labeled second antibody is used. The two antigen-specific antibodies are unlabeled. Advantages here include the use of monoclonal
antibodies "sensitive" to labeling as the liquid-phase antibody; disadvantages include the fact that both antigen-specific antibodies must come
from different species, e.g., rabbit/mouse.

4. CELIA: chemiluminescent immunoassay. This is analogous to the
conventional solid-phase RIA and is only to be recommended in systems
in which the samples are extracted prior to assaying, e.g., steroids. The
potential oxidation of label during incubation by endogenous peroxidases
or heme compounds must be taken seriously into account when developing CELIAs using aryl hydrazide labels.

Methods

1. Poly(phenylalanine–lysine) [poly(Phe-Lys) coating of solid phase].[5]
The following procedure is suitable for coating 1000 balls and has been
used with suitable factors to coat batches of up to 20,000 balls without loss
in precision. It is probable that the poly(Phe-Lys) must be warmed to 60–
70° before it dissolves, although it is manufacturer and batch dependent.

One thousand balls are transferred to a 400-ml beaker (1000 balls =
~230 ml) and are covered with 125 ml double-distilled water containing
2.5 mg poly(Phe-Lys). The balls are agitated to ensure coverage, this
process being repeated several times over the first 4–6 hr to remove
trapped gas bubbles. The coating procedure is complete within 48 hr at
ambient temperature, although coating times of up to 168 hr do not seem
to influence precision of coating.

After coating, the balls are washed once with distilled water and dried
under a stream of compressed air or nitrogen. The coated dry balls can be
kept when stored in a dry atmosphere for several months at ambient
temperature.

The drying process is necessary to optimize the hydrophobic bonding
between the benzene ring electrons of the styrene and phenylalanine moieties. If the drying step is omitted both coating precision and loading
capacity are decreased in the subsequent coupling of proteins to the ball
surfaces.

2. Covalent coupling of proteins to the poly(Phe-Lys) balls. a. Pen-
tane-1,5-dial activation. One thousand poly(Phe-Lys) balls are decanted

into a 400-ml beaker to which 125 ml double-distilled water containing 2.5 ml glutaraldehyde (250 g/liter, Sigma, Catalog No. G-5882) is added. The balls are activated for 40 min at ambient temperature during which time they are briefly stirred every 10 min to ensure uniform exposure to the pentane-1,5-dial and to release gas bubbles from the ball surfaces.

After activation, the balls are washed twice with 0.15 M NaCl and twice with distilled water. The activation step is controlled with Schiff's reagent (p-rosaniline treated with SO_2), a pink-lilac coloration on the ball surface denoting the presence of free aldehyde groups. A nontreated ball serves as a control, this showing little or no coloration with Schiff's reagent.

b. Protein coupling. A solution of the protein to be coupled (2–3 mg) in 125 ml distilled water is poured over the washed balls prepared in (a) above, briefly agitated, and left to stand for 20 min at ambient temperature before being adjusted to pH 8.2–8.8 with 5 ml 0.5 M K_2HPO_4. If amounts of protein (e.g., antigen or antibody) smaller than 2–3 mg are to be coupled to the balls, the total protein should be adjusted with bovine serum albumin or ovalbumin to give a final concentration of 2–3 mg/1000 balls, otherwise the coating quality is worse. The 20-min period before pH adjustment allows for a better dispersion of the protein to be coupled before the reaction of free aliphatic amino groups of the protein with the aldehyde groups on the balls takes place at around pH 8.5

This reaction is allowed to proceed overnight at ambient temperature or 4° after which 5 ml (25 g/liter) bovine serum albumin is added as saturation reagent. After a further 18–24 hr, 5 ml 0.5 M Tris–HCl, pH 8, is added to complete saturation, the balls once again being allowed to stand overnight before they are washed once with 0.15 M NaCl and once with distilled water before being covered for 5 min with 5 g/liter poly(vinyl alcohol) (PVA). The PVA solution is aspirated off and the balls dried under a stream of compressed air or nitrogen, after which they can be stored for several months at ambient temperature in a dry atmosphere (e.g., tightly stoppered bottles). The PVA coat serves both as bacteriostat and antimoisture agent, thus increasing the shelf life of the balls.

The time for the coupling procedure may appear long, but has shown itself to be far superior to purely adsorptive techniques for attaching proteins to polystyrene balls.

3. Labels and labeling. Three compounds have been used for labeling proteins, diazoluminol,[6] ABEI-H,[4,7] and ABEN-hemisuccinamide (ABEN-H).[7] Diazoluminol cannot be recommended where sensitive as-

[6] W. G. Wood and A. Gadow, J. Clin. Chem. Clin. Biochem. **21**, 789 (1983).

[7] H. R. Schroeder, R. C. Boguslaski, R. J. Carrico, and R. T. Buckler, this series, Vol. 57, p. 424.

says are to be developed, as the coupling at an alkaline pH of 9–10 for extended periods of time (24–72 hr) drastically affects the binding capabilities of the labeled antibodies.[6] This is seen by the low ratio of counts bound to counts added.

ABEI-H and ABEN-H are labels of choice, ABEN-H-labeled antibodies showing a light output of 5–10 times more than ABEI-H-labeled antibodies. The coupling is best carried out via an active ester which requires the conversion of the alkylamino group of ABEI or ABEN to the terminal carboxyl group of the respective hemisuccinamide using succinic anhydride, the reaction being carried out in dry dimethyl formamide (DMF). The active ester is formed with N-hydroxysuccinimide and dicyclohexylcarbodiimide (DCCD). The subsequent coupling to the protein is carried out in 0.05 M phosphate buffer, pH 7.5–8.2, making sure that the end concentration of DMF in the reaction mixture does not exceed 5%.

After labeling, the unreacted luminogen and labeled proteins are separated by gel filtration (e.g., Sephadex, Ultrogel, or Bio-Rad gels), the fractions with optimal binding characteristics (high specific binding and low unspecific binding) being pooled and portioned before being stored at $-20°$. The labels are stable for periods in excess of 2 years when stored between -20 and $-30°$. Full details for labeling are given below.

Labeling procedure for proteins using an active ester intermediate:

    a. Preparation of the hemisuccinamide of ABEI and ABEN:

        5 μmol ABEN or ABEI suspended in 0.5 ml dry dimethyl formamide (DMF)

        6 μmol succinic anhydride in 0.1 ml dry fresh distilled DMF

    Allow to react overnight at 4° in the dark

The hemisuccinamide is present in excess of 90% theoretical yield and is used as such.

    b. Preparation of the active ester: Take the hemisuccinamide and add

        5 μmol N-hydroxysuccinamide in 100 μl dry DMF

        7.5 μmol dicyclohexylcarbodiimide in dry DMF

    Allow to react as in (a) above

The active ester is used directly for labeling [see (c) below] or is portioned, lyophilized, and stored in a cool dry place until used; the stability is at least 4 weeks at 4°.

    c. Protein labeling with the active ester: Dissolve 8–12 mg protein in 1 mg 0.05 M phosphate buffer, pH 8. Add active ester with stirring to give a ratio protein : active ester 1 : 20

    Allow to react overnight as above in (a) and (b)

The incorporation is 2–5 mol label per mol protein. The mixture is separated by gel chromatography as stated in the text after which portions of 200–500 μl are stored frozen until use. The thawed portions are conserved by adding 10 μl 1.5 M NaN$_3$/500 μl. The stability of the thawed portions is 3–6 weeks at 4°.

4. Oxidation systems for aryl hydrazide luminescence. Many systems and substances have been published, each with its advantages and disadvantages.[8–10] The systems described here have been in routine use or under test for at least 3 months in comparative studies in luminescence immunoassays. The substances tested include hematoprotoporphyrin IX (HP-9), microperoxidase MP-11 (MP-11), horseradish peroxidase (HRP), cytochrome c (cyt c), and hemin. Stability over longer periods of time (up to 7 days) and drift phenomena have been studied on the "ready for use" reagents using assays with total measuring time over 3 hr ($\sim$360 tubes) for the latter.

The substances tested and listed above differ not only in light output of the system, but in the reaction kinetics, even when using identical luminogen concentrations. Differences are to be seen as to whether the tests are carried out with the luminogen in the liquid phase or bound in an antigen–antibody complex to the polystyrene ball. The standard oxidation system for the Berthold LB-950 luminometer is as follows:

Solution 1: 0.15 M NaCl containing 7.5 μM "catalyst" (these substances are not catalysts in the true sense of the word, as they are not "recycled" but perform a "one-off" reaction) (e.g., MP-11)

Solution 2: 0.33 M NaOH containing 3 mM hydrogen peroxide

300 μl solution 1 is added to the measuring cuvette containing the washed solid phase (polystyrene ball) and serves to keep it wet, thus improving precision

300 μl solution 2 is injected into the cuvette in the measuring position to initiate the light reaction; the light output is integrated over the period 0–12 sec

A single dispenser version[11] is used in all cases, the cofactor being added to the solid phase in the measuring cuvette before loading up the luminometer. The light emission is initiated with the injection of alkaline peroxide solution in the measuring position. Table I shows the main properties of the different systems with respect to light output, reagent stability (label degradation), and light-reaction kinetics.

The simplest way to change the light reaction kinetic ($t_{max}/2$) is to vary

---

[8] T. P. Whitehead, L. J. Kricka, T. J. N. Carter, and G. H. G. Thorpe, *Clin. Chem. (Winston-Salem, N.C.)* **25**, 1531 (1979).

[9] L. J. Kricka, P. E. Stanley, G. H. G. Thorpe, and T. P. Whitehead, eds. "Analytical Applications of Bioluminescence and Chemiluminescence," p. 149. Academic Press, Orlando, 1984.

[10] L. J. Kricka, *in* "Clinical and Biochemical Luminescence" (L. J. Kricka and T. J. N. Carter, eds.) Chapter 8, p. 153. Dekker, New York, 1982.

[11] W. G. Wood, U. Hantke, and A. J. Gross, *J. Clin. Chem. Clin. Biochem.* **23**, 47 (1985).

TABLE I

MAIN PROPERTIES OF THE OXIDATION SYSTEMS TESTED USING 50 μl OF A 1:2000
DILUTION OF ABEN-H ANTITRANSFERRIN

	System				
Component	MP-11	HP-9	HRP	Cyt c	Hemin
Signal (relative light units)[a]					
a. Reagent blank	14	4	10	5	16
b. 50 μl tracer	1,170	18	277	233	21,005
Ratio (b : a)	84	4.5	28	47	1,313
Kinetics $(t_{max}/2)^b$ (sec)	4.8	>12	>12	>12	1.8
Signal loss in 10 min (%)					
pH 7 liquid phase	80	—[c]	—	—	22
pH 3 liquid phase	5	—	—	—	2
pH 7 solid phase	3	—	—	—	0.5
pH 3 solid phase	0.5	—	—	—	0.05

[a] Relative light units = impulses integrated over the time 0–12 sec divided by 1,000, i.e.,
1,170 relative light units = 1,170,000 impulses/12 sec.
[b] $t_{max}/2$ represents the time taken in seconds for the peak light signal to fall to half its
height.
[c] The dashes (—) indicate that no measurements were made in these systems due to their
poor light output and kinetics.

the hydrogen peroxide concentration. In general, increasing the concentration increases $t_{max}/2$, the pH remaining around 13. The latter is important in helping to solublize the protein coat from the ball, and, because of this, the alternative way to change $t_{max}/2$, namely by changing the pH, is not given as the "first choice."

The reason for the rapid decay in the liquid phase at pH 7 is no doubt due to the heme compounds which can present "active oxygen" present in the solutions to the luminogen, thus leading to label oxidation. This effect is drastically reduced by lowering the pH to 3. The decay effects are hardly to be seen when the luminogen is bound to the solid phase in the form of a labeled antibody–antigen complex. This is probably due to the exclusion of the "active species" (cofactor–oxygen complex) by the hydrophobic interactions of the protein coat on the solid phase which reduces the diffusion capacity. This difference in reaction capability is of practical interest, especially when evaluating published oxidation systems!

5. Examples of solid-phase luminescence immunoassays to demonstrate the points described in the text. a. One component, three assay methods: the determinations of ceruloplasmin in serum. The determination of ceruloplasmin in serum has been measured using ILMA and

SPALT and using the oxidase activity of this protein using *p*-phenylene-diamine as substrate.[12] The assay flow schemes for SPALT and ILMA are as follows.

SPALT
   10 μl sample/standard (1 : 500 dilution)
   200 μl rabbit anticeruloplasmin
   Incubate 15 min at ambient temperature
   1 ceruloplasmin-coated ball
   Incubate 45 min on horizontal rotator (170–200 rpm) at ambient temperature
   Wash with 2 × 5 ml wash solution (0.15 *M* NaCl containing 3 ml/liter Tween 20)
   200 μl ABEI-H-labeled donkey anti-rabbit IgG
   Incubate 60 min on horizontal rotator at ambient temperature
   Wash as above, transfer washed ball to fresh measuring cuvette, prepare and measure as shown previously

ILMA
   10 μl sample/standard (1 : 1000 dilution)
   200 μl assay buffer 0.025 *M* Tris–HCl, 0.025 *M* phosphate, 0.08 *M* NaCl, 0.15 ml/liter Tween 20, pH 7.5
   1 rabbit anticeruloplasmin ball
   Incubate 60 min at ambient temperature on horizontal rotator
   Wash as for SPALT above
   200 μl ABEN-H-labeled rabbit anticeruloplasmin
   Incubate and wash as above, prepare and measure as for SPALT

The reference range for ceruloplasmin as determined in 86 healthy blood donors was similar for all three methods: SPALT, median 0.27 g/liter, range 0.09–0.71 g/liter; ILMA, median 0.25 g/liter, range 0.12–0.66 g/liter; and enzymatic method, median 0.28 g/liter, range 0.14–0.61 g/liter. The results from 50 patients suffering from surgically or histologically confirmed gynecological tumors were different: SPALT, median 0.78 g/liter, range 0.12–1.24 g/liter; ILMA, median 0.44 g/liter, range 0.15–1.42 g/liter; and enzymatic method, median 0.26 g/liter, range 0.18–0.41 g/liter. The ILMA appears to measure "intact" ceruloplasmin, the enzymatic method biologically active ceruloplasmin, and the SPALT all molecules containing the epitopes recognized by the antigen-specific antibody, i.e., intact ceruloplasmin together with fragments and/or aggregation products. This example demonstrates the use of different assays in

[12] H. A. Ravin, *Lancet* **1**, 726 (1956).

measuring biologically and immunologically active molecules as a further differential diagnostic aid, here in tumor diagnosis.

b. One assay, two labels: the α_2-macroglobulin ILMA. The results shown in Table II demonstrate the effect of substituting ABEN-H for ABEI-H as label for the antibody, the other assay details remaining the same. Changing the label did not change the values as seen by the regression data. Changing the label did however improve precision. This example is an extreme case of what can be achieved by changing the label; normally the improvement by such a substitution is not so marked.

c. One assay, two oxidation systems: the ferritin ILMA. Table III shows the results from two ferritin assays run at the same time, the only difference being the oxidation system used: on the one hand MP-11, on the other hemin. The molar concentrations of both were identical. Again, changing the oxidation system did not change the concentration of the analyte measured, as can be seen from the regression data derived from 6 assays set up in parallel.

TABLE II
EFFECT OF CHANGING THE LABEL IN THE
α_2-MACROGLOBULIN ILMA

Standards (g/liter) (1 : 500 dilutions)	Relative light units[a]	
	Assay a	Assay b
0	363/469	411/522
0.27	704/770	2908/2941
0.55	881/820	4108/3926
1.10	1079/1029	6559/6357
2.20	1345/1500	9457/9709
4.40	1923/1922	11230/11838
8.80	2327/2192	13685/13764

[a] Assay a, ABEI-H-labeled antibody; assay b, ABEN-H-labeled antibody. Mean precision of samples within the standard curve—155 serum and sputum samples in 6 parallel assays: assay a, 11.8%; assay b, 6.32%. Correlation data: assay a (x) vs assay b (y): $r = 0.982$, $a_{yx} = -0.07$ g/liter, $b_{yx} = 0.991$. Assay scheme: 10 μl sample/standard (1 : 500 dilution), 200 μl assay buffer (see ILMA assay), 1 rabbit anti-α_2-macroglobulin ball. Incubate and wash as for ceruloplasmin ILMA. 200 μl ABEN-H- or ABEI-H-labeled rabbit anti-α_2-macroglobulin. Incubate, wash, prepare, and measure as for ceruloplasmin assay.

TABLE III
FERRITIN ILMA USING HEMIN AND MP-11 AS
COFACTORS

Standard (μg/liter)	Relative light units[a]	
	Assay a	Assay b
0	217/275	85/88
5	473/484	95/92
20	971/866	147/152
50	2301/2307	249/293
200	7210/7509	763/755
500	14368/15389	1576/1555
1000	20352/21372	2908/3028

Control sera mean (range)	Measured concentrations (μg/liter)	
	Assay a	Assay b
39 (24–54)	45	36
70 (55–89)	73	88
193 (176–210)	194	194

[a] Assay a, with hemin; assay b, with MP-11. Correlation data: assay a (x) vs assay b (y) — regression line $y = a + bx$; number of samples = 223, $r = 0.993$, $a_{yx} = 1.23$, $b_{yx} = 0.996$. Assay flow scheme: 25 μl sample/standard, 200 μl assay buffer (see ILMA assay), 1 goat anti-liver ferritin ball. Incubate 150 min on horizontal rotator at ambient temperature. Wash as for ceruloplasmin (see SPALT assay); 200 μl ABEN-H-labeled rabbit anti-liver ferritin. Incubate 90 min as above, wash, process, and measure as ceruloplasmin (see SPALT assay).

d. An assay with problems: the SPALT assay for haptens. The effect of "bridge elements" and carrier proteins on hapten antiserum production and specificity has been well documented by Hunter.[13] As one normally uses labeled hapten as antigen, the antibodies to the bridge and carrier-protein entities play little or no part at all in the assays provided as kits. In the case of the SPALT, when labeled antibodies to the animal species IgG used in raising the antigen-specific antibodies are used, these

[13] W. M. Hunter, *in* "Radioimmunoassay and Related Procedures in Medicine," p. 3. IAEA, Vienna, 1982.

problems cannot be ignored for hapten assays, as antibodies to all three components mentioned above can react with the carrier-protein hapten conjugate bound to the solid phase, when the latter has a identical structure identical to the immunogen. This problem was seen in setting up SPALT assays for thyroxine[1] and gentamicin. In early experiments it was seen that when the same conjugate was used for immunization and ball coating [e.g., gentamicin–bovine serum albumin with 1-ethyl-3-(3-dimethylaminopropyl)carbodiimide (EDAC)] no displacement of label took place, no doubt as the antibodies to bovine serum albumin and EDAC rearrangement moieties[1] were in excess of the specific antibodies to gentamicin. However, when a gentamicin–transferrin conjugate made with DCCD was coupled to the solid phase, the assay was able to be set up very quickly as the only component common to both system halves was gentamicin. The same problems were seen with SPALT assays for cortisol and thyroxine. SPALT assay schemes for cortisol and gentamicin, the former being sensitive enough to be able to measure cortisol in saliva in a 10-μl sample, are outlined below.

    a. Cortisol in serum

        20 μl serum/standard (1 : 50 dilution) or 10 μl saliva

        200 μl rabbit anti-cortisol (the cortisol antiserum was raised against a bovine serum albumin–C_3-carboxymethyloxime conjugate; the cortisol conjugate for the balls also used a C_3-carboxymethyloxime for coupling)

        Incubate 10 min at ambient temperature

        1 cortisol–ovalbumin-coated ball

        Incubate 45 min on a horizontal rotator at ambient temperature

        Wash as for ceruloplasmin (see SPALT assay)

        200 μl ABEI-H-labeled donkey anti-rabbit IgG

        Incubate 60 min on a horizontal rotator at ambient temperature

        Wash, process, and measure as for ceruloplasmin (see SPALT assay)

    b. Gentamicin in serum

        10 μl sample/standard (1 : 450 dilution)

        200 μl rabbit antigentamicin (the gentamicin antiserum was raised against a gentamicin–EDAC–transferrin conjugate; the gentamicin–ovalbumin conjugate for the balls used DCCD as coupling agent)

        Incubate 15 min at ambient temperature

        1 gentamicin–ovalbumin-coated ball

        Proceed as for cortisol SPALT above in subsequent incubation and wash steps

The advantage of SPALT assays, as already stated, is that no analyte dilution with labeled antigen takes place, thus increasing the sensitivity of the system. Disadvantages include the need for excellent coating precision, especially when low analyte concentrations must be measured, i.e., where the maximal antibody concentration is allowed to react with the immobilized antigen after a prior liquid-phase incubation of antibody and antigen.

Concluding Remarks

This chapter has attempted to describe the problems and possible solutions for luminescence immunoassays designed to quantitate haptens and proteins in biological solutions, as well as fields of clinical relevance, the latter being in neonatology and cell culture media as well as in sputum, saliva, and lavage solutions.

The authors have favored solid-phase assays with separate incubation steps for sample and label in order to reduce possible oxidation of label during the incubation steps in assays without prior extraction. Despite the trend toward rapid one-step assays, acceptable precision and accuracy can only be attained using the "two-step" approach here described, at least at the present time and using the labels and systems described here. With the aid of automatic or semiautomatic wash apparatus the disadvantage of a two-step assay (seen as a function of lost time) is minimal.

It is possible to process over 1000 samples per day using the methods described, although an average of 600–700 samples per working day is more realistic. These figures are for the Berthold LB-950, but are possible with other semiautomated machines such as the LKB-1251, which although only having a 25-sample carousel, has the advantage of a shorter cycle time per sample.

The low detection limits for immunoluminometric assays, e.g., 500 attomol/tube for α2-macroglobulin and 75 attomol/tube for ferritin and thyroglobulin allow their use in situations where either low analyte concentration or small sample volumes are available, i.e., in the situations stated above.

Whether luminescence immunoassays will form part of the routine armamentarium in hospital laboratories remains to be seen, and is definitely dependent on commercialization of this sector. Until this time, such assays will remain a specialized tool of the laboratories developing and implementing them, rather than serving as an alternative to radioimmunoassay, which is the main aim of this laboratory.

[31] Immunoassays Using Acridinium Esters

By IAN WEEKS, MARIA STURGESS, RICHARD C. BROWN, and
J. STUART WOODHEAD

Introduction

During recent years, sensitive immunoassay procedures have found increasing applications in many areas of diagnostic medicine. With this increase in applications has come a diversification of methodologies far beyond the basic procedure of competitive binding radioimmunoassay. A major development has been the introduction of methods based on nonradioactive labels. While some of the new technologies such as therapeutic drug monitoring have relied almost exclusively on the use of nonisotopic labels, more established procedures such as those used for the assay of peptide hormones still use predominantly ^{125}I.

To some extent this reflects the conservative attitudes of laboratory medicine, which are understandable when new technologies require alternative and unfamiliar instrumentation. There is, however, the important practical consideration that most nonisotopic labels lack the sensitivity of detection and versatility of ^{125}I. In this chapter we describe recent studies using chemiluminescent acridinium esters as labels. In addition to providing the more obvious advantages of nonisotopic compounds such as stability, these molecules can provide superior immunoassay performance in the form of increased sensitivity when compared with radioisotopes.

Acridinium Ester Chemiluminescence

Historical

The synthesis and investigation of the properties of acridine-based compounds are not new; neither is the observation that certain classes undergo chemiluminescent reactions under the appropriate conditions. Indeed chemiluminescence is only one facet of what is a remarkable group of compounds whose properties enable them to be used in a wide variety of applications from dyestuffs to antibacterial agents.

Acridine-9-carboxylic acid which can be regarded as a precursor of chemiluminescent acridinium carboxylate esters was synthesized over a half-century ago.[1] At that time the chemiluminescence of lucigenin, a

---

[1] K. Lehmstedt and E. Wirth, *Ber. Dtsch. Chem. Ges.* **61,** 2044 (1928).

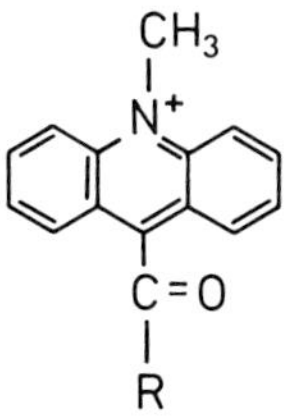

FIG. 1. The acridinium carbonyl nucleus.

bisacridinium salt, was observed[2]—the first synthetic molecule chemiluminescence to be studied since the work on luminol by Albrecht in 1928.[3]

Detailed studies have been reported on the chemiluminescence of 9-chlorocarbonylacridine hydrochloride,[4] 9-chlorocarbonyl-10-methylacridinium chloride,[4] and 9-cyano-10-methylacridinium nitrate.[5] It is seen that one requirement for chemiluminescence in these systems is the presence of a good leaving group at R, (Fig. 1) so that aryl esters exhibit the phenomenon more readily than do aliphatic esters.

Chemistry

Unlike luminol and its derivatives acridine chemiluminescence has no catalytic requirement for the production of high quantum yields. The chemiluminescence quantum yield Φ_{CL} is given by

$$\Phi_{CL} = \Phi_F \Phi_E \Phi_C$$

where Φ_F is the fluorescence quantum yield of the product molecule, Φ_E is the proportion of these molecules in an excited state, and Φ_C is the chemical yield of the particular reaction product (see Fig. 2).

Thus intense chemiluminescence requires the coordination of several processes and will occur only when both the mechanistic requirements of the chemistry and the quantum-mechanical selection rules are satisfied. To achieve this, side reactions must be minimal, the specific rate must be sufficiently high, and the electronic transitions must be spin and symmetry allowed. The energy involved is of the order of a few hundred kilojoules per mole.

[2] K. Gleu and W. Petsch, *Angew. Chem.* **58,** 57 (1935).
[3] H. O. Albrecht, *Z. Phys. Chem.* **136,** 321 (1928).
[4] M. M. Rauhut, D. Sheehan, R. A. Clarke, and A. M. Semsel, *J. Org. Chem.* **30,** 3587 (1965).
[5] F. McCapra and D. G. Richardson, *Tetrahedron Lett.* **43,** 3167 (1964).

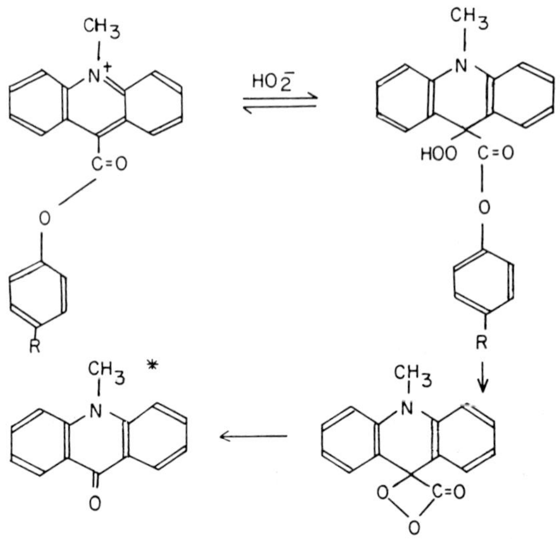

FIG. 2. Mechanism of the chemiluminescent reaction of acridinium esters.

The orbital symmetry conservation rules of the Woodward–Hoffman theory of electrocyclic reactions predict the formation of 1,2-dioxetanes as essential intermediates in these reactions. Acridinium chemiluminescence has therefore been postulated to occur via a concerted multiple bond cleavage mechanism involving a dioxetanone intermediate. The decomposition of this yields N-methylacridone in an excited singlet state which subsequently relaxes to its ground state with the emission of photons at a wavelength of approximately 430 nm.

An important property of acridinium salts is the formation of a corresponding carbinol in aqueous solution and the pH dependence of this equilibrium[6] (Fig. 3). The carbinol (or pseudo-base) does not undergo a chemiluminescence reaction though at high pH N-methylacridone formation will occur slowly via a dark reaction.[7]

Certain acridinium species are among the most highly chemiluminescent synthetic molecules currently known. The light emitting reaction is rather simpler than the luminol and isoluminol systems. First, it requires none of the many and varied catalysts of these phthalhydrazide mole-

[6] A. Hantzsch and M. Kalb, Ber. Dtsch. Chem. Ges. **32**, 3109 (1899).
[7] K. A. Zaklika, Ph.D. Thesis, University of Sussex, UK (1976).

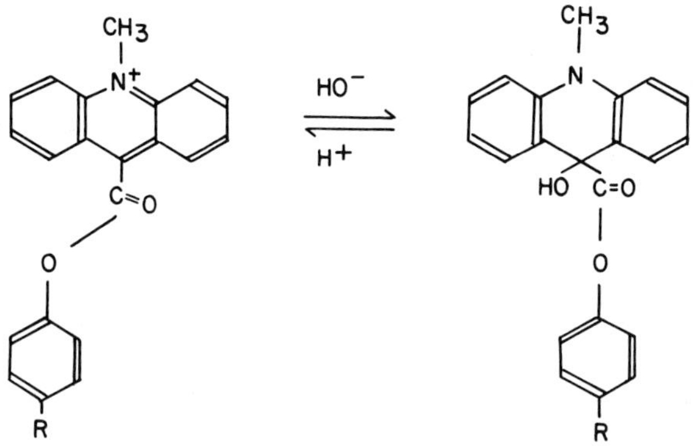

FIG. 3. The acridinium/carbinol base equilibrium.

cules[8] and second it does not proceed with as many "active oxygen" species as do these systems.[9]

Measurement

The sensitive measurement of visible photons emitted during the acridinium reaction requires the use of a photomultiplier tube (PMT) detector and is analogous to the measurement of radioactivity. In the latter, for example in β and γ counting, the emission frequencies are brought to within the spectral range of the PMT by energy transfer to a suitable fluorescent material, e.g., polyoxazole scintillation cocktails or a sodium iodide crystal.

In a sense therefore the measurement of acridinium chemiluminescence is more straightforward since the emission wavelength (~430 nm) is easily accessible to high efficiency photomultiplier tubes. Early purpose-built luminometers utilized analog measurement, that is the conversion of the PMT output to a photocurrent or photovoltage. However, it is perhaps more logical to perform luminescence measurements in a similar manner to radioactivity measurements as events per unit time. Digital photon counters are highly efficient and are now commercially available. Studies in this laboratory have been carried out with such systems manufactured by Laboratorium Professor Berthold, D7547, Wildbad, FRG.

[8] A. K. Campbell and J. S. A. Simpson, (*Tech. Life Sci. Metab. Res.*) **213**, 1 (1979).
[9] R. C. Allen, *in* "Chemical and Biochemical Generation of Excited States" (W. Adams and G. Cilento, eds.), p. 309. Academic Press, New York, 1982.

The rate of photon emission from the acridinium reaction depends upon several factors including reagent concentration and pH. Manipulation of the conditions enables the reaction rate to be easily controlled and reaction rates ranging from milliseconds to hours can be achieved.[4] The choice of reaction rate is dictated by the application since a very rapid reaction may have photon emission rates which exceed the resolving power of the photon counter and a slow reaction may not only be inconvenient but will also result in the accumulation of background emission. In either case the signal-to-noise ratio will be compromised; in the first instance due to loss of signal and in the second due to extreme accumulation of noise.

From the point of view of immunoassay it is desirable to have rapid quantitation of the signal such that high throughput is achievable at this stage in the procedure. Reaction times of less than 1 sec are easily measurable since digital photon counters have resolution within the order of nanoseconds. Of course such rapid reactions require initiation of the reaction while the chemiluminescent material is in front of the PMT. Although this was originally considered a significant problem it is now apparent that it can be achieved easily using simple reagent pumps and integrating the full photon count rate/time profile. Further, in this way all the available emission from the system is utilized, in contrast to long lived emissions where the practical constraints on counting time result in "wastage" of much of the emitted signal.

Acridinium Ester Labels

Background

Initial studies of the use of acridinium esters as labels in immunoassay were reported by Simpson et al.[10] Here, an aryl carboxyl derivative of an acridinium ester was coupled to protein by two methods, both involving activation of the carboxyl group. However, both the mixed anhydride reaction using isobutyl chloroformate and the use of carbodiimide succeeded only in producing labeled antibodies of low specific activity. Presumably this was due to either poor coupling efficiency or the destruction of the integrity of the label.

Further attempts in our laboratory to produce "activated" derivatives of acridinium salts in organic solvents prior to exposing them to an aque-

[10] J. S. A. Simpson, A. K. Campbell, J. S. Woodhead, A. Richardson, R. Hart, and F. McCapra, in "Bioluminescence and Chemiluminescence" (M. DeLuca and W. D. McElroy, eds.), p. 673. Academic Press, New York, 1981.

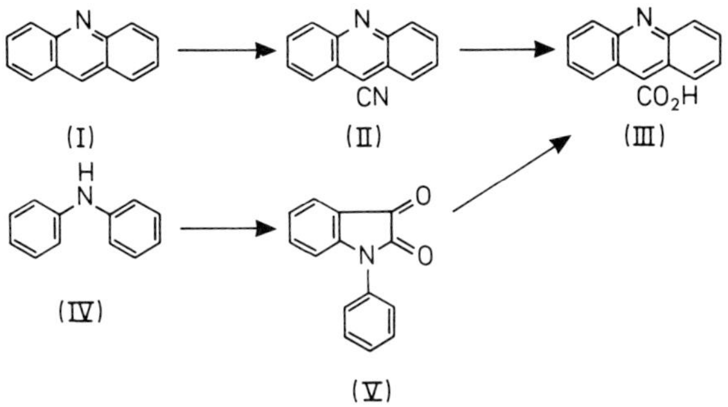

Fig. 4. Preparation of acridine-9-carboxylic acid.

ous environment also met with little success due to the labile nature of the molecule.

Synthesis of Dedicated Labels

The first acridinium ester label specifically designed for coupling to proteins was 4-(2-succinimidyloxycarbonylethyl)phenyl-10-methylacridinium-9-carboxylate fluorosulfonate.[11] One method of producing this involves using acridine (**I**) as the precursor (Fig. 4). Another method involves the production of acridine-9-carboxylic acid (**III**) from diphenylamine (**IV**) via N-phenylisatin (**V**) (Fig. 4).[12,13]

Synthesis of Acridine-9-carboxylic Acid (**III**)

From acridine. This is performed by the method of Lehmstedt and Wirth.[1] Potassium cyanide (5 g) is dissolved in water (30 ml) and a suspension formed by the addition of acridine (**I**) (2.3 g). Benzoyl chloride (7.5 g) is added dropwise with shaking. The reaction mixture becomes hot and is left to cool. The resultant solid is filtered off and dissolved in the minimum volume of boiling methanol and the solution is then adsorbed with charcoal. The mixture is filtered and allowed to cool to yield 9-cyanoacridine (**II**). Of this 1.5 g is heated with 90% (w/w) sulfuric acid (15 ml) for 2 hr on a boiling water bath with stirring. After cooling, 4.0 g of sodium nitrite is added in portions with stirring. After a further 2 hr on the boiling water bath, the mixture is poured into 250 ml of water and the resulting yellow precipitate filtered. The solid is dissolved in 0.5 M sodium carbonate and

[11] I. Weeks, I. Beheshti, F. McCapra, A. K. Campbell, and J. S. Woodhead, *Clin. Chem. (Winston-Salem, N.C.)* **29,** 1474 (1983).
[12] R. Stolle, *J. Prakt. Chem.* **105,** 137 (1922).
[13] M. S. Newman and W. H. Powell, *J. Org. Chem.* **26,** 812 (1961).

the solution filtered. The filtrate is acidified with sulfuric acid and the yellow solid isolated by filtration, washed with water, and dried under vacuum.

From Diphenylamine. Diphenylamine (**IV**) (10.5 g) is dissolved in redistilled carbon disulfide and added dropwise to a stirred, refluxing solution of oxalyl chloride (9.0 g) in 45 ml of carbon disulfide. The solution is refluxed for 1 hr and the solvent and excess oxalyl chloride removed under reduced pressure. One hundred milliliters of carbon disulfide is added followed by 26.6 g of anhydrous aluminum chloride in portions. Refluxing is continued for 1 hr and the solvent removed under reduced pressure. Crushed ice and 1 M hydrochloric acid are added and the solid material filtered and washed with water. The resulting solid is recrystallized from benzene–cyclohexane (3 : 5 v/v) to yield red crystals of *N*-phenylisatin (**V**). This is added to a 10% solution of potassium hydroxide in water (100 ml) and the mixture is allowed to cool and then poured into 500 ml of crushed ice/5 M hydrochloric acid. The yellow precipitate of acridine-9-carboxylic acid is filtered, washed with cold water, and dried *in vacuo.*

Synthesis of 4-(2-Succinimidyloxycarbonylethyl)phenyl-10-methylacridinium-9-Carboxylate Fluorosulfonate (Fig. 5)

Acridine-9-carboxylic acid (**III**) (5 g) is suspended in redistilled thionyl chloride (15 ml) and boiled under reflux for 3 hr. The solvent is then evaporated and dried under reduced pressure to yield orange crystals of acridine-9-carbonyl chloride (**VI**). Of this 2.3 g is suspended in anhydrous pyridine (35 ml), benzyl 4-hydroxyphenyl propanoate (9 nmol) is added, and the mixture stirred overnight at room temperature.

The mixture is poured into cooled, 1 M hydrochloric acid (250 ml) and the resulting yellow solid filtered, washed with water, and dried under reduced pressure. The 4-(2-benzyloxycarbonylethyl)phenyl-9-acridine carboxylate (**VII**) thus obtained is recrystallized from benzene/cyclohexane to yield yellow crystals (mp 135°).

Benzyl ester (**VII**) (0.46 g) is dissolved in 10 ml of hydrogen bromide/acetic acid mixture (45/55 w/w) and the solution stirred for 2 hr at 50–55°. The solution is poured into 100 ml of water and the resulting yellow solid filtered, washed with water, and dried under reduced pressure. The solid is recrystallized from acetonitrile/chloroform (1/1 v/v) to yield 4-(2-carboxyethyl)phenyl-9-acridine carboxylate (**VIII**) as yellow needles (mp 270–273°).

N-Hydroxysuccinimide is recrystallized from ethyl acetate by the ad-

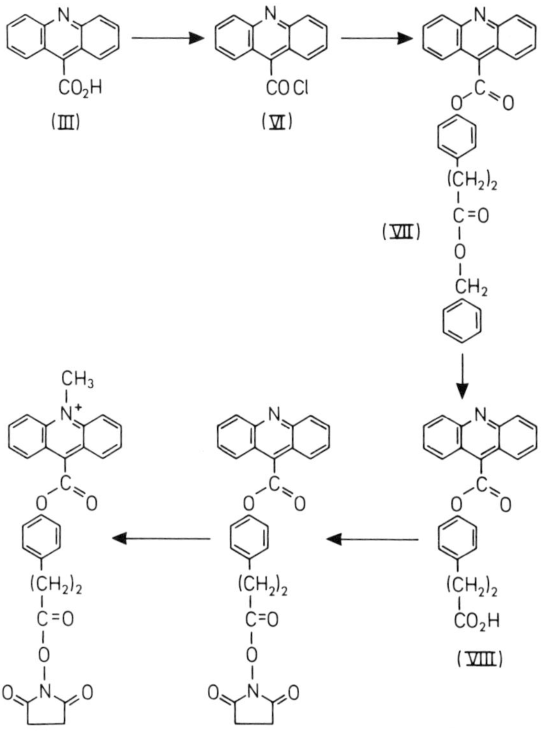

FIG. 5. Synthesis of 4-(2-succinimidyloxycarbonylethyl)phenyl-10-methylacridinium 9-carboxylate fluorosulfonate (**X**).

dition of diisopropyl ether and 62 mg of the purified material dissolved in 5 ml of anhydrous dimethylformamide together with 0.2 g of (**VIII**) and the mixture cooled to $-20°$ in CO_2/methanol. N,N-dicyclohexylcarbodiimide (123 mg) is added and the mixture stirred for 2 hr at $-20°$ and then overnight at room temperature. One drop of glacial acetic acid is then added and the mixture left for a further 30 min. The dicyclohexylurea is removed by filtration and the material obtained by evaporation of the liquor is recrystallized from benzene/cyclohexane to yield 4-(2-succinimidyloxycarbonylethyl)phenyl-9-acridine carboxylate (**IX**) as pale yellow crystals.

Two hundred and thirty-four milligrams of (**IX**) is dissolved in anhydrous chloroform (15 ml) and 0.5 ml of methyl fluorosulfonate is added. The precipitate which forms after stirring for 18 hr at room temperature is filtered and washed with anhydrous benzene to yield yellow crystals of 4-(2-succinimidyloxycarbonylethyl)phenyl-10-methylacridinium-9-carboxylate-fluorosulfonate (**X**).

Acridinium Ester Labeling Reaction

The use of *N*-succinimidyl esters in protein chemistry is well established.[14] Since they react with primary and secondary aliphatic amines to couple via an amide linkage, (**X**) can therefore be made to couple to any suitable amine species to form an acridinium-labeled conjugate. Moreover the reaction is capable of proceeding rapidly under mild conditions. Figure 6 shows the rate of incorporation of the label into guinea pig immunoglobulins under aqueous conditions at pH 8.0. Following mixing of a solution of the label in dimethyl formamide with the protein solution, samples were removed at suitable time intervals and applied to small columns of Sephadex G-50. Aliquots of the fractions were measured luminometrically. The increase in void volume chemiluminescence thus reflects uptake of label and it is seen that the reaction is complete after only 10 min.

Acridinium Ester Labeling of Purified Monoclonal Antibodies

Recent advances in high-performance liquid chromatography make it possible to purify monoclonal antibodies from ascitic fluid in high yields. Two types of column are normally used, consisting of either an anion-exchange matrix or a hydroxylapatite matrix.

Since the labeling procedure described here relies upon the conjugation of the acridinium ester label to antibody lysyl residues, amine containing buffers such as Tris–HCl must be avoided. The purification of the antibodies on a hydroxylapatite column eluted in phosphate buffer is therefore preferred in this case. A typical protocol is as follows. Five hundred microliters of ascitic fluid is centrifuged for 20 min at 30,000 g to remove lipids and loaded onto the hydroxylapatite column. The column is eluted using a sodium phosphate gradient from 10 to 400 mM, pH 6.8, in the presence of 10 μM $CaCl_2$. Figure 7 shows a typical profile for the purification of a monoclonal antibody to human prolactin (hPR). Immunoreactivity in the fractions is assessed by taking 5-μl aliquots and incubating with ^{125}I-labeled hPR (10,000 cpm) in 400 μl phosphate-buffered saline (PBS, 0.1 M, pH 7.4) and 100 μl of normal rabbit serum. After incubation for 2 hr at room temperature 1 ml of polyethylene glycol (MW 6000, 20%

---

[14] G. W. Anderson, J. E. Zimmerman, and F. Callahan, *J. Am. Chem. Soc.* **86**, 1839 (1964).

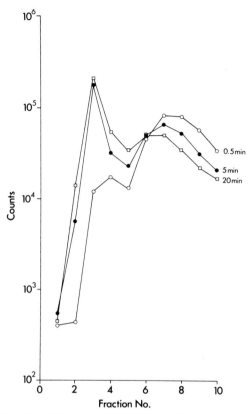

FIG. 6. Incorporation of acridinium ester label **(X)** into IgG as measured by monitoring chemiluminescence in fractions from a gel exclusion column. Samples from a reaction mixture of IgG and acridinium label **(X)** were taken at appropriate time intervals and loaded onto small columns of Sephadex G-50.

w/w) is added and the tubes mixed thoroughly. The precipitate is isolated by decantation of the supernates following centrifugation at 2500 g for 15 min and the radioactivity quantified using a gamma-counter. The fractions exhibiting immunoreactivity are then pooled and dialyzed against 0.1 M sodium phosphate buffer (pH 8.0) such that a final pH of 8.0 is obtained in the dialysis tube. The protein concentration is estimated by optical density measurement at 280 nm and a 50-μg aliquot removed for labeling.

One hundred micrograms of 4-(2-succinimidyloxycarbonylethyl)-phenyl-10-methyl-9-acridinium carboxylate fluorosulfonate is dissolved in 400 μl of dry acetonitrile and 10 μl of this removed to a glass vial. Fifty micrograms of the IgG is dissolved in 200 μl of phosphate buffer (0.1 M,

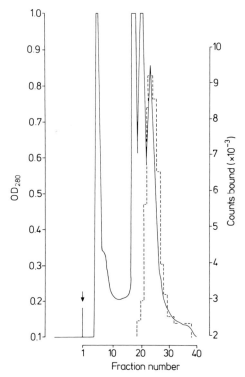

FIG. 7. Purification of mouse monoclonal antibodies to human prolactin using hydroxyl-apatite HPLC. The area within the dotted line represents the immunoreactivity profile represented by the binding of ^{125}I-labeled hPR. Injection is denoted by the arrow.

pH 8.0) and the solution added to the acridinium solution. The mixture is left to stand for 15 min at room temperature after which time 100 μl of a solution of lysine monohydrochloride (10 mg/ml) in pH 8.0 phosphate buffer is added and left for a further 15 min. The mixture is then applied to a column of Sephadex G-25 (medium grade) which is equilibrated and eluted with PBS (pH 6.3) containing 0.01% sodium azide, 0.1% human serum albumin, and 20 mg/liter of normal horse IgG. Five hundred microliter fractions are collected. Ten-microliter aliquots are removed from each fraction and added to 1 ml of the elution buffer, 10-μl aliquots then being removed for luminometry. Chemiluminescence is initiated by programming the luminometer to inject 200 μl of an aqueous solution of 0.05 M sodium hydroxide containing 0.05% hydrogen peroxide (100 volume) and the emitted photon counts integrated over 10 sec.

The void volume fractions possessing chemiluminescence activity are pooled and stored in aliquots at $-20°$.

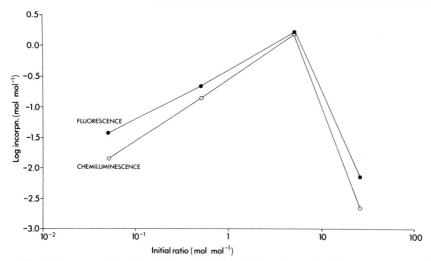

FIG. 8. Incorporation of acridinium label into IgG as determined by measurement of chemiluminescence and *N*-methylacridone fluorescence. The apparent decrease at high initial ratios is an artifact of loss of protein.

Previous work involving the labeling of proteins with luminol and isoluminol has shown a decrease in apparent chemiluminescence quantum yield upon modification of the label.[15,16] It is apparent that the acridinium ester label described in the previous section does not suffer this problem as can be seen by inspection of Fig. 8. This shows the relationship between the initial and final label/protein ratios for a given set of conditions of pH, incubation time, and temperature. The incorporation ratio was deduced in two ways. First, the luminometer was calibrated with known amounts of pure label and the emission intensity of the purified labeled protein translated into molar quantities of label. Second, following oxidation of the sample with alkaline hydrogen peroxide the fluorescence intensity of the dissociated *N*-methylacridone product was measured fluorimetrically (λ_{ex} = 395 nm, λ_{em} = 430 nm) using a spectrophotofluorimeter previously calibrated with known amounts of *N*-methylacridone. Protein concentrations were measured using Coomassie Blue G250 which was not influenced by the acridinium or the acridone. The fact that the incorporation ratios as calculated independently by these two methods were the same within experimental error suggests that no

[15] J. S. A. Simpson, A. K. Campbell, M. E. T. Ryall, and J. S. Woodhead, *Nature (London)* **279**, 646 (1979).
[16] H. R. Schroeder, C. M. Hines, D. D. Osborn, R. C. Moore, R. L. Hartle, F. F. Wogoman, R. W. Rogers, and P. O. Vogelhut, *Clin. Chem. (Winston-Salem, N.C.)* **27**, 1378 (1981).

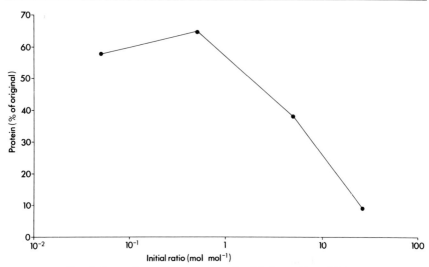

FIG. 9. Loss of protein upon excessive labeling with acridinium.

change in chemiluminescence quantum yield occurs upon coupling to protein.

At high concentration ratios however considerable loss of labeled protein is observed (Fig. 9).

Properties of Acridinium-Labeled Antibodies

It is desirable when producing chemiluminescent antibodies that both the quantum yield of the label and the immunoreactivity of the antibody are preserved as far as possible. Loss of quantum efficiency has in the past been a feature of chemiluminescent systems based on luminol and isoluminol.[17] Acridinium esters of the type dealt with here suffer less from such loss because of the dissociative nature of the chemiluminescent reaction (see above).

Retention of antibody immunoreactivity is crucially dependent upon the level of incorporation of the label which, if excessive, may even lead to protein loss by precipitation (see above). The susceptibility of the antibody to such effects depends upon the chemistry of the immunoglobulin and hence varies from antibody to antibody. However, for the majority of antibodies it is possible to incorporate up to 3 molecules of label into each antibody molecule without deleterious consequences. Figure 10 shows an antibody dilution curve for a monoclonal antibody to α_1-fetopro-

---

[17] R. B. Brundrett and E. H. White, *J. Am. Chem. Soc.* **96,** 74 (1974).

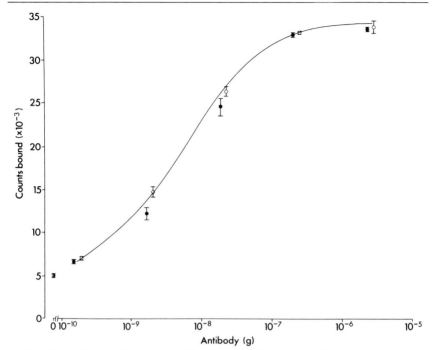

Fig. 10. Antibody dilution curve of mouse monoclonal (anti-α_1-fetoprotein) antibodies binding to ^{125}I-labeled AFP. The solid line represents the dilution curve for unlabeled antibodies, (●) are points for acridinium-labeled antibodies having an incorporation of 0.3 mol/mol, and (○) are for those having an incorporation of 2.9 mol/mol.

tein (AFP) in which ^{125}I-labeled AFP is bound to antibody which is unlabeled and also antibody labeled to two different average incorporation ratios. It can be seen that the curves are superimposable even when each mole of antibody is labeled with approximately 3 mol of acridinium ester.

Chemiluminescence Immunoassays

Polypeptide Immunoassays Utilizing Acridinium Esters

In order to take advantage of the high potential specific activities which can be achieved by labeling proteins with acridinium salts, it is necessary to design suitable immunochemistry. As long ago as 1968, Miles and Hales[18] drew attention to the potential advantages of reagent excess, labeled antibody assay systems when compared with conven-

---

[18] L. E. M. Miles and C. N. Hales, *Nature (London)* **219**, 186 (1968).

tional radioimmunoassays. They argued that assays employing excess binding reagent were kinetically more favorable than systems where the antibody is used at high dilution, with the result that immunoradiometric technology can provide significant improvements in analytical speed and sensitivity. More recently, theoretical analyses[19] have supported the concept that reagent excess systems would benefit from the utilization of high specific activity labels, since the sensitivity of many immunoradiometric systems has undoubtedly been restricted by the detection limit of ^{125}I itself together with the radiolytic damage it induces in antibody molecules.

The full potential of labeled antibody immunoassay systems has been slow to be realized. This is largely the result of the practical difficulties involved in the affinity purification of antibody from crude antisera prior to the labeling procedure. This difficulty has been overcome by the use of monoclonal antibodies. It should also be appreciated, however, that it is only in the last few years that nonisotopic labels together with suitable detection systems have been able to offer tangible improvements over ^{125}I-based methods.

The following assay procedures have been carefully designed to take full advantage of the benefits of labeled antibody methodology and also exploit the unique potential of acridinium esters in providing high specific activity labels. An additional feature has been the utilization of magnetizable particles to provide a rapid and convenient separation system. As a result, the performance of many of the assays developed so far shows considerable improvements compared with ^{125}I-based methods.

Preparation of Solid Phase Antibody

Polyclonal antiserum is mixed with an equal volume of sodium phosphate buffer (0.1 M, pH 7.4) containing 0.15 M sodium chloride. Anhydrous sodium sulfate is then added (180 mg/ml solution) with stirring until it is dissolved. After 30 min, the precipitate is collected by centrifugation and redissolved in sodium phosphate buffer (0.1 M, pH 7.4). The solution of immunoglobulins is dialyzed against the same buffer overnight at 4°.

Magnetic particles (Advanced Magnetics, Cambridge, MA 02138) are suspended in distilled water at a concentration of 50 mg/ml. After two washes in distilled water, followed by one wash in sodium phosphate buffer (0.1 M, pH 7.4), the particles are suspended in a 5% solution of glutaraldehyde in phosphate buffer and agitated for 2 hr at room temperature. The particles are then washed with phosphate buffer until there is no smell of glutaraldehyde.

[19] R. P. Ekins, in "Monoclonal Antibodies and Developments in Immunoassay" (A. Albertini and R. P. Ekins, eds.), p. 3. Elsevier/North-Holland Biomedical Press, Amsterdam, 1981.

The immunoglobulin solution is then added using a particle/protein ratio of 6/1 (w/w) and the suspension agitated overnight at room temperature. After three washes in phosphate buffer, a solution of 0.2 M glycine in phosphate buffer is added and the suspension agitated for a further 2 hr. The solid phase antibody is then washed three times with phosphate buffer and then with the same buffer containing 1 M sodium chloride. Finally the preparation is suspended at a concentration of 50 mg/ml in phosphate buffer (0.1 M, pH 6.3) containing 0.15 M sodium chloride, 0.1% human serum albumin, and 0.01% sodium azide (assay buffer). The solid phase antibody is stored at 4°. These sequential washing procedures can be carried out most conveniently in flat-sided tissue culture flasks utilizing a flat-bed magnet.

Immunochemiluminometric Assay of α_1-Fetoprotein

Measurement of circulating α_1-fetoprotein (AFP) can be useful in the diagnosis of several pathological states. In maternal serum it provides a screening test for the antenatal detection of fetal neural tube defects. Further, serum AFP can be a useful tumor marker in the diagnosis and management of certain hepatomas or teratomas. Assays of AFP must therefore provide long-term reliability as well as a facility for high throughput and rapid turnaround. The AFP ICMA is carried out as follows.

Solid phase antibody is diluted 1/10 with assay buffer (see above) so that 100 μl contains approximately 0.5 mg solid phase. Monoclonal antibody labeled to an average incorporation of 0.33 mol acridinium ester/mol is diluted in assay buffer so that 100 μl contains approximately 3.8 ng of antibody protein. Standards, calibrated against a reference preparation of AFP (72/227, National Institute for Biological Standards and Control, Holly Hill, London NW3, UK), are prepared in horse serum.

Samples and standards (100 μl) are mixed with 100 μl labeled antibody and 100 μl solid phase antibody and the reaction is allowed to proceed for 1 hr at room temperature. One milliliter of a washing solution consisting of 0.01 M sodium dihydrogen orthophosphate containing 0.15 M sodium chloride, 0.01% sodium azide, 0.1% human serum albumin, and 1% Triton X-100 is added to each tube. The tubes are placed on a magnetic separator (Corning Medical, Medfield, MA 02052) and after 3 min the liquid is decanted and the tubes allowed to stand inverted for a further 1 min. The wash cycle is then repeated.

Pellets are resuspended in 200 μl distilled water; this can be done automatically when using the LB950 luminometer. The luminescent reaction is initiated by addition of 200 μl 0.05 M sodium hydroxide containing 0.05% (100 vol) hydrogen peroxide. Luminescent counts are integrated

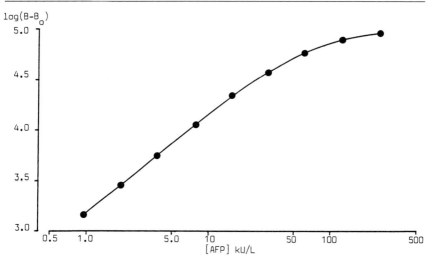

FIG. 11. Dose–response curve of a two-site immunochemiluminometric assay for α_1-fetoprotein.

over a 5-sec period and results plotted as $\log[B - B_0]$ against $\log[\text{AFP}]$ where B is the counts bound at each dose and B_0 the counts bound at zero dose.

A typical dose–response relationship is shown in Fig. 11. A precision profile based on duplicate determinations shows an interassay coefficient of variation (CV) of 10% over the range 10–300 kU/liter. It should be noted that if necessary sensitivity can be increased substantially by increasing the incorporation of label up to 3 mol acridinium ester/mol antibody.[20]

Immunochemiluminometric Assay of Thyrotropin

The importance of circulating thyrotropin (TSH) measurement in the assessment of thyroid function has increased significantly during recent years. In addition to the introduction of regional screening tests for congenital hypothyroidism, the diagnostic potential of TSH assays has been extended by the introduction of a number of sensitive immunoradiometric procedures capable of differentiating the suppressed hormone levels found in thyrotoxic patients from normal levels.[21]

The most sensitive TSH assay developed so far has been a two-site

[20] I. Weeks, A. K. Campbell, and J. S. Woodhead, *Clin. Chem.* (*Winston-Salem, N.C.*) **29**, 1480 (1983).
[21] J. S. Woodhead and I. Weeks, *Ann. Clin. Biochem.* **22**, 455 (1985).

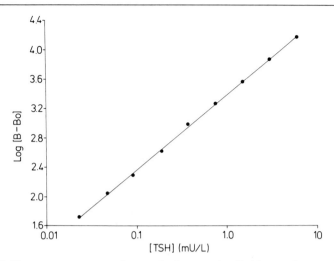

FIG. 12. Dose–response curve of a two-site immunochemiluminometric assay for thyrotropin.

immunoassay procedure utilizing cellulose-coupled polyclonal antibody and acridinium ester-labeled monoclonal antibody.[22] Clinical studies carried out so far have confirmed the importance of this procedure as a first line test of thyroid function. The method outlined below is based on the original procedure with the modification that cellulose particles, which require separation by centrifugation, are replaced by magnetizable particles.

Monoclonal antibody labeled to an activity of 3 mol acridinium ester/mol antibody is diluted in assay buffer so that 100 μl contains 1.5 ng protein. Standards are prepared in horse serum and are calibrated against a reference preparation of TSH (68/38, National Institute for Biological Standards and Control, Holly Hill, London NW3, UK). To 100 μl sample or standard is added 100 μl labeled antibody and the reaction proceeds for 2 hr at room temperature. At this stage 100 μl of a 1/20 dilution of antibody suspension is added equivalent to 0.25 mg solid phase/tube. After a further 1 hr, 1 ml of wash buffer is added and the magnetic particles separated and washed further as in the previous assay.

Luminometry is carried out as before using the LB 950 Automatic Luminescence Analyzer.

Figure 12 shows a dose–response curve for the range 0.025 to 6 mU/liter. In practice, the dose–response is linear as far as 60 mU/liter and the

[22] I. Weeks, M. Sturgess, K. Siddle, M. K. Jones, and J. S. Woodhead, *Clin. Endocrinol.* (*Oxford*) **20**, 489 (1984).

between assay CV is $<10\%$ for the range 0.1–60 mU/liter for duplicate tests.

Competitive Immunochemiluminometric Assay for Free Thyroxine

The success of any nonisotopic immunoassay technology is dependent on its application to the measurement of a wide range of analytes. In the endocrine laboratory, for example, this involves the availability of suitable methodologies for peptides, steroids, and other low-molecular-weight analytes such as thyroid hormones and cyclic nucleotides. While the two-side methodology is clearly inappropriate for low-molecular-weight compounds, the simplicity of labeling antibodies rather than a range of antigens each with individual properties has led us to explore the potential use of acridinium ester-labeled antibodies in the assay of thyroid hormones.

In principle, the method involves a competitive reaction between the analyte in a sample and a solid-phase antigen derivative for a limited quantity of labeled antibody. In practice a method has been developed in which the competitive reaction is carried out in the liquid phase following introduction into a sample of acridinium ester-labeled antibody and a soluble hapten–carrier complex. Assay separation is achieved by the introduction of excess solid phase anti-carrier antibody. Since the competitive reaction is dependent on the free analyte concentration at any given time, the following procedure has been developed to provide a rapid, direct assay for free thyroxine (free T_4).

A T_4–rabbit immunoglobulin (IgG) conjugate is prepared by first converting N-glutaryl T_4 to an active ester by reaction with N-hydroxysuccinimide. To do this, 300 mg glutaryl T_4 is dissolved in 1 ml dimethyl formamide and 35 mg N-hydroxysuccinimide (first recrystallized from ethyl acetate) and 40 mg dicyclohexylcarbodiimide is added. The mixture is stirred overnight at room temperature and crystals of dicyclohexylurea are removed by centrifugation. A 5-μl aliquot of the product is added to 45 mg rabbit IgG dissolved in 500 μl 0.1 M phosphate buffer (pH 7.4) containing 0.15 M sodium chloride. After 30 min reaction at room temperature, solid material is removed by filtration and the product purified by gel filtration on a 1 × 10-cm column of Sephadex G-25. The product should have a molar incorporation ratio of approximately 2 : 1, glutaryl T_4 : rabbit IgG. The conjugate is stored in aliquots in assay buffer at $-20°$ and diluted in assay buffer before use so that 100 μl contains approximately 15 ng conjugate.

Monoclonal antibody to a T_4–albumin conjugate[23] is labeled to an

[23] C. N. Mpoko, D. B. Gordon, I. Laing, G. Corbitt, and C. C. Storey, Clin. Chim. Acta 146, 215 (1985).

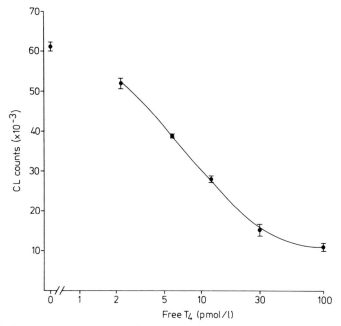

FIG. 13. Dose–response curve of an immunochemiluminometric assay for free thyroxine.

incorporation level of 1 mol acridinium ester/mol antibody and diluted in assay buffer so that 100 μl contains approximately 200 pg antibody protein.

Labeled monoclonal antibodies (100 μl) and T_4–rabbit IgG conjugate (100 μl) are added to 100 μl serum samples or standards. After 30 min at room temperature 100 μl sheep (anti-rabbit IgG) antibody linked to magnetizable particles (diluted 1/20 to give approximately 0.25 mg solid phase/tube) is added. After 10 min at room temperature the particles are separated, washed, and assayed luminometrically as in the methods described above.

Figure 13 shows a standard curve for free T_4 obtained with this procedure. Studies carried out so far have indicated that there is no interaction between the T_4–rabbit IgG conjugate and endogenous binding proteins so that the method does not suffer the artifactual problems encountered in analog tracer procedures as a result of such interaction.[24]

An example of the potential advantage of this system is shown in Fig. 14 which compares results obtained on patients with chronic renal failure, many of whom have significantly reduced albumin levels, using the

[24] R. P. Ekins, *J. Clin. Immunoassay* **7**, 163 (1984).

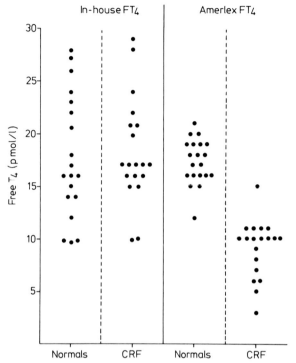

FIG. 14. Measurements of free thyroxine (FT$_4$) in normal subjects and patients with chronic renal failure using the ICMA and a single-step radiolabeled analog RIA.

luminometric method and an analog tracer procedure (Amerlex free T$_4$, Amersham International, Amersham, UK). It will be noted that while the analog method yields significantly low results on these patients the labeled antibody method reveals normal free T$_4$ values which are consistent with the euthyroid status of these patients.

Conclusions

In addition to the well-recognized advantages of nonisotopic immunoassay systems, assays based on acridinium ester-labeled antibodies have demonstrated two outstanding features. First, it has been possible to develop procedures for polypeptide measurement which are among the most sensitive immunoassays so far described. Second, the technology is not restricted to any one class of molecular species. As demonstrated above it has been possible to develop sensitive, rapid, and convenient methods for the measurement of low-molecular-weight analytes. The realization that these methods can provide valid measurements of free analytes in serum will lead to many new applications. The areas most likely to benefit from these high sensitivity procedures will be the analysis of free hormones as well as the detection of low concentrations of tumor markers and infectious agents. Acridinium ester-based immunoassay offers not only a logical alternative to radioisotope procedures but also a real improvement in analytical performance.

[32] Surface Chemiluminescent Immunoassays of Steroids

By F. Kohen, J. De Boever, and J. B. Kim

Research in endocrinology, as well as the diagnosis of endocrine disorders, and monitoring therapy, require the determination of steroidal hormones or of specific steroid metabolites in body fluids. In blood plasma, the steroid to be measured may be present in picomolar concentration, often in the presence of much higher concentrations of chemically related but hormonally inert substances. Hence, highly sensitive and specific methods are required. To be clinically useful, these should also be cheap, rapid, reliable, and safe. Radioimmunoassay (RIA) seemed to provide the most practical approach to achieve this, and specific RIA of steroid hormones is now widely used in clinical and pharmacological laboratories. However, the very success and widespread use of RIA has raised several problems which include (1) shelf-life and stability of radiolabeled compounds, (2) high cost of radioactive waste disposal, and (3) health hazards as a result of exposure to the use of radiolabeled hormones, and to the solvents necessary for liquid-scintillation counting.

To avoid these drawbacks, while retaining the specificity of an immunoassay, we explored the use of chemiluminescence as an end point in the assay. As a chemiluminescent marker, derivatives of isoluminol possessing alkyl chains of an optional length of 2–6 carbon atoms and terminating with a primary amino group were chosen since these compounds could be measured upon oxidation at picomolar levels.[1,2] Furthermore, the feasibility of using these isoluminol derivatives in protein binding reactions has

[1] H. R. Schroeder and F. M. Yeager, *Anal. Chem.* **50,** 1114 (1978).
[2] H. R. Schroeder, R. C. Boguslaski, R. J. Carrico, and R. T. Buckler, this series, Vol. 57, p. 424.

been demonstrated for biotin[3] and for thyroxine.[4] Chemiluminescent conjugates of steroids or steroid glucuronides were prepared by conjugating carboxy derivatives of steroids or steroid glucuronides to primary amino derivatives of isoluminol.[5,6] These conjugates then served as markers in immunoassay procedures. Two types of formats were adopted: (1) assays that do not require physical separation of bound and free hormone, so-called "homogeneous," and (2) assays that require a separation step, so-called "heterogeneous." In the "homogeneous" type of assays, oxidation of the chemiluminescent marker conjugate in the presence of homologous antibody[7-9] or of binding protein[3] produces enhancement of chemiluminescence. In binding reactions this light enhancement is inhibited by the addition of homologous ligand in a dose-dependent manner. Assays based on this principle have been developed for progesterone,[7] estriol,[8] estriol 16α-glucuronide,[9] and more recently for total estrogens.[10] Although sensitivity similar to RIA (5–10 pg/tube) is obtained, and the assay can be easily automated, the homogeneous assay procedure can be affected by interference from luminescent compounds present in extracts of plasma or diluted urine samples. For instance, it has been shown that the presence of materials of biological origin can modify the reaction kinetics of the chemiluminescent label upon oxidation and consequently the light output.[5] It is thus advisable to use an on-line computer analysis of the chemiluminescent reactions to improve the reliability of these assays.[11]

In the heterogeneous type of assays bound and free hormone are separated using dextran-coated charcoal or solid-phase methods. The use of dextran-coated charcoal can be readily applied to plasma or urinary

[3] H. R. Schroeder, P. O. Vogelhut, R. J. Carrico, R. C. Boguslaski, and R. T. Buckler, *Anal. Chem.* **48,** 1933 (1976).

[4] H. R. Schroeder, F. M. Yeager, R. C. Boguslaski, and P. O. Vogelhut, *J. Immunol. Methods* **25,** 275 (1979).

[5] F. Kohen, M. Pazzagli, M. Serio, J. de Boever, and D. Vandekerckhove, *in* "Alternative Immunoassays" (W. P. Collins, ed.), p. 103. Wiley, New York, 1985.

[6] F. Kohen, E. A. Bayer, M. Wilchek, G. Barnard, J. B. Kim, W. P. Collins, I. Beheshti, A. Richardson, and F. McCapra, *in* "Analytical Applications of Bioluminescence and Chemiluminescence" (L. J. Kricka, P. E. Stanley, G. H. Thorpe, and T. P. Whitehead, eds.), p. 149. Academic Press, New York, 1984.

[7] F. Kohen, M. Pazzagli, J. B. Kim, H. R. Lindner, and R. C. Boguslaski, *FEBS Lett.* **104,** 201 (1979).

[8] F. Kohen, J. B. Kim, and H. R. Lindner, *in* "Bioluminescence and Chemiluminescence" (W. D. McElroy and M. A. DeLuca, eds.), p. 357. Academic Press, New York, 1981.

[9] F. Kohen, J. B. Kim, G. Barnard, and H. R. Lindner, *Steroids* **36,** 405 (1980).

[10] G. Messeri, A. L. Caldini, G. F. Bolelli, M. Pazzagli, A. Tommasi, P. L. Vannuchi, and M. Serio, *Clin. Chem. (Winston-Salem, N.C.)* **30,** 653 (1984).

[11] A. Tommasi, M. Pazzagli, M. Damiani, R. Salerno, G. Messeri, A. Magini, and M. Serio, *Clin. Chem. (Winston-Salem, N.C.)* **30,** 2597 (1984).

extracts of steroids such as progesterone,[12] cortisol,[13,14] testosterone,[15] and urinary steroid metabolites such as testosterone 17-glucuronide.[16] Sensitivity and precision similar to RIA were achieved. However, these assays are affected by the presence of materials of biological origin and dextran-coated charcoal may cause stripping of the bound fraction.

In order to avoid the drawbacks of dextran-coated charcoal, we examined solid phase separation techniques.[17] In these assays, the solid phase support consists of antibody-coated tubes or polystyrene balls or of first or second antibody covalently linked to polyacrylamide beads or cellulose.

In the immunoadsorption technique the adsorbed antibody-bound fraction can be washed several times by simple addition of buffer, followed by aspiration. This washing step permits removal of factors which might quench chemiluminescence. This approach is simple and does not require a centrifugation step. Using this approach, chemiluminescence-based immunoassays have been developed for plasma steroids such as progesterone,[18] estradiol,[19] testosterone,[20] and for urinary steroid glucuronide such as estriol 16α-glucuronide,[21] pregnanediol 3α-glucuronide,[22] and estrone 3-glucuronide.[23]

Although this method does not require a centrifugation step and can be

[12] M. Pazzagli, J. B. Kim, G. Messeri, G. Martinazzo, F. Kohen, F. Francheschetti, A. Tommasi, R. Salerno, and M. Serio, *Clin. Chim. Acta* **225,** 287 (1981).

[13] M. Pazzagli, J. B. Kim, G. Messeri, F. Kohen, G. F. Bolelli, A. Tommasi, R. Salerno, G. Monetti, and M. Serio, *J. Steroid Biochem.* **14,** 1005 (1981).

[14] M. Pazzagli, J. B. Kim, G. Messeri, F. Kohen, G. F. Bolelli, A. Tommasi, R. Salerno, and M. Serio, *J. Steroid Biochem.* **14,** 1181 (1981).

[15] M. Pazzagli, M. Serio, P. Munsun, and D. Rodbard, *in* "Radioimmunoassay and Related Procedures in Medicine," p. 747. IAEA, Vienna, 1982.

[16] M. Pazzagli, G. Messeri, A. L. Caldini, G. Monetic, G. Martinazzo, and M. Serio, *J. Steroid Biochem.* **19,** 407 (1983).

[17] F. Kohen and H. R. Lindner, *in* "Luminsecenza" (M. Pazzagli, ed.), p. 83. OIC Medical Press, Firenze, 1983.

[18] F. Kohen, J. B. Kim, H. R. Lindner, and W. P. Collins, *Steroids* **38,** 73 (1981).

[19] J. B. Kim, G. J. Barnard, W. P. Collins, F. Kohen, H. R. Lindner, and Z. Eshhar, *Clin. Chem. (Winston-Salem, N.C.)* **28,** 1120 (1982).

[20] W. P. Collins, G. J. Barnard, J. B. Kim, D. A. Weerasekera, F. Kohen, Z. Eshhar, and H. R. Lindner, *in* "Immunoassays for Clinical Chemistry: A Workshop Meeting" (W. M. Hunter and J. E. T. Corrie., eds.), p. 373. Churchill-Livingstone, Edinburgh and London, 1983.

[21] G. Barnard, W. P. Collins, F. Kohen, and H. R. Lindner, *J. Steroid Biochem.* **14,** 941 (1981).

[22] Z. Eshhar, J. B. Kim, G. Barnard, W. P. Collins, S. Gilad, H. R. Lindner, and F. Kohen, *Steroids* **38,** 89 (1981).

[23] D. A. Weerasekera, J. B. Kim, G. J. Barnard, W. P. Collins, F. Kohen, and H. R. Lindner, *Acta Endocrinol.* **101,** 254 (1982).

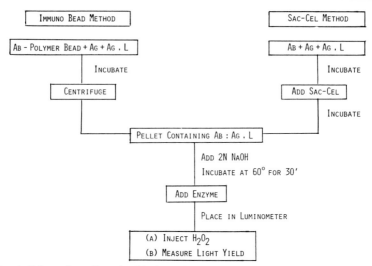

FIG. 1. Schematic outline of solid-phase chemiluminescence procedures for haptens. Ab, antibody; Ag, antigen (hapten); Ag.L, chemiluminescent antigen.

considered a single tube assay, it has some disadvantages: (1) the technique is antibody consumptive, (2) the limited amount of antibody that can be bound to the tube wall or ball, (3) possible loss of immunoreactivity of the adsorbed antibodies, and (4) considerable batch to batch variation in the plastic surface (tubes or balls) with respect to behavior of adsorbed antibody.[24] On the other hand, the use of a primary specific monoclonal or polyclonal antibody or of a second antibody directed against the first antibody covalently coupled to a solid support (e.g., small polyacrylamide beads 5–10 μm in diameter or cellulose) seemed to improve the reliability of the assay.[24]

In the approach described here the following features are incorporated: (1) a steroid-chemiluminescent marker conjugate serves as the labeled ligand, (2) specific polyclonal or monoclonal antibodies are utilized either in the liquid phase or in the solid phase covalently coupled to polymer beads (Fig. 1, Method A), (3) a solid-phase second antibody directed against the first antibody is utilized when the binding reaction is carried out in the liquid phase (Fig. 1, Method B), (4) a centrifugation step after the binding reaction is used to remove interfering luminescent compounds along with unbound steroid, (5) dissociation of the label is achieved by incubation of the antibody bound fraction at pH 13 for 30 min at 60°, and (6) luminometry is performed by oxidation of the label with a H_2O_2–microperoxidase system (see Fig. 1). These assays have enabled

[24] F. Kohen, H. R. Lindner, and S. Gilad, *J. Steroid Biochem.* **19,** 413 (1983).

the development of simple, reliable assays for the measurement of plasma steroids, e.g., estradiol,[25] testosterone,[24] urinary steroid metabolites, e.g., pregnanediol 3α-glucuronide[24] and estrone 3-glucuronide,[5] and therapeutic drugs such as digoxin.[17] Furthermore, the use of solid-phase techniques enabled the development of a direct immunoassay for progesterone in unextracted serum[26] and in saliva.[27] Results of various types of solid phase immunoassays will be reported here.

Reagent Solutions

Stock Solutions of Steroid-Chemiluminescent Marker Conjugates. Stock solutions (1 µg to 100 µg/ml) of these derivatives are prepared in spectral grade ethanol and stored at −20°. The working solution is prepared freshly on the day of assay by diluting the stock solution to 1 µg/ml with assay buffer.

Assay Buffer (PBS). The buffer consists of 0.05 M sodium phosphate buffer (pH 8.0) containing 9 g/liter NaCl (analytical grade), 0.1 g/liter bovine serum albumin (RIA grade, Sigma), and 1 g/liter sodium azide.

Wash Solution. Dissolve 9 g of NaCl, 1 g of sodium azide, and 0.5 ml of Tween 20 [polyoxyethylene sorbitan monolaureate, Sigma] in doubly distilled water and dilute to 1 liter.

Heme Catalysts. A stock solution of 1 mg/ml of microperoxidase (MP-11, Sigma) is prepared in PBS (10 mmol/liter pH 7.4). This stock solution can be stored for up to 2 months at 4°. A working solution is prepared freshly before each assay by adding 100 µl of the stock solution to 4.9 ml of distilled water.

Oxidant Solution. A working solution is prepared freshly by mixing 0.1 ml of 30% hydrogen peroxide solution (Merck, Darmstadt, FRG) with 15 ml of distilled water.

Procedures

Preparation of Steroid-Chemiluminescent Marker Conjugates

These conjugates were synthesized in two steps, involving the formation of a peptide bond between the carboxylic group of the steroid derivative and the terminal amino group of the isoluminol derivative.

[25] J. De Boever, F. Kohen, and D. Vanderkerckhove, *Clin. Chem. (Winston-Salem, N.C.)* **29**, 2068 (1983).

[26] J. De Boever, F. Kohen, D. Vandekerckhove, and G. Van Maele, *Clin. Chem. (Winston-Salem, N.C.)* **30**, 1637 (1984).

[27] J. De Boever, F. Kohen, and D. Vandekerckhove, *Clin. Chem. (Winston-Salem, N.C.)* **32**, 763 (1986).

Preparation of the Steroid Active Ester Derivatives. In the first step of the reaction activated *N*-succinimide esters of carboxy derivatives of steroids (hemisuccinate, carboxymethyloxime, or glucuronide) were prepared by dissolving the steroid (5 to 10 mg) in dry dimethyl formamide (0.5 ml). Carbodiimide (3 to 6 mg) and *N*-hydroxysuccinimide (1.5 to 3 mg) were added to the solution, and the reaction mixture was incubated overnight at room temperature. The resulting urea was filtered, and the elute containing the activated *N*-succinimide ester steroid derivative was used in the next step without any further purification. As a representative example, the structure of the *N*-hydroxy ester derivative of progesterone 7α-carboxyethyl thioether (compound **2**) is shown in Fig. 2.

Preparation of Isoluminol Derivatives. The isoluminol derivative possessing a short alkyl chain of two carbon atoms and terminating with a primary amino group (Fig. 2, compound **3a**) [6-*N*-(2-aminoethyl)-*N*-eth-

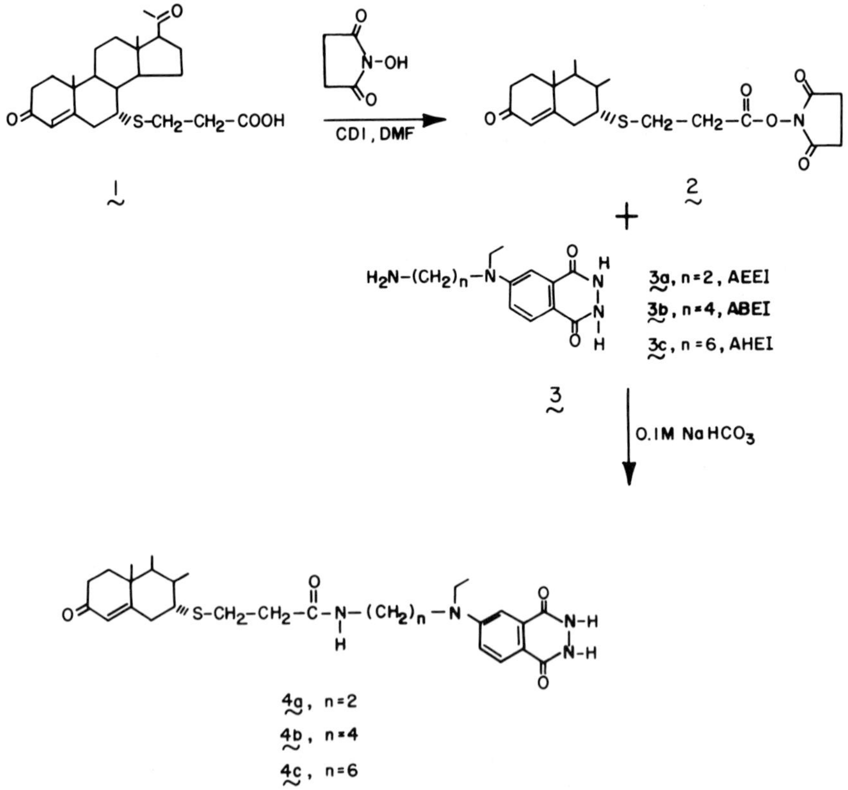

FIG. 2. A synthetic scheme for the preparation of chemiluminescent markers of progesterone 7α-carboxyethylthioether.

yl]amino-2,3-dihydropthalazine-1,4-dione (aminoethylethyl isoluminol, AEEI) was synthesized according to the procedure developed by Schroeder *et al.*[1,2] for the synthesis of aminobutylethylisoluminol (Fig. 2, compound **3b,** ABEI) and of aminohexylethylisoluminol (Fig. 2, compound **3c**). Aminoethylethylisoluminol (**3a**) was characterized by electron impact ionization mass spectrometry.[27] A molecular ion peak (M^+) at 248 was observed. The UV spectrum of AEEI (**3a**) in 0.01 N NaOH showed peaks at 220 nm (ε 17,200), 290 nm (ε 25,000), and 320 nm (ε 12,100).

Conjugation of Chemiluminescent Marker to Activated Steroids. The isoluminol derivatives (**3a, 3b,** or **3c**, 2 to 5 mg) in $NaHCO_3$ solution (0.5 ml, 0.13 M) was added slowly to the reaction containing the activated steroid ester (as shown above). The mixture was mixed for 1 hr and was then acidified to pH 2 with 1 N HCl and extracted with ethyl acetate. The organic phase was washed with water, and 0.1 M $NaHCO_3$ solution, dried with anhydrous Na_2SO_4, filtered, and taken to dryness under reduced pressure. The residue was purified by thin-layer chromatography (TLC) on silica gel 60 plates using chloroform–methanol (80:20) as the developing solvent. The band with an R_f 0.6 to 0.7, showing absorption at the long wave region and in the short wave region, as in the case with the Δ^4-3-ketosteroids, was scratched off from the plate. The steroid-chemiluminescent marker conjugates were then eluted from the silica with spectral grade ethanol. The concentrations of the conjugates was then determined by using $\varepsilon = 25,000$ at 285 nm. The UV absorption spectra of the conjugates were superimposable on those of equimolar mixtures of the steroid derivative and the isoluminol marker.[28] The molar extinction coefficients of isoluminol derivatives and steroid-chemiluminescent marker conjugates are listed in Table I. The steroid-chemiluminescent marker conjugates were further characterized by mass spectrometry using field desorption (FD) and fast atom bombardment (FAB) techniques.[29] These methods gave an indication to the molecular ion peak and fragmentation pattern of the conjugates. Figure 2 shows representative examples of the proposed structures of progesterone 7α-carboxyethyl thioether isoluminol conjugates (**4a, 4b,** and **4c**).

Preparation of Antibodies to Steroids

Specific polyclonal antibodies to steroids conjugated to bovine serum albumin were raised in rabbits and characterized in terms of titer, affinity, and specificity by RIA procedures. The hybridoma technique of Köhler

[28] M. Pazzagli, J. B. Kim, G. Messeri, G. Martinazzo, F. Kohen, F. Franceschetti, G. Moneti, R. Salerno, A. Tommasi, and M. Serio, *Clin. Chim. Acta* **115**, 277 (1981).
[29] M. Pazzagli, G. Messeri, A. L. Caldini, G. Monetic, G. Martinazzo, and M. Serio, *J. Steroid Biochem.* **19**, 407 (1983).

TABLE I

ULTRAVIOLET SPECTRUM CHARACTERISTICS OF VARIOUS ISOLUMINOL
DERIVATIVES AND STEROID-CHEMILUMINESCENT MARKER CONJUGATES

Compound	(nm) (ε)		
AEEI	220 (22,700)	280 (24,500)	320 (8,600)
ABEI	220 (19,500)	290 (27,500)	320 (14,400)
AHEI	220 (17,600)	290 (25,100)	320 (12,900)
P-11-HS-AEEI	236 (25,500)	290 (26,000)	320 (12,250)
P-11-HS-ABEI	238 (27,700)	291 (27,500)	320 (13,700)
P-11-HS-AHEI	240 (27,400)	291 (26,900)	320 (13,800)
P-7-CET-AEEI	234 (26,000)	285 (22,000)	320 (12,000)
P-7-CET-ABEI	232 (27,000)	285 (21,000)	320 (12,000)
P-7-CET-AHEI	230 (28,000)	285 (20,000)	320 (11,000)
T-3-CMO-ABEI	250 (29,100)	290 (29,000)	320 (14,900)
Cortisol-3-CMO-ABEI	250 (29,000)	290 (29,300)	320 (14,000)
Cortisol-21-HS-ABEI	237 (28,200)	291 (26,270)	320 (13,400)
E_2-6-CMO-ABEI	235 (29,700)	290 (31,250)	320 (18,300)
Pregn-3-gluc-AHEI	230 (28,000)	285 (21,000)	320 (10,000)
E_3-3-gluc-ABEI	230 (29,000)	285 (24,000)	320 (11,000)

and Milstein[30] was used to generate monoclonal antibodies to steroids.[31] Polyclonal and monoclonal antibodies belonging to the IgG class were purified by affinity chromatography on Sepharose-Protein A (Pharmacia, Uppsala, Sweden).[17] The purified antibodies were stored in PBS at $-20°$ at a concentration of 1 mg protein/ml.

Preparation of the Immunoadsorbant

Two milligrams of purified IgG fraction of the monoclonal or poly- clonal antibody was dialyzed against buffer phosphate (3 mmol/liter, pH 6.3, the "coupling buffer") overnight, then mixed with 40 mg of Immuno- bead matrix (Bio-Rad, Richmond, CA) in a final volume of 4 ml of cou- pling buffer. To this suspension was added 15 mg of 1-ethyl-3-(3-dime- thylaminopropyl)carbodiimide and the mixture was incubated overnight at 4°. The reaction mixture was then treated as directed by the supplier (Bio-Rad). The antibodies so prepared were stored at 4° in phosphate buffer (5 mmol/liter, pH 7.2) containing, per liter, 10 g of bovine serum albumin and 10 g of sodium azide, at a concentration of 10 mg of immu- noadsorbant per milliliter. Before immunoassay, the immunoadsorbent is

---

[30] G. Köhler and C. Milstein, *Eur. J. Immunol.* **6,** 511 (1976).
[31] F. Kohen and S. Lichter, *in* "Monoclonal Antibodies: Basic Principles, Experimental and Clinical Applications in Endocrinology" (G. Forti, M. B. Lipsett, and M. Serio, ed.), p. 87. Raven Press, New York, 1986.

diluted 100- to 1000-fold, depending on the titer of the conjugated antibody, with assay buffer containing uncoupled immunobead matrix at a concentration of 1 mg/ml.

Light Measurements

Measurements of light emission were made with a Lumac Luminometer Model 2080 (Lumac Systems, Basel, Switzerland) using the automatic injection and integration modes of the instrument, and Lumacuvette P polystyrene test tubes (12 × 50 mm) as reaction vessel. The Luminometer was also connected to a storage oscilloscope (Type 5111, Tektronix, Beaverton, OR) in order to observe the kinetics of light emission. When the Luminometer was used in the automatic injection mode, readings on the Luminometer started immediately after initiation of the light reaction. The light emission was then integrated for 10 sec, and the total light production (TLP) was recorded as arbitrary light units.

Evaluation of Steroid-Chemiluminescent Marker Conjugates

The various steroid-chemiluminescent marker conjugates were evaluated as suitable labels in the development of chemiluminescent-based immunoassays of plasma steroids and their urinary metabolites. Two factors were investigated: (1) their light efficiency, and (2) their binding affinity for the homologous antibody. The results obtained are described below.

Light Yield of the Steroid-Chemiluminescent Marker Conjugate. Conjugation of aminoalkylisoluminol derivatives to carboxy steroids did not significantly affect the chemiluminescent characteristics of the isoluminol molecule. Under optimal pH and oxidizing conditions, the detection limits of the different steroid-chemiluminescent conjugates were in the order of 0.1–0.4 fmol/tube.[23] The detection limit was influenced by two factors: (1) the length of spacer between the steroid and isoluminol, and (2) the pH of the oxidizing system. For instance, the detection limit, defined twice the mean of the background chemiluminescent signal, was 126 fg for estrone 3-glucuronide aminoethylethylisoluminol, the conjugate having an alkyl chain of two carbons between the steroid and isoluminol, and 218 fg for estrone 3-glucuronide aminobutylethylisoluminol, the conjugate having an alkyl chain of four carbons between the steroid and isoluminol.[23]

The other factor affecting the rate of the chemiluminescent reaction and consequently the total light production (TLP) of the chemiluminescent reaction was the pH of the oxidizing system. For instance, when oxidation of progesterone 11α-hemisuccinate aminobutylethylisoluminol

TABLE II
EFFECT OF pH ON LIGHT EMISSION

Duration of integration (sec)	pH 8.0		pH 13.0	
	Arbitrary light units	Total light production (%)	Arbitrary light units	Total light production (%)
0–60	4,119	100	207,857	100
0–2	2,937	71.3	34,528	16.6
2–12	442	10.7	84,790	40.7
12–30	369	8.9	55,431	26.6
30–60	371	9.1	33,108	16.1

conjugate was carried out at pH 8.0, more than 70% of the light yield occurred within 2 sec after initiation of the reaction. When the oxidation was performed at pH 13.0, however, 16.6% of the light emission occurred within the first 2 sec after the initiation of the reaction. Moreover, the TLP of the conjugate at pH 13 was 50 times higher than that obtained at pH 8.0. The results are shown in Table II. It was concluded that the optimal pH to carry the light reaction was 13.

Binding Affinity of the Steroid-Chemiluminescent Marker Conjugates for the Homologous Antibody

The binding affinity of the various marker conjugates for the homologous antibody was investigated since the sensitivity and specificity of the immunoassay depend largely on the structure of the conjugate and on the type of antibody used, monoclonal or polyclonal, in the system. The ability of the steroid-chemiluminescent marker conjugates to compete for the binding sites of the homologous antibody was evaluated in a radioimmunoassay procedure using tritiated homologous steroid as the tracer and charcoal as a separation system.[28] The results indicated that the affinity of the conjugate can be higher, lower, or similar to that of unaltered ligand depending on the ability of the antibody to recognize the bridge connecting the steroid to the isoluminol molecule. In general, conjugates possessing shorter alkyl chains such as aminoethylethylisoluminol (AEEI) or aminobutylethylisoluminol (ABEI) yielded steeper standard curves than those conjugates possessing longer alkyl chain, aminohexylethylisoluminol (AHEI) (Fig. 3). In addition, the use of monoclonal antibodies in the system improved the sensitivity of the method, and eliminated the problem of bridge recognition (Fig. 4).

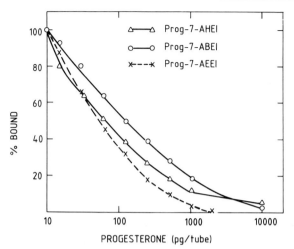

FIG. 3. Dose–response curves for progesterone using different progesterone chemiluminescent marker conjugates, monoclonal antibodies to progesterone 7-carboxyethylthioether BSA, and Sac-Cel as a separation method.

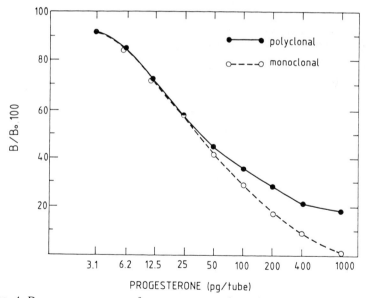

FIG. 4. Dose–response curves for progesterone using progesterone 11α-hemisuccinylaminobutylethylisoluminol as marker, polyclonal (●) or monoclonal (○) antibodies, and Sac-Cel as a separation method.

Plasma Estradiol-Extraction Assay Immunobead Separation

The evaluation of the chemiluminescence immunoassay for plasma estradiol has been fully described in several papers.[19,20,25] We report here the technique used in our laboratory using Immunobead separation system (Fig. 1, Method A).

Reagents. 6-(*N*-Ethyl-4-[estradiol-6-(*O*)-carboxymethyloxime]butyl)-amino-2,3-dihydrophthalazine-1,4-dione (estradiol–6-CMO–ABEI conjugate) (E$_2$–CMO–ABEI) and monoclonal antibodies were prepared as previously described.[25] Assay buffer and reagents for initiating the light reaction are described in this chapter in the section entitled "Reagent Solutions."

Sample Collection and Extraction. Blood samples are obtained from women by venipuncture and transferred into tubes containing heparin. After centrifugation, the plasma is removed and stored at −20° until assayed for estradiol content. Plasma, 0.25 to 2 ml, adjusted to 2 ml with distilled water, is extracted with 10 ml of diethyl ether. The ether phase is evaporated under nitrogen and the dry residue is dissolved in 0.5 ml of assay buffer. It is necessary to include duplicate ether blanks in each assay by extracting 0.25 ml of double distilled water by the same procedure.

Immunoassay Procedure. Monoclonal antibody, coupled to Immunobeads, is diluted 1000-fold to give a suspension of 1 mg of Immunobead matrix per milliliter of assay buffer (see Section "Preparation of the Immunoadsorbent") and 0.1 ml of this dilution is added per Lumacuvette. Reconstituted plasma extracts or standards in 0.1 ml of assay buffer are added, followed by 100 pg of E$_2$–CMO–ABEI conjugate in 0.1 ml of assay buffer. The contents are mixed, incubated overnight at 4°, 1 ml of wash solution is added, and the tubes are centrifuged (10 min, 2000 *g*, room temperature). The supernatant fluid is decanted, and sodium hydroxide (2 *N*, 200 µl) is added to each tube. The tubes are incubated at 60° for 30 min. After cooling to room temperature, 0.1 ml of diluted microperoxidase is added to each individual tube which is immediately placed in the luminometer. Chemiluminescence is initiated by the rapid injection of 100 µl of oxidant solution (0.3% of H$_2$O$_2$), and the signal is recorded for 10 sec.

Evaluation. A typical dose–response curve and within-batch precision profile are shown in Fig. 5. The sensitivity of the method (defined as the reagent blank minus 2 SD) is routinely less than 5 pg per tube. The intraassay coefficient of variation was 7.2, 9.4, and 8.4% for samples containing, respectively, 152, 697, and 2188 pg estradiol/ml (*n* = 9).

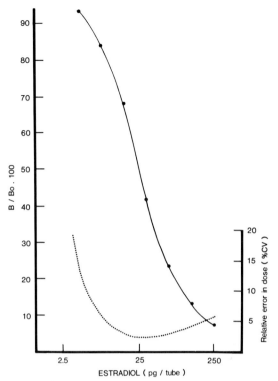

FIG. 5. Estradiol chemiluminescence immunoassay: dose–response curve (●) and smoothed precision profile (· · ·).

The estradiol values obtained by chemiluminescence immunoassay were compared with those determined with two different radioimmunoassay methods, one involving the same monoclonal antibody plus iodinated estradiol, ^{125}I-RIA, the other a polyclonal antibody and tritiated estradiol, ^{3}H-RIA. In the latter RIA, dextran-coated charcoal was used to separate bound and free ligand. Values, determined by the three methods agreed well: for chemiluminescence vs ^{125}I-RIA, $R = 0.93$ ($n = 149$); for chemiluminescence vs ^{3}H-RIA, $r = 0.94$ ($n = 105$); for ^{125}I-RIA vs ^{3}H-RIA, $r = 0.94$ ($n = 111$). The three methods were similar in sensitivity, precision, and specificity. The present chemiluminescence immunoassay method is suitable for clinical application, e.g., for monitoring ovulation induction, and can provide satisfactory nonisotopic alternatives to RIA.[25]

Plasma Progesterone-Direct Assay-Immunobead Separation

The introduction of simple, direct immunoassays for the measurement of steroids in unextracted serum or plasma represents an important advance in routine clinical analysis. Extraction with an organic solvent is avoided by the use of synthetic steroids (e.g., danazol) or reagents (e.g., 8-anilino-1-naphthalenesulfonic acid) which block the binding to, or displace steroids from, serum binding proteins. The measurement of steroids in plasma by a direct assay requires (1) an antiserum with high avidity and specificity, (2) a good separation technique which would remove any component that would enhance or reduce the signal, and (3) a set of standards in an appropriate matrix. We report here a solid phase chemiluminescence immunoassay in unextracted serum.[26]

Reagents. Progesterone 11α-hemisuccinyl-aminobutylethylisoluminol (the chemiluminescent ligand) and homologous polyclonal antiprogesterone IgG covalently coupled to Immunobeads were prepared as described previously.[26] Danazol was a gift from Winthrop Labs, Brussels, Belgium. Steroid-free serum was prepared by mixing 100 ml of male serum with a 1 g of charcoal at 4° overnight, followed by centrifugation (3 times at 20,000 g).

Immunoassay Procedure. Polyclonal antiprogesterone 11α-hemisuccinyl-BSA IgG, coupled to Immunobeads, is diluted 100-fold with a suspension of Immunobead matrix containing 1 mg of matrix beads/ml buffer, and 0.1 ml of this suspension is added to each cuvette. Serum samples (50 μl) are diluted with 0.2 ml assay buffer containing 500 ng of danazol. Standards contained, per ml, 0.8 ml of assay buffer, 0.2 ml of normal male serum, 2 μg of danazol, and 62.5 pg to 8 ng of progesterone. Duplicate 50-μl aliquots of diluted serum samples or of standards are added to Lumacuvettes containing 0.1 ml of diluted solid-phase antibody. The contents are mixed and incubated at 4° for 1 hr. Subsequently, progesterone 11-ABEI conjugate (50 pg) in 0.1 ml assay buffer is added to all tubes. The tubes are mixed, incubated overnight at 4°, 0.9 ml of wash solution is added, and the tubes are centrifuged (10 min, 2000 g). The supernatant is decanted, and the wash procedure is repeated. The light yield of the chemiluminescent marker bound to the solid-phase antibody is assayed using the conditions described for the estradiol assay (see section on Plasma Estradiol Extraction Assay).

Evaluation. A typical dose–response curve and within-batch precision profile are shown in Fig. 6. The sensitivity of the assay is 2 pg/tube. The intraassay coefficient of variation was 9%, and the interassay coefficient of variation was 10.6%. For 13 unknown serum samples assayed four to nine times (mean six times) in consecutive assays, interassay coefficient of variation were 16.9, 10.2, and 8.7 over the range 3–6.5, 7–40, and 41–

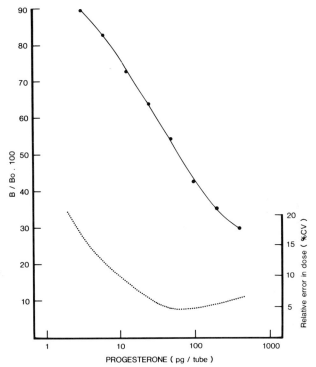

FIG. 6. Progesterone chemiluminescence immunoassay: dose–response curve (●) and smoothed precision profile (· · ·).

80 nmol/liter. The specificity and accuracy of this method are similar to those of a conventional radioimmunoassay for progesterone in which a radioligand of tritiated progesterone and serum extraction is used. Progesterone values obtained by this direct assay method agreed with those obtained by radioimmunoassay ($r = 0.987$, $n = 208$). The results of this study indicate that the assay is analytically valid and can provide a convenient nonisotopic alternative to radioimmunoassay of progesterone for routine use in a clinical laboratory.[26]

Estimation of Serum Digoxin–Sac-Cel Separation

Measurement of digoxin in plasma for drug monitoring is a routine procedure in hospital laboratories since digoxin has a low therapeutic index. We have developed a rapid assay for determination of plasma digoxin levels, using donkey anti-rabbit cellulose suspension (Sac-Cel) for separation of bound and free hormone. The methodology is given below.

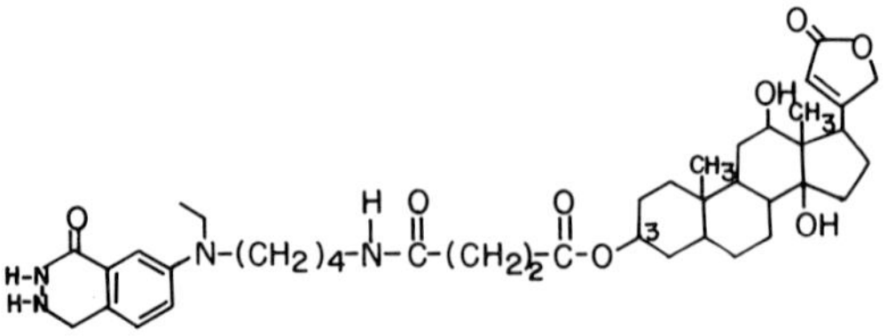

FIG. 7. Proposed structure of digoxigenin hemisuccinylaminobutylethylisoluminol conjugate.

Reagents. The labeled marker, digoxigenin hemisuccinate aminobutylethylisoluminol conjugate (Fig. 7) is prepared by coupling digoxigenin hemisuccinate to aminobutylethylisoluminol. Polyclonal antidigoxin antibodies are prepared in rabbits and an IgG fraction is prepared by chromatography on Sepharose-Protein A. Assay buffer consists of PBS, pH 7.4, containing 50 mM EDTA and 0.1% azide. The standards are prepared by dissolving digoxin in ethanol to a concentration of 1 mg/ml: this solution is diluted with a pool of normal human sera. The calibration curve extends from 0.3 to 4.3 ng digoxin/ml. Donkey anti-rabbit cellulose suspension (Sac-Cel) was purchased from Wellcome Lab. (UK).

Immunoassay Procedure. Polyclonal antidigoxin IgG antibodies are diluted 1:1000 with assay buffer, and 0.1 ml of this dilution is added per Lumacuvette. Serum samples or standards (50 μl) are diluted with 0.8 ml of assay buffer. Duplicate 0.8 ml aliquots of diluted serum samples or of standards are added to Lumacuvettes containing 0.1 ml of diluted antibody. The tubes are incubated at room temperatures for 1 hr. Subsequently, digoxigenin–ABEI conjugate (100 pg) in 0.1 ml assay buffer is added to all tubes. The incubation is continued for another hour, and 50 μl of Sac-Cel suspension is added to all tubes. The tubes are incubated for 30 min, and 1 ml of wash solution is added. The tubes are centrifuged (15 min, 2000 g), and the supernatant is decanted. The light yield of the conjugate bound to the pellet containing the antibody-bound fraction is measured using the conditions described for the estradiol assay.

Assessment of the Method. A typical dose–response curve for digoxin is shown in Fig. 8. The sensitivity of the assay is 5 pg/tube. The antibody to digoxin used in the assay showed minimal cross-reaction with digitoxin (5%), and nonsignificant cross-reaction (<0.01%) with prednisolone, spironolactone, progesterone, and cortisol. The intraassay variation was 8.5% ($n = 10$) and the interassay variation on 20 assays was 11.2%. Digoxin values obtained by this method agreed well with those obtained by a fluorescence polarization immunoassay[32] method ($r = 0.94$, $n = 98$).

[32] F. Kohen, B. Gayer, J. Ausher, B. Rosenkranz, U. Klotz, and J. C. Frölich, unpublished data from this laboratory.

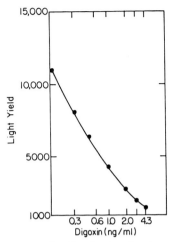

FIG. 8. Dose–response curve for digoxin chemiluminescence immunoassay.

The results of this study indicate that the chemiluminescence immunoassay for plasma digoxin is comparable in performance to a fluorescence polarization immunoassay method routinely used in clinical laboratories.

Remarks. The work reviewed here indicates that derivatives of isoluminol are useful alternatives to isotopic labels for the measurement of steroids in biological fluids. The procedures are comparable to established RIA in terms of sensitivity, specificity, accuracy, and precision. However, the setting up of the present technology in the routine clinical laboratory is more complex than the establishment of RIA procedures. This limitation is due to several components of the assay, such as the structure and binding affinity of the steroid-chemiluminescent marker conjugate and the type of antibody (polyclonal or monoclonal) used in the assay. It is important to match the antibody used with the particular marker conjugate (see Figs. 3 and 4).

Another limitation of assays using isoluminol derivatives is the final oxidation of the label at high pH. Sodium hydroxide is added to the solid-phase antibody-bound label for the following reasons: (1) to dissociate the label from the antibody-bound complex, and (2) to raise the pH so that the light yield of the marker is enhanced (see Table II), and the sensitivity of the assay is improved. The enhancement of the light emission necessitates one extra incubation step (see Fig. 1). Despite these limitations, chemiluminescence-based immunoassays do provide a feasible alternative to RIA, and the methods avoid the use of a radioactive label.

[33] Photographic Detection of Chemiluminescent and Bioluminescent Reactions

By LARRY J. KRICKA and GARY H. G. THORPE

In chemiluminescent and bioluminescent assays light emission is usually measured with a photomultiplier tube or a silicon photodiode. A less expensive and simpler alternative is photographic film. The combination of chemiluminescence or bioluminescence and photographic film has a number of advantages.

1. Totally chemical assays can be devised in which light produced in a chemical reaction is detected by means of the chemical reactions occurring in the film emulsion.
2. Multiple assays can be performed simultaneously using low cost, portable instrumentation which requires no power source.
3. A permanent visual record of results is obtained.
4. Rapid assays are possible, particularly if an "instant" rather than a conventional "wet" development type of film is used.

The major limitations of using photographic film as a detector are that it is less sensitive than a photomultiplier tube and the photographic results are semiquantitative rather than quantitative. However, a further degree of quantitation can be achieved using, for example, neutral density filters or a densitometer. On balance these limitations are outweighed by the advantages of the photographic detection technique.

Types of Photographic Film

Early work was performed with conventional black and white film, but more recently instant film has been preferred because of the convenience of automatic processing. Four types of instant film have been used, Polaroid Land Types 47 (ASA 3000), 57 (ASA 3000), 410 (ASA 10,000), and 612 (ASA 20,000) black and white print film. These films are panchromatic, and although there are differences in the relative spectral sensitivity between films, it has not proved necessary to match films to a particular chemi- or bioluminescent emission spectrum.

The sensitivity of a film is represented by a speed number, based on either the ASA or DIN film speed system. The higher the number the more sensitive the film. Polaroid Type 612 has a speed of ASA 20,000 and is the most sensitive instant film generally available. Some improvements in film sensitivity may be possible by prefogging the film, but this would serve to complicate what is essentially a simple detection procedure.

The density response to exposure of a film is quantified by the slope of the characteristic curve (a plot of image density versus log exposure). This indicates the contrast of a film. A medium contrast film (e.g., Type 47, slope 1.40) produces a graded response in image density across a wide range of exposure, while a high contrast film such as Type 410 (slope 2.30) exhibits a very rapid change in image density across a narrow range of exposure. Ideally, a film intended for use as a detector of chemiluminescence or bioluminescence should combine high sensitivity with a low contrast as this would both detect very low light levels and give a graded response in image density across a wide range of exposure. Such a combination is not available and a very high speed film such as Type 612 (ASA 20,000) has a high rather than low contrast.

Exposure times in photographic assays vary from a few seconds to days. Generally, the more recent assays based on Polaroid Type 612 instant film use very short exposures (e.g., 5–30 sec) and so the final detection step, including exposure and developing, can be completed in less than 2 min.

Instrumentation

The kinetics of light emission (i.e., a flash versus a prolonged emission) are important factors in the design of instrumentation for photographic assays. In reactions that produce a prolonged emission, the use of film poses no major problems because the reactions can be initiated prior to exposure to the film. However, for reactions that produce a flash of light, the reaction must be initiated directly in front of the film if the light emission is to be recorded and this requirement complicates the design and construction of suitable instrumentation.

Many of the original photographic chemiluminescent assays were performed in a dark room by placing assay tubes in contact with a piece of film and initiating the reaction directly on top of the film.[1] Alternatively, the film can be wrapped around the assay tube.[2] Although this protocol is suitable for reactions which produce a flash or a glow of light, it is inconvenient and several instruments have been designed to enable photographic assays to be performed away from a dark room.

[1] U. Isacsson and G. Wettermark, *Anal. Chim. Acta* **68**, 339 (1974), and references to the early Russian work cited therein (Tables IV–VI).
[2] W. R. Seitz and M. P. Neary, *Anal. Chem.* **46**, 188A (1974).

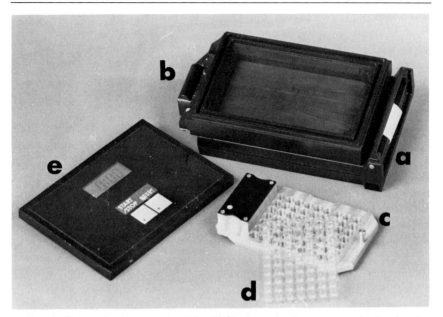

FIG. 1. Camera luminometer (a, Polaroid film back; b, shutter; c, mask; d, microtiter plate; e, lid and timer).

Camera Luminometer

Two versions of this instrument are illustrated in Figs. 1 and 2. In its simplest form it is a light-tight box located on top of a film holder (Polaroid Type CB103).[3,4] The interior of the camera is finished in matt black and a loose fitting labyrinth ensures a light-tight fit between the lid and the body of the instrument. A slidable shutter is interposed between the film holder and the body of the instrument. In the closed position the shutter supports the reaction vessel holder, and in the open position the holder drops onto the film. Various types of reaction vessel holder can be inserted into the instrument and Fig. 1 shows a holder for a microtiter plate. This holder is made in the form of an array of holes to separate the light emission from individual wells and the lower edges are beveled to enable the shutter to be reinserted smoothly.

Another version of this instrument incorporates a multiple pipet into the lid of the instrument (Fig. 2). The pipet consists of 63 disposable pipet

[3] T. P. Whitehead, G. H. G. Thorpe, L. J. Kricka, J. E. C. Gibbons, and R. A. Bunce, European Patent 110610 (1984).
[4] R. A. Bunce, G. H. G. Thorpe, J. E. C. Gibbons, P. R. Killeen, G. Ogden, L. J. Kricka, and T. P. Whitehead, *Analyst* **110,** 657 (1985).

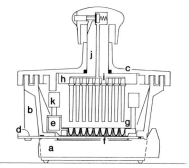

FIG. 2. Cross section of the camera luminometer and the multiple pipet (a, Polaroid film holder; b, body of camera luminometer; c, top assembly; d, slidable shutter; e, MTP holder containing light emitting diodes and battery; f, film; g, 7 × 9 matrix of microtiter plate wells; h, manifold plate; i, chamber connecting pipet to the valve; j, multiple pipet; k, stop that limits travel of pipet). Reproduced with permission from Bunce *et al.*[4]

tips fitted into a manifold plate which is connected to a finger-operated valve. The assembly can be moved up and down inside the luminometer so that the pipet tips can be positioned in the reaction vessels held in the mask. The multiple pipet is filled by simply immersing the pipet tips in a reservoir of reagent. Greater control of the reproducibility of filling of the pipet is achieved using a combined reagent reservoir and loader (Fig. 3). The loader guides the pipet into the reservoir so that all of the tips are immersed to the same depth. Once filled, the valve is closed and the pipet slowly withdrawn from the reservoir and then carefully transferred into the camera luminometer. Reagent is dispensed simultaneously into reaction vessels lying in contact with the film by lowering the pipet tips into the vessels and then opening the valve. The precision of this pipet at 200 μl was 0.9% (coefficient of variation), but in most instances luminescent

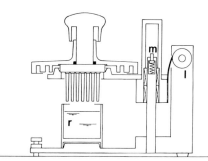

FIG. 3. Multiple pipet mounted in combined reagent loader reservoir (l, constant force spring; m, spring-loaded restriction; r, reservoir). Reproduced with permission from Bunce *et al.*[4]

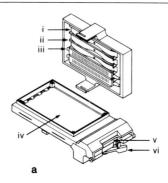

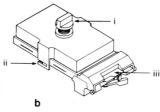

FIG. 4. (a) The separated MAST photocassette lower member and lid: (i) platform spring; (ii) platform; (iii) test chamber; (iv) phosphor screen (this is removed for use with the chemiluminescent assay); (v) film slot; (vi) control lever. (b) Assembled MAST photocassette: (i) control knob; (ii) latch; (iii) film envelope. Reproduced with permission from Miller *et al.*[5]

reactions can be triggered by an excess of reagent and so pipet accuracy is not critical.

Operation of the camera luminometer with and without the multiple pipet is described in a subsequent section.

Photocassette

This instrument, illustrated in Fig. 4, monitors light emission from a MASTpette reaction vessel (Fig. 5).[5,6] It comprises a light-tight chamber located on top of a Polaroid instant film pack, and a lid which holds up to 5 MASTpettes. Once assembled and loaded with glowing MASTpettes the film envelope is removed and the MASTpettes are pushed into contact

[5] S. P. Miller, V. A. Marinkovich, D. H. Riege, W. J. Sell, D. L. Baker, N. T. Eldredge, J. W. Dyminski, R. G. Hale, J. M. Peisach, and J. F. Burd, *Clin. Chem. (Winston-Salem, N.C.)* **30,** 1467 (1984).
[6] C. R. Brown, K. W. Higgins, K. Frazer, L. K. Schoelz, J. W. Dyminski, V. A. Markinovich, S. P. Miller, and J. F. Burd, *Clin. Chem. (Winston-Salem, N.C.)* **31,** 1500 (1985).

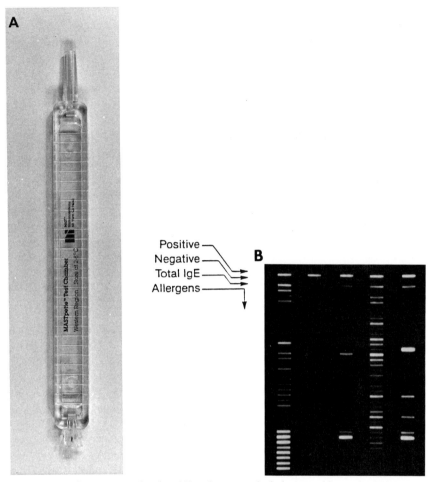

Positive
Negative
Total IgE
Allergens

FIG. 5. MASTpette test chamber (A) and some typical photographic results (B) showing the IgE-specific and total IgE response of one nonatopic and four atopic individuals.

with the film using a control knob in the lid. After exposure the instant film is developed in the normal way. The photocassette is only suitable for reactions which produce a steady light emission as there is no facility for initiating reactions in front of the film.

Modified Polaroid Camera

Modification of a camera to monitor luminescent reactions has been reported.[7] The lens was removed from a Polaroid camera and a sliding

[7] H. R. Schroeder and P. O. Vogelhut, European Patent Appl. 0071859 (1983).

metal plate attached as a shutter. A thin clear plastic plate and an aluminum block with an array of holes which accommodated 12 reaction tubes was positioned above the shutter, and the whole assembly covered by a light-tight lid. The lid was pierced by holes above each reaction tube and the holes plugged with rubber septa. In use, the tubes were filled with sample and reagent, the shutter removed, and the luminescent reaction initiated by injecting a coreactant through the septa into the tubes. The underneath of the aluminum block is recessed and the area of film exposed (spot diameter) under the reaction tube is proportional to the intensity of light emission. This instrument is suitable for reactions which produce a flash of light or a prolonged light emission.

Flow-Through Systems

In the patent literature an instrument for detecting atmospheric pollutants has been disclosed which consists of a light-tight enclosure containing photographic film in close proximity to a disc containing a chemiluminescent coreactant.[8] Air is drawn through the system past the film and coreactant by a pump. Light emitted as a result of the interaction of pollutants (e.g., ozone) with the coreactant (rhodamine B) is detected by the film. In one embodiment the film is advanced continuously to produce a running record of the amount of analyte present in the air sample.

Cross-Talk between Reaction Vessels

The exposure of a piece of film by a glowing reaction vessel can sometimes be influenced by light emission from adjacent reaction vessels, an effect known as "cross-talk." This effect can be minimized or eliminated by isolating the reaction vessels in one of two ways. Reaction vessels can be either located in a mask similar to that used in the camera luminometer[3,4] or a dye solution can be used as a chemical mask. The latter approach is employed in the MASTpette to isolate light emission from adjacent glowing threads.[6] Dye solution between the threads ensures that each thread only exposes film directly beneath it since the optically dense solution prevents lateral transmission of light. Cross-talk may also occur in arrays of connected reaction vessels such as a clear plastic microtiter plate. Light emitted by a brightly glowing well can be transmitted through the plastic to adjacent wells (a light piping effect) and thus falsely elevate the observed light emission from such wells. This effect can be prevented by using microtiter plates fabricated from colored plastic. Light reaching the film through the thickness of the plastic is only marginally

---

[8] L. A. Cavanagh, U.S. Patent 3,923,462 (1975).

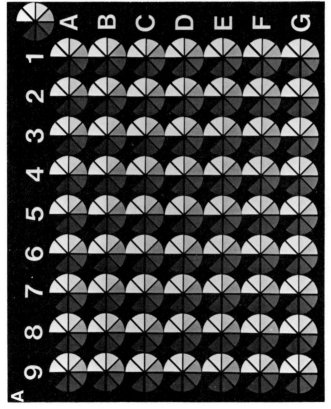

FIG. 6. Array of stepped neutral density filters and sample identification system (A) and some typical results from a chemiluminescent assay for luminol using instant film and neutral density filters (B) (a, 50 and b, 5 pmol, c, 500, d, 50, and e, 5 fmol). Reproduced with permission from Bunce *et al.*[4]

attenuated whereas transmission of light between wells is reduced due to the extended pathlength.[4]

Assessment and Quantitation of Photographic Results

In their simplest form the results of photographic assays are assessed by visual comparison of exposures (black through grey to white) produced by samples and standards. Further quantitation can be achieved if the reaction vessel is held in a cylindrical cavity above the film.[7] The area of film exposed beneath the cavity is dependent upon the intensity of light emission, and intensity is proportional to the diameter of the exposed zone. Alternatively, semiquantitative results can be obtained by interposing neutral density filters between the glowing reaction vessels and film.[4,8] A stepped neutral density filter or optical step wedge, is illustrated in Fig. 6 together with some typical results illustrating the response of the filter to increasing light intensity.

Photographic results can also be quantitated using a densitometer. This method is employed to determine allergen-specific IgE levels from photographic prints obtained in MAST chemiluminescent enzyme immunoassays.[6]

Photographic Assays

Tables I[9–14] and II[15–18] summarize the range of analytes detected using photographic assays. The majority of the assays listed in Table I are liquid phase reactions performed in glass or plastic tubes positioned above a piece of film. Solid phase reactions can also be successfully adapted to a photographic format. Enzymes (e.g., bacterial and firefly luciferase) re-

[9] K. Green, L. J. Kricka, G. H. G. Thorpe, and T. P. Whitehead, *Talanta* **31,** 173 (1984).

[10] B. J. McCarthy and L. J. Kricka, *J. Soc. Dyers Colour.* **101,** 228 (1985).

[11] T. J. N. Carter, T. P. Whitehead, and L. J. Kricka, *Talanta* **29,** 529 (1982).

[12] J. E. Becvar, *in* "Bioluminescence and Chemiluminescence" (M. A. Deluca and W. D. McElroy, eds.), p. 583. Academic Press, New York, 1981.

[13] G. H. G. Thorpe and L. J. Kricka, unpublished data.

[14] N. M. Lukovskaya and M. I. Gerasimenko, *J. Anal. Chem. USSR* **26,** 1462 (1971).

[15] G. H. G. Thorpe, S. B. Moseley, L. J. Kricka, R. A. Stott, and T. P. Whitehead, *Anal. Chim. Acta* **170,** 101 (1985).

[16] I. Sampson, J. A. Matthews, G. H. G. Thorpe, and L. J. Kricka, *Anal. Lett.* **18B,** 1307 (1985).

[17] H. X. Wang, J. C. Hall, G. H. G. Thorpe, G. G. Nickless, J. George, and L. J. Kricka, *Med. Lab. Sci,* **43,** 145 (1986).

[18] G. H. G. Thorpe, T. P. Whitehead, R. Penn, and L. J. Kricka, *Clin. Chem. (Winston-Salem, N.C.)* **30,** 806 (1984).

solved by gel electrophoresis have been detected directly in the gel by placing the gel on top of a piece of film and applying the appropriate luminescent coreactants to the gel surface. This bioluminescent staining process has been termed "autolumography."[12] The immunoassays collected in Table II are examples of solid phase chemiluminescent reactions involving horseradish peroxidase conjugates which have been immobilized onto the surface of a plastic microtiter plate, a nitrocellulose membrane, or a cellulose thread. It has also proved possible to perform photographic assays using suspensions of immobilized reagents and pads which have been impregnated with reagents.[11] Glucose can be detected in this manner using a tube containing a paper pad impregnated with glucose oxidase and horseradish peroxidase in contact with a transparent membrane impregnated with luminol. Hydrogen peroxide formed by the action of glucose oxidase on glucose reacts with peroxidase and luminol at the interface between the paper pad and membrane, and the emitted light exposes the film directly below the tube. In the patent literature more complex integral test devices comprising layers of reagents physically associated with the photoresponsive layer have been disclosed, but none is available commercially.[19]

Gas-phase chemiluminescent reactions can also be monitored photographically. An automated system for the detection of ozone and other pollutants in air has been disclosed by Cavanagh.[8]

Photographic Enhanced Chemiluminescent Immunoassay for Cytomegalovirus (CMV)-Specific Antibody Monitored Using a Camera Luminometer

Reagents and Controls

A 96-well poly(vinyl chloride) U-bottom M24 microtiter plate (Dynatech Laboratories, Billinghurst, UK) was coated with 200 μl of a 1 : 200 dilution in carbonate–bicarbonate buffer (0.05 M, pH 9.6) of either CMV CF antigen (Flow Laboratories, Rickmansworth, UK) or CMV negative control antigen (Flow Laboratories). The plates were covered with clingfilm and incubated at 4° overnight. The coated plate was emptied of unbound antigen and washed 3 times with phosphate-buffered saline (PBS) (Oxoid, Hants, UK) containing 0.5% Tween 20 (BDH Chemicals Ltd). The plate was shaken out vigorously between each wash to remove excess PBS and at the end of the third wash the plate was blotted dry.

---

[19] P. O. Vogelhut, European Patent Appl. 0019786 (1980).

TABLE I
CHEMILUMINESCENT AND BIOLUMINESCENT PHOTOGRAPHIC ASSAYS

Analyte	Assay system	Detection limit	Type of film	Reference
Alcohols				
Ethanol	Bacterial luciferase–NADH : FMN oxidoreductase[a]–alcohol dehydrogenase	1 pmol	Polaroid Type 612	9
Amino acids				
Cysteine	Luminol–iodine	0.01 µg	—	1
Anions				
Sulfide	Luminol–iodine	5 ng	—	1
Biomass				
Microbial growth	Firefly luciferase–luciferin	—	Polaroid Type 612	10
Carbohydrates				
Glucose	Glucose oxidase–horseradish peroxidase–luminol	28 nmol	Polaroid Type 410	11
Cofactors				
ATP	Firefly luciferase–luciferin	2 pmol	Polaroid Type 612	9
NADH	Bacterial luciferase–NADH : FMN oxidoreductase	200 pmol	Polaroid Type 612	9
Cyclic hydrazides				
7-[N-(4-Aminobutyl)-N-ethyl]-aminonaphthalene-1,2-dicarboxylic acid hydrazide	Microperoxidase–peroxide	0.94 nmol	Polaroid Type 410	7
Luminol	Horseradish peroxidase–peroxide	500 fmol	Polaroid Type 612	4
Enzymes				
Bacterial luciferase	NADH : FMN oxidoreductase–NADH-FMN-decanal	20 ng	Kodak Royal-X pan 4166	12
Firefly luciferase	ATP–luciferin	—	Kodak Royal-X pan 4166	12
Horseradish peroxidase	Luminol–peroxide–p-iodophenol	0.24 fmol	Polaroid Type 612	13
Metal ions				
Bismuth	Lucigenin–peroxide	5 µg	—	1
Cerium(IV)	Luminol–peroxide–copper(II)	0.1 µg	—	1

Cobalt(II)	Luminol–peroxide	2 ng	—	1
Copper(II)	Luminol–peroxide	0.03 µg	—	1
Lead	Lucigenin–peroxide	0.5 µg	—	1
Manganese(II)	Siloxene	0.1 µg	—	1
	Lucigenin–peroxide	1 µg	—	1
	Lucigenin–peroxide–organic amine	0.1 µg	—	1
Mercury	Luminol–peroxide	0.2 µg	—	1
Silver	Lucigenin–peroxide	0.1 µg	—	1
Thorium(IV)	Luminol–peroxide–copper(II) or hemin	1 µg	—	1
Vanadium	Luminol–peroxide–Co[(NH$_3$)$_4$(NO$_2$)$_2$]Cl	0.04 µg	—	1
Zirconium	Luminol–peroxide–copper(II)	—	—	1
Naphthols				
1-Nitroso-2-naphthol	Luminol–peroxide–cobalt(II)	20 µg	Isoortho plates	14
Oxidants				
Chlorine–hypochlorite	Luminol–peroxide	0.5 µg	—	1
Hydrogen peroxide	Luminol–copper(II)	0.1 µg	—	1
	Luminol–Co[(NH$_3$)$_4$(NO$_2$)$_2$]Cl	2 ng	—	1
	Lucigenin	14.7 µmol	Polaroid Type 410	11
	Microperoxidase–7-[N-(4-aminobutyl)-N-ethyl]aminonaphthalene-1,2-dicarboxylic acid hydrazide	5.6 nM	Polaroid Type 410	7
Iodine	Luminol	1 µg	—	1
Ozone	Rhodamine B	—	Polaroid Type 57, Type 410. Kodak Type 103aF	8
Potassium ferricyanide	Luminol	0.3 µg	—	1
Steroids				
Cholylglycine	Bacterial luciferase–NAD(P)H : FMN oxidoreductase–7α-hydroxysteroid dehydrogenase	50 pmol	Polaroid Type 612	9

^{a} NADH dehydrogenase (FMN).

TABLE II

PHOTOGRAPHIC IMMUNOASSAYS BASED ON HORSERADISH PEROXIDASE CONJUGATES

Analyte	Solid support	Detection system	Type of film	Reference
Cytomegalovirus-specific IgG	PVC microtiter plate	Luminol–peroxide–6-hydroxybenzothiazole	Polaroid Type 612	15
	Nitrocellulose membrane	Luminol–peroxide–p-iodophenol	Polaroid Type 612	16
Factor VIII re-lated antigen	PVC microtiter plate	Luminol–peroxide–p-iodophenol	Polaroid Type 612	17
Ferritin	PVC microtiter plate	Luminol–peroxide–firefly luciferin	Polaroid Type 612	18
IgE	Cellulose threads	Cyclic diacyl hydrazide–peroxide	Polaroid Type 57	6
Malaria-specific IgG	PVC microtiter plate	Luminol–peroxide–p-iodophenol	Polaroid Type 612	13
Rubella-specific IgG	PVC microtiter plate	Luminol–peroxide–p-iodophenol	Polaroid Type 612	13

Sera which had been previously tested for CMV antibodies by complement fixation and immunofluorescence were used as positive and negative controls.

Procedure

Samples or controls were diluted 1 : 20 with PBS, and 200 μl added to each well using an 8-place multipipettor (Dynatech Laboratories). The plate was then incubated for 30 min at room temperature in an enclosed box. After incubation with sample the plate was emptied and washed three times with PBS.

Goat anti-human IgG horseradish peroxidase conjugate (Miles Laboratories, Slough, UK) was diluted 1 : 25,000 with PBS and 200 μl dispensed into each well. After 30 min incubation at room temperature the HRP conjugate was removed and the plate washed 3 times leaving PBS in the wells for 1 min during each rinse. Peroxidase conjugate bound to the surface of the wells was then quantitated using a chemiluminescent assay.

Chemiluminescent Assay

The luminol–hydrogen peroxide–p-iodophenol substrate mixture was prepared by adding 1 ml p-iodophenol (40.9 mM in dimethyl sulfoxide) to 100 ml of a luminol (1.25 mM), hydrogen peroxide (2.7 mM) mixture prepared in Tris buffer (0.1 M, pH 8.5).[20,21] Immediately after the final wash procedure and after the plate had been cut to an appropriate size, chemiluminescent reactions were initiated by addition of 200 μl of substrate solution to each well. The glowing plate was then positioned in the mask and placed in the camera luminometer which had been loaded with Polaroid Type 612 film. The lid was replaced and the shutter withdrawn allowing the mask to drop onto the film. After 15 sec the shutter was inserted. The instant film was then pulled from the film holder and after 35 sec (automatic processing) the film backing was removed to reveal the developed photograph on which the degree of exposure was proportional to the intensity of light emission. In order to standardize interpretation, sera with known antibody levels were included on each plate positioned at the beginning and end of each batch of specimens. Each print could be inserted into a clear plastic holder premarked for identification of individual wells. As the print is a mirror-image of the glowing plate it was important to adhere to a careful identification system. The exposure time could

[20] L. J. Kricka, G. H. G. Thorpe, and T. P. Whitehead, European Patent 116454 (1983).
[21] G. H. G. Thorpe, L. J. Kricka, S. B. Moseley, and T. P. Whitehead, *Clin. Chem.* (*Winston-Salem, N.C.*) **31**, 1335 (1985).

be adjusted for minor differences in reagents, or incubation times, thus permitting the reference specimens to be developed to the same degree with each batch of assays performed. Some typical results are shown in Fig. 7.

Photographic Chemiluminescent Assay for Chromium(III) Using a Camera Luminometer with an Integral Multiple Pipet

Reagents and Standards

A stock chromium(III) chloride hexahydrate solution (0.01 M) was prepared in freshly distilled, deionized water and standards prepared by serial dilution.

A solution of luminol (0.1 M) was prepared in sodium hydrogen carbonate solution (0.05 M) containing EDTA (0.17 mM). The pH of the solution was adjusted to pH 11.0 using sodium hydroxide (1 M) and hy-

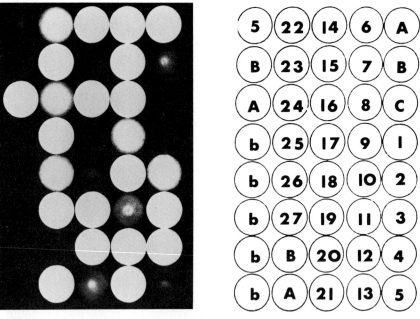

FIG. 7. Photographic results of an enhanced chemiluminescent enzyme immunoassay for cytomegalovirus-specific IgG on a series of 27 samples (standards: A, positive; B, weakly positive; C, negative; blank wells, b). [Reproduced with permission from G. G. Nickless, G. H. G. Thorpe, L. J. Kricka, T. P. Whitehead, L. J. Wells, and F. A. Ala, *J. Virological Methods* **12**, 313 (1985).

drogen peroxide added to give a final concentration of 0.01 *M*. The luminol–EDTA–peroxide assay reagent was allowed to stand for 1 hr prior to use. It can be used immediately but background light emission is lower if it is allowed to mature.

Procedure

Aqueous samples (50 μl) were added to the wells of a 9 × 7 portion of a poly(vinyl chloride) microtiter plate. This was then positioned in the mask and placed in the camera luminometer. The reservoir of the reagent loader was filled with the luminol–EDTA–peroxide reagent. The pipet valve was opened and the pipet lowered into the reagent to a predetermined depth equivalent to a dispensed volume of 200 μl. The valve was then closed and the pipet slowly withdrawn and transferred to the camera luminometer. The shutter was withdrawn, allowing the microtiter plate and its mask to drop onto the film. The pipet was then lowered into the wells and the reagent dispensed by releasing the valve. After a suitable exposure time, generally less than 1 min, the pipet was raised clear of the plate and the

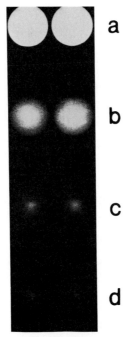

FIG. 8. Photographic results of a chemiluminescent assay for chromium(III) (a, 5 pmol; b, 500 fmol; c, 50 fmol; d, 5 fmol). Reproduced with permission from Bunce *et al.*[4]

shutter reinserted. The instant film was pulled from its holder and after 35 sec the film backing was removed to reveal the developed photograph.[4] Figure 8 shows some typical results for an assay of aqueous chromium(III) standards.

Commercial Systems

The MAST chemiluminescent assay system for total IgE and allergen-specific IgE (MAST Immunosystems Inc, Mountain View, CA 94043) is currently the only commercially available assay system which combines chemiluminescence and photography. The assay is performed in a test chamber which contains a parallel array of allergen-sensitized cellulose threads (Fig. 5). The threads are incubated successively with patient's serum and an anti-IgE peroxidase conjugate. Bound conjugate is then detected by adding hydrogen peroxide, a cyclic diacyl hydrazide, and a dye mixture which acts as a chemical mask. The glowing threads are exposed to Polaroid Type 57 film for 30 min. The amount of bound allergen-specific IgE is determined from densitometric measurements on the developed print (immunograph).[6]

Conclusions

The combination of photography and either chemiluminescence or bioluminescence provides a simple and convenient system for assaying a wide range of substances. As yet this type of analytical system is largely unexplored but the recent commercialization of a photographic immunoassay will act as a spur to more extensive studies of photographic assays.

[34] On-Line Computer Analysis of the Kinetics of Chemiluminescent Reactions: Application to Luminescent Immunoassays

By MARIO PAZZAGLI and MARIO SERIO

Introduction

Chemiluminescent immunoassay (CIA) procedures have been described for steroids, protein hormones, and other biological compounds. The main difference between CIA and radioimmunoassay (RIA) methods

METHODS IN ENZYMOLOGY, VOL. 133

is the measurement of the tracer. The measurement of the chemiluminescent (CL) tracer is more critical than the measurement of radioactivity for
three reasons.

1. The tracer must be activated through an additional step, i.e., an
oxidation reaction, before the analytical signal of the CL tracer can be
measured, whereas radioisotopes spontaneously emit β or γ radiations;
the oxidation reaction can be an additional source of variation in the
measurement of the CL tracer.

2. The analytical signal obtained by the CL tracer during the measurement changes with time, whereas radioactive tracers give a constant signal, which is more easily quantifiable.

3. The signal of a CL reaction is not strictly related to the number of
emitted photons and thus to the number of reacted molecules[1]; the
luminometers used for measurements of luminescence record values in
arbitrary light units, whereas microprocessor controlled β and γ radiation
counters can automatically evaluate the counting efficiency and convert
counts per minute into disintegrations per minute and then into moles of
radiolabeled tracer.

These characteristics of CL tracers can negatively influence the accurate measurement of light emission. Therefore we have connected a microcomputer on-line with a luminometer. This system allows both a detailed analysis of the time course of the output light signal, and the
simultaneous computation of several parameters of both the light emission and the "shape" of CL reaction kinetics.

In particular this system has been used to select the most suitable
parameter of light emission, to study the interfering effects of biological
samples on the measurements of chemiluminescence, to evaluate the ability of various phase separation systems routinely used in the immunological procedures to remove these interfering effects, and to investigate the
possibility of revealing individual samples with significant changes in the
kinetic light emission.

Light Emission Measurements

The present methods for absolute measurement of light emission of
luminescent reactions are very laborious.[1] Consequently commercial
luminometers usually record light emission in arbitrary light units. Moreover the analytical signal is not constant, but changes with time and thus
is more difficult to quantify. A typical light signal of a CL reaction is

---

[1] H. H. Seliger, this series, Vol. 57, p. 560.

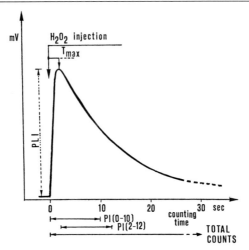

FIG. 1. Time course of the light signal obtained upon oxidation of a CL tracer with hydrogen peroxide–microperoxidase system at pH 13.9: PLI, peak light intensity; T_{max}, time interval from the oxidant injection to PLI; PI, portion integration for a fixed time interval.

illustrated in Fig. 1, with the most common[2,3] measurement parameters including (1) TL (total light), the integrated value of the photon emission that reaches the detector over the total time of emission. This is the most representative measure of the number of photons emitted in the course of the CL reaction. (2) PLI (peak light intensity), a very fast measure of CL reactions, often (but not always) related to TL. (3) PI (portion integration), the integrated value of counts for a fixed time interval, usually 0–10 or 2–10 sec. This measure is the most widely used for routine applications because it is faster than TL and more reliable than PLI.[3]

Shape Analysis of the Kinetics of Chemiluminescent Reactions

None of above described parameters of light emission, however, takes into consideration the "shape" of the CL reaction kinetics, whereas interfering compounds usually modify both the light efficiency and the pattern of the kinetics. Figure 2 illustrates an example of the effects of biological interferents on CL reaction kinetics. Therefore, for every CL reaction, not only the light emission, but also a "shape" index, should be measured to reveal modifications in the kinetics pattern and thus possible disturbances in the light-production efficiency of the CL reaction.

[2] A. K. Campbell and J. S. A. Simpson, *in* "Techniques in the Life Science," Vol. B213, p. 1. Elsevier/North Holland, 1979.
[3] M. Pazzagli, J. B. Kim, G. Messeri, F. Kohen, G. F. Bolelli, A. Tommasi, R. Salerno, G. Moneti, and M. Serio, *J. Steroid Biochem.* **14**, 1005 (1981).

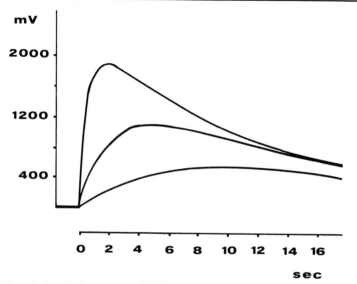

	PI 0-10 sec.	Tmax sec	5/10 Ratio
A (CL-tracer) _	117473	1.8	.54
B (" with 10 μl of urine)_	72416	3.5	.43
C (" with 20 μl of urine)_	38668	6.8	.33

FIG. 2. Interfering effects of different amounts of a urine sample on the CL tracer reaction and on PI (0–10 sec), T_{max}, and ratio 5/10 values.

Recently Wampler et al.[4] have described the use of computer analysis for the kinetic profiles of bioluminescent reactions for monitoring assay performance of a bioluminescent assay for hydrogen peroxide in the presence of extraneous materials. In their system, they calculated a shape index based on the generation of reduced moments and computation of mean, skewness, and kurtosis for every kinetics as briefly summarized in Fig. 3.

We have followed their approach and, moreover, we have investigated two other "shape" indices for CL reactions: T_{max} (in seconds), which

[4] J. E. Wampler, M. G. Mulkerrin, and E. S. Rich, Jr., Clin. Chem. (Winston-Salem, N.C.) **25**, 1628 (1979).

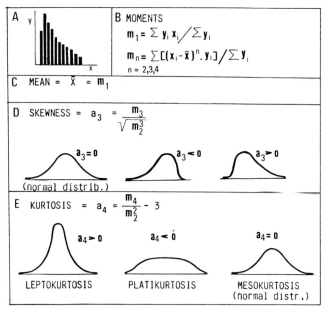

FIG. 3. Shape analysis of CL reactions using the procedure described by Wampler *et al.*[4] (A) The CL reaction is divided in sample intervals (i.e., Fig. 4) of x (fixed time interval) and y (integrated value of counts for that time interval). (B) Equations for the computation of first, second, third, and fourth reduced moments. (C, D, and E) Equations for computation of mean, skewness, and kurtosis and meaning of these parameters in the shape analysis.

represents the time interval from the oxidant injection to the PLI (see Fig. 1), and the 5/10 ratio, i.e., the ratio of the PI values during 0–5 and 0–10 sec. The ability of the 5/10 ratio to identify changes in the kinetic shape is based on the shift of the count distribution vs time in CL reactions undergoing interference (see Fig. 2).

Instrumentation

We measured light emission with an Auto Picolite Packard (model 6500, Packard Instruments, Downers Grove, IL) using 12×47 mm Lumacuvette P Polystyrene (Lumac System, Basel, Switzerland) as reaction tubes. We connected the analog output of the luminometer to a storage oscilloscope (type 5111; Tektronix, Beaverton, OR) to observe the pattern of light emission. In addition we interfaced the output signal from the photomultiplier tube of the luminometer to a microcomputer (General Processor, Model T/08, dual 5¼ in. single-density floppy disk drive, 48 kbyte read/write memory) as described in Tommasi et al.[5]. Two different programs have been developed for this system.

1. DECODE. This compiled program from a BASIC source file and a driver in assembly code is necessary for rapid real-time acquisition of data from the luminometer (via the interface) to reduce the time interval between successive samples. The light emitted by the CL tracer is divided into sample intervals from 8 to 3000 msec and into variable numbers of samples per reaction from 1 to 255. Data are stored on disk. An example of acquisition of a CL reaction is reported in Fig. 4.

2. GRAFST2. This program in Microsoft BAS80 language provides data processing and printing. It allows computation of the data and printing of a table (one row per sample), in which light emission parameters and "shape" indices are reported together (e.g., Table I).

Reagents

We obtained microperoxidase (MP-11), bovine serum albumin (Cohn fraction V; BSA), dextran-coated charcoal, and human γ-globulins (hIgG) from Sigma, St. Louis, MO, 30% hydrogen peroxide solution and sodium hydroxide, from Merck, Darmstadt, FRG, and Sac-Cel donkey anti-mouse antibody-coated cellulose suspension from Wellcome, Beckenham, England.

The assay buffer was a borate buffer 0.1 M, pH 8.6, containing 9 g of NaCl, 80 mg of BSA, and 80 mg of hIgG per liter.

Dextran-coated charcoal was prepared as described elsewhere,[6] and diluted in assay buffer to 1.0 mg/0.2 ml.

Steroid-CL tracers were linked covalently to 6-[N-(4-aminobutyl)-N-ethyl] isoluminol (ABEI) and then purified and characterized as already described.[7]

All the other reagents, solutions, and procedures for CL reactions were as previously described[3,6] unless specified otherwise.

---

[5] A. Tommasi, M. Pazzagli, M. Damiani, R. Salerno, G. Messeri, A. Magini, and M. Serio, Clin. Chem. (Winston-Salem, N.C.) **30**, 1597 (1984).

[6] M. Pazzagli, J. B. Kim, G. Messeri, F. Kohen, G. F. Bolelli, A. Tommasi, R. Salerno, and M. Serio, J. Steroid Biochem. **14**, 1181 (1981).

[7] M. Pazzagli, G. Messeri, A. L. Caldini, G. Moneti, G. Martinazzo, and M. Serio, J. Steroid Biochem. **19**, 407 (1983).

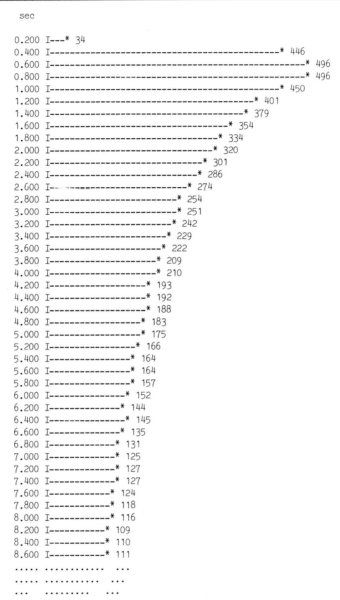

FIG. 4. Example of acquisition of data of a typical CL reaction by the computer–lumino-meter system.

TABLE I
EXAMPLE OF A TABLE CREATED BY THE PROGRAM GRAFST2[a]

N	PLI (counts/sec)	Total counts	PI 0–10 sec	PI 2–10 sec	T_{max} (sec)	Ratio 2/10	Ratio 5/10	Mean (sec)	Skewness	Kurtosis
Blank	120	423	423	316	0.600	36.26	68.81	3.77	0.605	−0.806
Nonspecific binding	180	645	645	451	0.500	36.30	69.05	3.69	0.646	−0.713
Total	7,300	44,041	44,041	32,217	0.600	37.72	70.27	3.64	0.667	−0.747
Zero	5,230	32,764	32,764	24,297	0.500	37.63	70.79	3.61	0.665	−0.687
7.8 pg	5,120	31,592	31,592	23,175	0.500	38.55	71.98	3.55	0.708	−0.616
15.6 pg	4,880	29,735	29,735	21,806	0.400	38.03	71.12	3.57	0.696	−0.631

[a] Light emission parameters and shape indices are reported together (one row per kinetics). For abbreviations see text.

Results

Choice of the Light Emission Parameter of the CL Tracer

We investigated the reproducibility of the light emission parameters in the range of the dose–response curve values for several CIA methods such as cortisol,[6] progesterone,[8] testosterone,[9] testosterone 17β-D-glucuronide,[10] and total urinary estrogens.[11] The results are reported in Fig. 5. The PI (2–10 sec) value was found significantly more reproducible than PI (0–10 sec) and PLI measurements and therefore it was used for the calculations of the assay results.

Effect of Plasma or Urine on the CL Reaction

The study of the interferences from biological samples on the measurement of chemiluminescent reactions has shown that such effects are usually associated with modifications of the shape of the light emission kinetics (Fig. 2). These results suggest that a simultaneous evaluation of the shape of a chemiluminescent reaction and measurement of light emission can be combined to assess luminescent immunoassays as an internal control of the interferences in measurements of the chemiluminescent tracer.

We have studied the effect of plasma or urine on the measurement of CL tracer; to do so, we oxidized pooled plasma or urine at various dilutions in the absence (blanks) or the presence of 100 fmol of cortisol–3–carboxymethyloxime–aminobutylethylisoluminol (cortisol–3–CMO–ABEI) under the following conditions.

a. Diluted plasma or urine (0.1 ml each) (with an amount of biological fluid ranging from 0.156 to 20 μl per sample), NaOH (2 M), microperoxidase (10 μM), and hydrogen peroxide (a 133-fold dilution of the stock with double distilled water).

b. Same as a, except the microperoxidase solution was 20 μM.

c. Same as a, except the hydrogen peroxide solution (a 35-fold dilution of the stock).

[8] M. Pazzagli, J. B. Kim, G. Messeri, G. Martinazzo, F. Kohen, F. Franceschetti, A. Tommasi, R. Salerno, and M. Serio, *Clin. Chim. Acta* **115**, 287 (1981).

[9] M. Pazzagli, M. Serio, P. Munson, and D. Rodbard, *in* "Radioimmunoassay and Related Procedures in Medicine," p. 747. IAEA, Vienna, 1982.

[10] P. L. Vannucchi, G. Messeri, G. F. Bolelli, M. Pazzagli, A. Masala, and M. Serio, *J. Steroid Biochem.* **18**, 625 (1983).

[11] G. Messeri, A. L. Caldini, G. F. Bolelli, M. Pazzagli, A. Tommasi, P. L. Vannucchi, and M. Serio, *Clin. Chem. (Winston-Salem, N.C.)* **30**, 653 (1984).

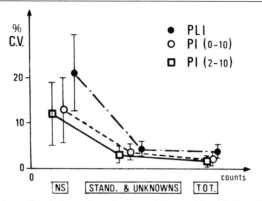

Fig. 5. Coefficient of variation values (mean ± SD; $n = 20$) of light emission parameters at different levels of CL tracer concentration: NS (nonspecific binding) are values at background levels; standard and unknowns and total activity are values in the typical range of CIA methods for steroids.

d. Same as a, except the plasma or urine samples were pretreated with dextran-coated charcoal before the CL reaction.

The results (Figs. 6 and 7) can be summarized as follows.

1. Both plasma and urine interfere with the CL reaction, in the blank reaction and in the measurement of CL tracer, but with opposite effects, i.e., the light emission of the blank increases, simulating a CL effect as described by Kohen et al.,[12] and the measurement of the CL tracer is quenched.[6]

2. In standard conditions of oxidation reactions (Fig. 6, a), 0.312 μl of pooled plasma or 1.25 μl of pooled urine could induce a significant variation in the light emission measurements; moreover these variations were not significantly modified by changing the oxidation reaction conditions (Fig. 6, b and c).

3. The pretreatment of plasma with charcoal did not significantly modify the interfering content of the samples whereas using urine, the pretreatment with charcoal substantially decreased both the blank and the quenching interferences; therefore the light measurement parameters became similar to the values obtained in the absence of urine (Fig. 6, d).

4. In all the experiments the "shape" indices indicated the presence of biological fluid interferences associated with the variations in the light emission parameters (Fig. 7).

[12] F. Kohen, J. B. Kim, J. Barnard, and H. R. Lindner, Steroids 36, 405 (1980).

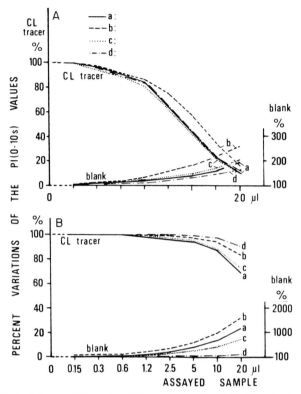

FIG. 6. Effect of plasma (A) or urine (B) samples on the PI (2–10 sec) values (expressed as percentage variation) measured in the absence (blank) or presence (CL tracer) of 100 fmol per tube of cortisol–3-CMO–ABEI. For experimental conditions see text.

From these data, we can take some suggestions when the development of new CIA methods have to be performed, in particular for the choice of extraction procedures or for the use of suitable solid-phase separations systems as summarized in Fig. 8.

Calculations for the Shape Test

We performed the "shape" test within every assay which included a set of data of the dose–response curve points and those of unknown biological samples. We first computed the mean ± SD of the various shape parameters of the dose–response curve points (without interferences by definition). Then we calculated the comparison index (I) for both dose response curve points and unknowns (I_u) according to the formula

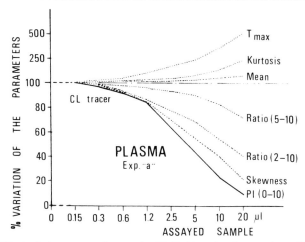

FIG. 7. Effect of plasma samples on the PI (2–10 sec) values and on different shape indices (expressed as percentage variation of the noninterfered values) in the oxidation reaction conditions of experiment a in Fig. 6.

reported in Table II in which two examples of computation of shape indices are reported.

The maximum value of I observed in the group of noninterfered kinetics (I_{max}) was used for the "shape" test: if $I_u \leq I_{max}$ the PI (2–10 sec) value of the unknown sample was accepted and then used to determine the amount of analyte in the assay tube; if $I_u > I_{max}$ the sample was rejected.

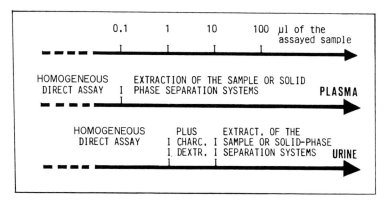

FIG. 8. Interfering effects of biological samples on the CL tracer measurement and possible procedures for the development of CIA methods.

TABLE II

COMPUTATION OF THE COMPARISON INDEX I^a

(A)[b]

$$I = \sum \left(\frac{|\overline{P_i} - P_i|}{\sigma_i} \right) \Big/ N$$

(B)

N	PI		T_{max} (sec)	Ratio		Mean (sec)	Skewness	Kurtosis
	0–10 sec	2–10 sec		2/10	5/10			
Blank	423	316	0.600	36.26	68.81	3.77	0.605	−0.806
Nonspecific binding	645	451	0.500	36.30	69 05	3.69	0.646	−0.713
Total	44,041	32,217	0.600	37.72	70.27	3.64	0.667	−0.747
Zero	32,764	24,297	0.500	37.63	70.79	3.61	0.665	−0.687
7.8 pg	31,592	23,175	0.500	38.55	71.98	3.55	0.708	−0.616
15.6 pg	29,735	21,806	0.400	38.03	71.12	3.57	0.696	−0.631
....	....	....	....	....	....	....	....	....
500 pg	2,490	1,836	0.500	37.51	70.73	3.66	0.689	−0.693
Mean			0.510	38.01	71.09	3.62	0.676	−0.670
±SD			0.050	0.54	0.44	0.03	0.016	0.028

(C) Sample 15.6 pg/tube
     Use of the "shape" parameter: mean

$$I = \frac{|3.62 - 3.57|}{0.03} = 1.66$$

(D) Use of the "shape" parameters: mean, skewness, kurtosis

$$I = \left(\frac{|3.62 - 3.57|}{0.03} + \frac{|0.676 - 0.696|}{0.016} + \frac{|0.670 - 0.631|}{0.028} \right) \Big/ 3 = 1.46$$

[a] Following the procedure described by Wampler et al.[4] using different shape parameters.
[b] (A) Equation used for computation of I. (B) Example of standard curve data of a CIA method for urinary free cortisol. (C) and (D) Computation of I using two different shape parameters.

     An example of application of the shape test to a set of cortisol CIA data is reported in Table III. The graphical distribution of the I values computed using different shape parameters for both standard kinetics or kinetics of unknown biological samples is reported in Fig. 9.

     For routinary applications, we used the 5/10 ratio values to compute I and I_{max} (Fig. 9) because this parameter is more sensitive than T_{max} in detecting interferences. In addition the 5/10 ratio value seems to have the same sensitivity as the index used by Wampler et al.,[4] based on the generation of the second, third, and fourth reduced moments (Fig. 9). The computation of the 5/10 ratio value, in comparison to the parameters used

TABLE III
EXAMPLE OF SHAPE TEST[a]

N	PI 0–10 sec	2–10 sec	T_{max} (sec)	Ratio 5/10	I (I_s or I_u)	Shape test
Standard curve data						
Zero	32,768	24,297	0.500	70.79	0.681	
7.8 pg	31,592	23,175	0.500	71.98	2.022	
15.6 pg	29,735	21,806	0.400	71.12	0.047	
....	....	....	....	....	....	
500 pg	2,490	1,836	0.500	70.73	0.818	
Mean				71.09		
±SD				0.44		
I_{max}				2.022		
Unknown samples data						
a	21,292	16,138	0.500	71.31	0.500	Accepted
b	6,535	4,727	0.500	71.20	0.250	Accepted
c	17,026	12,523	0.500	72.35	2.863	Rejected

[a] Performed on a typical set of data of a CIA method for urinary free cortisol, which includes data of the standard curve points, and data of unknown biological samples.

by Wampler et al.,[4] in fact, induced a substantial decrement in the computation time (1–2 sec per kinetics versus 20–25 sec per kinetics, using the computer luminometer system described here) and therefore increased the applicability of the "shape" test to a large number of samples.

Discussion

In this paper we have briefly reviewed some of the problems connected to CL tracer measurements and how these problems become relevant when CL measurements are performed in the presence of biological samples. In an attempt to improve the present methods of measurement of CL reactions we have developed an on-line computer analysis system following the approach described by Wampler et al.[4] for bioluminescent reactions. This system first allows the storage on disk of the data from the luminometer (via the interface) and their subsequent analysis and computation. The microcomputer–luminometer system has then been utilized for the choice of the most reproducible light emission parameter and for studying the interfering effects of biological samples on the CL reaction. The results indicate that (1) plasma and urine samples strongly interfere with the CL reaction both in the blank and in the CL tracer measurement; (2) these interferences cannot be removed by changing the CL reaction

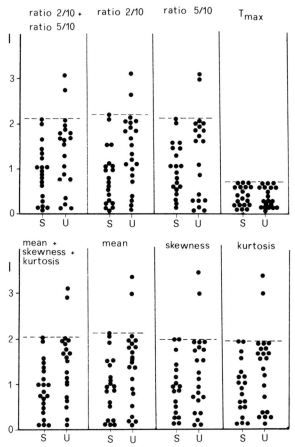

FIG. 9. I values obtained using different shape indices in a set of data from a CIA method for urinary free cortisol. S = I values of 10 points in duplicate of the dose–response curve; U = I_u values of 10 urine unknown samples in duplicate. The dotted line represents the maximum value of I in the S group of kinetics (I_{max}) which is used for the shape test. Note that one urine sample in duplicate shows I_u values higher than I_{max} in all the methods used for the shape analysis except for T_{max}.

oxidation conditions, but the use of suitable solid-phase separation systems such as Sac-Cel (Wellcome, England) or the urine pretreatment with charcoal markedly reduced their interfering content; and (3) the reliability of CL measurements is improved if we perform a contemporaneous evaluation of the "shape" of CL reaction kinetics; "shape" indices in fact have been shown to reflect the efficiency of the CL reaction.

The 5/10 ratio value can be used for this purpose because its ability to detect interferences in the CL reaction is comparable to the procedure described by Wampler *et al.*[4]; the computation time is, on the contrary, substantially reduced.

It should be considered, however, that the "shape" analysis can detect interferences which modify the kinetic profile of the CL reaction; some other kinds of interference (e.g., absorption of light by a colored sample) would not be detected by the proposed system.

This approach has been applied in our laboratory to the development and assessment of several CIA methods for steroids in both plasma and urine samples. The results indicate that the reliability achieved by the CIA methods approaches that of RIA only when an automatic control of nonspecific interferences coming from biological samples is introduced in the procedure.

Acknowledgments

This work is supported by a grant from the Italian Research Council (Biomedical Technology Program: Immunochemistry project).

[35] Chemiluminescence Detection System for High-Performance Liquid Chromatography

By Kazuhiro Imai

Chemiluminescence (CL) reactions have been used for the analysis of minute amounts of substances. The following CL reactions have been reported: gallic acid–hydrogen peroxide CL reaction for Co^{2+},[1,2] luminol–hydrogen peroxide CL reaction for Co^{2+},[3] luminol–hydrogen peroxide-Co^{3+} CL reaction for proteins,[4] zinc and cadmium,[5] luminol–borate CL reaction for alkyl phosphate,[6] luminol–peroxidase CL reaction for hydrogen peroxide[7] and in immunoassay,[8–12] luminol–Cu^{2+}(phenanthroline)–

[1] S. Nakahara, M. Yamada, and S. Suzuki, *Anal. Chim. Acta* **141,** 255 (1982).

[2] M. Yamada, T. Komatsu, S. Nakahara, and S. Suzuki, *Anal. Chim. Acta* **155,** 259 (1983).

[3] J. L. Burguera, A. Townshend, and S. Greenfield, *Anal. Chim. Acta* **114,** 209 (1980).

[4] T. Hara, M. Toriyama, and K. Tsukagoshi, *Bull. Chem. Soc. Jpn.* **57,** 289 (1984).

[5] J. L. Burguera, M. Burguera, and A. Townshend, *Anal. Chim. Acta* **127,** 199 (1981).

[6] U. Fritsche, *Anal. Chim. Acta* **118,** 179 (1980).

[7] B. Olson, *Anal. Chim. Acta* **136,** 113 (1982).

[8] T. P. Whitehead, L. J. Kricka, T. J. N. Carter, and G. H. G. Thorpe, *Clin. Chem.* (*Winston-Salem, N.C.*) **25,** 1531 (1979).

[9] T. P. Whitehead, G. H. G. Thorpe, T. J. N. Carter, C. Groucutt, and J. Kricka, *Nature* (*London*) **305,** 158 (1983).

glucose oxidase CL reaction for glucose,[13-15] luminol–oxidase–peroxidase CL reaction for D-glucose[16] and for polyamines,[17] luminol–lipoxygenase–linoleic acid CL reaction for lipoxygenase,[18] electrogenerated CL reaction of luminol for Co^{2+} [19] and for other metals ions,[20] isoluminol–hydrogen peroxide CL reaction for amidase[21] and for immunoassay,[22-24] electrogenerated CL reaction of tris(bipyridyl)ruthenium complex for oxalate,[25] electrochemiluminescence for aromatic hydrocarbons,[26] lophine–hydrogen peroxide CL reaction for Co^{2+},[27] 1,10-phenanthroline–hydrogen peroxide CL reaction for Cu^{2+},[28] flavine mononucleotide(FMN)–hydrogen peroxide CL reaction for Cu^{2+},[29] hydrogen peroxide–peroxidase CL reaction for S^{2-},[30] hypobromide–azure B CL reaction for S^{2-},[31] hypochlorite–fluorescein CL reaction for S^{2-},[32] hypochlorite–hydrogen peroxide CL for Cl^- [33] and for benzo[a]pyrene-7,8-diol,[34] permanganate-FMN CL reaction for So_3^{2-},[35] permanganate CL

[10] T. Olsson, K. Bergström, and A. Thore, *Clin. Chim. Acta* **138**, 31 (1984).

[11] Y. Ikariyama, S. Suzuki, and M. Aizawa, *Anal. Chim. Acta* **156**, 245 (1984).

[12] M. Maeda and A. Tsuji, *Anal. Chim. Acta* **167**, 241 (1985).

[13] D. Pilosof and T. A. Nieman, *Anal. Chem.* **54**, 1698 (1982).

[14] P. J. Worsfold, J. Farrelly, and M. S. Matharu, *Anal. Chim. Acta* **164**, 103 (1984).

[15] C. Ridder, E. H. Hansen, and J. Ruzicka, *Anal. Lett.* **15**, 1751 (1982).

[16] T. Hara, M. Toriyama, and M. Imaki, *Bull. Chem. Soc. Jpn.* **55**, 1854 (1982).

[17] S. Kamei, F. Mashige, A. Ohkubo, M. Yamanaka, M. Okada, Y. Yoshimura, and K. Imahori, *Jpn. J. Clin. Chem.* **13**, 66 (1984).

[18] E.-M. Lilius and S. Laakso, *Anal. Biochem.* **119**, 135 (1982).

[19] K. E. Haapakka, *Anal. Chim. Acta* **139**, 229 (1982).

[20] K. E. Haapakka and J. Kankare, *Anal. Chim. Acta* **138**, 253 (1982).

[21] B. R. Branchini, J. D. Hermes, F. G. Salituro, N. J. Post, and G. Claeson, *Anal. Biochem.* **111**, 87 (1981).

[22] E. H. J. M. Jansen, R. H. Van Den Berg, G. Zomer, R. Both-Miedema, C. Enkelaar-Willemsen, and R. W. Stephany, *Anal. Chim. Acta* **170**, 21 (1985).

[23] E. H. J. M. Jansen, C. A. Laan, R. H. Van Den Berg, R. W. Stephany, and G. Zomer, *Anal. Chim. Acta* **170**, 29 (1985).

[24] J. de Boever, F. Kohen, D. Dhont, D. Vandekerckhove, and G. Van Maele, *Anal. Chim. Acta* **170**, 117 (1985).

[25] I. Rubinstein, C. R. Martin, and A. J. Bard, *Anal. Chem.* **55**, 1580 (1983).

[26] H. Schaper, *J. Electroanal. Chem.* **129**, 335 (1981).

[27] D. F. Marino and J. D. Ingle, Jr., *Anal. Chem.* **53**, 292 (1981).

[28] M. Yamada and S. Suzuki, *Anal. Lett.* **17**, 251 (1984).

[29] M. Yamada and S. Suzuki, *Chem. Lett.* p. 1747 (1982).

[30] J. L. Burguera and A. Townshend, *Talanta* **27**, 309 (1980).

[31] J. Teckentrup and D. Klockkow, *Talanta* **28**, 653 (1981).

[32] J. L. Burguera, A. Townshend, and S. Greenfield, *Anal. Chim. Acta* **114**, 209 (1980).

[33] D. F. Marino and J. D. Ingle, Jr., *Anal. Chim. Acta* **123**, 247 (1981).

[34] A. Thompson, H. H. Seliger, and G. H. Posner, *Anal. Biochem.* **130**, 498 (1983).

[35] M. Yamada, T. Nakada, and S. Suzuki, *Anal. Chim. Acta* **147**, 401 (1983).

reaction for SO^{2-} [36] and for humic acid,[37] $S_2O_8^{2-}$ CL reaction for indole,[38] tetrakis-*N*-dialkylaminoethylene CL reaction for oxygen,[39,40] aryl oxalate–hydrogen peroxide CL reaction for fluorescer,[41–43] lucigenin CL reaction for vitamin C,[44] for heparin[45] and for immunoassay,[46–48] and rhodospin CL reaction for Ca^{2+}.[49]

The growing interest in CL analysis in recent studies might be derived from the fact that CL analysis has some advantages over photo-induced fluorescence analysis. In principle, CL emerges on a dark existing background permitting possible enlargement of a signal (an emitted light) with a detector such as a photomultiplier. An extremely high signal-to-noise ratio (S/N) and large dynamic ranges (5 orders of magnitude) of signals can be attained. The luminol CL reaction[50] can be used to detect 10–9 *M* hydrogen peroxide,[51] 2×10^{-9} *M* vitamin B_{12},[52] 70 fg Co^{2+},[53] 10 pg hematin,[54] 100 pg hemoglobin,[54] and 100 pg myoglobin[54]. Other examples include the ozone CL reaction to detect 50 ng/liter NO_2[55] and 8 ng/liter phosphorus.[56] Glucose could also be determined in certain cases in the $10^{-8}–10^{-4}$ *M* ranges.[57,58]

Another advantage of CL analysis is that it does not require a sophisticated light source and it needs only a rather simple detection device having a photomultiplier or photon counter. The development of HPLC

[36] F. X. Meixner and W. A. Jaeschke, *Int. J. Environ. Anal. Chem.* **10,** 51 (1981).
[37] D. F. Marino and J. D. Ingle, Jr., *Anal. Chim. Acta* **124,** 23 (1981).
[38] H. Imai, H. Yoshida, T. Masujima, and T. Ohwa, *Bunseki Kagaku* **33,** 54 (1984).
[39] T. M. Freeman and W. R. Seitz, *Anal. Chem.* **53,** 98 (1981).
[40] B. F. MacDonald and W. R. Seitz, *Anal. Lett.* **15,** 57 (1982).
[41] P. A. Sherman, J. Holzbecher, and D. E. Ryan, *Anal. Chim. Acta* **97,** 21 (1978).
[42] V. K. Mohant, J. N. Miller, and H. Thakrar, *Anal. Chim. Acta* **145,** 203 (1983).
[43] K. Honda, J. Sekino, and K. Imai, *Anal. Chem.* **5,** 940 (1983).
[44] L. L. Klopf and T. A. Nieman, *Anal. Chem.* **56,** 1539 (1984).
[45] R. A. Steen and T. A. Nieman, *Anal. Chim. Acta* **155,** 123 (1983).
[46] I. Weeks, I. Beheshti, F. McCapra, A. K. Campbell, and J. S. Woodhead, *Clin. Chem. (Winston-Salem, N.C.)* **29,** 1474 (1983).
[47] I. Weeks, A. K. Campbell, and J. S. Woodhead, *Clin. Chem. (Winston-Salem, N.C.)* **29,** 1480 (1983).
[48] I. Weeks and J. S. Woodhead, *Clin. Chim. Acta* **141,** 275 (1984).
[49] A. K. Campbell and M. B. Hallett, *J. Physiol. (London)* **338,** 537 (1983).
[50] H. O. Albrechet, *Z. Phys. Chem.* **136,** 321 (1928).
[51] W. R. Seitz and M. P. Neary, *Anal. Chem.* **46,** 188A (1974).
[52] T. L. Sheehan and D. M. Hercules, *Anal. Chem.* **49,** 446 (1977).
[53] M. Yamada and S. Suzuki, *Chem. Lett.* p. 783 (1983).
[54] H. A. Neufeld, C. J. Conklin, and R. D. Towner, *Anal. Biochem.* **12,** 303 (1965).
[55] R. D. Cox, *Anal. Chem.* **52,** 332 (1980).
[56] K. Matsumoto, K. Fujiwara, and K. Fuwa, *Anal. Chem.* **55,** 1665 (1983).
[57] D. T. Bostick and D. M. Hercules, *Anal. Chem.* **47,** 447 (1975).
[58] J. P. Auses, S. L. Kook, and J. T. Maloy, *Anal. Chem.* **47,** 244 (1975).

techniques has enabled us to measure at high resolution a variety of mixtures of chemical substances. Precise solvent delivery utilizing a pumping system has also been well suited for detecting reproducibly the emitted CL intensity which often decays very fast. Thus, the combination of this high resolving power and the precise delivering system of solutions in HPLC and the sensitive detection ability of CL analysis as well as the cheap mechanics would be one of the plausible techniques for a sensitive, precise, and facile determination of trace levels of compounds in the complex mixture of substances in biochemical and biomedical fields.

In the present state of knowledge, the peroxyoxalate CL reaction,[59,60] ozone CL reaction,[61] hypochlorite–hydrogen peroxide CL reaction,[62] electrochemiluminescence,[63] luminol–Co^{2+}–peroxide,[64] and lucigenin CL reaction[65] were the only examples for use as detection systems for HPLC. Fluorescers such as dansyl (DNS) derivatives,[66–73] polyaromatic hydrocarbons,[74–79] aminopyrene,[80] and rhodamine B,[81] and nonfluorescent compounds, such as amino acids,[81] 1,2-dichloroethylene,[81] butyl disulfide,[81] fructose,[83] vitamin C,[84] nitrosamine,[85] glucuronic acid,[86] protein,[87] and

[59] M. M. Rauhut, L. J. Bollyky, B. G. Roberts, M. Loy, R. H. Whitman, A. V. Iannotta, A. M. Semsel, and R. A. Clarke, *J. Am. Chem. Soc.* **89,** 6515 (1967).
[60] M. M. Rauhut, *Acc. Chem. Res.* **2,** 80 (1969).
[61] A. Fontijn, A. J. Sabadell, and R. J. Ronco, *Anal. Chem.* **42,** 575 (1970).
[62] L. Mallet, *C. R. Hebd. Seances Acad. Sci.* **185,** 352 (1972).
[63] S. Park and D. A. Tryk, *Rev. Chem. Intermed.* **4,** 43 (1980).
[64] T. G. Bruno and W. R. Seitz, *Anal. Chem.* **47,** 1639 (1975).
[65] J. R. Totter, *Photochem. Photobiol.* **22,** 203 (1975).
[66] K. Kobayashi and K. Imai, *Anal. Chem.* **52,** 424 (1980).
[67] G. Mellbin, *J. Liq. Chromatogr.* **6,** 1603 (1983).
[68] K. Miyaguchi, K. Honda, and K. Imai, *J. Chromatogr.* **316,** 501 (1984).
[69] K. Miyaguchi, K. Honda, and K. Imai, *J. Chromatogr.* **303,** 173 (1984).
[70] G. Mellbin and B. E. F. Smith, *J. Chromatogr.* **312,** 203 (1984).
[71] G. J. de Jong, N. Lammers, F. J. Spruit, U. A. Th. Brinkman, and R. W. Frei, *Chromatographia* **18,** 129 (1984).
[72] K. Kobayashi, J. Sekino, K. Honda, and K. Imai, *Anal. Biochem.* **112,** 99 (1981).
[73] T. Koziol, M. L. Grayeski, and R. Weinberger, *J. Chromatogr.* **317,** 355 (1984).
[74] K. W. Sigvardson and J. W. Birks, *Anal. Chem.* **55,** 432 (1983).
[75] K. W. Sigvardson, J. M. Kennish, and J. W. Birks, *Anal. Chem.* **56,** 1096 (1984).
[76] M. L. Grayeski and A. J. Weber, *Anal. Lett.* **17,** 1539 (1984).
[77] K. W. Sigvardson and J. W. Birks, *J. Chromatogr.* **316,** 507 (1984).
[78] B. Shoemaker and J. W. Birks, *J. Chromatogr.* **209,** 251 (1981).
[79] C. Blatchford and D. J. M.-Lawes, *J. Chromatogr.* **321,** 227 (1985).
[80] R. Weinberger, C. A. Mannan, M. Cerchio, and M. L. Grayeski, *J. Chromatogr.* **288,** 445 (1984).
[81] J. W. Birks and M. C. Kuge, *Anal. Chem.* **52,** 897 (1980).
[82] A. McDonald and T. A. Nieman, *Anal. Chem.* **57,** 936 (1985).
[83] M. S. Gandelman and J. W. Birks, *J. Chromatogr.* **242,** 21 (1982).

transition metals,[88] were sensitively determined by those combined systems.

In this chapter, the CL detection system for HPLC of fluorescers using the aryl oxalate CL reaction is mainly described. Since the basic assembly of the CL detection system for HPLC is similar, only a slight modification to the instrumentation for the system due to the flow rates of the effluent and the reagent and type of CL reactions permits the sensitive detection of substances. This is also true for the bioluminescence detection system for HPLC.[89]

Peroxyoxalate Chemiluminescence Detection System

Aryl oxalate and hydrogen peroxide react with each other to give an unstable, energy-rich intermediate, 1,2-dioxetanedione (around 70 kcal/mol), which in turn excites the fluorescer to the singlet excited state. The excited fluorescer returns to the ground state releasing light, the maximum wavelength of which is the same as that in fluorescence emission. The fluorescer is again excited by the residual intermediate if present in the medium. Light continues to be generated until the generation of the intermediate terminates. This overall reaction is called for the peroxyoxalate CL reaction (Fig. 1).

Only hydrogen peroxide reacts with aryl oxalate to give an intermediate (Table I).[59] Table II summarizes the use of flow injection analysis for the detection of hydrogen peroxide itself[90,91] or for hydrogen peroxide generated by glucose oxidase,[92,93] galactosidase,[93] uricase,[94] aldehyde oxidase,[95] cholesterol oxidase,[96] and amino acid oxidase.[97] Experimental de-

[84] R. L. Veazey and T. A. Nieman, *J. Chromatogr.* **200**, 153 (1980).

[85] R. C. Massey, C. Crews, D. J. McWeeny, and M. E. Knowles, *J. Chromatogr.* **236**, 527 (1982).

[86] L. L. Klopf and T. A. Nieman, *Anal. Chem.* **57**, 46 (1985).

[87] T. Hara, M. Toriyama, and T. Ebuchi, *Bull. Chem. Soc. Jpn.* **58**, 109 (1985).

[88] M. P. Neary, R. F. Seitz, and D. M. Hercules, *Anal. Lett.* **7**, 583 (1974).

[89] W. D. Bostick, M. S. Denton, and A. Tsuji, *Clin. Chem.* (*Winston-Salem, N.C.*) **31**, 430 (1985).

[90] P. Van Zoonen, D. A. Kamminga, G. Gooljer, N. H. Velthorst, and R. W. Frei, *Anal. Chim. Acta* **167**, 249 (1985).

[91] D. C. Williams, III, G. F. Huff, and W. R. Seitz, *Anal. Chem.* **48**, 1478 (1976).

[92] D. C. Williams, III, G. F. Huff, and W. R. Seitz, *Anal. Chem.* **48**, 1003 (1976).

[93] V. I. Rigin, *J. Anal. Chem. USSR* (*Engl. Transl.*) **34**, 619 (1979).

[94] G. Scott, W. R. Seitz, and J. Ambrose, *Anal. Chim. Acta* **115**, 221 (1980).

[95] V. I. Rigin, *J. Anal. Chem. USSR* (*Engl. Transl.*) **36**, 1111 (1981).

[96] V. I. Rigin, *J. Anal. Chem. USSR* (*Engl. Transl.*) **33**, 1265 (1983).

[97] V. I. Rigin, *J. Anal. Chem. USSR* (*Engl. Transl.*) **38**, 1328 (1983).

TABLE I
EFFECT OF PEROXIDES ON THE QUANTUM YIELD OF
DNPO CHEMILUMINESCENCE[a][59]

Peroxide	Conc. (M)	Quantum yield, einsteins mol$^{-1} \times 10^2$
Hydrogen peroxide	0.082	7.3
t-Butyl hydroperoxide	0.20	0.01
Tetralin hydroperoxide	0.10	0.008
Peroxybenzoic acid	0.10	0.003
Benzoyl peroxide	0.10	Very low

[a] Reactions with 0.010 M bis(2,4-dinitrophenyl) oxalate (DNPO) and 4.16×10^{-4} M 9,10-diphenylanthracene in dimethylphthalate at 25°.

TABLE II
USE OF THE PEROXYOXALATE CL REACTION FOR THE FLOW INJECTION ANALYSIS OF HYDROGEN
PEROXIDE ITSELF OR GENERATED BY ENZYMATIC REACTION

Analyte (detection limit)	Enzyme	CL reagent[a]	Fluorescer (catalyst)	Reference
Hydrogen peroxide (1.2 pmol)	—	TCPO	Perylene	90
Glucose (7×10^{-8} M)	Immobilized glucose oxidase	TCPO	Perylene (trimethylamine)	92
NADH (2×10^{-7} M) and LDH activity	Coupling with methylene blue	TCPO	Perylene (trimethylamine)	91
Glucose (33 pmol) Lactose (35 pmol)	Immobilized glucosidase β-galactosidase	TCPO	9,10-Diphenylanthracence (trimethylamine)	93
Uric acid (1×10^{-6} M)	Uricase	CPPO	Perylene (trimethylamine)	94
Formaldehyde (1.67×10^{-9} M)	Immobilized aldehyde oxidase	TCPO	9,10-Diphenylanthracene (trimethylamine)	95
Cholesterol ester Cholesterol (200 fmol)	Immobilized cholesterol hydrolase, cholesterol oxidase	TCPO	9,10-Diphenylanthracene (trimethylamine)	96
L-Amino acids, valine, leucine tyrosine (200 fmol)	Immobilized L-amino acid oxidase	TCPO	9,10-Diphenylanthracene (trimethylamine)	97

[a] TCPO, bis(2,4,6-trichlorophenyl) oxalate; CPPO, bis (2,4,5-trichloro-6-pentoxycarbonyl) oxalate.

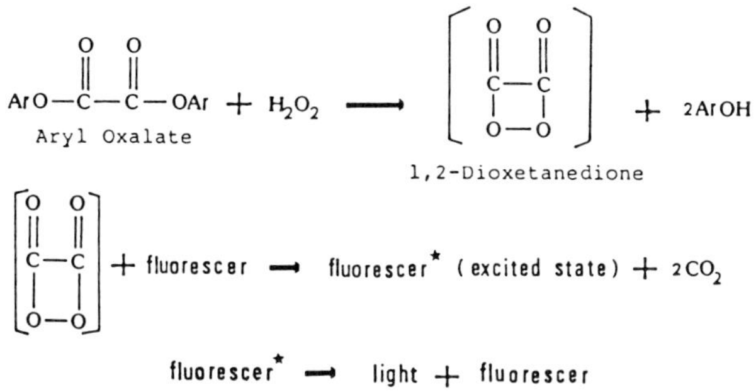

FIG. 1. Peroxyoxalate CL reaction system.

tails regarding flow injection analysis can be found in Seitz.[98] However, no reports on its application to HPLC detection of hydrogen peroxide generated by enzyme reactions have appeared.

This technique was also applied to chemiluminescence enzyme immunoassay.[99,100]

Fluorescers which are excited by the peroxyoxalate CL reaction must be easily oxidized and excited by intermediates such as rubrene, rhodamine, perylene, and N,N-dimethylaminonaphthalene. The CL intensities are also dependent on their fluorescence quantum yield. For examples of the differences in CL intensities among fluorescers see Fig. 4. For the detailed information on the kind of fluorescers that would be excited by the peroxyoxalate reaction system, the reader is referred to other articles.[75,101,102]

The peroxyoxalate CL reaction was first applied to the determination of DNS-amino acids by HPLC.[66] Table III lists many fluorescers used in the HPLC–CL detection system.

Determination of Fluorescers

This chapter describes the detection of DNS- and NBD-amino acids by isocratic elution HPLC–CL as well as detection of DNS-amino acids by gradient elution HPLC–CL.

[98] W. R. Seitz, this series, Vol. 57, p. 445.
[99] H. Arakawa, M. Maeda, and A. Tsuji, *Clin. Chem. (Winston-Salem, N.C.)* **31,** 430 (1985).
[100] H. Arakawa, M. Maeda, and A. Tsuji, *Chem. Pharm. Bull.* **30,** 3036 (1982).
[101] K. Imai, K. Honda, and K. Miyaguchi, *in* "Bioluminescence and Chemiluminescence: Instrumentations and Applications" (K. Van Dyke, ed.), Vol. 2, p. 65. CRC Press, Boca Raton, Florida, 1985.
[102] K. Honda, K. Miyaguchi, and K. Imai, *Anal. Chim. Acta* **177,** 111 (1985).

TABLE III
HPLC–Peroxyoxalate CL Detection of Fluorescers

Fluorescers (detection limit)	HPLC condition	CL reactions and reagents[a]	Reference
DNS-amino acids (10 fmol)	Column μBondapak C_{18} (10 μm, 300 × 4 mm diameter) Eluent 50 mM Tris–HCl (pH 7.7)–acetonitrile (16:9, v/v) Flow rate. 0.18 ml/min	5 mM TCPO (ethyl acetate) Flow rate: 0.15 ml/min 0.5 M H_2O_2 (acetone) Flow rate: 1.2 ml/min	66
DNS-amino acids (0.5 fmol)	Column Nucleosil C_{18} (5 μm, 200 × 5 mm diameter) Eluent for DNS-amino acids 50 mM Tris–HCl (pH 7.7)–acetonitrile (35:65, v/v)	For DNS-amino acids 5 mM TCPO (ethyl acetate) Flow rate: 0.5 ml/min 0.5 M H_2O_2 (acetone) Flow rate: 1.3 ml/min	67
DNS-epinephrine (6 fmol) DNS-norepinephrine (16 fmol)	Eluent for DNS-catecholamines 50 mM Tris–HCl (pH 7.7)–acetonitrile (15:85, v/v) Flow rate: 1 ml/min	For DNS-catecholamines 5 mM TCPO (ethyl acetate) Flow rate: 0.3 ml/min 0.5 M H_2O_2 (acetone) Flow rate: 0.5 ml/min	
DNS-amino acids (0.28 fmol)	Column MBC-ODS-2 (10 μm, 250 × 1 mm diameter) Eluent 0.1 M imidazole (pH 7.0)–acetonitrile (7:3, v/v) Flow rate: 0.03 ml/min	1 mM TCPO (ethyl acetate)–0.1 M H_2O_2 (acetone) (1:3, v/v) Flow rate: 0.6 ml/min	68
DNS-amino acids (2–5 fmol)	Column TSK-GEL Type ODS-120A (5 μm, 150 × 4.6 mm diameter) Eluent A 0.1 M imidazole-HNO_3 (pH 7.0)–acetonitrile (90:10, v/v) Eluent B 0.1 M imidazole-HNO_3 (pH 7.0)–acetonitrile (55:45, v/v) Flow rate: 0.3 ml/min	1 mM TCPO (ethyl acetate)–0.1 M H_2O_2 (acetone) (1:3, v/v) Flow rate: 1.8 ml/min	69

TABLE III (*continued*)

Fluorescers (detection limit)	HPLC condition	CL reactions and reagents[a]	Reference
	Gradient program linear gradient from 30%B (70%A) to 99% (1%A) over 256 min		
DNS-amines (0.8–14 fmol) NBD-amines (19–270 fmol) OPA-amines (94–580 fmol)	Column Nucleosil C_{18} (5 μm, 200 × 5 mm diameter) Eluent for DNS- and NBD-amines 0.05 M Tris–HCl (pH 7.7)–acetonitrile (25 : 75, v/v) Eluent for OPA-amines 0.05 M Tris–HCl (pH 7.7)–acetonitrile (15 : 85, v/v) Flow rate: 1 ml/min	5 mM TCPO (ethyl acetate) Flow rate: 0.3 ml/min 0.5 M H_2O_2 (acetone) Flow rate: 0.5 ml/min	70
DNS-drug (1–10 pg)	Column Zorbax ODS (7 μm, 50 × 4.6 mm diameter) Eluent 2.5 mM imidazole buffer (pH 7.0)–THF 2 : 1, v/v) Flow rate: 1 ml/min	10 mM TCPO (ethyl acetate) or 10 mM DNPO (acetonitrile) Flow rate: 1 ml/min 1.5 M H_2O_2 (THF) Flow rate: 2 ml/min	71
Fluorescamine-labeled norepinephrine, dopamine (25 fmol)	Column TSK LS 160 (5 μm, 230 × 5 mm diameter) Column temperature: 60° Eluent 50 mM imidazole–HCl (pH 7.3) containing 0.16 mM EDTA–acetonitrile (4 : 1, v/v)	1 mM TCPO (ethyl acetate) Flow rate: 1 ml/min 0.8 M H_2O_2 (acetone) Flow rate: 2.2 ml/min	72
DNS-hydrazone of fluorocortin butyl (168 fmol)	Precolumn ODS-3 RAC II (5 μm, 100 × 4.6 mm diameter) Analytical column Partisil 5 ODS-3 (5 μm, 250 × 4.6 mm diameter) Column temperature: 30° Eluent 9.9 mM Tris–HNO_3 in H_2O–acetonitrile (1 : 4, v/v) (pH 7.35)	6.88 mM TCPO (ethyl acetate–isopropanol, 1 : 4, v/v) Flow rate: 0.6 ml/min 0.1 M H_2O_2 (2-propanol–ethyl acetate, 4 : 5, v/v) Flow rate: 1.2 ml/min Reaction temperature: 30°	73

(*continued*)

TABLE III (continued)

Fluorescers (detection limit)	HPLC condition	CL reactions and reagents[a]	Reference
Polycyclic aromatic hydrocarbons (0.77 pg–72 ng)	Column Resolve C_{18} (5 μm, 150 × 3.9 mm diameter) Eluent 75% acetonitrile–Tris–HNO_3 (pH 7.4) Flow rate: 0.5 ml/min	5 mM TCPO (acetate) Flow rate: 0.34 ml/min 0.8 M H_2O_2 (acetone) Flow rate: 0.3 ml/min	74
Polycyclic aromatic hydrocarbons (0.56–18 pmol)	Column Microsphere ODS (5 μm, 250 × 1 mm diameter) Eluent 5 mM phosphate buffer–methanol–ethyl acetate (25 : 45 : 30, v/v/v) Flow rate: 0.05 ml/min	4 mM TCPO (ethyl acetate) Flow rate: 0.6 ml/min 1.01 M H_2O_2 (methanol) Flow rate: 0.6 ml/min	76
Polycyclic aromatic amines (0.11 pg–120 ng)	Column Zorbax ODS (250 × 4.6 mm diameter) Eluent 2 mM Tris–HNO_3 (pH 7.4)–acetonitrile (20 : 80–40 : 60, v/v) Flow rate: 0.5–1.0 ml/min	13.4 mM TCPO (ethyl acetate) Flow rate: 0.6 ml/min 1.15 M H_2O_2 (ethyl acetate) Flow rate: 1.2 ml/min	75
Nitropolycyclic aromatic hydrocarbons (0.25–8.5 pg)	Column Zorbax ODS (250 × 4.6 mm diameter) Eluent 50 mM Tris–HCl (pH 6/5)–acetonitrile (20–30 : 80–70, v/v) Flow rate: 1 ml/min	Reduction column Zinc particle (40–80 μm diameter)–glass beads (~40 μm diameter) (1 : 1) (35 × 3.2 mm diameter) CL reagents 13.4 mM TCPO (ethyl acetate) Flow rate: 0.6 ml/min 1.15 M H_2O_2 (ethyl acetate) Flow rate: 1.2 ml/min	77
Aminopyrene (0.058 fmol)	Column μBondapak C_{18} Eluent 8.25 mM Tris–HNO_3 (pH 7.2–7.4)–acetonitrile (25 : 75, v/v) Flow rate: 1.0 ml/min	1.34 mM TCPO (ethyl acetate) Flow rate: 0.6 ml/min 1.0 M H_2O_2 (isopropanol–ethyl acetate, 100 : 125, v/v) Flow rate: 1.2 ml/min	80

[a] TCPO, bis(2,4,6-trichlorophenyl) oxalate; DNPO, bis(2,4-dinitrophenyl) oxalate.

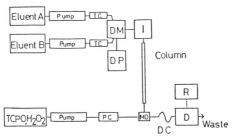

Fig. 2. Flow diagram for HPLC–peroxyoxalate CL detection system. TC, trapping column; DM, dynamic mixer; I, injector; DP, damper; CPO, bis(2,4,6-trichlorophenyl) oxalate; H_2O_2, hydrogen peroxide; PC, dammy column; MD, mixing device; DC, delayed coil; D, detector; R, recorder.

Reagents and Solvents

Because the sensitivity of the detection system is so high, trace amounts of contaminated fluorescers in the reagents and solvents result in a high background. Therefore, the purity of the reagents and solvents is very important to obtain a high signal-to-noise ratio. (2,4,6-Trichlorophenyl)oxalate (TCPO) was crystallized from ethyl acetate (mp 193–195°, lit. 190–192°,[103] which is commercially available (Wako Pure Chemical Industries, Co., Osaka, and Fluka AG, Buchs, Switzerland). Imidazole is crystallized from spectrograde ether [mp 88–89°, 90–91° (Merck Index)]. Water is first deionized, distilled, and purified by passing through a column of ODS. However, HPLC grade distilled water (commercially available) can be used. The purification of hydrogen peroxide and nitric acid is rather difficult and still not successful. Other HPLC-grade solvents, such as acetonitrile, acetone, and ethyl acetate, are used without further purification.

Instrumentation

A flow diagram for the HPLC–CL system is shown in Fig. 2. The fluorescers separated on the columns are mixed in a rotating flow mixing device with the CL reagent solution. The mixture is placed in a spiral type flow cell generating CL. Chemiluminescence is detected by a photomultiplier tube (PMT) and the CL intensity is recorded.

The ODS trapping columns are necessary for removing fluorescent impurities present in the eluting solvents, especially in case of gradient elution HPLC in order to suppress the baseline shift.[69]

The pumps used for delivering column eluent and the reagent solutions are the usual reciprocating type with concomitant use of a damper and a

[103] A. G. Mohan and N. J. Turro, *J. Chem. Educ.* **51**, 528 (1974).

dammy column for back pressure. They should be pulseless, i.e., a syringe-type or air pressure-type pump.[67,70] Originally, two pumps, one for aryl oxalate (TCPO in the usual case) and the other for hydrogen peroxide solution, were used for the CL reagent solution.[66] However, the mixed solution of TCPO and hydrogen peroxide in ethyl acetate and acetone (1:3, v/v) was found to retain 90% of its original activity after 4 hr storage at room temperature. So, the system shown in Fig. 2 has one pump for the CL reagent solution.

An instantaneous mixing of the two solutions is important for obtaining a stable baseline since the CL reaction occurs instantaneously and the intensity declines very rapidly. For thorough mixing in the flow stream a rotating mixing device has been developed (Fig. 3). The column effluent and the CL reagent solution go into the bottom of the vessel from the opposite sides and are mixed well by rotation. As the mixture rises, it is removed from the top of the vessel. The diameters of the inlets are important to control the flow stream into the vessel of the two solutions to be mixed. A stream flow ratio from 1:4 to 1:10 is preferred for thorough mixing. A delay tube attached to the device is necessary to maintain the appropriate reaction time. Usually 5–40 cm by 0.25 mm diameter PTFE tubing is used.

A spiral type flow cell is suitable for collecting the generated CL effectively. In this experiment a flow cell with a 120 μl volume is made from glass tube (0.6 mm diameter). Ideally, a flow cell made by hollowing out a groove on a stainless-steel plate with two holes in each end and covering the groove with a glass would be preferrable for convenient handling.

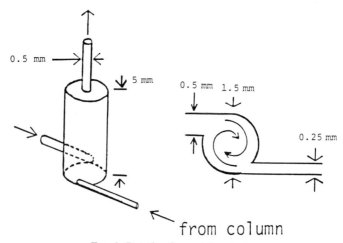

FIG. 3. Rotating flow mixing device.

A PMT or photon counter[67] is good for detecting chemiluminescence. However, because of the high background derived from the fluorescent impurities in the solvents and/or reagents a PMT was used in this experiment.

Isocratic Elution HPLC

Examples of an isocratic elution of DNS-amino acids and NBD-amino acids on a semimicrobore column (2.1 mm diameter) HPLC are now described.

Eluent: 0.1 M imidazole nitrate buffer (pH 7.0):acetonitrile (8:2, v/v) for separation of DNS-amino acids; 0.1 M imidazole nitrate buffer (pH 7.0):acetonitrile (9:1, v/v) for NBD-amino acids

Pump for eluent: model 302 (Gilson Co. Villiers-Le-Bell, France)

Trapping column for impurities: preparative C_{18} (55–105 μm, 20 × 3.9 mm diameter, Waters Assoc., Milwaukee, WI)

Analytical column: Hypersil ODS (5 μm, 150 × 2.1 mm diameter, Shandon Co., U.K.)

Eluent flow rate: 0.15 ml/min

CL reagent solution: 0.1 M hydrogen peroxide in acetone:1 mM TCPO in ethyl acetate (3:1, v/v)

Pump for CL reagent: Shimadzu LC 5A (Shimadzu Seisakusho Co., Tokyo, Japan)

Dammy columns for back pressure: a TSK LS 120T and a TSK LS 120A (ODS column, 250 × 4.6 mm diameter, Toyo Soda Co., Tokyo, Japan)

Reagent flow rate: 0.60 ml/min

Rotating flow mixing device: 8.8 μl inner volume

Delay tube: 50 × 0.25 mm diameter stainless-steel tubing

CL reaction time before the CL monitor: about 2.0 sec

CL monitor: Luminomonitor I (Atto Co., Tokyo, Japan)

CL monitor range: ×100

Recorder range: 20 mV FS

Then 0.5 μl of 80 nM DNS-Gly and DNS-Pro and 1.6 μM NBD-Gly and NBD-Pro in the column eluent were injected, separated, and detected. Figure 4a and b shows the chromatograms of 40 fmol DNS-GLY and DNS-Pro and 800 fmol NBD-Gly and NBD-Pro, respectively. Detection limits for DNS derivatives are around a few femtomoles and for NBD derivatives 50 fmol.

Gradient Elution HPLC

Gradient elution of 18 DNS-amino acids is described.

Column: Hypersil ODS (5 μm, 150 × 2.1 mm diameter)

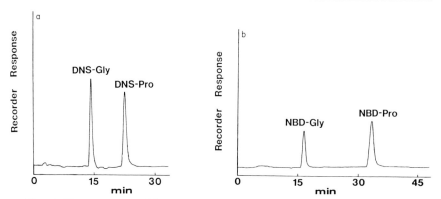

FIG. 4. Chromatograms of fluorescers detected by peroxyoxalate CL detection system (isocratic elution). (a) DNS-amino acids (40 fmol each); (b) NBD-amino acids (800 fmol each). Separation and detection conditions are described in the text.

Eluent A: 0.1 M imidazole nitrate (pH 7.0)–acetonitrile (9:1, v/v)
Eluent B: 0.1 M imidazole nitrate (pH 7.0)–acetonitrile (55:45, v/v)
Pumps for eluents: pumps of model 302 and 303 (Gilson Co.)
Gradient program: gradient elution from A:B (8:2) to A:B (2:8) over 100 min (System Controller MS-4, Gilson Co., Villiers-Le-Bell, France)
Eluent flow rate: 0.15 ml/min

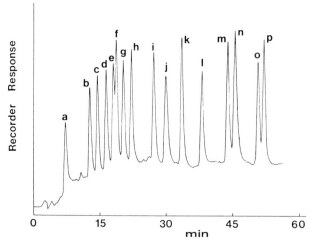

FIG. 5. Chromatogram of DNS-amino acids (gradient elution). (a) Asp, (b) Asn, (c) Gln, (d) Ser, (e) Arg, (f) Thr, (g) Gly, (h) Ala, (i) Pro, (j) Lys, (k) Val, (l) Met, (m) Ile, (n) Leu, (o) Trp, (p) Phe (50 fmol each). Separation and detection conditions are described in the test.

The other conditions are the same as in the isocratic elution. Figure 5 shows a chromatogram of 18 DNS-amino acids. As shown in Fig. 5, impurities of unknown origin result in an elevated baseline and they alter the sensitivity increment. More detailed investigations on the cause of the impurities will be necessary for the attomol detection of fluorescers.

In a similar manner, the other fluorescers depicted in Table III can be determined sensitively.

Comments

The HPLC-CL detection system can be routinely used as a sensitive and selective method as the instrumentation for the detection systems (such as pulseless pumps, luminescence detectors with flow cells, or special parts for CL generation system[78,79,82,85] becomes more readily available.

[36] Phagocytic Leukocyte Oxygenation Activities and Chemiluminescence: A Kinetic Approach to Analysis

By ROBERT C. ALLEN

Identification of infecting microbes by the immune system is the first in a series of steps leading to phagocyte microbicidal action. From the primative alternative pathway complement cascade to the highly evolved antigen-specific immunoglobulins, humoral immune mechanisms provide information linking the target microbe to killer phagocytes such as the polymorphonuclear neutrophil (PMNL) and eosinophil leukocytes, blood monocytes, and monocyte-derived tissue macrophages. Recognition of the opsonified microbe results in phagocytosis, activation of the respiratory burst metabolism, and generation of oxygenating agents.[1-5] As demonstrated in chronic granulomatous disease, a condition in which the patient's PMNL and monocytes can phagocytose microbes but are unable

[1] A. J. Sbarra and M. L. Karnovsky, J. Biol. Chem. 234, 1355 (1960).
[2] F. Rossi, D. Romeo, and P. Patriarca, Res, J. Reticuloendothel. Soc. 12, 127 (1972).
[3] J. A. Badwey and M. L. Karnovsky, Annu. Rev. Biochem. 49, 695 (1980).
[4] R. C. Allen, in "The Reticuloendothelial System. II. Biochemistry and Metabolism" (A. J. Sbarra and R. Strauss, eds.), p. 309. Plenum, New York, 1980.
[5] B. M. Babior, in "The Reticuloendothelial System. II. Biochemistry and Metabolism" (A. J. Sbarra and R. Strauss, eds.), p. 339. Plenum, New York, 1980.

to activate redox metabolism, oxygenation activity is necessary for effective microbicidal action.[6,7]

Chemiluminescence (CL) is an energy product of phagocyte oxygenation activity.[8–10] Measurement of this luminescence is nondestructive and allows continuous monitoring of phagocyte oxygenation activity. As an end product of the microbe–humoral–phagocyte interaction, phagocyte luminescence may also be used for assessing the functional interrelationships of the various reacting components. For example, when stimulus is not limiting, CL is proportional to the functional capacity of the phagocytes present. Likewise, if the number of phagocytes is not limiting, microbe opsonification kinetics can be studied as the rate of stimulated phagocyte oxygenation activity yielding CL. Thus the CL approach can provide an *in vitro* model system for analysis of the humoral immune mechanisms.[11–15]

Both native CL and chemiluminigenic probe-dependent CL techniques are successfully employed for assessment of PMNL and monocyte–macrophage oxygenation activities in a number of laboratories.[16,17] Unfortunately, a major hindrance to the more widespread use of the CL approach is the perception that it is exotic, difficult to measure, and impossible to understand. Hopefully, the following description of methodology and discussion with illustrative examples will help dispel these notions and serve as evidence for the validity and utility of luminescence testing.

Oxygen, Biology, and Hund's Maximum Multiplicity Rule

In the interest of semantic clarity, oxidation is here defined as the removal of one or more electrons or reducing equivalents from a substrate atom or molecule, i.e., dehydrogenation. The combination of oxygen with

[6] E. L. Mills and P. G. Quie, *Rev. Infect. Dis.* **2**, 505 (1980).
[7] R. C. Allen, R. L. Stjernholm, M. A. Reed, T. B. Harper, III, S. Gupta, R. H. Steele, and W. W. Waring, *J. Infect. Dis.* **136**, 510 (1977).
[8] R. C. Allen, R. L. Stjernholm, and R. H. Steele, *Biochem. Biophys. Res. Commun.* **47**, 679 (1972).
[9] M. A. Trush, M. E. Wilson, and K. Van Dyke, this series, Vol. 57, p. 462.
[10] R. C. Allen, *Front. Biol.* **48**, 197 (1979).
[11] V. G. Hemming, R. T. Hall, P. H. Rhodes, A. O. Shigeoka, and H. R. Hill, *J. Clin. Invest.* **58**, 1379 (1976).
[12] R. C. Allen, *Infect. Immun.* **15**, 828 (1977).
[13] P. Stevens and L. S. Young, *Infect. Immun.* **16**, 796 (1977).
[14] R. C. Allen and M. M. Lieberman, *Infect. Immun.* **45**, 475 (1984).
[15] C. S. Via, R. C. Allen, and R. C. Welton, *J. Rheumatol.* **11**, 745 (1984).
[16] P. G. Quie, E. L. Mills, and B. Holmes, *Prog. Hematol.* **10**, 193 (1977).
[17] M. H. Grieco and D. K. Meriney, "Immunodiagnosis for Clinicians: Interpretation of Immunoassays." Year Book Med. Publ., Chicago, Illinois, 1983.

a substrate can be defined as either oxidation or oxygenation. Even when there is no absolute loss of electrons from the substrate, the large electronegativity of oxygen predicts a relative dehydrogenation of the substrate oxygenated. Oxidation is the general term; oxygenation is more specific. Oxygenations are oxidations, but not all oxidations are oxygenations. During metabolism, glucose is oxidized to CO_2 and water in a series of dehydrogenations as it passes through the glycolytic and tricarboxylic acid pathways. The product reducing equivalents enter the mitochondrial electron transport chain where ultimate reaction with O_2 is catalyzed by cytochrome oxidase, the terminal enzyme of that redox chain. When sufficient activation energy is available, glucose will burn in the presence of O_2. The net reaction as well as the change in standard free energy is the same for either burning or metabolism. However, in metabolism, glucose undergoes piecemeal dehydrogenation; it is oxidized but not directly oxygenated. In burning, glucose reacts directly with O_2.

The biomedical literature is not without misinformation related to the reactivity of free radicals in general, or to the chemistry of the superoxide anion in particular. It is therefore prudent to consider which positions and propositions are valid and which are not. As a first consideration, biological molecules are with few exceptions nonradical and of singlet spin multiplicity. The few exceptions include the flavoproteins and metalloproteins that participate in one electron redox metabolism. In obedience to Hund's rule of maximum multiplicity,[18,19] the oxygen we breathe is a paramagnetic diradical molecule of triplet multiplicity ($^3\Sigma_g^-$ O_2). Likewise, the superoxide anion is a paramagnetic radical but of doublet multiplicity ($\cdot O_2^-$). Although reactions of O_2 and $\cdot O_2^-$ with biological substrates are thermodynamically favored, living and nonliving matter, both composed of singlet multiplicity substrate, continue to survive in an atmosphere of 21% O_2.

There are two distinct aspects to any reaction. The first is related to the direction and magnitude of the chemical change. This aspect is defined by the equilibrium constant, and as such reflects the thermodynamic feasibility of the reaction occurring. The second aspect is related to the rate of reaction. If the change in free energy is negative, reaction is favored, but the rate at which a possible reaction occurs cannot be determined from thermodynamic parameters alone.

Factors influencing rate may be considered with respect to the empirically derived Arrhenius equation:

$$k = Ae^{-E_a/RT} \qquad (1)$$

---

[18] F. Hund, *Z. Phys.* **63**, 719 (1930).
[19] G. Herzberg, "Molecular Spectra and Molecular Structure. I. Spectra of Diatomic Molecules." Van Nostrand-Reinhold, Princeton, New Jersey, 1965.

where k is the rate constant, A is the "frequency factor," and in the exponential term, E_a is the energy of activation, R is the gas constant, and T is the absolute temperature.[20]

Rate theories based upon collision and statistical mechanics have been advanced to more completely explain this empirical relationship.[21,22] According to the collision theory,

$$k = ZPe^{-E_a/RT} \qquad (2)$$

A is the product of Z, the "collision number," times P, the "probability" or "steric" factor. The latter term is a correction factor and as such provides a measure of excursion from the rate of reaction expected based on collision alone.

The more rigorous theory of absolute reaction rates is developed from consideration of the equilibrium relationship between the reactants and the activated complex expressed in terms of partition functions.[21] Accordingly the equation of rate becomes

$$k = \kappa(kT/h)K^* \qquad (3)$$

where κ is the transmission coefficient, the term (kT/h) is equivalent to the effective rate of crossing the energy barrier to the activated complex, and K^* is the partition function equilibrium constant. In the constant (kT/h), k is the Boltzmann entropy constant, T is the absolute temperature, and h is Planck's constant, and, thus, (kT/h) equals approximately 10^{12} sec^{-1} at 25°.

The reaction rate of organic substrate with either O_2 or $\cdot O_2^-$ is very much lower than predicted by collision alone. Thus, the value of the steric factor P must be very small. This might be explained by difficulty in orbital overlap between radical and nonradical reactants.[23] The low reactivity of a singlet substrate with either triplet O_2 and doublet $\cdot O_2^-$ can also be considered with respect to the preservation of state symmetry.[24,25] The unfavorable nature of such singlet–triplet and singlet–doublet reactions is reflected in very low values for the transmission coefficient, as described in Eq. (3), and proportionately low reaction rates.

[20] K. J. Laidler, "Chemical Kinetics." McGraw-Hill, New York, 1950.
[21] S. Glasstone, K. J. Laidler, and H. Eyring, "The Theory of Rate Processes." McGraw-Hill, New York, 1941.
[22] H. Eyring, S. H. Lin, and S. M. Lin, "Basic Chemical Kinetics." Wiley, New York, 1980.
[23] W. H. Koppenol and J. Butler, *FEBS Lett.* **83**, 1 (1977).
[24] E. Wigner and E. E. Witmer, *Z. Phys.* **51**, 859 (1928).
[25] L. Salem, "Electrons in Chemical Reactions: First Principles." Wiley (Interscience), New York, 1982.

Radicals tend to react with other radicals. The probability for overlap between the partially filled orbitals of radical reactants favors reaction. With regard to collision theory, the value of the "probability" factor for such reactions approximates unity. In terms of statistical mechanics, doublet–doublet annihilation reactions yielding singlet products are spin allowed, and the transmission coefficient approximates unity. Consider the disproportionation of $\cdot O_2^-$. As the pH approaches 4.8, the pK_a of hydrodioxylic acid ($\cdot O_2H$), anion–anion repulsion is minimized and reaction rate is maximized.[26] The multiplicity, J, of a molecule is equal to the absolute value of the product of the sum of electron spins, S, multiplied by two, plus one, i.e., $J = |2S| + 1$. Therefore, $\cdot O_2^-$ will have doublet multiplicity, i.e., $|2(\frac{1}{2})| + 1 = 2$. The Wigner–Witmer correlation rules predict that a doublet–doublet annihilation redox reaction such as

$$\cdot O_2^- + \cdot O_2H \rightarrow O_2H^- + {}^1O_2 \tag{4}$$

can have a potential energy surface of singlet, $|2(1/2 - 1/2)| + 1 = 1$, or triplet, $|2(1/2 + 1/2)| + 1 = 3$, multiplicity. Since the reaction requires overlap of the partially filled orbitals of the reacting radicals, the singlet surface will be favored, and the products O_2H^- and O_2 will be of singlet multiplicity.[27,28] However, if disproportionation involves intermediate reaction with a metal or metalloprotein, such as the reaction catalyzed by superoxide dismutase, the product molecules can be of mixed multiplicity.

The technique of cytochrome c reduction commonly used for measurement of $\cdot O_2^-$ [29] can also be considered with regard to the spin conservation rules. In the reaction

$$\text{Cyt } c(Fe^{3+}) + \cdot O_2^- \rightarrow \text{Cyt } c(Fe^{2+}) + O_2 \tag{5}$$

ferric iron of quartet multiplicity, $|2(3/2)| + 1$, reacts with doublet multiplicity, $|2(1/2)| + 1$, $\cdot O_2^-$ to yield ferrous iron of quintet multiplicity, $|2(3/2 + 1/2)| + 1$, and triplet, $|2(3/2 - 1/2)| + 1$, molecular oxygen.

The maximum multiplicity and spin correlation rules explain why biological systems are protected against O_2. On the other hand, the phagocytes are capable of realizing and directing the thermodynamic potential of O_2 for microbicidal action. This feat is accomplished through the enzyme-directed redox reactions of respiratory burst metabolism. As will be fully considered, phagocyte redox metabolism provides multiple reaction

[26] D. Behar, G. Czapski, J. Rabini, L. M. Dorfman, and H. A. Schwarz, *J. Phys. Chem.* **74**, 3209 (1970).

[27] J. Stauff, H. Schmidkeenz, and G. Hartmann, *Nature (London)* **198**, 281 (1963).

[28] A. U. Khan, *Science* **168**, 476 (1970).

[29] J. M. McCord and I. Fridovich, *J. Biol. Chem.* **244**, 6049 (1969).

pathways and mechanisms to overcome Hund's multiplicity rule. Phagocytosis is associated with activation of NAD(P)H oxidase.[1,2] This
NADPH : O_2 oxidoreductase catalyzes the univalent reduction

$$O_2 + H^+ + e^- \rightarrow \cdot O_2H \underset{}{\overset{pK_a = 4.8}{\rightleftharpoons}} \cdot O_2^- + H^+ \tag{6}$$

in which oxygen is converted from a triplet diradical (O_2) to a doublet
radical ($\cdot O_2^-$).[4,5] The reaction may also be responsible for acidification of
the phagolysosome to a pH of approximately 4.8, the pK_a of $\cdot O_2H$.[10] At
pH 4.8, disproportionation, as described in Eq. (4), will be maximal. The
product H_2O_2 serves as substrate for myeloperoxidase (MPO) of PMNL
and monocytes,[30–32] as well as eosinophil peroxidase (EPO).[33] The initial
step is a haloperoxidation,[34,35]

$$H_2O_2 + Cl^- \rightarrow H_2O + OCl^- \tag{7}$$

yielding singlet multiplicity hypohalite. The halide can be Cl^-, Br^-, or I^-,
but Cl^- is of physiologic significance. In the acid environment of the
phagolysosome, generation of chlorine is also possible via the reaction[36]

$$HOCl + H^+ + Cl^- \rightleftharpoons H_2O + Cl_2 \tag{8}$$

The products of halide oxidation can react with additional peroxide in the
reaction

$$OCl^- + H_2O_2 \rightarrow {}^1O_2 + Cl^- + H_2O \tag{9}$$

As predicted by the Wigner–Witmer correlation rules, these reactions
proceed via the singlet surface and all of the products are of singlet multiplicity.[34,37–40] Thus, the MPO and EPO systems provide a mechanism for
overcoming Hund's multiplicity barrier.

The ${}^1O_2$ generated in Eqs. (4) and (9) can react directly with singlet
multiplicity organic molecules,[41,42]

[30] S. J. Klebanoff and R. C. Clark, "The Neutrophil: Function and Clinical Disorders."
North-Holland Publ., Amsterdam, 1978.
[31] J. M. Zgliczynski, in "The Reticuloendothelial System. II. Biochemistry and Metabolism" (A. J. Sbarra and R. Strauss, eds.), p. 255. Plenum, New York, 1980.
[32] H. Rosen and S. J. Klebanoff, J. Biol. Chem. 252, 4803 (1977).
[33] A. J. Bos, R. Wever, M. N. Hamers, and D. Roos, Infect. Immun. 32, 427 (1981).
[34] R. C. Allen, Biochem. Biophys. Res. Commun. 63, 675 (1975).
[35] R. C. Allen, Biochem. Biophys. Res. Commun. 63, 684 (1975).
[36] J. E. Harrison and J. Schultz, J. Biol. Chem. 251, 1371 (1976).
[37] M. Kasha and A. U. Khan, Ann. N.Y. Acad. Sci. 171, 5 (1970).
[38] A. U. Khan and D. E. Kearns, Adv. Chem. 77, 143 (1968).
[39] J. R. Kanofsky, J. Biol. Chem. 258, 5991 (1983).
[40] A. U. Khan, Biochem. Biophys. Res. Commun. 122, 668 (1984).
[41] C. S. Foote and S. Wexler, J. Am. Chem. Soc. 86, 3879 (1964).
[42] E. J. Corey and W. C. Taylor, J. Am. Chem. Soc. 86, 3882 (1964).

$$^1O_2 + Sub \rightarrow Sub-O_2 \tag{10}$$

yielding dioxygenated products such as dioxetanes and dioxetanones.[43,44] As an alternative possibility, hypochlorite or chlorine may directly chlorinate biological substrates,[31,36]

$$HOCl\ (Cl_2) + Sub \rightarrow Sub-Cl + H_2O\ (HCl) \tag{11}$$

and the chlorinated product may then react with peroxide,

$$Sub-Cl + H_2O_2 \rightarrow Sub-O_2 + H^+ + Cl^- \tag{12}$$

to yield the dioxetane intermediate, and ultimately the excited carbonyl product.[45,46] Note that the reactants and products of Eqs. (7) through (12) are all of singlet multiplicity, but that $^1O_2$ is not directly involved as an intermediate in dioxygenation by Eqs. (11)–(12) pathway.

The above mechanisms do not rule out a more direct reactive role for radicals in microbe killing. Several investigators have suggested hydroxyl radical (·OH) as a phagocyte-generated microbicidal agent.[47,48] They propose that ·OH is generated in the Haber–Weiss reaction[49]

$$H^+ + \cdot O_2^- + H_2O_2 \rightarrow \cdot OH + H_2O + O_2 \tag{13a}$$
$$\cdot OH + H_2O_2 \rightarrow H_2O + \cdot O_2^- + H^+ \tag{13b}$$

The previously discussed difficulty of radical–nonradical orbital overlap should exert a rate-limiting effect. In fact, Eq. (13a) cannot be demonstrated in well-defined reacting systems over a pH range of 6.0 to 10.6.[50,51] However, reaction rate can be significantly enhanced by the multiplicity-manipulating effect of metals catalysts.

Radical reactions, such as the Haber–Weiss, are also subject to termination via radical–radical annihilation,

$$\cdot O_2H + \cdot OH \rightarrow H_2O + {}^1O_2 \tag{14}$$

which should proceed via the favorable singlet, $|2(1/2 - 1/2)| + 1$, surface to yield H_2O and $^1O_2$ as products.[52,53] This reaction is essentially diffusion controlled with a rate constant of $10^{10}\ M^{-1}\ sec^{-1}$.[54]

[43] F. McCapra, *J. Chem. Soc., Chem. Commun.* p. 155 (1968).
[44] K. R. Kopecky, *in* "Chemical and Biological Generation of Excited States" (W. Adam and G. Cilento, eds.), p. 85. Academic Press, New York, 1982.
[45] E. A. Chandross, *Tetrahedron Lett.* p. 761 (1963).
[46] W. H. Richardson and V. F. Hodge, *J. Am. Chem. Soc.* **93**, 3996 (1971).
[47] S. J. Weiss, G. W. King, and A. F. LoBuglio, *J. Clin. Invest.* **60**, 370 (1977).
[48] A. I. Tauber and B. M. Babior, *J. Clin. Invest.* **60**, 374 (1977).
[49] F. Haber and J. Weiss, *Proc. R. Soc. London, Ser. A* **147**, 332 (1934).
[50] G. J. McClune and J. A. Fee, *FEBS Lett.* **67**, 294 (1976).
[51] B. B. Halliwell, *FEBS Lett.* **72**, 8 (1976).
[52] R. M. Arneson, *Arch. Biochem. Biophys.* **136**, 352 (1970).
[53] E. W. Kellogg and I. Fridovich, *J. Biol. Chem.* **250**, 8812 (1975).
[54] K. Sehested, O. L. Rasmussen, and H. Fricke, *J. Phys. Chem.* **72**, 626 (1968).

The abundant generation of H_2O_2 and the ubiquitous presence of metal suggest that at least some ·OH might be generated by the phagocyte. The great electronegativity of this oxidized water appears sufficient to overcome orbital overlap limitations, and, as such, the reaction

$$\cdot OH + Sub \rightarrow \cdot Sub^+ + OH^- \tag{15}$$

may proceed by a doublet surface to yield a doublet multiplicity product. In the absence of limiting conditions such as charge repulsion, radicals react readily with other radicals. Thus, the variety of reactions possible will depend on the type and quantity of radical available. For example, reaction of substrate radicals

$$\cdot Sub^+ + \cdot Sub \rightarrow Sub-Sub^+ \tag{16}$$

would yield a covalently crosslinked product. Whereas, the anion–cation annihilation reaction with superoxide

$$\cdot Sub^+ + \cdot O_2^- \rightarrow Sub-O_2 \tag{17}$$

would yield a dioxygenated product.[55]

The ultimate product is the same for the reactions described by Eqs. (14) + (10) and for the reactions described by Eqs. (15) + (17). Substrate dioxygenation, by whatever combination of mechanisms, is in compliance with the fundamental rules of chemistry. First, the energy available must be sufficient to override Hund's multiplicity barrier. With regard to molecular oxygen, the difference between the triplet ground state and the lowest singlet excited state is ~23 kcal/mol.[37] Second, total state symmetry must be conserved.[19,24,25]

Chemiluminigenic Probing of Phagocyte Oxygenation Activities

The method of initial rates[56] can be modified and applied to kinetic analysis of stimulated phagocyte oxygenation activity.[14] Phagocytes respond to stimuli by activating "respiratory burst" metabolism

$$Phagocyte\ (resting) + stimulus \rightarrow phagocyte\ (activated) \tag{18}$$

This stimulated redox metabolism is required for the generation of oxygenating agents

$$Phagocyte\ (activated) \rightarrow oxygenating\ agents \tag{19}$$

[55] R. C. Allen, in "Chemical and Biological Generation of Excited States" (W. Adam and G. Cilento, eds.), p. 309. Academic Press, New York, 1982.
[56] I. H. Segel, "Enzyme Kinetics." Wiley (Interscience), New York, 1975.

which are directed to microbe killing. At least a portion of the resulting substrate oxygenations yield electronically excited products

$$\text{Oxygenating agents} + \text{Sub} \rightarrow \text{Sub}-O_2^* \rightarrow h\nu \qquad (20)$$

and a portion of these excited molecules relax by photon emission.[8,10] Luminescence is an energy product of the microbe–humoral–phagocyte interaction,[14] and, as such, can be described by the rate equation

$$v = dh\nu/dt = k'[\text{Phagocyte}]^p[\text{Stimulus}]^s[\text{Sub}]^c \qquad (21)$$

where v is the luminescence velocity ($dh\nu/dt$) expressed as photons per minute, k' is the proportionality constant, the exponential p is the reaction order with respect to the type and concentration of phagocyte, s is the order with respect to the type and concentration of stimulus, and c is the order with respect to the type and quantity of substrate available for oxygenation. In the sections that follow, the neutrophil leukocyte (PMNL) is employed as phagocyte. The stimuli include complement-opsonified zymosan (OZ) and the chemical agent phorbol myristate acetate (PMA).

In the native system of phagocyte luminescence the substrate presented for oxygenation is for the most part defined by the microbe phagocytosed. Since the quantum yield, i.e., the ratio of photons to oxygenation events, depends on the nature of the substrate oxygenated, native phagocyte CL is dependent on the type of microbe phagocytosed. Thus, control of substrate presents a major difficulty with regard to quantifying the relationship described by Eq. (21). Phagocytes function by killing microbes, and the accompanying luminescence is a product of this activity. As such, the quantum yield of these oxygenations is very poor in comparison with the high efficiency of bioluminescence systems. However, through the introduction of substrates susceptible to oxygenation and of high chemiluminescence yield, it is possible to mimic the high efficiency, and, to some extent, the specificity of bioluminescence systems.[55] The chemiluminigenic probes (CLPs), luminol and lucigenin (DBA) have similar CL quantum yields but very different reactivities. Thus, selective use of CLPs provides increased sensitivity as well as specificity with regard to measurement of oxygenation activity.

Preparation of Buffered Media, Stimuli, and Chemiluminigenic Probes

After 15 years of trial and error, I have not found an ideal buffered medium for phagocyte luminescence studies; however, a large number of media can and have been successfully used. Best results are obtained using isotonic media containing physiologic concentrations of calcium and magnesium, D-glucose at approximately 100 mg/dl, and at least a trace phosphate. Cell function and viability, as well as reproducibility of the luminescence response, are greatly improved by adding at least 100 mg/dl high purity, immunoglobulin-free albumin or gelatin; immunoglobulins, especially if denatured, can activate phagocytes and should be avoided. The medium should also be free of indicator dyes and practically colorless to visible light.

Preparation of Phosphate-Buffered Saline with Glucose and Albumin. Phosphate-buffered saline with D-glucose and albumin (PBSGA) is prepared by dissolving 8.0 g NaCl, 0.2 g KCl, 0.62 g KH_2PO_4, and 1.14 g Na_2HPO_4 in 800 ml water. After the salts are in solution, 10 ml of 10% (w/v) D-glucose and 20 ml of 5% (w/v) albumin are added; the pH is adjusted to 7.3, and the volume is adjusted to 1 liter. The medium is then filtered to sterility and kept refrigerated (4°) until used. Where indicated sodium ethylenediaminetetraacetate (Na_2EDTA) is also added to a final concentration of 50 μM.

Preparation of Complete Veronal Buffer. Complete veronal buffer (CVB) is prepared by dissolving 7.6 g NaCl, 0.33 g KCl, and 1.0 g sodium 5,5-diethylbarbiturate (veronal) to 800 ml water. With constant mixing 5.6 ml 1.0 N HCl, 5 ml 0.1 M $MgCl_2$, 10 ml 10% (w/v) D-glucose, 20 ml 5% (w/v) albumin, and finally 15 ml 0.03 M $CaCl_2$ are added. The medium is adjusted to a pH of 7.3, and the volume is adjusted to 1 liter. The medium is then filtered to sterility and kept refrigerated until used. This medium is similar to the veronal buffer routinely employed for studies of the complement system.[57]

Preparation of Opsonified Zymosan. Zymosan A, a preparation of *Saccharomyces cerevisiae* cell wall,[57] is suspended in normal (0.85% w/v) saline to a concentration of 250 mg/dl, and heated in a boiling water bath for 20 min. After cooling to 22°, the suspension is centrifuged at a relative centrifugal force (RCF) of 300 for 10 min, and the supernatant is discarded. Zymosan is then opsonified by resuspending the pellet in 200 ml fresh-frozen (−70°) pooled sera. Following gentle rotation at 22° for 20 min, the suspension is again centrifuged as described above. The supernatant is discarded, an additional 200 ml of sera is added to the pellet, and the incubation and centrifugation steps are repeated. The pellet of opsonified zymosan is then washed with 500 ml of normal saline, centrifuged as described above, and the supernatant discarded. This saline wash is repeated twice in order to remove residual protease activity. The washed OZ is then adjusted to the original concentration, aliquoted to storage

[57] M. M. Mayer, *in* "Experimental Immunochemistry" (E. A. Kabot, ed.), p. 133. Thomas, Springfield. Illinois, 1971.

tubes, and frozen at $-70°$ until used. This OZ suspension contains 600 ± 200 zymosan particles per microliter.

Preparation of Phorbol Myristate Acetate. PMA, a cocarcinogen extracted from croton oil, is known to cause specific degranulation of PMNL and activation of redox metabolism.[58–60] A 5 mM stock solution of phorbol 12-myristate 13-acetate (PMA) is prepared in spectral grade dimethyl sulfoxide (DMSO). This stock solution is further diluted with water to attain the desired concentration of PMA used for stimulation.

Preparation of Chemiluminigenic Probes. Two different CLPs are employed in the experiments presented. Luminol (5-amino-2,3-dihydro-1,4-phthalazinedione), a cyclic hydrazide with a reported quantum yield of 0.01, is prepared as a 20 mM stock concentration in DMSO. The stock is kept refrigerated in the dark until diluted with water or buffer for testing. The concentration of aqueous luminol is assayed spectrophotometrically based on a millimolar extinction coefficient of 7.63 at 347 nm.[61]

Lucigenin (10,10'-dimethyl-9,9'-biacridinium dinitrate, DBA) is a water-soluble acridinium salt with a quantum yield comparable to that of luminol. A 5 mM solution is prepared in water and kept refrigerated in the dark. The concentration of the working solution is assayed spectrophotometrically based on millimolar extinctions of 37.3 and 9.65 at 369 and 430 nm, respectively.[62] Lucigenin has chemical characteristics in common with the viologens, and as such, caution should be exercised when preparing and using this CLP.

Preparation of Polymorphonuclear Neutrophil Leukocytes

Isolation of Leukocytes from Whole Blood. Blood, obtained from healthy volunteers, is collected in 7 ml capacity sterile evacuated tubes containing 10 mg K_3EDTA and 1 ml 6% (w/v) dextran (70,000 MW). Heparin may also be employed as the anticoagulant. EDTA is suggested as anticoagulant when complement activation is to be avoided. After gentle mixing the erythrocytes are allowed to settle at $37°$ for 1 hr. The leukocyte-rich plasma is gently removed and centrifuged at an RCF of 200 for 10 min. The plasma is discarded and the leukocyte pellet is resuspended in PBSGA. The centrifugation step is repeated and the cells resus-

[58] J. G. White and R. D. Esensen, *Am. J. Pathol.* **75**, 45 (1974).
[59] J. E. Repine, J. G. White, C. C. Clawson, and B. M. Holmes, *J. Lab. Clin. Med.* **83**, 911 (1974).
[60] L. R. DeChatelet, P. S. Shirley, and R. B. Johnston, Jr., *Blood* **47**, 545 (1976).
[61] J. Lee and H. H. Seliger, *Photochem. Photobiol.* **15**, 227 (1972).
[62] J. R. Totter, *Photochem. Photobiol.* **3**, 231 (1964).

pended in 1 ml of PBSGA. If erythrocyte contamination is heavy, hypotonic lysis with 0.3% (w/v) saline can be employed at this stage. Use of ammonium chloride for erythrocyte lysis is not recommended.

Purification of PMNL by Gradient Centrifugation. The leukocyte fraction containing PMNL can be further purified by isopyknic focusing on a continuous gradient of colloidal poly(vinylpyrrolidone)-coated silica (Percoll, Pharmacia Chemicals). The gradient is prepared by adding 19 ml of 10× concentrated PBSGA, 1.4 mg Na$_2$EDTA, and 50 ml of 0.02 N HCl to 130 ml of stock isoosmotic Percoll. After mixing, the pH is adjusted to 7.2, 20 ml of the gradient material is added per centrifuge tube, and the tubes are kept refrigerated until used. After venipuncture while the whole blood is settling, two gradient tubes are centrifuged at an RCF of 20,000 for 20 min to generate the continuous gradient.

The leukocyte suspension (1 ml) is then applied to the top of the continuous gradient of one tube, and 1 ml of mixed density marker beads (Pharmacia) suspended in PBSGA is applied to the balance tube to calibrate density. The tubes are then centrifuged at an RCF of 500 for 20 min. The PMNL band is located at an approximate buoyant density of 1.09. Immature, dead, and damaged granulocytes are of lesser density and thus are avoided. This band is removed taking care to avoid contamination with the slightly more dense trace erythrocyte band lying below. The isopyknic technique helps ensure PMNL uniformity for testing. The PMNL are washed with 15 ml PBSGA and centrifuged at an RCF of 200 for 10 min. The supernatant is discarded and the wash–centrifugation step repeated. The PMNL are resuspended in 2 ml fresh PBSGA, total cell counts are made, and cytospin slides are Wright stained for differential counting. This procedure routinely yielded a PMNL purity of approximately 95%. The other 5% of cells are eosinophils. The PMNL concentration is adjusted to 1000 PMNL/μl. Just prior to testing the required number of PMNL in PBSGA is added to sterile, siliconized vials containing 1.8 ml CVB with the stated concentration of luminol or DBA. For the temperature variable studies of Figs. 29–32, the conditions are the same except that 0.8 ml CVB with CLP is used. Stimulus is added at time zero, and the final volume is either 2.0 or 1.0 ml as indicated.

Photon Counting

Liquid Scintillation Counters and Photon Counters. Even relatively weak sources of luminescence can be readily measured using the photon counting capacity of standard liquid scintillation counters.[63–65] Scintilla-

[63] E. Tal, S. Dikstein and F. G. Sulman, *Experientia* **20,** 652 (1964).
[64] E. Schram, *Arch. Int. Physiol. Biochim.* **75,** 897 (1967).
[65] P. E. Stanley and S. G. Williams, *Anal. Biochem.* **29,** 381 (1969).

tion counters are actually two photon counters operated in coincidence. The coincidence circuitry is designed to measure the shower or pulse of photons resulting from the near simultaneous relaxation of the multitude of fluorescent molecules excited by a single ionizing radiation event, e.g., β emission, and to filter out background chemiluminescence. However, in the out-of-coincidence mode the instrument is a photon counter capable of measuring single photon events, i.e., chemiluminescence. The newer model scintillation counters do not require refrigeration, and most are equipped with high sensitivity, bialkali spectral response photomultiplier tubes. A Beckman Instruments LS200 series scintillation counter equipped with bialkali spectral response photomultiplier tubes, operated in the out of coincidence mode at the tritium channel settings was employed in all of the ambient (23–25°) temperature experiments presented. A Berthold Instruments LB950 luminometer specifically designed for photon counting was employed for the temperature control experiment depicted in Figs. 29 through 32.

Calibration. The luminescence velocity measurements, in relative counts per minute (CPM), were converted to blue (luminol equivalent) photons per minute by multiplying the relative CPM by a photon conversion factor. This conversion factor was established by calibrating the counter with an established blue photon emitting standard prepared by Seliger.[66,67] The value of this count-to-photon conversion factor typically is in the range of 10 to 20.

Luminol as Chemiluminigenic Probe

According to the rate relationship described in Eq. (21), luminescence velocity is proportional to the concentration and quantum yield of the substrate available for dioxygenation. Introduction of CLPs allows control over the previously described problem of substrate variability. In addition, the probe approach greatly increases sensitivity while providing information with regard to the nature of the oxygenation activity measured. Figure 1 depicts the temporal traces of CL velocity obtained from a constant number of PMNL stimulated with a fixed quantity of OZ. The variable in this experiment is the concentration of luminol present. The CL product of luminol dioxygenation by stimulated granulocytes[55,68]

$$\text{Luminol} + O_2 \xrightarrow{\text{PMNL (activated)}} \text{aminophthalate} + N_2 + h\nu \qquad (22)$$

[66] H. H. Seliger, this series, Vol. 57, p. 560 (1978).
[67] H. H. Seliger, *in* "Liquid Scintillation Counting: Recent Applications and Developments" (C. T. Peng, D. L. Horrocks, and E. L. Alpens, eds.), Vol. 2, p. 281. Academic Press, New York, 1980.
[68] R. C. Allen and L. D. Loose, *Biochem. Biophys. Res. Commun.* **69**, 245 (1976).

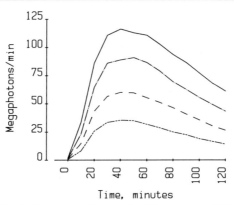

Fɪɢ. 1. Effect of luminol concentration on the CL response of OZ-stimulated PMNL. Conditions: 20,000 PMNL in 50 μl PBSGA were added to 1.9 ml of CVB with luminol concentration as the variable. The stimulus, 100 μg OZ (50 μg/ml final), was added at time zero. The top through the bottom curves depict the CL responses with luminol at 80, 40, 20, and 10 μM final concentration respectively. Each curve is on the average of triplicate measurements taken at 23°.

is dependent on the generation of oxygenation agents by the PMNL and on the concentration of luminol present. As the concentrations of luminol is increased, the reaction begins to approach the zero-order condition with respect to CLP. Thus, the CLP–PMNL–CL relationship approximates a substrate–enzyme–product condition. The stimulus–phagocyte system is complex in comparison with simple isolated enzymes. Whereas simple enzymes rapidly respond to substrate, the peak velocity of the stimulus–phagocyte system is temporally delayed. Despite the time lag, the phagocyte can be described by the Henri–Michaelis–Menten equation[69,70]

$$v = V_{max}[CLP]/(K_s + [CLP]) \qquad (23)$$

where v is the velocity, V_{max} is the maximum velocity, [CLP] is the concentration of the CLP, and K_s, the apparent Michaelis constant, is the concentration of CLP yielding half-maximum velocity.

Comparison of OZ and PMA as Stimuli. When luminol is used as CLP the PMNL CL response is very much dependent on the type of stimulus employed. The temporal kinetics of the CL response to OZ, a phagocytosable particle, is similar to that observed using either complement-opsonified or IgG-opsonified microbes as well as immune complexes (data

[69] V. Henri, *C. R. Hebd. Seances Acad. Sci.* **135,** 916 (1902).
[70] L. Michaelis and M. L. Menten, *Biochem. Z.* **49,** 333 (1913).

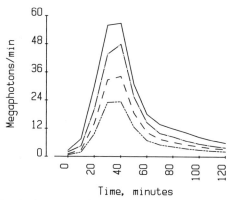

FIG. 2. Effect of luminol concentration on the CL response of PMA-stimulated PMNL. Conditions were the same as for Fig. 1 except that 150 nM PMA was the stimulus.

not shown). Contact with OZ or opsonified microbes results in recognition via complement and/or IgG Fc receptors of the phagocyte, activation of redox metabolism, phagocytosis, specific (secondary) and azurophilic (primary) degranulation, and formation of phagolysosomes.[71,72] On the other hand, the chemical stimulus PMA causes specific degranulation with full activation of granulocyte redox metabolism, but little or no azurophilic degranulation and no phagolysosome formation.[58-60] Figure 2 depicts the effect of luminol concentration on the CL response following stimulation with PMA. Note that the luminol-dependent response of PMNL to PMA stimulation differs from that obtained using OZ with respect to both magnitude and temporal pattern. Whereas the initial response is similar using either stimulus, peak velocity is reached at approximately 35 min with PMA and is followed by relatively rapid deceleration. With OZ the peak velocity is reached at 42 min and then gradually declines.

Henri–Michaelis–Menten Relationship. Regardless of whether OZ or PMA is used as stimulus, the Henri–Michaelis–Menten relationship holds if CLP is considered as substrate. The double reciprocal plots of CL velocity against luminol concentration using OZ and PMA as stimuli are presented in Fig. 3, and the values for K_s and V_{max}, calculated by the statistical method of Wilkinson are presented in Table I.[73] Note that both the V_{max} and the K_s are significantly effected by the choice of stimulus.

[71] P. M. Henson, *J. Immunol.* **107**, 1535 (1971).

[72] D. F. Bainton, *J. Cell Biol.* **58**, 249 (1973).

[73] G. N. Wilkinson, *Biochem. J.* **80**, 324 (1961).

TABLE I
KINETIC PARAMETERS OF THE PMNL LUMINESCENCE RESPONSE

PMNL[a]	CLP[b]	Stimulus[c]	$K_s{}^d$	$V_{max}{}^d$	$T_n{}^e$
20,000	Luminol	OZ	35 ± 1	168 ± 3	42.4 ± 0.6
20,000	Luminol	PMA	22 ± 1	73 ± 2	34.9 ± 0.2
20,000	DBA	OZ	44 ± 4	4 ± 0	78.1 ± 1.2
20,000	DBA	PMA	85 ± 14	57 ± 5	108.2 ± 0.5

[a] Number of PMNL (96% neutrophils; 4% eosinophils) in 2.0 ml total volume CVB.
[b] The concentrations of the chemiluminigenic probes luminol or DBA were varied described in Figs. 1–6.
[c] Final concentrations of stimuli: 50 μg/ml OZ; 150 nM PMA.
[d] The apparent Michaelis constant, expressed as μM CLP ± the standard error (SE), and the maximum velocity, expressed as megaphotons/minute (10^6 photons/min) ± the SE, were calculated according to Wilkinson.[73]
[e] Nodal time: the mean time interval from stimulus addition to estimated peak CL velocity, in mintues ± the SE.

These differences may reflect the nature and location of the oxygenation activities measured.

Role of Myeloperoxidase. With OZ as stimulus, there is true phagocytosis, azurophilic degranulation, and formation of an acidified phagolysome space.[10,74–77] This environment is optimal for MPO microbicidal function.[78,35] When presented with Cl^- and H_2O_2 at an acid pH, cell-free MPO oxidizes native substrate yielding CL.[32,34,35] Under such conditions, addition of luminol results in an increased luminescence by a factor of greater than 10^4 (data not shown). Although the products are the same, i.e., aminophthalate and a photon, the MPO-catalyzed dioxygenation of luminol differs in mechanism from the classical base-catalyzed reactions yielding CL.[79,55] The pK_a of H_2O_2 is 10.6, and, thus, base is required for proton dissociation. In alkaline solution, O_2H^- with trace metal as catalyst, serves as oxidant in the luminol CL reaction.[80,81] However, in the absence of MPO, the rate of luminol reaction with H_2O_2 is very slow at pH 5.

[74] B. J. Bentwood and P. M. Henson, *J. Immunol.* **124**, 855 (1980).
[75] M. S. Jensen and D. F. Bainton, *J. Cell Biol.* **56**, 379 (1973).
[76] S. Ohkuma and B. Poole, *Proc. Natl. Acad. Sci. U.S.A.* **75**, 3327 (1978).
[77] R. van Zweiten, R. Wever, M. N. Hamers, R. S. Weening, and D. Roos, *J. Clin. Invest.* **68**, 310 (1981).
[78] S. J. Klebanoff, *J. Bacteriol.* **95**, 2131 (1968).
[79] H. O. Albrecht, *Z. Phys. Chem.* **136**, 321 (1928).
[80] K. D. Gundermann, *Top. Curr. Chem.* **46**, 61 (1974).
[81] E. H. White and D. F. Roswell, *Acc. Chem. Res.* **3**, 54 (1970).

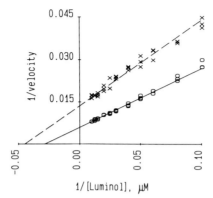

FIG. 3. Double reciprocal plot of peak measured CL velocity, expressed as megaphotons/minute (i.e., 10^6 photons/min), against luminol concentration with OZ (O) or PMA (×) as stimuli. The data were derived from experiments as described in Figs. 1 and 2.

When PMA is used as stimulus there is activation of redox metabolism without true phagocytosis. Degranulation is for the most part confined to the specific granules. Although vacuolization occurs, there is no true phagolysosome formation, and the degree of acidification is not known. Therefore, if MPO is involved in the oxygenation activity of PMA-stimulated PMNL, its contribution would be suboptimal relative to its role in true phagocytosis. PMA uncouples the structure–function relationship required for efficient microbicidal function, and in so doing, allows escape of redox intermediates, such as $\cdot O_2^-$, to the extracellular space where their presence can be measured.

DBA (Lucigenin) as Chemiluminigenic Probe

In alkaline solution, the acridinium salt DBA reacts with H_2O_2 to yield electronically excited N-methylacridone which in turn relaxes by photon emission.[62,82,83] The quantum yield of this reaction, 0.01–0.02, is comparable to that of luminol. As such, DBA has been employed as a CLP for measurement of phagocyte oxygenation activity,[55,84,85] but as will be demonstrated, the physical properties and chemical reactivity of DBA differ greatly from those of luminol.

[82] K. Gleu and W. Petsch, *Angew. Chem.* **48,** 57 (1935).
[83] J. R. Totter, V. J. Medina, and J. L. Scoseria, *J. Biol. Chem.* **235,** 238 (1960).
[84] R. C. Allen, *in* "Bioluminescence and Chemiluminescence: Basic Chemistry and Analytical Applications" (M. A. DeLuca and W. D. McElroy, eds.), p. 63. Academic Press, New York, 1981.
[85] R. C. Allen and B. A. Pruitt, Jr., *Arch. Surg. (Chicago)* **117,** 133 (1981).

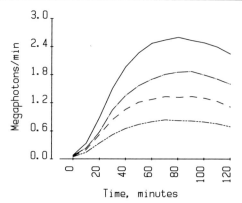

FIG. 4. Effect of DBA concentration on the CL response of OZ-stimulated PMNL. The conditions were as described for Fig. 1 except that the top through the bottom curves depict the CL responses with DBA (lucigenin) at 100, 50, 25, and 12.5 μM final concentration, respectively.

Figures 4 and 5 depict the DBA-dependent CL responses of OZ and PMA stimulated PMNL respectively. The conditions of testing are identical to those presented for Figs. 1 and 2 except that DBA is used as CLP in place of luminol. The relationship of CL velocity to DBA concentration is depicted by the double reciprocal plot of the data presented in Fig. 6. The calculated values for K_s and V_{max} are presented in Table I.

Luminol and DBA Measure Different Oxygenation Activities

Both luminol and DBA may be employed as CLPs for measurement of the respiratory burst metabolism that follows phagocyte stimulation.

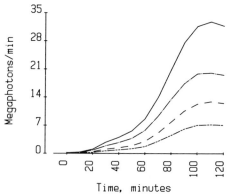

FIG. 5. Effect of DBA concentration on the CL response of PMA-stimulated PMNL. Conditions were the same as for Fig. 4 except that 150 nM PMA was used as stimulus.

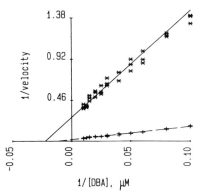

FIG. 6. Double reciprocal plot of peak measured CL velocity in megaphotons/minute versus DBA concentration with OZ (∗) and PMA (+) as stimuli. The data were derived from experiments as described in Figs. 4 and 5.

However, a comparison of the raw CL data illustrates that each probe measures a temporally different activity. The time interval from addition of OZ to peak CL velocity is 42 min with luminol and 78 min with DBA. Furthermore, with OZ as stimulus, the CL V_{max} with luminol is over 40-fold higher than with DBA. The temporal response following PMA stimulation shows an even more dramatic CLP dependence. Luminol detects a relatively early oxygenation activity with a peak velocity at 35 min followed by a rapid deceleration. DBA detects a late oxygenation activity with peak velocity being reached at 108 min. In fact when PMA is stimulus, the temporal relation of luminol and DBA responses is reciprocal. These differences may reflect the involvement of different oxygenating agents, difference in the physical environment of the activity, or both.

Although both luminol and DBA luminescence can be elicited with alkaline peroxide, the mechanism of CLP dioxygenation differs with respect to each CLP. As shown in Fig. 7, the net reaction responsible for luminol CL is a dioxygenation, i.e., luminol reacts with oxygen to yield the dioxygenated product aminophthalate and a photon. The net reaction for DBA luminescence is also a dioxygenation, but in addition, DBA is reduced by two electrons. Thus, the DBA luminescence reaction is a reductive dioxygenation. These differences may explain why the EPO and MPO–halide–peroxide systems yield high-intensity CL with luminol and no CL with DBA (data not shown). For these acid haloperoxidase systems, H_2O_2 is the oxidant and Cl^- is the reductant as described in Eq. (7). The CLP may react with $OOCl^-$ or $^1O_2$, the intermediate and product of Eq. (9), to yield dioxygenated product as in Eq. (10), or alternatively, CLP oxygenation might involve an intermediate halogenation as described in Eqs. (11) and (12). Luminol can serve as CLP under such conditions; DBA cannot.

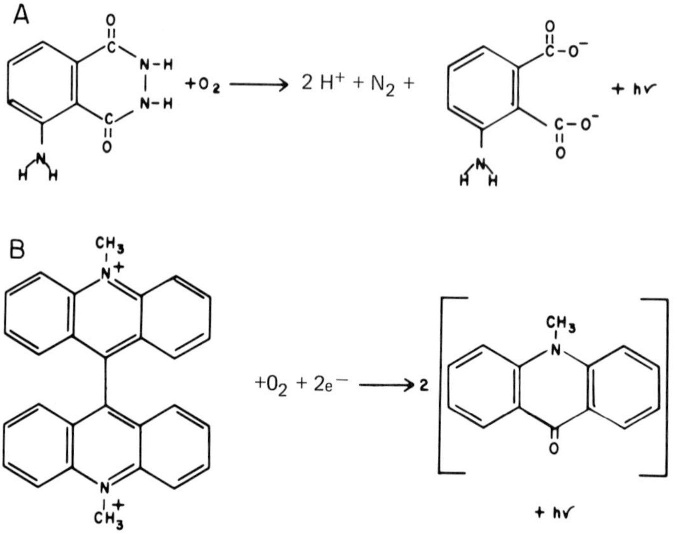

FIG. 7. Luminol versus DBA dioxygenation. (A) Generalized luminol reaction: dioxygenation; (B) generalized lucigenen reaction: reductive dioxygenation. Reprinted from Allen.[55]

There is no evidence to suggest that radicals are directly involved in the luminol dioxygenation reaction catalyzed by MPO or EPO. However, nonperoxidase radical mechanisms may be responsible for a small portion of the luminol-dependent PMNL luminescence, and possibly a large portion of luminol-dependent macrophage luminescence. Radical dehydrogenation of luminol by a metal–peroxide complex or by the $\cdot$OH radical, as described in Eq. (15), would yield a doublet multiplicity luminol intermediate that could react with $\cdot O_2^-$ in a radical–radical annihilation to yield dioxygenated product as described in Eq. (17). With regard to DBA luminescence, the net reaction is a two-electron reduction plus a dioxygenation. As such, the sequence of radical dehydrogenation followed by radical reduction described for luminol could not apply to DBA.

The mechanistic possibilities for reductive dioxygenation of DBA are presented in Fig. 8. Reaction with alkaline peroxide is the classical method for DBA luminescence. Since H_2O_2 is a divalently reduced molecular oxygen, reaction with DBA to yield two molecules of N-methylacridone would satisfy the stoichiometry of a reductive dioxygenation. With regard to biological significance, DBA is only weakly reactive with H_2O_2 over the pH range from acid to neutral. The second alternative involves univalent reduction of DBA. The product, a radical doublet multiplicity cation, could then react with $\cdot O_2^-$ in a radical–radical, anion–cation anni-

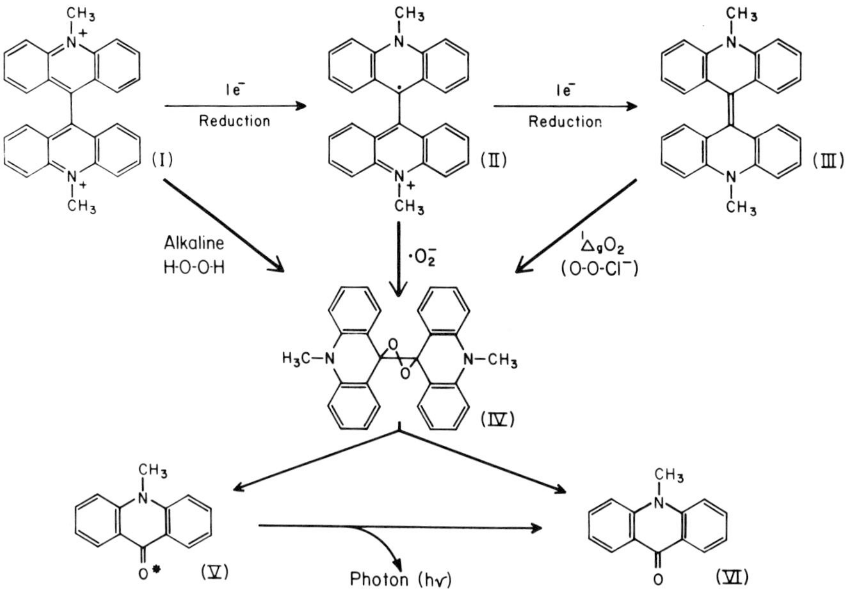

FIG. 8. Pathways to DBA dioxygenation. Reprinted from Allen.[55]

hilation reaction to yield the moloxide (dioxetane) intermediate which disintegrates yielding one excited and one ground state N-methylacridone. Univalent reduction could be metal or metalloprotein catalyzed with $\cdot O_2^-$ serving as the initial as well as the secondary reductant. The third possibility for dioxygenation requires divalent reduction of DBA followed by reaction with $^1O_2$.[86] The second pathway involving $\cdot O_2^-$ is supported by the superoxide dismutase inhibition studies presented in a following section.

Effect of Azide on CLP-Dependent CL

Azide is a potent inhibitor of MPO, but it does not inhibit the oxidases responsible for the respiratory burst.[87,88] As such, azide has been employed for studying the contribution of MPO to PMNL microbicidal action. In this regard, the microbicidal activity and luminescence of azide-treated PMNL closely approximates that obtained from PMNL of patients with hereditary myeloperoxidase deficiency.[89] However, in assess-

[86] F. McCapra and R. A. Hann, J. Chem. Soc., Chem. Commun. p. 442 (1969).
[87] S. J. Klebanoff, Science 169, 1095 (1970).
[88] R. J. McRipley and A. J. Sbarra, J. Bacteriol. 94, 1425 (1967).
[89] H. Rosen and S. J. Klebanoff, J. Clin. Invest. 58, 50 (1976).

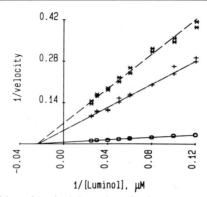

FIG. 9. Azide inhibition of luminol-dependent CL from OZ-stimulated PMNL. Double reciprocal plot of peak CL velocity in megaphotons/minute versus luminol concentration. Conditions: 20,000 PMNL in 50 μl PBSGA were added to 1.9 ml of CVB containing the final concentrations of luminol indicated. The PMNL suspensions were tested in the absence ($\bigcirc$) and in the presence of 1.25 μM (+) and 5.0 μM (∗) azide. The stimulus, 100 μg of OZ (50 μg/ml final), was added at time zero, and velocity measurements were taken in duplicate at 10 min intervals for 2 hr.

ing the action of azide on CLP-dependent CL, it should be realized that azide is not exclusively an MPO inhibitor. It is to varying degrees inhibitory to many other metalloenzymes.[90] Furthermore, azide is a quenching agent for $^1O_2$.[91,92]

Luminol as CLP. The effect of azide can be analyzed with regard to the Henri–Michaelis–Menten relationship linking CL velocity to CLP concentration.[56] Figure 9 depicts the double reciprocal plot of CL velocity, measured from OZ-stimulated PMNL, against the concentration of luminol employed as CLP. The composite data obtained in the absence and in the presence of 1.25 and 5.0 μM azide illustrate that azide inhibition is potent and noncompetitive, i.e., the V_{max} changes but the K_s remains relatively constant. Figure 10 depicts the results of a study in which PMA was employed as stimulus. The conditions were otherwise identical to those described for Fig. 9. Again inhibition is essentially noncompetitive. The calculated values for K_s and V_{max} are presented in Table II. Note that the magnitude of inhibition with PMA is relatively less than observed using OZ as stimulus.

DBA as CLP. If luminol, for the most part, measures MPO-dependent oxygenation activity, and if azide inhibits such activity, then the results

[90] A. Rigo, R. Stevanato, P. Viglino, and G. Rotilio, *Biochem. Biophys. Res. Commun.* **79,** 776 (1977).

[91] D. E. Kearns, W. Fenical, and P. Radlick, *Ann. N.Y. Acad. Sci.* **171,** 34 (1970).

[92] C. S. Foote, T. T. Fujimoto, and Y. C. Chang, *Tetrahedron Lett.* p. 45 (1972).

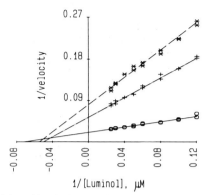

FIG. 10. Azide inhibition of luminol-dependent CL from PMA-stimulated PMNL. Conditions were as described for Fig. 9 except that 150 nM PMA was employed as the stimulus.

are as expected. However, DBA-dependent CL is in a direct sense MPO independent. The double reciprocal plots of CL velocity, measured from OZ-stimulated PMNL, against the concentration of DBA, in the absence and presence of 0.5 and 5.0 μM azide are presented in Fig. 11, and the values of K_s and V_{max} are given in Table II. Instead of inhibition, there is a relatively large noncompetitive amplification of DBA-dependent CL in the presence of azide. Inhibition of MPO, and possibly to some extent catalase and superoxide dismutase, would result in the accumulation of H_2O_2 and $\cdot O_2^-$, the substrates of such enzymes. These accumulated intermediates of redox metabolism might therefore be available for reaction with DBA by the mechanisms previously discussed. However, interpreta-

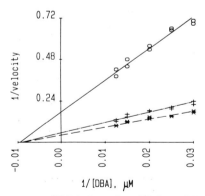

FIG. 11. Azide enhancement of DBA-dependent CL from OZ-stimulated PMNL. The conditions were as described for Fig. 9 except that DBA was employed as CLP, and azide was tested at 0.5 μM (+) and 5.0 μM (∗) final concentrations.

TABLE II
EFFECTS OF AZIDE AND SUPEROXIDE DISMUTASE ON KINETIC PARAMETERS OF THE
PMNL LUMINESCENCE RESPONSE

PMNL[a]	CLP[b]	Stimulus[c]	Inhibitor[d]	K_s[e]	V_{max}[e]
20,000	Luminol	OZ	None	37 ± 4	178 ± 13
20,000	Luminol	OZ	Azide, 1.25 μM	44 ± 6	22 ± 2
20,000	Luminol	OZ	Azide, 5.00 μM	42 ± 5	14 ± 1
20,000	Luminol	PMA	None	14 ± 1	49 ± 2
20,000	Luminol	PMA	Azide, 0.50 μM	20 ± 2	19 ± 2
20,000	Luminol	PMA	Azide, 5.00 μM	18 ± 2	12 ± 1
20,000	DBA	OZ	None	105 ± 25	6 ± 1
20,000	DBA	OZ	Azide, 0.50 μM	110 ± 37	18 ± 4
20,000	DBA	OZ	Azide, 5.00 μM	114 ± 31	24 ± 4
20,000	DBA	PMA	None	197 ± 65	103 ± 28
20,000	DBA	PMA	Azide, 1.25 μM	111 ± 46	17 ± 5
20,000	DBA	PMA	Azide, 5.00 μM	169 ± 46	19 ± 4
20,000	Luminol	OZ	None	29 ± 2	213 ± 9
20,000	Luminol	OZ	SOD, 50 U/ml	33 ± 2	230 ± 9
20,000	Luminol	OZ	SOD, 200 U/ml	39 ± 6	238 ± 22
20,000	Luminol	PMA	None	19 ± 2	90 ± 4
20,000	Luminol	PMA	SOD, 50 U/ml	20 ± 2	85 ± 4
20,000	Luminol	PMA	SOD, 200 U/ml	23 ± 2	96 ± 4
20,000	DBA	OZ	None	67 ± 13	4 ± 0
20,000	DBA	OZ	SOD, 50 U/ml	64 ± 7	3 ± 0
20,000	DBA	OZ	SOD, 200 U/ml	49 ± 7	2 ± 0
20,000	DBA	PMA	None	170 ± 42	232 ± 45
20,000	DBA	PMA	SOD, 50 U/ml	83 ± 23	64 ± 10
20,000	DBA	PMA	SOD, 200 U/ml	70 ± 16	5 ± 1

[a] Number of PMNL in 2.0 ml total volume CVB.

[b] The concentrations of chemiluminigenic probes, luminol and DBA, were varied as described for Figs. 9 through 16.

[c] Final concentration of stimuli: 50 μg/ml OZ; 150 nM PMA.

[d] Sodium azide and bovine erythrocyte superoxide dismutase were tested at the final concentrations indicated.

[e] The apparent Michaelis constant, expressed as μM CLP ± the standard error (SE), and the maximum velocity, expressed as megaphotons/minute (i.e., 10^6 photons/min) ± the standard error, were calculated according to the method of Wilkinson.[73]

tion is clouded by data such as those presented in Fig. 12. The figure depicts the results of a DBA study in which PMA was employed as stimulus in the absence and in the presence of 1.25 and 5.0 μM azide. Although the pattern is complex, azide inhibits the temporally delayed

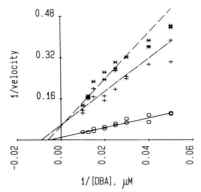

FIG. 12. Azide inhibition of DBA-dependent CL from PMA-stimulated PMNL. The conditions were as described for Fig. 10 except that DBA was the CLP.

DBA-dependent CL characteristically associated with PMA stimulation of PMNL.

Effect of Superoxide Dismutase on CLP-Dependent CL

At the near neutral physiologic pH of the extracellular environment, anion–anion repulsion limits the rate of superoxide disproportionation as described in Eq. (4).[26,93] Consequently, $\cdot O_2^-$ can accumulate in the extracellular space where it can be measured as superoxide dismutase (SOD) inhibitable cytochrome c reducing activity as described in Eq. (5).[29] SOD can catalyze disproportionation in neutral to basic pH environments because the reaction is stepwise. The first step is reduction of the metal prosthetic group by $\cdot O_2^-$; the second step is reduction of a second $\cdot O_2^-$ to HO_2^- by reaction with the reduced SOD. Inhibition of cytochrome c reduction by SOD is considered proof for the generation of $\cdot O_2^-$ during PMNL redox metabolism.[5,94] In the same manner, SOD can be used to investigate the role of $\cdot O_2^-$ in CLP-dependent CL activities of phagocytes.

OZ as Stimulus. The effect of SOD on OZ-stimulated PMNL is depicted in the data of Fig. 13 and the values for K_s and V_{max} are presented in Table II. The figure presents the plot of reciprocal CL velocity versus reciprocal luminol concentration in the absence and in the presence of 100 and 400 units of bovine erythrocyte SOD. Although there is essentially no change in V_{max}, the change in K_s suggests that SOD causes a relatively

[93] B. H. J. Bielski and A. O. Allen, *J. Phys. Chem.* **81,** 1048 (1977).
[94] I. Fridovich, *in* "Free Radicals in Biology" (W. A. Pryor, ed.), Vol. 1, p. 239. Academic Press, New York, 1976.

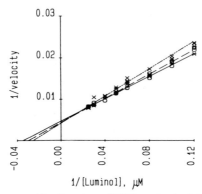

FIG. 13. Effect of superoxide dismutase (SOD) on luminol-dependent CL from OZ-stimulated PMNL. Conditions were as described for Fig. 9 except that measurements were in the absence (○) and in the presence of 50 units/ml (∗) or 200 units/ml (×) SOD.

small competitive inhibition. Figure 14 depicts the results obtained under the same conditions of testing with the exception that DBA was employed as the CLP. SOD does inhibit DBA-dependent CL in a pattern suggesting a mixture of noncompetitive and uncompetitive inhibition.[56] These data are consistent with the proposal that $\cdot O_2^-$ is a reactant in the DBA-dependent CL of OZ-stimulated PMNL.

PMA as Stimulus. The effects of SOD on PMA-stimulated PMNL using luminol and DBA as CLPs are presented in Figs. 15 and 16, respectively. The conditions of testing and the concentrations of SOD employed are those previously described, and the values for K_s and V_{max} are presented in Table II. With luminol as CLP, SOD did not exert a significant

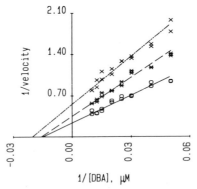

FIG. 14. Superoxide dismutase inhibition of DBA-dependent CL from OZ-stimulated PMNL. Conditions were as for Fig. 13 except that DBA was employed as CLP.

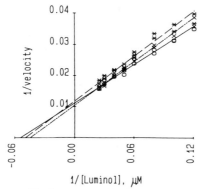

FIG. 15. Effect of superoxide dismutase on luminol-dependent CL from PMA-stimulated PMNL. Conditions were as for Fig. 13 except that 150 nM PMA was employed as stimulus.

effect on V_{max}; however, the small effect on K_s suggests a relatively small competitive inhibition similar to that previously described using OZ. With DBA, a strong inhibition was obtained with 400 units (200 units/ml) SOD. The pattern suggests a mixture of noncompetitive and uncompetitive inhibition.

The results of the azide and the SOD inhibition studies suggest that luminol-dependent CL results primarily, but not exclusively, from the action of MPO in PMNL. In contrast, DBA-dependent CL is not related to MPO activity, and is inhibited by SOD suggesting the involvement of $\cdot O_2^-$. Thus, probing with luminol and DBA allows differential analysis of the peroxidase- and oxidase-associated oxygenation activities that follow phagocyte stimulation.

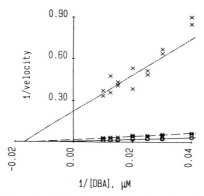

FIG. 16. Superoxide dismutase inhibition of DBA-dependent CL from PMA-stimulated PMNL. Conditions were as for Fig. 15 except that DBA was employed as CLP.

Stimulus Dependence of Phagocyte Activation

When the concentration of CLP is sufficient to approximate the zero-order condition, the previously described relationship of Eq. (21) simplifies to

$$v = k'[\text{Phagocytes}]^P[\text{Stimulus}]^s \qquad (24)$$

By increasing the phagocyte concentration and/or measuring only the initial CL velocity so as to approximate the zero-order condition with respect to phagocytes, the relationship of Eq. (24) in turn simplifies to

$$v = k'[\text{Stimulus}]^s \qquad (25)$$

As such, the kinetics of stimulation can be directly studied with regard to the nature and concentration of the stimulus employed.

Opsonified Zymosan as Stimulus

Luminol as CLP. Figure 17 depicts the temporal CL responses measured from a constant number of PMNL with a constant concentration of luminol as CLP, and four different concentrations of OZ as stimulus. Figure 18 presents the results obtained using one-fourth the number of PMNL. The experimental conditions are otherwise the same as in Fig. 17. Note that the temporal geometry of the CL responses varies with the concentration of OZ employed, and, to a lesser extent, with the concentration of PMNL employed. These results are in marked contrast to the

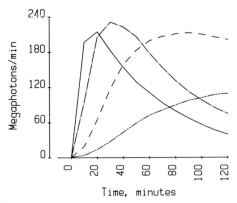

FIG. 17. Effect of OZ concentration on the luminol-dependent CL response of PMNL. Conditions: 40,000 PMNL in 50 μl PBSGA were added to 1.9 ml of CVB containing 40 μM luminol. The four curves from left to right depict the CL responses to 250, 62.5, 15.6, and 3.9 μg/ml final concentration of OZ, respectively. Measurements were in duplicate.

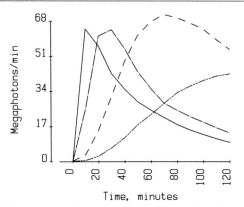

FIG. 18. Effect of OZ concentration on the luminol-dependent CL response using different OZ:PMNL ratios. Conditions were as described for Fig. 17 except that 10,000 PMNL were employed.

symmetric CL responses depicted in Fig. 1 where the concentration OZ is held constant and luminol concentration is the variable. The temporal pattern of phagocyte activation is exquisitely sensitive to the type and concentration of stimulus employed. Although CLPs increase the sensitivity for detecting stimulated phagocyte oxygenation activity, these probes do not affect the temporal kinetics.

DBA as CLP. Figure 19 presents the composite results obtained using a constant concentration of DBA as the CLP, and three different concen-

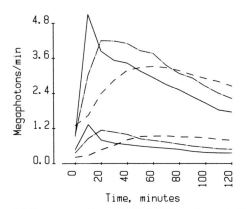

FIG. 19. Effects of OZ concentration and OZ:PMNL ratio on the DBA-dependent CL response of PMNL. Conditions were as described for Figs. 17 and 18 except that DBA was employed as CLP. The top three curves depict the responses of 40,000 PMNL to 250, 62.5, and 15.6 μg/ml OZ. The bottom three curves depict the responses of 10,000 PMNL to 250, 62.5, and 15.6 μg/ml OZ.

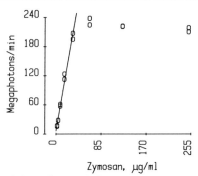

Fɪɢ. 20. Plot of luminol-dependent peak CL velocity measured during the initial 30 min poststimulation interval versus the concentration of OZ employed as stimulus. The data were derived from experiments as described in Fig. 17.

trations of OZ as stimulus for each of the two different concentrations of PMNL tested. The concentrations of OZ tested are the same as those depicted by the curves of the same line type in Figs. 17 and 18. Likewise, the PMNL concentration of the upper set of curves is four times that of the bottom set. Although the luminol-dependent CL responses are magnitudinally greater than those obtained using DBA, the temporal patterns are similar using either CLP.

Estimating Reaction Order. The method of initial rates can be modified for analysis of the kinetics of phagocyte stimulation. In Fig. 20, the peak CL velocity measured during the initial 30 min poststimulation time interval, as depicted in Fig. 17, is plotted against the concentration of OZ. Note that at the lower concentrations of OZ the reaction is essentially first order with respect to the stimulus OZ, but at higher OZ concentrations

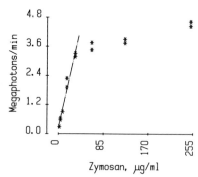

Fɪɢ. 21. Plot of DBA-dependent peak CL velocity measured during the initial 30 min poststimulation interval versus the concentration of OZ employed as stimulus. The data were derived from experiments as described in Fig. 19.

(>50 μg/ml) the reaction is zero order with respect to stimulus even during this initial test period. Figure 21 presents the results obtained under the same experimental conditions except that DBA was employed as the CLP. Although the velocity using DBA is one-fiftieth that obtained with luminol, the relationship of peak CL velocity to OZ concentration is essentially the same for both CLPs.

The data presented in Figs. 20 and 21 illustrate the rate relationship described by Eq. (25). The calculated values for the proportionality constant and reaction order with respect to stimulus are presented in Table III. It should be appreciated that given sufficient time PMNL will become

TABLE III

REACTION ORDER WITH RESPECT TO STIMULUS AND NODEL TIME WITH RESPECT TO
STIMULUS : PMNL RATIO

			CL velocity[a] $= k'$[Stimulus]^s			
PMNL[b]	CLP[c]	Time int. (min)[d]	[Stimulus][e]	Constant (k')	Order (s)	r^2[f]
40,000	Luminol	0–30	OZ: 0–50 μg/ml	8.14	0.94	0.96
40,000	DBA	0–30	OZ: 0–50 μg/ml	0.18	0.86	0.98
40,000	Luminol	0–30	PMA: 0–100 nM	0.56	1.31	0.76
40,000	DBA	0–30	PMA: 0–100 nM	0.34	0.86	0.98

		Nodal time [g] $= k''$([Stimulus]/[PMNL])$^{-t}$			
		Stimulus, range[e]	Constant (k'')	Order (t)	r^2
10,000	Luminol	OZ: 7–250 μg/ml	475	−0.57	0.99
40,000	Luminol	OZ: 7–250 μg/ml	488	−0.55	0.98
10,000	DBA	OZ: 15–250 μg/ml	428	−0.59	0.95
40,000	DBA	OZ: 7–250 μg/ml	549	−0.61	0.97
10,000	Luminol	PMA: 4–120 nM	151	−0.42	0.99
		120–600 nM	108	−0.27	0.93
40,000	Luminol	PMA: 4–120 nM	148	−0.42	0.99
		120–600 nM	128	−0.29	0.97
10,000	DBA	PMA: 37–600 nM	204	−0.18	0.92
40,000	DBA	PMA: 18–600 nM	182	−0.17	0.96

[a] CL velocity is expressed as megaphotons/minute.
[b] Number of PMNL in 2.0 ml total volume CVB.
[c] Chemiluminigenic probes, final concentrations: luminol, 40 μM; DBA, 50 μM.
[d] Initial poststimulation time interval during which peak CL velocity was measured.
[e] The range of stimulus concentrations used for calculations.
[f] The coefficient of determination (r^2) is presented for each set of experimental data.
[g] Nodal time (T_n) is expressed as minutes.

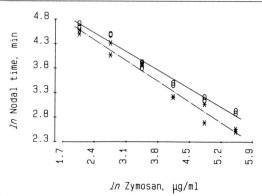

In Zymosan, μg/ml

FIG. 22. Double log plot of nodal time (T_n) versus OZ concentration. The data were derived from the luminol (○)- and DBA (∗)-dependent CL responses as described in Figs. 17 and 19, respectively.

rate limiting even when the ratio of stimulus to phagocyte is relatively low.

Nodal Time. The time required for a set number of PMNL to reach peak CL velocity is dependent on the type and concentration of stimulus employed. The nodal time, T_n, is defined as the point in time separating acceleration and deceleration phases of the phagocyte CL response, i.e., the estimated time interval from stimulus addition to peak velocity.[14] When the number of PMNL is held constant, T_n is inversely related to the stimulus concentrations. When a set quantity of OZ is employed as stimulus and the number of PMNL is varied as described in Table IV, T_n is directly related to PMNL concentration. As such, T_n reflects the attainment of a limiting condition of phagocyte stimulation, and can be described by the relationship

$$T_n = k''([\text{Stimulus}]/[\text{PMNL}])^{-t} \tag{26}$$

As the ratio of stimulus to phagocyte is increased, the time to peak velocity is decreased. Figure 22 depicts the double natural log (ln) plot of T_n versus the concentration of OZ using luminol and DBA as CLPs. The values for nodal time were calculated from the experiments described in Figs. 17 and 19. Although the CL velocity using luminol is magnitudinally greater than that obtained using DBA, the relationship between T_n and the concentration of OZ is well maintained regardless of the CLP employed.

PMA as Stimulus

Luminol as CLP. The temporal pattern of luminol-dependent Cl following stimulation of PMNL with PMA is very different from that ob-

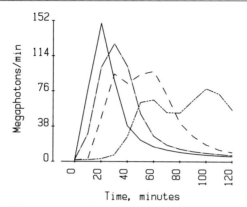

Fig. 23. Effect of PMA concentration on the luminol-dependent CL response of PMNL. Conditions were as described for Fig. 17 except that PMA concentration was the variable. The four curves from left to right (top to bottom) depict the responses to 600, 150, 38, and 9 nM PMA.

served with OZ. Except for the use of PMA as stimulus, the CL responses presented in Figs. 23 and 24 are obtained under the same conditions of testing as previously described for Figs. 17 and 18, respectively. Note the biphase response obtained with lower concentrations of PMA, suggesting the possibility of two separable activities.

DBA as CLP. The results obtained using a fixed concentration of DBA as the CLP and three different concentrations of PMA for each of two concentrations of PMNL tested are presented in Fig. 25. Except for the stimulus, the conditions are as described for Fig. 19. The biphasic re-

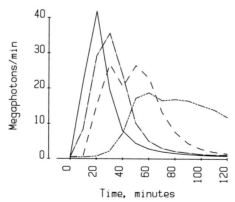

Fig. 24. Effect of PMA concentration on the luminol-dependent CL response using different PMA:PMNL ratios. Conditions were as described for Fig. 23 except that 10,000 PMNL were employed.

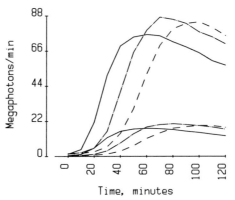

FIG. 25. Effects of PMA concentration and PMA : PMNL ratio on the DBA-dependent CL responses of PMNL. Conditions were as described for Figs. 23 and 24 except that 50 μM DBA was employed as CLP. The top three curves from left to right depict the responses of 40,000 PMNL to 600, 150, and 38 nM PMA, respectively. The bottom curves depict the responses of 10,000 PMNL to 600, 150, and 38 nM PMA.

sponse is not observed using DBA as the CLP. The composite plot of peak CL velocity, during the initial 30 min interval using luminol or DBA as CLP, versus the concentration of PMA employed for stimulation is presented in Fig. 26. Eq. (25) is applicable at the lower PMA to PMNL ratios.

Nodal Time. The relationship of nodal time to stimulus concentration, as described in Eq. (26), also applies with PMA as stimulus, but in comparison to OZ, the choice of CLP has a large effect on the results obtained from PMA-stimulated PMNL. The double natural log plot presented in Fig. 27 was constructed from the experimental results using luminol and

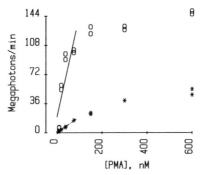

FIG. 26. Composite plot of peak luminol (○)- and DBA (∗)-dependent CL velocities, measured during the initial 30 min poststimulation interval, versus the concentration of PMA employed as CLP. The data were derived from experiments as described in Figs. 23 and 25.

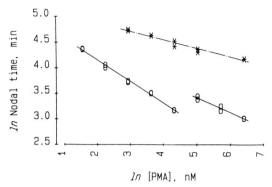

FIG. 27. Double log plot of nodal time (T_n) versus PMA concentration. The data were derived from the luminol (○)- and DBA (*)-dependent CL experiments as described in Figs. 23 and 25.

DBA as described in Figs. 23 and 25, respectively. The biphasic nature of the luminol-dependent response is reflected in the break observed at approximately 100 nM PMA; at PMA concentrations below 100 nM, peak velocity is associated with the second phase. The DBA-dependent responses occur relatively late poststimulation, and at lower concentrations of PMA, the T_n was not obtained during the 2-hr period of measurement. With DBA as CLP, the stimulus–T_n relationship was continuous throughout.

Effect of Phagocyte Concentration

Measurement of CL velocity in the initial poststimulation period ensures that the rate is relatively independent of PMNL. However, given sufficient time, the zero-order condition with respect to stimulus is reached even at low stimulus to PMNL ratios. When PMNL are held constant, the nodal time is inversely related to the quantity of stimulus employed. Since the nodal time is the time required to reach peak CL velocity, it is also the temporal measure of saturation with regard to stimulated phagocyte oxygenation activity. This concept is illustrated by the relative effects of PMNL number and stimulus concentration on CL response. At peak CL velocity, the response has reached the zero-order condition with respect to stimulus, and, as such, peak velocity is related to phagocyte concentration by the equation

$$v = k'[\text{Phagocytes}]^P \tag{27}$$

Based on the peak velocity measurements presented in Figs. 17, 18, 19, 23, 24, and 25, CLP-dependent PMNL luminescence is essentially first order with respect to PMNL regardless of the stimulus or CLP employed.

Table IV and Fig. 28 present the results of an experiment specifically designed to assess reaction order with respect to PMNL. Figure 28a depicts the plot of peak CL velocity versus the final concentration of PMNL tested. OZ or PMA are used as stimuli and luminol is employed as CLP at a final concentration of 60 μM. With regard to Eq. (27) v is expressed as peak velocity in megaphotons/minute, and phagocyte concentration is expressed as PMNL/microliter. The value of k' is 9.4 for PMA and 18.8 for OZ, and the order of the reaction, p, is 0.93 and 0.85 using PMA and OZ, respectively. Three separate sets of runs were required to obtain these data, and, as such, a small cell-aging artifact is imposed by the 4-hr period separating the first run from the last.

TABLE IV
KINETIC PARAMETERS WITH RESPECT TO PMNL CONCENTRATION

PMNL[a]	Time (hr), PV[b]	CLP[c]	Stimulus[d]	$K_s{}^e$	$V_{max}{}^e$	$T_n{}^f$
5,000	3.0	Luminol	PMA	33 ± 3	33 ± 2	27.8 ± 0.2
10,000	4.5	Luminol	PMA	30 ± 3	64 ± 4	31.2 ± 0.1
15,000	6.0	Luminol	PMA	34 ± 4	92 ± 6	33.3 ± 0.1
20,000	6.0	Luminol	PMA	26 ± 2	103 ± 3	33.8 ± 0.3
30,000	4.5	Luminol	PMA	22 ± 2	161 ± 5	31.6 ± 0.2
40,000	3.0	Luminol	PMA	20 ± 1	203 ± 4	28.7 ± 0.2
5,000	3.5	Luminol	OZ	83 ± 9	105 ± 7	22.7 ± 0.1
10,000	5.0	Luminol	OZ	97 ± 7	193 ± 10	21.7 ± 0.1
15,000	6.5	Luminol	OZ	80 ± 6	227 ± 11	24.2 ± 0.2
20,000	6.5	Luminol	OZ	70 ± 6	254 ± 14	25.7 ± 0.4
30,000	5.0	Luminol	OZ	65 ± 5	417 ± 19	24.1 ± 0.1
40,000	3.5	Luminol	OZ	34 ± 2	410 ± 10	29.2 ± 0.5

$$V_{max} = k'[\text{PMNL}/\mu l]^p$$

	k'	p	r^2
With PMA as stimulus	15.4	0.86	0.99
With OZ as stimulus	55.5	0.72	0.97

[a] Number of PMNL in 2.0 ml CVB. Count determined by hemocytometer.

[b] Time interval from venipuncture to stimulation in hours; postvenipuncture age of PMNL.

[c] Each of the twelve runs was in duplicate using eight different concentrations of luminol ranging from 5 to 60 μM.

[d] Final concentrations of stimuli: 100 μg/ml OZ; 150 nM PMA.

[e] The apparent Michaelis constant, expressed as μM luminol ± the standard error (SE), and the maximum velocity, expressed as megaphotons/minute ± the SE, were calculated according to Wilkinson.[73]

[f] Nodal time: the mean time interval from stimulus addition to estimated peak CL velocity, in minutes ± the SE.

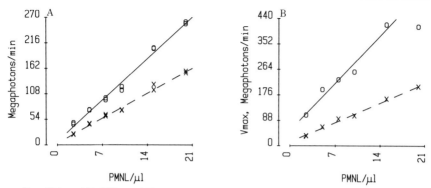

FIG. 28A and B. Effect of PMNL concentration on peak CL velocity (A) and estimated V_{max} (B). Conditions: concentrations ranging from 5,000 to 40,000 PMNL in 50 μl PBSGA were added to 1.9 ml of CVB with 60 μM luminol as CLP (A) or with variable concentrations of luminol ranging from 5 to 60 μM (B). Stimulation was with 100 μg/ml OZ ($\bigcirc$) or 150 nM PMA ($\times$).

This set of experiments also included luminol as a variable. Fig. 28b depicts the plot of estimated V_{max} versus PMNL/μl tested. Table IV presents the kinetic parameters K_s, V_{max}, and T_n derived from these studies. For either stimulus, the K_s decreases with increasing PMNL concentration, but the V_{max} is essentially in the first-order condition with respect to PMNL. When the concentration of OZ is held constant, the value of T_n increases with PMNL concentration reflecting a relative decrease in the OZ : PMNL ratio. Although the data are not presented, a similar effect was observed using aggregated immunoglobulin. However, this effect was not observed using PMA as stimulus. Instead the T_n was observed to increase with the postvenipuncture age of the cell.

Effect of Temperature

As described in Eq. (1), rate processes increase in velocity with increasing temperature. In physiologic chemistry the ratio of reaction velocities at two temperatures 10° apart is referred to as the Q_{10} of the reaction.[56] Actually, the Q_{10} is proportional to the activation energy, E_a, of the Arrhenius equation. For example, a Q_{10} of 2 is equivalent to an activation energy of approximately 12.6 kcal/mol over the region from 25 to 35°.

All of the experiments described thus far were conducted within a 23 to 25° temperature range. Although the modified scintillation counters used for photon counting were fan ventilated to disperse instrument heat and ensure interrun temperature stability throughout the experiment, the

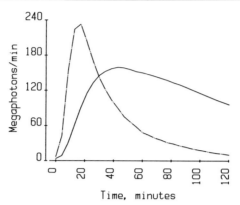

FIG. 29. Effect of temperature on the luminol-dependent CL response of OZ-stimulated PMNL. Conditions: 40,000 PMNL in 100 μl PBSGA were added to 800 μl of CVB with 30 μM luminol as CLP. The stimulus, 100 μl of 100 μg OZ (100 μg/ml final) was added at time zero. The lower (solid) and the higher curves depict the mean of quadruplet measurements at temperatures of 23 and 37° respectively.

operating temperature was under ambient control. Recently single photon counters with temperature control have become commercially available. The plots of CL velocity versus time presented in Figs. 29, 30, 31, and 32 were obtained using such an instrument. In each graphic the responses at 23 and 37° are depicted by the solid and broken curves, respectively. Although run using half the final reaction volume (1 ml) previously employed, and in an instrument requiring a different sample geometry, the results obtained at 23° are very similar to those previously presented. As

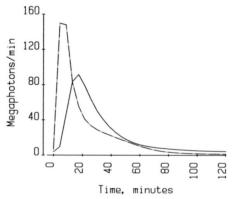

FIG. 30. Effect of temperature on the luminol-dependent CL response of PMA-stimulated PMNL. The conditions were as described for Fig. 29 except that 150 nM PMA was the stimulus.

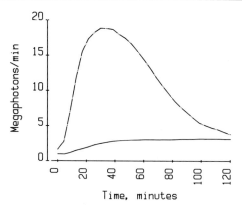

FIG. 31. Effect of temperature on the DBA-dependent CL response of OZ-stimulated PMNL. The conditions were as described for Fig. 29 except that 120 μM DBA was employed as CLP.

expected, a relatively marked acceleration in the initial CL response is obtained by increasing the temperature to 37°.

Luminol as CLP. The luminol-dependent CL responses following stimulation of PMNL with either OZ or PMA are presented in Figs. 29 and 30, respectively. The increase in initial velocity and the shortened T_n observed at the higher temperature are consistent with more rapid expression of PMNL oxygenation capacity. However, the finite nature of this capacity is indicated by the similarity in total integral CL response at either temperature. PMNL effect microbicidal action at 23 and at 37°. It is

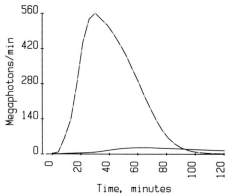

FIG. 32. Effect of temperature on the DBA-dependent CL response of PMA-stimulated PMNL. The conditions were as described for Fig. 30 except that 120 μM DBA was employed as CLP.

the rate of microbicidal activity that is increased by increasing temperature.

DBA as CLP. The DBA-dependent CL responses to OZ or PMA stimulation, as presented in Figs. 31 and 32, respectively, are very different from those obtained with luminol as CLP. These preliminary data suggest that the Q_{10} for DBA-dependent CL activity is relatively higher than the Q_{10} observed using luminol. This difference might reflect the nature of the different oxygenation activities measured by each CLP. As an alternative explanation, higher temperature might favor change in membrane fluidity and structural integrity resulting in greater exposure of DBA to the products of oxidase activity. These possibilities are presently being explored.

Thermal Inactivation of PMNL. In biochemical systems, reaction rate increases with temperature up to a maximum and then drops precipitously due to enzyme denaturation. Exposure to 56° for 15 min completely inhibits the PMNL metabolic response to stimuli, generation of oxygenating agents, and luminescence. Although heat treatment denatures a key enzyme or enzymes required for metabolism, there is no direct effect on MPO activity. This observation can be employed as a control to ensure that luminescence is a consequence of PMNL respiratory burst metabolism.

When streptococci or other bacteria lacking a cytochrome system are added to chronic granulomatous disease PMNL or to MPO in the absence of viable PMNL, both native and luminol-dependent CL activities are observed.[95] This luminescence results from the accumulation of the streptococcal metabolic by-products, acid and H_2O_2, which can fulfill the pH optimum and substrate requirements of MPO. Likewise, heat-inactivated PMNL yield CL on exposure to acid and H_2O_2.

Direct Measurement of Phagocyte Activity in Unseparated Whole Blood

The relatively short lifetime as well as the tendency for specific degranulation and activation of redox metabolism present major difficulties in PMNL research. Even under ideal conditions leukocyte preparation is a time-consuming process, and although great care is usually taken to avoid cell damage and activation during preparation and purification, there is always the potential for introducing artifact. One possibility for obviating or minimizing these problems is direct analysis of phagocyte function in unseparated whole blood.

Problem of Hemoglobin Absorption. A major limitation with regard to

[95] R. C. Allen, E. L. Mills, T. R. McNitt, and P. G. Quie, *J. Infect. Dis.* **144,** 344 (1981).

the direct measurement of phagocyte function in unseparated whole blood is the relatively high extinction coefficient of hemoglobin in the blue region of the visible spectrum. A significant portion of the CL emission spectrum of both luminol and DBA overlaps with the absorption spectrum of hemoglobin, and, thus, absorption of CL will be proportional to the hemoglobin present.

Dilution of whole blood decreases the number of erythrocytes and the concentration for hemoglobin, but also decreases the number of leukocytes. To be effective, the whole blood dilution should be sufficient to decrease hemoglobin to a concentration that is not readily detected by vision. This typically requires a final dilution of approximately 1 : 1000 or greater with a proportionate decrease in leukocyte concentration. Fortunately, the ultrasensitivity of the CLP approach for quantifying oxygenation activity allows direct measurement of phagocyte function from such highly diluted, unseparated whole blood specimens.

Preparation of Whole Blood. Whole blood is collected into evacuated tubes containing K_3EDTA as anticoagulant as described under "Preparation of PMNL." Within 30 min of testing, a small portion of the well mixed blood (50 μl) is mixed with PBSGA (4.95 ml) to achieve a 1 : 100 dilution. An aliquot of the diluted specimen (50 μl) is then added to CVB containing CLP (1.9 ml), and the CL response is measured following addition of stimulus at time zero. The remainder of the undiluted whole blood is used for a complete blood count with differential leukocyte count. Data with regard to type and number of leukocytes present is required for calculating specific CL activities.

Use of EDTA as anticoagulant ensures against complement activation and associated phagocyte stimulation. Phagocyte viability and function are well preserved in EDTA whole blood with minimal loss of activity even after 8 hr at room temperature (22°). Adding the diluted blood specimen to CVB just prior to testing reverses the effect of EDTA. The CVB contains sufficient Ca^{2+} and Mg^{2+} to nullify the effect of any remaining EDTA present in the relatively small aliquot of diluted whole blood added.

In the experiments presented, whole blood is tested at a final dilution of 1 : 4000. In previous studies we reported the use of 10 μl (1 : 200 dilution) and 1 μl (1 : 2000 dilution) equivalents of whole blood for testing.[85,96] The use of 5 μl of whole blood with correction of hemoglobin absorption has also been suggested.[97] However, use of very high concentrations of

[96] R. C. Allen, *Proc. Int. Congr. Clin. Chem., 11th, 1981* p. 1043 (1982).
[97] P. DeSole, S. Lippa, and G. P. Littarru, *J. Clin. Lab. Autom.* **3**, 391 (1983).

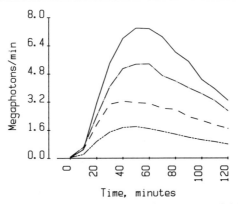

FIG. 33. Effect of luminol concentration on the CL response of OZ-stimulated phagocytes in diluted, unseparated whole blood. Conditions: 0.5 μl of blood containing 2,600,000 erythrocytes, 105,000 platelets, 1,900 segmented and 100 band neutrophils, 200 eosinophils, 300 monocytes, 700 lymphocytes, and 0.28 μl plasma in 50 μl PBSGA was added to 1.9 ml CVB with luminol as the variable. The stimulus, 100 μg OZ (50 μg/ml final), was added at time zero. The top through the bottom curves depict the responses with luminol at 80, 40, 20, and 10 μM final concentration, respectively. Each curve is the average of triplicate measurements taken at 23°.

whole blood should be avoided if accurate determination of specific activity is desired.[98,99]

Comparison of CL Activities of Purified and Whole Blood Phagocytes. The effectiveness of the whole blood approach for analysis of stimulated phagocyte oxygenation activity is demonstrated by the data presented in Figs. 33–36. These experiments were conducted on the same day using blood from the same donor as described for Figs. 1 and 2 and 4 and 5, respectively. Activity is measured from a 0.5 μl equivalent of EDTA-anticoagulated whole blood as described above. The kinetic parameters of the whole blood CL response, presented in Table V, can be compared to those of the purified PMNL presented in Table I.

The temporal kinetics of the stimulus and CLP-dependent CL responses using whole blood are in good agreement with those obtained from purified PMNL. The whole blood data are also in agreement with respect to magnitude, i.e., the specific luminescence activities (V_{max}/phagocyte) are relatively well maintained across the magnitudinal differ-

[98] H. Fischer, M. Ernst, F. E. Maly, T. Kato, H. Wokalek, M. Heberer, D. Maas, B. Peskar, E. T. Rietschel, and H. Staudinger, *in* "Luminescence Assays: Perspectives in Endocrinology and Clinical Chemistry" (M. Serio and M. Pazzagli, eds.), p. 229. Raven Press, New York, 1982.

[99] L. R. DeChatelet and P. S. Shirley, *Clin. Chem. (Winston-Salem, N.C.)* **27**, 1739 (1981).

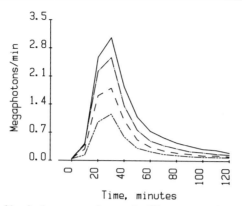

FIG. 34. Effect of luminol concentration on the CL response of PMA-stimulated phago-cytes in diluted whole blood. Conditions were the same as for Fig. 33 except that 150 nM PMA was the stimulus.

ence in phagocyte concentration and in stimulus : phagocyte ratio. When luminol is the CLP and OZ is the stimulus, the estimated K_s is consistently higher for whole blood compared with purified PMNL. This increase may be related to the decreased concentration of PMNL used in whole blood testing, or it may reflect competitive binding of luminol by additional components present in the whole blood specimen. However, the K_s is approximately the same for whole blood or PMNL when luminol is the CLP and PMA is employed as nonphagocytic stimulus.

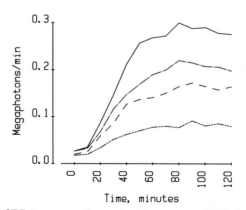

FIG. 35. Effect of DBA concentration on the CL response of OZ-stimulated phagocytes in diluted whole blood. The conditions were as described for Fig. 33 except that the top through the bottom curves depict the responses with DBA at 100, 50, 25, and 12.5 μM final concentration, respectively.

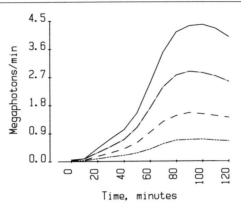

FIG. 36. Effect of DBA concentration on the CL response of PMA-stimulated phagocytes in diluted whole blood. Conditions were as described for Fig. 35 except that 150 nM PMA was the stimulus.

When DBA is employed as CLP with either OZ or PMA as stimulus, the specific activity of whole blood is higher than predicted by the purified PMNL studies. This increase may be related to the additional presence of monocytes in the whole blood specimen. The MPO content of monocytes is less than half that of PMNL. However, both PMNL and monocytes can respond to OZ and PMA by activation of respiratory burst metabolism.

TABLE V

KINETIC PARAMETERS OF THE WHOLE BLOOD PHAGOCYTE LUMINESCENCE RESPONSE

Whole blood[a] (μl)	CLP[b]	Stimulus[c]	K_s[d]	V_{max}[d]	T_n[e]
0.5	Luminol	OZ	58 ± 4	12.8 ± 0.5	49.2 ± 1.1
0.5	Luminol	PMA	22 ± 2	3.8 ± 0.1	27.6 ± 0.1
0.5	DBA	OZ	37 ± 4	0.4 ± 0.0	69.3 ± 2.3
0.5	DBA	PMA	156 ± 26	10.8 ± 1.4	93.9 ± 0.8

[a] The equivalent of whole blood added to 1.9 ml CVB for a final dilution of 1:4000. The 0.5 μl equivalent contained approximately 2,600,000 erythrocytes, 105,000 platelets, 1900 segmented and 100 band neutrophils, 200 eosinophils, 300 monocytes, 700 lymphocytes, and 0.28 μl plasma.

[b] The concentrations of the CLPs luminol and DBA were varied as described for Figs. 1–6.

[c] Final concentration of stimuli: 50 μg/ml OZ; 150 nM PMA.

[d] The apparent Michaelis constant, expressed as μM CLP ± the standard error (SE), and the maximum velocity, expressed as megaphotons/minute ± the SE, were calculated as described for Table I.

[e] Nodal time: the mean time interval ± the SE from stimulus addition to estimated peak CL velocity, in minutes.

Thus monocytes oxidase activity, measured as DBA-associated CL, should be proportionally higher than monocyte MPO activity, measured as luminol-associated CL.

Although these data are not shown, whole blood phagocytes are also susceptible to inhibition by azide and SOD. Inhibition is dependent on the combination of stimulus and CLP employed as described for the PMNL studies. Regardless of the CLP-stimulus combination, whole blood luminescence is completely inhibited by heating to 56° for 15 min.

The CLP method for directly measuring phagocyte oxygenation activity in unseparated whole blood is highly sensitive. Use of selective CLP–stimulus–inhibitor combinations also allows relative specific and differential measurement of phagocyte-generated oxygenating agents.

Acknowledgments

The opinions or assertions presented herein are the private views of the author and are not to be construed as reflecting the views of the Department of the Army or the Department of Defense of the U.S.A.

I wish to thank Drs. Ruth R. Benerito, David F. Jadwin, and Basil A. Pruitt, Jr. for their review and comments.

[37] Cellular Chemiluminescence Associated with Disease States

By KNOX VAN DYKE and CHRIS VAN DYKE

Introduction

Cellular chemiluminescence (CL) can be defined as light produced from cells that has a chemical origin. Generally, this light is in the visible spectrum (400–600 nm). Cellular CL under discussion here is produced from single cells such as neutrophils, monocytes, or macrophages (alveolar or peritoneal). These cells are phagocytic, i.e., they can engulf particles. If the particles are living biological entities such as bacteria, yeast, or virus, they may be killed by the various chemical processes available to phagocytic cells. This killing of invading organisms forms a portion of the natural defense mechanism available to humans and other animal species (see Figs. 1 and 2).

If this natural cellular defense mechanism is defective or is increased

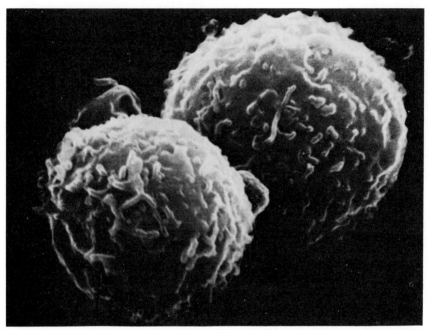

Fig. 1. Scanning electron micrographs of human neutrophils. Cells were fixed in 2% glutaraldehyde for 30 min and sequentially dehydrated in various concentrations of ethanol from 50 to 100%. They were subjected to critical point drying and stained with gold. ×3700.

or decreased by disease, these changes may be detected by measuring cellular CL. Therefore, the quantitation of light from human neutrophils has been found to be useful in the detection of genetic deficiencies, and studies of inflammatory diseases, infection, degenerative diseases, and cancer. The main findings involving genetic diseases are in the diagnosis of the neutrophil abnormalities, i.e., chronic granulomatous disease and myeloperoxidase deficiency. Studies related to cell CL in inflammatory disease include arthritis, exercise-induced asthma, and pollen-induced allergy. Infection by bacteria, virus, mycoplasma, and yeasts can be followed using CL of human neutrophils. Diabetes, renal dialysis, and cancer (including leukemias) have been subjected to important research using cellular CL. The humoral defense mechanism involving the coating of particles with recognition factors (either antibody or complement) known as opsonization will not be covered in this chapter. Comments will be confined to either detection or analysis of disease states by changes in cellular mechanisms as measured by cellular CL.

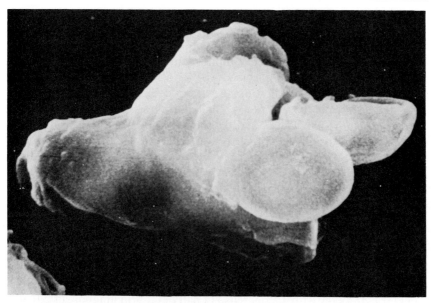

FIG. 2. Scanning electron micrograph of human neutrophils engulfing opsonized zymosan particles. Cells were prepared as in Fig. 1. ×3350.

Genetic Abnormalities

Chronic Granulomatous Disease (CGD). CGD was the first disease investigated using cellular CL.[1] It is a disorder associated with a high susceptibility to catalase-positive bacterial infection[2,3] and chronic inflammation. The disease has been partially related to defects in the function of granulocytes. These cells fail to kill catalase-positive bacteria even though there is no apparent decrease in granulocytic ingestion of bacteria.[2] Evidence indicates that this disease is related to an abscence of the "respiratory burst" associated with phagocytosis of normal granulocytes. Indeed, granulocytes from patients with CGD fail to produce antibacterial substances such as superoxide anion or hydrogen peroxide in response to bacterial exposure. In addition, CGD neutrophils fail to generate CL or display the increased hexose monophosphate shunt activity in response to

[1] R. L. Stjernholm, R. C. Allen, R. H. Steele, W. W. Waring, and J. A. Harris, *Infect. Immun.* **7,** 313 (1973).
[2] A. I. Tauber, N. Borregaard, E. Simons, and J. Wright, *Medicine (Baltimore)* **62,** 286 (1983).
[3] E. L. Mills and P. G. Quie, *Rev. Infect. Dis.* **2,** 505 (1980).

bacteria.[4] The genetic heterogeneous complexity of this disease is probably reflective of a variety of point mutations in the sequence of events coupled to neutrophilic oxidative metabolism. Fundamental defects of CGD neutrophils include glucose-6-phosphate dehydrogenase deficiency, glutathione peroxidase, abnormal NADPH affinity for its substrate, abnormal activation of NADPH oxidase from defective triggering of membrane potential, and dysfunction of an electron transport system including flavoproteins, cytochrome b, and quinones.[5] CGD disease in males is familial and its transmission is clearly X-linked. The X-linked inheritance in females can be followed by luminol-dependent CL[6,7] and females that are heterozygous for the CGD trait produce CL intermediate between noncarriers and X-linked male or female carriers. The simplest method used to detect CGD is to do a luminol-dependent whole blood assay with phorbol myristate acetate (PMA) as a soluble stimulant[8] (see section on assay).

Myeloperoxidase Deficiency. The deficiency of myeloperoxidase can be detected easily using luminol-dependent CL. Myeloperoxidase deficiency does not generally cause a prominent reduction of host body defense. It carries only a 50% decrease in native CL (no luminol added) and a 95% decrease in luminol-dependent CL activated by opsonized bacteria or zymosan.[9] Confirmation of myeloperoxidase deficiency is accomplished by staining for peroxidase.

Inflammation

Arthritis. A study by James *et al.*[10] indicated that joint fluid from individuals with rheumatoid arthritis produces soluble factors which stimulate luminol-dependent CL from neutrophils from joints or peripheral blood. They indicated that the soluble factors may be immune complexes. Other studies[11] in arthritics indicated that platelet activating factor (PAF) or *N*-formylmethionylleucylphenlalanine (nFMLP) stimulated luminol-dependent chemiluminescence (LDCL) 6-fold relative to normals. Drug treatments using low dose methotrexate or gold appear to decrease this stimulation of LDCL in certain individuals only.

[4] R. C. Allen, R. L. Stjernholm, M. A. Reed, T. B. Harper, S. Gupta, R. H. Steele, and W. W. Waring, *J. Infect. Dis.* **136**, 510 (1977).
[5] J. I. Gallin, *Ann. Intern. Med.* **99**, 657 (1983).
[6] E. L. Mills, K. S. Rholl, and P. G. Quie, *J. Clin. Microbiol.* **12**, 52 (1980).
[7] E. L. Mills, K. S. Rholl, and P. G. Quie, *J. Clin. Invest.* **66**, 332 (1980).
[8] L. R. DeChatelet and P. S. Shirley, *Clin. Chem.* (*Winston-Salem, N.C.*) **27**, 739 (1981).
[9] P. Stevens, D. J. Winston, and K. Van Dyke, *Infect. Immun.* **22**, 41 (1978).
[10] D. W. James, W. H. Betts, and L. G. Cleland, *J. Rheumatol.* **10**, 184 (1983).
[11] M. Thornton, K. Van Dyke, C. Van Dyke, and A. DiBartolomeo, *Fed. Proc., Fed. Am. Soc. Exp. Biol.* **44**, 1839 (1985).

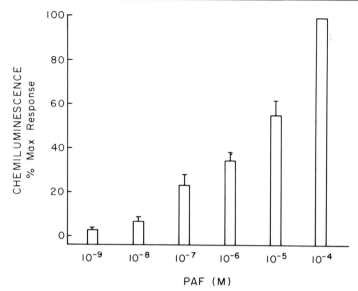

FIG. 3. Effect of various doses of platelet activating factor (PAF) on whole blood chemiluminescence. Whole blood was assayed as in luminol-dependent CL (ref. 15). This assay was accomplished using a liquid scintillation counter.[40]

Exercise-Induced Asthma. Recent findings indicate that a phospholipid substance, platelet-activating factor (PAF), is released from lungs of individuals with exercise-induced asthma. Once this substance is released, its concentration increases dramatically in the blood.[12] Morley *et al.*[13] have proposed that an exacerbation of asthma symptoms arises as a consequence of PAF formation within the lung. They suggested that inflammation arises due to exposure to PAF including an action on platelets and products which causes a reduced threshold to various stimuli and induction of hypertrophy in smooth muscle. It has been shown by Poitevin *et al.*[14] and Edinboro *et al.*[15] that PAF causes LDCL of human neutrophils (see Fig. 3). This CL from neutrophils can increase up to 50-fold in exercise-induced asthmatics.[16] PAF receptor inhibitors decrease the CL from PAF stimulation of human neutrophils.[16] One of these sub-

[12] K. E. Grandel, E. Schwartz, D. Greene, M. L. Wardlow, R. N. Pinckard, and R. S. Farr, *Fed. Proc., Fed. Am. Soc. Exp. Biol.* **42,** 1026 (1983).
[13] J. Morley, S. Sanjar, and C. P. Page, *Lancet* **2,** 1142 (1984).
[14] B. Poitevin, R. Roubin, and J. Benveniste, *Immunopharmacology* **7,** 135 (1984).
[15] L. Edinboro, K. Van Dyke, D. Peden, V. Castranova, and D. Wierda, *Microchem. J.* (in press).
[16] K. Van Dyke, unpublished observations (1984).

stances is a natural product from a chinese traditional remedy known as kadsurenone.[17]

Pollen-Induced Cellular CL. Lindberg *et al.* demonstrated that pollen grains produced cellular CL with isolated neutrophils from allergic and nonallergic individuals.[18] Serum and neutrophils from nonallergic persons produced highly specific and reproducible CL. But these same substances from an allergic person produced less CL. They speculated that the inhibition of cellular CL was due to IgE.

Infection

Bacteria. Bacterial infections generally elevate host cellular CL. This has been noted in animal and human studies.[19] Luminol-dependent CL has been used as an indicator of germ warfare using *Franciella tularensis*.[20] Virulent *Salmonella typhi* were found to depress cellular CL, so apparently increases are not always found associated with infection, but this is an exception rather than rule.

Virus. Pox virus has been shown to produce luminol-dependent CL when infecting mouse spleen cells.[21] This CL reached a peak in 60–90 min postinfection and declined thereafter. Irradiated virus still produced CL but heat-inactivated virus does not, indicating that a viral protein is important in producing light.

Influenza virus induces CL in the absence of antiviral antibody as demonstrated in mouse spleen cells[22] and human neutrophils.[23,24]

Sendai virus, which causes respiratory tract infections in humans, mice, and swine, has been shown to trigger CL in mouse spleen cells.[23]

Mycoplasma are wall-less bacteria that frequently contaminate *in vitro* cultures. It has been shown that such contamination can produce cellular CL irrespective of other modes of CL production such as soluble or particle stimulants and virus, etc.[25] Cultures must be free of mycoplasma

---

[17] T. Y. Shen, S. B. Hwang, M. N. Chang, T. W. Doebber, M.-H. T. Lam, M. S. Wu, X. Wang, G. Q. Han, and R. E. Li, *Proc. Natl. Acad. Sci. U.S.A.* **31**, 261 (1985).

[18] R. E. Lindberg, J. L. Pinnas, and J. F. Jones, *J. Allergy Clin. Immunol.* **69**, 388 (1982).

[19] J. P. McCarthy, R. S. Bodroghy, P. D. Jarhling, and P. L. Sobocinski, *Infect. Immun.* **30**, 824 (1980).

[20] R. E. Kossack, R. L. Guerrant, P. Densen, J. Schadelin, and G. L. Mandell, *Infect. Immun.* **31**, 674 (1981).

[21] E. Peterhans, J. Mundy, and C. R. Parish, *Eur. J. Immunol.* **10**, 477 (1980).

[22] E. Peterhans, *in* "Cellular Chemiluminescence" (K. Van Dyke and C. Castranova, eds.), CRC Press, Boca Raton, Florida, 1986.

[23] E. Peterhans, *Virology* **105**, 445 (1980).

[24] E. L. Mills, Y. Debets-Ossenkopp, and H. A. Verbrugh, *Infect. Immun.* **32**, 1200 (1981).

[25] E. Peterhans, G. Bertoni, P. Koppel, R. Wyler, and R. Keller, *Eur. J. Immunol.* **14**, 201 (1984).

in order to attribute cellular CL to the added stimulant. Cellular CL with mycoplasma can be used as a handy assay for its presence, but positive proof of its presence is shown by colony formation on mycoplasmic agar plates.

Cancer

Chemotherapy for cancer is toxic for a variety of cells including human neutrophils. Since many of the anticancer agents cause bone marrow suppression, it would not be surprising to find an effect on neutrophilic cellular function because the cells mature in marrow. This could and does increase susceptibility to infection. In one study 6 weeks of chemotherapy (postunilateral mastectomy) caused a 50% decrease in cellular CL of granulocytes[26] (see Fig. 4). In a study with cells obtained from patients with acute leukemia, the following observations were made: (1) acute lymphoid leukemia cells were negative regarding the production of CL and (2) acute myeloid leukemia cells of M1, M2, or M3 subgroups of Fab receptor were negative for production of CL. From 13 patients with M4 Fab classification, 10 were positive for CL and phagocytosis. One of 5 patients with M5 Fab classification gave cellular CL. The M classification with Fab subgroups is defined as follows: M1 cells are mostly abnormal myeloblasts; M2 cells contained myeloblasts and a few promyelocytes; M3 cells were heavily granulated promyelocytes; M4 (myelomonocytic leukemia) cells were monocytes with a few promyelocytes; and M5 was pure monocytic leukemia.[27]

It is clear that maturation of neutrophils is an important consideration in producing cellular CL. This may explain the 50% decrease using chemotherapeutic agents to treat cancer in the previous study, i.e., drugs affect maturation of neutrophils.

Miscellaneous Diseases with Risk of Infection

Chronic Hemodialysis. Leukocytes from chronic renal dialysis patients have a higher resting level of light emission and a reduced response to soluble and particulate stimuli compared to normals.[28]

Renal Transplant. Leukocytes from renal transplant patients have a markedly reduced response to soluble and particulate stimuli compared to cells from normals.[28]

[26] M. Matamoros, B. K. Walker, K. Van Dyke, and C. J. Van Dyke, *Pharmacology* **27**, 29 (1983).
[27] M. Hirano and T. Matsui, *in* "Cellular Chemiluminescence" (K. Van Dyke and V. Castranova, eds.). CRC Press, Boca Raton, Florida, 1986.
[28] S. V. Shah, *in* "Cellular Chemiluminescence" (K. Van Dyke and V. Castranova, eds.). CRC Press, Boca Raton, Florida, 1986.

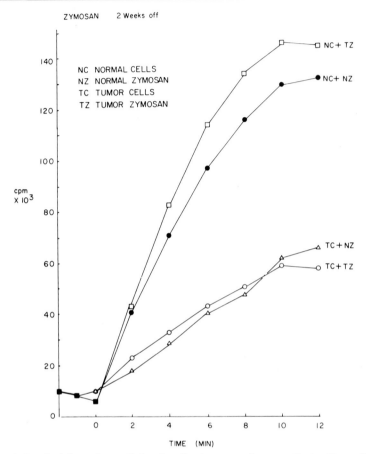

FIG. 4. Luminol-dependent cellular chemiluminescence from a patient with a unilateral mastectomy using cancer chemotherapy drugs cyclophosphamide, methotrexate, 5-fluorouracil, vincristine, and prednisone.[26] One can observe a large difference in CL between the normal cells (with zymosan opsonized from normal or tumor serum) and tumor cells with zymosan opsonized with either serum. This was accomplished in a liquid scintillation counter.

Diabetic Patients. Leukocytes from diabetic patients have a reduced response of cellular CL to soluble and particulate stimuli. Resting levels of cell CL are decreased possibly due to serum factors.[28]

Secondary Hypothroidism. Cellular CL of human neutrophils increased in secondary hypoparathyroidism.[29] It is thought that this is due to serum factors.

[29] S. N. Tuma, R. R. Martin, L. E. Mallette, and G. Eknoyan, *J. Lab. Clin. Med.* **97,** 291 (1981).

Sarcoidosis

Patients with acute sarcoidosis produce more monocytic cellular CL with opsonized zymosan than the comparable cells from normal patients. Kanegasaki *et al.*[30] believe that the disease activates the monocytes to particulate stimuli. When patients with active sarcoidosis receive glucocorticoids the enhanced monocytic CL returns to normal.

Assay Systems

Discussion

Native Cellular CL. The original detection of native cellular CL was by Robert Allen *et al.*[31] Opsonized particles (coated with antibody/complement) were added to human neutrophils in a suitable buffer to produce light. There were no added chemicals in this system to produce light more efficiently.

Luminol-Assisted Cell CL. A major problem with doing native CL studies is that many neutrophils are needed to produce detectable CL (5×10^6–10^7). Therefore a chemical bystander molecule (luminol or 3-aminophthalhydrazide) is added to the preparation of cells to allow efficient detection of blue light.[32] In a variety of studies luminol has mirrored similar activity as native CL in most cases.[33-36] Luminol is synonomous with peroxidase activity.[34] The minimum number of neutrophils usable in this assay is 5×10^4.

Lucigenin-Assisted Cell CL. Lucigenin (bis-*N*-methyl acridinium nitrate) has been used as a chemiluminigenic probe for the measurement of neutrophil oxidation–reduction activity.[37] It is a water-soluble substance eliminating the need of adding solvents. Lucigenin has not been used as

[30] S. Kanegasaki, J. Y. Homma, H. Homma, and M. Washizaki, *Int. Arch. Appl. Immunol.* **64,** 72 (1981).

[31] R. C. Allen, R. L. Stjernholm, and R. H. Steele, *Biochem. Biophys. Res. Commun.* **47,** 67 (1972).

[32] R. C. Allen and L. D. Loose, *Biochem. Biophys. Res. Commun.* **69,** 245 (1976).

[33] K. Van Dyke, M. A. Trush, M. A. Wilson, P. Stealey, and P. Miles, *Microchem. J.* **22,** 463 (1977).

[34] P. Stevens, D. J. Winston, and K. Van Dyke, *Infect. Immun.* **22,** 41 (1978).

[35] M. E. Wilson, M. A. Trush, K. Van Dyke, J. M. Kyle, M. D. Mullett, and W. A. Neal, *J. Immunol. Methods* **23,** 315 (1978).

[36] P. R. Miles, P. Lee, M. A. Trush, and K. Van Dyke, *Life Sci.* **20,** 165 (1977).

[37] R. C. Allen and T. F. Lint, *in* "Analytical Applications of Bioluminescence and Chemiluminescence" (E. Schram and P. Stanley, eds.), p. 589. State Publishing and Printing, Westlake Village, Caifornia, 1979.

widely as luminol but apparently mirrors the production of superoxide ($\cdot O_2^-$) fairly well.

7-Dimethylaminonaphthalene-1,2-dicarbonic Acid Hydrazide (DMNH)-Assisted Cell CL. DMNH has been found to be more efficient than luminol in measuring cellular CL from monocytes stimulated with opsonized particles.[38]

Assays

All assays are performed as indicated. In all cases a blank (without stimulant) should be accomplished using the same conditions as with stimulant. The integrated CL value from blank should be subtracted from CL obtained with stimulant in those cases where integration of CL is used.

Native CL. Native CL from human neutrophils can be assayed as follows: $5 \times 10^6 - 1 \times 10^7$ granulocytes in 7 ml Krebs–Ringer bicarbonate (KRB) solution (pH 7.4) are added to a standard liquid scintillation vial (dark adapted). Heat-killed opsonized bacteria, 0.6 mg dry weight, are added and CL is followed for 1 hr in a liquid scintillation counter[39] set in out of coincidence mode using ^3H settings.

Leukocytes (granulocytes) are isolated by settling blood in 5% KRB-dextran (200,000 MW) for 45 min at 37° and collecting plasma. The leukocyte-rich plasma is centrifuged at 120 *g* for 4 min. Supernatant fluids are decanted and sedimented leukocytes washed by resuspension in KRB buffer and recentrifuged as before. Cells are resuspended at $5 \times 10^6/7$ ml buffer.

Opsonization: 10^9 colony-forming units of bacteria are opsonized by incubation in 1 ml of 50% serum at 37° for 30 min. Bacteria should be washed once to remove serum. A ratio of bacteria to granulocytes of 200 : 1 is effective.[34]

Luminol-Dependent CL. Isolated neutrophils. Isolated neutrophils ($10^5 - 10^6$) are placed in 0.5 ml HEPES buffer (containing Ca^{2+}). The composition for this buffer is 145 m*M* NaCl, 5 *M* KCl, 10 m*M* Na-HEPES, and 5.5 m*M* glucose, pH 7.4. HEPES buffer (1.5 ml) containing luminol is added[39] at a concentration of 1×10^{-7} *M* (final concentration). [Luminol was dissolved as 10^{-3} *M* in 1% dimethyl sulfoxide HEPES (with luminol dissolved in DMSO first, then diluted in HEPES).] Either opsonized bacteria in HEPES (20 μl CFU $= 10^9$) or zymosan (20 μl of 1 mg/ml) in HEPES is added to initiate phagocytosis and CL. Chemiluminescence is assessed in a standard or a mini-LSC vial as before. Soluble stimulants

[38] C. Eschenbach, W. Hahm, and U. Adrian, *Klin. Wochenschr.* **63,** 364 (1985).
[39] M. A. Trush, M. E. Wilson, and K. Van Dyke, this series, Vol. 57, p. 462.

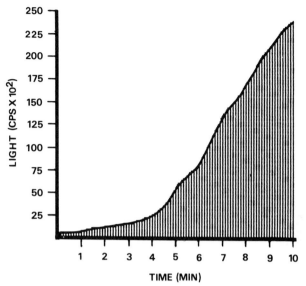

FIG. 5. Luminol-dependent cellular CL with whole blood activated by zymosan.[45] This assay was accomplished in a Bethold 9500T luminometer at 37°.

such as 10^{-6} M phorbol myristate acetate (PMA), 10^{-5} M nFMLP, 10^{-6} M ionophore A23187, or 10^{-5} M PAF can be used.

Whole blood assay. Whole blood assay is accomplished according to the following conditions: human blood is drawn into a Vacutainer containing heparin (green top). It is diluted in HEPES buffer in the ratio 9 parts buffer : 1 part blood. Each LSC vial contains a final volume of 2 ml. It contains the following ingredients: 1 ml of 1 : 10 diluted blood, 200 μl of 10^{-5} M luminol (diluted 1 : 100 with HEPES), and to initiate the reaction 200 μl PMA 10^{-5} M[40] (PMA was dissolved in DMSO at 2 mg/ml).

The CL is accomplished in a liquid scintillation counter (LSC) for 60 min. When CGD blood is compared to normal it produces no light[40] (see Fig. 5).

Lucigenin-Dependent CL. Isolated neutrophils. Neutrophils (10^4–10^5) are placed in 1.5 ml HEPES buffer in an LSC vial with 200 μl of 50 nM (25 μM final concentration) lucigenin in HEPES. The reaction is initiated with 300 μl of a particle (opsonized zymosan, 50 g). Alternatively, PMA at 0.25 M final concentration (0.05 μM) in 300 μl is added. PMA produces lucigenin-dependent CL efficiently.[41]

[40] L. E. Edinboro, K. Van Dyke, D. Peden, V. Castranova, and D. Wierda, *Microchem. J.* **31,** 26 (1985).
[41] R. C. Allen, *in* "Bioluminescence and Chemiluminescence" (M. DeLuca and W. D. McElroy, eds.), p. 63. Academic Press, New York, 1981.

Whole blood assay. Whole blood assay is accomplished according to the following conditions: heparinized whole blood is diluted 1 part blood : 199 parts buffer (10 μl blood in 2 ml HEPES) and is added to 0.5 μM lucigenin (final concentration). Twenty microliters of PMA (25 nM) or 50 μg of opsonized zymosan is added to initiate the CL reaction.[42]

DMNH-Assisted CL. Partially isolated monocytes (10^5) in 100 μl phosphate-buffered saline (PBS) was added to a 3-ml test tube. One hundred seventy-five microliters of PBS and 175 μl DMNH solution containing 1.02 μg DMNH are added to the preparation. The CL assay was accomplished in a Berthold 9500 luminometer.[43] [DMNH solution is made by dissolving 5.1 mg DMNH in 2 ml of dimethyl sulfoxide and diluted with 8 ml of phosphate-buffered saline. Phosphate-buffered saline is made 0.15 M, pH 7.2.]

Instrumentation

There are three different instruments available to measure cellular chemiluminescence. They are the liquid scintillation counter (set in out of coincidence mode or with one photomultiplier disconnected), luminometer (with or without computer assistance), and photometer. The concentration of reagents and volume of reaction mixture can vary with the instrument.

Liquid Scintillation Counter (LSC)

The most available instrument to do cellular CL is the liquid scintillation counter. However, the instrument should be an ambient one or, if refrigerated, the refrigeration should be turned off for several days to allow thermal equilibration with room temperature. Second, the coincidence circuitry must be shut off to count photons, otherwise the counts generated from each tube (if not coincident) will be subtracted from each other. Alternatively, one photomultiplier (PM) tube can be shut off or disconnected. With more recent LSCs, this is difficult to accomplish because of a tie-in with computer systems. In any case, a call to the service representative from the company that manufactured the instrument would be useful. Tritium settings for the LSC-PM windows are effective for cellular CL.[44]

Cellular CL generally works far more efficiently in a heated chamber (37°). There is no simple method to accomplish this in the standard LSCs. Therefore, luminometers with temperature control were developed.

[42] R. C. Allen and B. A. Pruitt, *Arch. Surg. (Chicago)* **117**, 133 (1982).
[43] C. Eschenbach, W. Hahm, and U. Adrian, *Klin. Wochenschr.* **63**, 364 (1985).
[44] K. Van Dyke, *Packard Tech. Bull.* **20**, 1 (1974).

Luminometers

There are a variety of commercial luminometers available including instruments from Berthold (Wildbad, Germany), Packard (Downers Grove, IL), and LKB (Turku, Finland). With all of this instrumentation, the option is provided to have computer linkage. Software is available to record, store, and analyze data making handling of data a much simpler task. The Berthold systems are generally linked to Apple computer systems (II-E); Packard uses IBM-PC as the computer and LKB uses their own system. A more detailed description has appeared.[45]

Photometers

There are three commercial photometers available that could be useful for measurement of cellular CL. In general, these are not the most sensitive detectors of cellular light. They will detect bystander-assisted (luminol or lucigenin) cellular CL. However, more of the chemiluminigenic substance must be added to get reproducible measurements. Examples of these instruments are from Turner Designs (Mountain View, CA), SML/Aminco (Urbana, IL), and Aloka (Tokyo, Japan).

Helpful Hints

Although the subject of isolation of cells has been covered in *Methods in Enzymology* previously, a few helpful hints may be useful. Neutrophils tend to lose biological activity over a matter of 3–4 hr. Therefore, a technique which isolates the cells quickly without possible physiological damage is most useful. Neutrophils isolated within 30 min with monopoly resolving medium (Flow Laboratories, McLean, VA) using Ficoll–Hypaque of density 1.114 are quite viable. If the medium is quickly washed free from the cells, the biological activity is quite good for 3–4 hr postisolation. With some nutritional/disease conditions (ingestion of fish oil for 6 weeks) or with diabetes/kidney dialysis or a combination, the density of the cells changes and the density-based isolation system does not separate these cells.

The concentration of the bystander molecule (BM-luminol or lucigenin) relative to the number of phagocytic cells can be important. If the BM concentration is too large (generally $>10^{-6} M$), the metabolic burst of isolated cells can be activated without a stimulant. Therefore, it is important to measure the cell CL without stimulant to make sure the baseline is relatively constant. When doing whole blood CL, higher amounts of BM

[45] K. Van Dyke, M. R. Van Scott, and V. Castranova, this series, Vol. 132.

concentration can be used. However, a relatively constant baseline is still quite important.

Once cells are isolated, it is easier to keep them in a solution that does not contain added calcium until assay. Routinely, we find it best to keep cells at room temperature rather than on an ice bath and add calcium back at the time of assay.

Calculations

There are three methods to analyze the time versus light curve generated by typical chemiluminescent reactions. These are integration, peak height measurement, and slope analysis. Of these three, we have found integration to be the best representation. Peak height measurement is simple but not particularly descriptive. Slope analysis is seldom used and is omitted here. It is, however, useful for rate of reaction studies.

Peak height measurement involves measuring a peak (y axis) and associating it with a time (x axis). Multiple peaks within the same graph may be measured and these numbers compared to the results of other experiments. This technique is useful if the magnitude and the kinetics from several experiments are very similar (results will be nearly the same as integration) or if a more qualitative approach is appropriate.

Integration is a technique that represents a graph in terms of the area it encloses. There are three approximations that we have used in our lab. These are Simpson's rule, the trapezoidal rule, and cutting and weighing the area. The latter requires no mathematical manipulations. Instead, one obtains an x–y plot of all data to be compared on sheets of paper with uniform mass distribution. Photocopying is adequate in this regard. Then, the graphs are cut out and weighed on an analytical balance. Given that the paper is uniform, then this mass is directly proportional to the area.

Usually, the most exact method of integration is Simpson's rule. This system uses parabolas to approximate a curve. Data must be in a specific form to apply this rule. Points must be paired so that the x axis (time) is divided into an even number of intervals of the same length, e.g., 0, 5, or 10 sec. This length is referred to as t (time) below. The proper y axis (light) value must be associated with the x axis. These y values are numbered y_1, y_2, ... , y_n where n is equal to the total number of data pairs. The area under the curve is then given by

$$\text{Area} = (\Delta t/3)\,(y_1 + 4y_2 + 2y_3 + 4y_4 + 2y_5 + \ldots + 2y_{n-2} + 4y_{n-1} + y_n)$$

A less exacting method is the trapezoidal rule. This mathematical technique uses trapezoids to approximate the area a curve encloses. Although generally less accurate (accuracy is dependent on the shape of the curve), the approximation is still good and is quite useful. The data must be in the same form as that of Simpson's rule with the exception that there need not be an even number of data pairs. The area under the curve is given by

$$\text{Area} = \Delta t([y_1/2] + y_2 + y_3 + \ldots + y_{n-1} + [y_n/2])$$

Computers have helped in the above mathematical calculations. Some luminometers have computer links available. The Berthold and Packard luminometers have on-line integration and provide printouts. Software programs are available that give prompts for the input of data and, having obtained the data, will perform the area calculation.[46]

Conclusions

It has been documented that cellular CL can be a useful assay in a variety of diseases that can affect the biological activity of phagocytic cells. There are a variety of choices of types of assays of cellular CL including "native," luminol, lucigenin, and DMNH dependent. Instruments available to do cellular CL include the liquid scintillation counter, luminometer, and photometer. The calculations are simple using either slope analysis, peak height, or total integration; the latter is generally the most useful.

[46] R. J. Tallarida and R. B. Murray, p. 47. Springer-Verlag, Berlin and New York, 1981.

[38] Luminescent Bioassays Based on Macrophage Cell Lines

By Patrick De Baetselier and Eric Schram

Cells of the monocyte/macrophage lineage serve several essential functions for immune responses. In the first instance, macrophage cells display effector functions such as a receptor-mediated phagocytosis and macrophage-mediated tumor lysis resulting in the destruction or catabolism of the target antigen. Second, they represent an important accessory cell for a variety of regulatory immune functions and they produce a series of nonspecific molecules which modulate immune functions, e.g., interleukin 1 (IL-1) and prostaglandins. Furthermore, these different functions subserved by macrophage cells are controlled by immunoregulatory molecules secreted by activated T lymphocytes (collectively designated as lymphokines).

Considering this central role of macrophage cells in the immune system, assay systems had to be developed to probe accurately the functional activity of macrophage cells as well as their regulation. An extreme sensitive method to monitor macrophage activity, at least when studying the process of antigen catabolism, is provided by chemiluminescent measurements. Indeed, the process of phagocytosis and subsequent killing of ingested pathogens in the phagolysosomes is associated with the production and release of highly reactive oxygen metabolites such as superoxide anion (O_2^-) hydrogen peroxide (H_2O_2), singlet oxygen (O_2^Δ), and hydroxyl radicals ($OH\cdot$). This activity, termed the respiratory burst,[1] appears to represent the major mechanism by which macrophage cells exert their antimicrobial and antitumor activity as well as tissue injury that may accompany inflammatory reactions. The degeneration of these labile oxygen components results in light emission, i.e., chemiluminescence (CL),[2-4] that can be amplified through addition of exogenous luminescent substrates such as luminol or lucigenin (chemiluminescent probes or CLP).[5-7] CL measurements thus allow measurement of the biological activity of macrophage cells in different individuals in a given disease state such as chronic granulomatous disease or in response to drugs.[8]

Besides these diagnostic applications, this assay system also allows testing of the effect of different agents on macrophage cell activation and function, implying that CL measurements could be adopted to screen analytically various immunological and pharmacological agents. Yet, in order to relate quantitatively macrophage-mediated CL emission to the presence of either a triggering or a modulating agent, the quantity and functional capacity of the responding macrophage cells should be held constant. This obligatory requirement is difficult to fulfill, primarily be-

---

[1] B. M. Babior, *N. Engl. J. Med.* **298**, (1978).
[2] R. C. Allen, R. L. Stjernholm, and R. H. Steele, *Biochem. Biophys. Res. Commun.* **47**, 679 (1972).
[3] R. C. Allen, S. J. Yevich, R. W. Orth, and R. H. Steele, *Biochem. Biophys. Res. Commun.* **60**, 909 (1974).
[4] M. A. Thrush, M. E. Wilson, and K. S. Van Dyke, this series, Vol. 57, p. 462.
[5] R. C. Allen and L. D. Loose, *Biochem. Biophys. Res. Commun.* **69**, 245 (1976).
[6] P. Stevens, D. J. Winston, and K. S. Van Dyke, *Infect. Immun.* **22**, 41 (1978).
[7] K. Van Dyke, C. Van Dyke, and J. Udeinya, *Clin. Chem. (Winston-Salem, N.C.)* **25**, 1655 (1979).
[8] B. R. Bloom, B. Diamond, R. Muschel, N. Rosen, J. Schneck, G. Damiani, O. Rosen, and M. Scharff, *Fed. Proc., Fed. Am. Soc. Exp. Biol.* **37**, 2765 (1978).

cause of the extreme heterogeneity of the cell types analyzed, the various stages of differentiation of primary macrophages, and the inability of primary macrophages to grow in culture. Hence the standardization of macrophage-dependent chemiluminescence for analytical purposes is hampered by the extreme heterogeneity of the macrophage cells analyzed. One approach to circumvent this problem, inherent in the analysis of distinct subsets of functional cells, would be the development of immortalized macrophage cell lines. Indeed the availability of uniform, cloned populations of macrophage cell lines might simplify to a great extent the study of macrophage-mediated CL emission and the modulation of this emission.

For the past several years different investigators have developed and studied cloned macrophage-like cell lines since macrophage-related cell lines offer many advantages for the study of macrophage functions.[8] These advantages range from convenience in obtaining large numbers of cells of relatively homogeneous nature by growth in culture, to purity of the cell population ensuring that experimental results obtained are due only to macrophages. A variety of cell lines and variants exist which stably differ in degree of maturation, sensitivity to inducing agents, and extent of mature characteristics. Finally since the cell lines are growing rapidly, it is relatively easy to synchronize or select cells at different stages of the cell cycle for study of physiological functions throughout the cell division process. Most studied macrophage lines are derived from murine macrophage tumors and serve as model systems for understanding macrophage functions. Yet these few macrophage-like tumor cells may not represent the whole array of macrophage phenotypes and functions. For instance, in the human system, few if any human cell lines exist that exhibit the general characteristics of macrophages. So far two relatively undifferentiated promonocytic cell lines, U937 and HL-60, have been reported which can be induced to differentiate into nondividing cells with macrophage-like properties.[9,10] To circumvent the limitations imposed by the restricted availability of macrophage cell lines, different attempts were made to immortalize primary human or murine macrophages. Two methods were mainly adopted: transfection with viral DNA[11] and somatic cell hybridization.[12]

We have adopted the somatic hybridization technology for the genera-

[9] L. T. Clement and J. E. Lehmeyer, *J. Immunol.* **130**, 2763 (1983).
[10] R. F. Todd, J. D. Griffin, J. Ritz, L. M. Nadler, T. Abrams, and S. R. Schlossman, *Leuk. Res.* **5**, 491 (1981).
[11] Y. Nagata, B. Diamond, and B. R. Bloom, *Nature (London)* **306**, 597 (1984).
[12] E. Tzehoval, S. Segal, N. Zinberg, and M. Feldman, *J. Immunol.* **132**, 1741 (1984).

tion of continuous cell lines expressing macrophage phenotypes and macrophage functions. This technology is based on cell fusion between a normal somatic partner and a malignant cell partner, resulting in a hybrid which will manifest different properties derived from the two parents. A successful hybrid will inherit the malignant properties of the tumoral partner (infinite growth *in vitro* and *in vivo*) and functional properties of the normal somatic partner. This technology was developed initially for the immortalization in culture of antibody-secreting lymphocytes[13] and of functional T cells.[14] A prerequisite for the generation of stable functional hybrids that can be maintained in long-term cultures is the identification of a stable, highly fusogenic tumor cell line. Furthermore, since compelling evidence has been provided that a phenotypic identity is required between the two fusion partners for efficient functional immortalization, macrophage-like cell lines were screened for their fusogenic capacity. We have isolated such a macrophage-like fusogenic cell line, derived from the murine lymphosarcoma J744 clone. This 8-azaguanine-resistant cell line (designated J774-C2E2-HAT) appeared to be a very efficient tumor cell line for PEG-mediated somatic hybridization with different murine macrophages (i.e., splenic macrophages, peritoneal macrophages). The macrophage hybridomas thus generated exhibited, in contrast to the parental tumor fusion partner, a variety of macrophage functions such as phagocytosis, monokine secretion, tumoricidal effects, and production of reactive oxygen metabolites. In particular this last function has been extensively analyzed in order to identify macrophage cell lines that can be adopted as cellular reagents for CL probing.[15,16]

Principle of the Bioassay System

The cellular bioassay described here is based on the assumption that provided the quantity and functional capacity of the macrophage cell population are held constant, using a homogeneous macrophage cell line, the degree of stimulated macrophage oxygenation activity and resulting CL emission will reflect the stimulating capacity of a triggering agent as

[13] G. Kohler and C. Milstein, *Nature (London)* **256**, 495 (1975).
[14] O. Irigoyen, P. V. Rizzolo, Y. Thomas, L. Rogozinski, and L. Chess, *J. Exp. Med.* **154**, 1827 (1981).
[15] P. De Baetselier, L. Brys, E. Vercauteren, L. Mussche, R. Hamers, and E. Schram, *in* "Analytical Applications of Bioluminescence and Chemiluminescence" (L. J. Kricka, P. E. Stanley, G. H. G. Thorpe, and T. P. Whitehead, eds.), p. 287. Academic Press, New York, 1984.
[16] P. De Baetselier, L. Brys, L. Mussche, L. Remels, E. Vercauteren, and E. Schram, *in* "Cellular Bioluminescence" (K. Van Dyke, ed.). CRC Press, Boca Raton, Florida, 1986 (in press).

described by the following equation:

Macrophage + triggering agent → macrophage oxygenation +   CLP   →   CL
                                           activity   (1)
(constant)   (variable)   (variable)   (constant) (variable)

As such, the triggering capacity of a particular agent is proportional to the macrophage oxygenation activity and therefore to the measured CL. Furthermore, if the triggering agent is held constant, one is able to measure the activity of other agents which modulate the activity of either the responding macrophage cells or the inherent stimulating activity of different triggering agents according to the following equation:

Macrophage + triggering agent + modulating agent →
 (constant)   (constant)   (variable)
                                    macrophage oxygenation +   CLP   →   CL
                                              activity
                                            (variable)   (constant) (variable) (2)

Hence, the modulating activity of a particular agent is proportional to the macrophage oxygenation activity and therefore to the measured CL. Using Eqs. (1) and (2) we have adopted macrophage-mediated CL emission as an assay system to analyze qualitatively and quantitatively different triggering and modulating agents. As will be illustrated below, this cellular assay system has enabled us to monitor the following substances: macrophage-activating factors such as lymphokines and endotoxins, opsonizing compounds such as antibodies, complement, and immune complexes, macrophage-inhibiting substances such as drugs, and macrophage-specific antibodies.

Bioassay Procedure

Cell Lines

The generation of functional macrophage hybridomas through somatic hybridization and the characteristics of the cell lines thus generated have been described elsewhere.[15,16] Briefly, the macrophage hybridoma cells were cultured in RPMI 1640 medium supplemented with 10% fetal calf serum (Gibco, Grand Island, NY), 2 mM glutamine, penicillin (100 U/ml), and streptomycin (100 μg/ml). The macrophage cell lines were maintained in culture by serial passage. All cultures were incubated at 37° in a humidified atmosphere containing 5% CO_2.

Chemiluminometric Measurements

The methodology used for chemiluminometric measurements using macrophage hybridoma cell lines is depicted in Fig. 1. This procedure

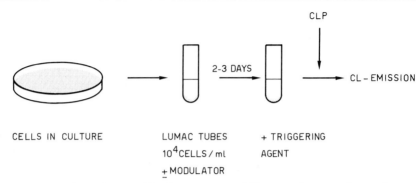

FIG. 1. Experimental protocol for the assessment of CL responsiveness using continuous macrophage cell lines.

allows direct measurement of the oxygenation activity of macrophage cells in the test tubes without additional manipulation of the macrophage cells such as scraping, harvesting, and cell transfer, which might be deleterious to their activity. Furthermore, the oxygenation activity of the macrophage cells can be measured either constitutively or following incubation with modulating agents.

For determinations of chemiluminescence, macrophage hybridoma cells were transferred from tissue culture petri dishes to lumacuvettes (PST cuvettes, Lumac 4960) at a concentration of 1 to 2×10^4 cells in 1 ml of culture medium in the presence or absence of modulators (see below). The lumacuvettes were subsequently incubated for 2 to 3 days. The medium in the lumacuvettes was removed and replaced with veronal-buffered solution containing Ca^{2+} (2×10^{-4} M), Mg^{2+} (5×10^{-4} M), albumin (BSA, 0.2 mg/ml), and glucose (0.2 mg/ml). CL emission after addition of a CLP (10^{-4} M luminol, 5'-amino-2,3-dihydro-1,4-phthalazinedione or 10^{-4} M lucigenin, 9,9'-bis-N-methylacridinium nitrate; Boehringer Mannheim) and a triggering substance was recorded in a 6-channel Biolumat apparatus LB9505 from the Berthold company (Wildbad, FGR). The kinetics of the light emission were recorded in 6 parallel experiments over a total of half an hour. Which method of data presentation is used will be determined, in part, by the response observed and the experimental conditions. In general the CL response is represented as the peak of maximum CL response (CL_{max}) and/or the stimulation index. The stimulation index (SI) is often used as a measure of the ability of modulating agents to induce responsivity in resting macrophage cell lines.

The induction of CL emission can be obtained by a variety of substances (collectively designated as triggering agents). Furthermore, the biological oxygenation activity of macrophages can be modulated in an either positive (enhancement, induction of CL emission) or negative way (inhibition of CL emission) by other substances (collectively designated as modulating agents). The preparation of various triggering and modulating agents has been reported elsewhere.[16] As triggering agents we use chemical triggering agents such as phorbol myristate acetate (PMA), opsonized particles (zymosan, *Micrococcus luteus, Plasmodium chabaudi*) and anti-macrophage antibodies. Currently used modulating agents consisted of lipopolysaccharides (LPS), lymphokine supernates of activated T lymphocytes (LK-SN), and recombinant immune interferon (IFN-γ).

The Use of Chemiluminescence as an Index to Study the Regulation of
    Macrophage Activation

On the basis of functional criteria we have selected two macrophage hybridoma cell lines (i.e., 2C11-12 and LA5-9) for chemiluminescent assays.[15,16] Regarding luminol-aided chemiluminescence, only the LA5-9 cell line was found to emit light constitutively following stimulation with various triggers such as zymosan, microbes, and PMA. The other line produced either a weak or no CL in response to the different triggering agents tested. However, when culturing these cells in a medium containing soluble factors released by activated T lymphocytes (LK-SN), a significant luminescence was produced following stimulation with triggering agents (Fig. 2). These results indicate that the macrophage hybridomas can be activated to differentiate by soluble factors released by activated lymphocytes. Several interpretations might be made concerning the induction or augmentation of respiratory activity in these macrophage hybridoma cell lines. Indeed, there is compelling evidence that T cell lymphokines influence both the differentiation and activation of monocyte/macrophage cells.[17,18] Hence, the induction of chemiluminescent responsiveness in C211-12 cells could reflect the maturation of these cells from their undifferentiated state to a more mature cell, functionally resembling a circulating monocyte. Alternatively, the chemiluminescent responsiveness enhanced in LA5-9 cells by these lymphokines could be analogous to the enhanced respiratory burst observed with the activation of mature macrophages.[19] Whatever the mechanism underlying this lymphokine-mediated induction or activation of CL responsiveness, our data indicate that these cell lines may be an appropriate model for the study of

[17] C. F. Nathan, M. L. Karnovsky, and J. R. Davis, *J. Exp. Med.* **133,** 1456 (1971).
[18] C. F. Nathan, N. Nogueira, C. Juangbhanich, J. Ellis, and Z. A. Cohn, *J. Exp. Med.* **149,** 1056 (1979).
[19] A. Nakagawara, N. M. Desantis, N. Nogueira, and C. F. Nathan, *J. Clin. Invest.* **70,** 1042 (1982).

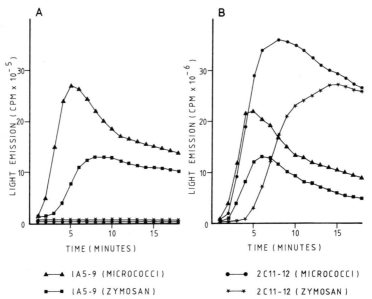

Fig. 2. CL responsiveness of macrophage cells lines in the absence (A) or presence (B) of 5% LK-SN. The CL response assssed against zymosan or micrococci.

the molecular events involved in the generation of a respiratory burst by macrophage cells, as well as for the study of the role of various macrophage-activating substances in this process. In particular, according to Eq. (2), macrophage-activating agents can be considered as modulating agents that induce macrophages to respond to various stimulating agents by CL emission. In such an assay system, the measured CL will be proportional to the quality and quantity of a particular macrophage activating agent tested as illustrated below:

Macrophage   + modulating agent → activated macrophage + stimulating agent →
(unresponsive)   (macrophage   (responsive)   (constant)
                activating
                factor)

$$\text{macrophage oxygenation} + \quad \text{CLP} \quad \rightarrow \quad \text{CL} \tag{3}$$
                     activity
                     (variable)   (constant)   (variable)

Using the relation described in Eq. (3) we have adopted these macrophage hybridoma cell lines as a cellular reagent for the quantitative and qualitative analysis of the macrophage-activating capacity of endotoxins and lymphokines.

CL Detection of Endotoxins

Since small amounts of bacterial endotoxin have been described to prime mononuclear phagocytes[20] and neutrophils[21] for enhanced stimulated release of O_2^- and H_2O_2, we have analyzed whether endotoxins induce or enhance the CL responsiveness of the macrophage hybridoma cell lines. Hence the macrophage cell lines 2C11-12 and LA5-9 were cultured for 24 or 48 hr in the presence of *Escherichia coli* endotoxin concentrations ranging from 1 ng to 1 pg. These time intervals for the 2C11-12 and LA5-9 cell lines were selected on the basis of preliminary experiments aimed at determining the kinetics of the optimal induction of chemiluminescent capability. Following incubation with these modulators, the luminol aided CL responsiveness was assessed toward triggering substances such as PMA, opsonized zymosan, and opsonized micrococci. The results compiled in Table I indicate that LPS induces a concentration-dependent CL responsiveness toward the different triggering agents. In this context it should be mentioned that the quantification of extremely low concentrations of endotoxins using the above described test system may offer an alternative to currently used tests for pyrogen activity. Indeed, exogenous pyrogens such as endotoxins stimulate mononuclear phagocytes to produce an internal pyrogen thought to be the primary mediator of fever. So far the most currently used assay systems for pyrogenicity testing are the cumbersome rabbit test and the *Lumilus* amebocyte lysate (LAL) test.[22] Since 1 pg/ml LPS(E) could be detected using an appropriate macrophage cell line (LA5-9 triggered with micrococci, Table I), one might consider this assay system as a candidate for routine pyrogenicity testing.

CL Detection of Macrophage-Activating Factors

Using these macrophage hybridoma cell lines we have used the chemiluminometric assay system to analyze the macrophage-activating factor (MAF) of various lymphokine preparations. As a source of lymphokines we used supernates of either mouse, human, or rabbit lymphocytes stimulated with Con A (LK-SN). The results shown in Table II,A indicate that, after cultivation of LA5-9 or 2C11-12 cells in conditioned medium containing murine LK-SN for 24 or 48 hr, a significant CL response is recorded after stimulation with various triggering agents. Conditioned me-

[20] M. J. Pabst, H. B. Hedegaard, and R. B. Johnston, *J. Immunol.* **128,** 123 (1982).

[21] L. A. Guthrie, L. C. McPhail, P. M. Henson, and R. B. Johnston, *J. Exp. Med.* **160,** 1656 (1984).

[22] F. C. Pearson and W. Weary, *BioScience* **30,** 461 (1980).

TABLE I
INDUCTION OR ACTIVATION OF CL RESPONSIVENESS BY ENDOTOXINS

		CL response following stimulation with[b]					
		PMA		Zymosan		Micrococci	
Cell line	Modulator[a]	CL_{max}	SI	CL_{max}	SI	CL_{max}	SI
LA5-9	—	810	—	180	—	176	—
	1 ng LPS	11,574	14.2	1,245	6.9	1,811	10.2
	1 pg LPS	2,347	2.9	451	2.5	1,259	7.1
2C11-12	—	26	—	23	—	41	—
	1 ng LPS	640	24.6	313	9.48	584	14.2
	1 pg LPS	35	1.3	60	1.8	113	2.7

[a] LA5-9 or 2C11-12 cells were incubated for 24 or 48 hr, respectively, with 1 ng or 1 pg of lipopolysaccharide from *E. coli* (LPS).
[b] The luminol-aided response is represented as the maximum CL emission ($CL_{max} \times 10^{-3}$) or the stimulation index (SI = CL with modulator divided by background CL).

dium derived from Con A-activated human or rabbit lymphocytes did not induce a significant activation or differentiation of the 2C11-12 or LA5-9 cell lines. Hence these results indicate that this chemiluminometric assay can be adopted to probe for the presence of MAF lymphokines in various biological fluids. Furthermore, the results demonstrate a species specificity for expression of MAF activity in this murine assay, since supernates from stimulated rabbit or human lymphocytes did not exhibit CL-inducing activity toward the murine macrophage cell lines 2C11-12 and LA5-9.

Since many recent reports emphasized the role of IFN-γ as the major or sole MAF lymphokine,[23,24] we have analyzed the potency of various interferon preparations to induce CL responsiveness. As shown in Table II,B, the amount of recombinant IFN-γ added to the cell lines equivalent to an IFN-γ activity of 10 U/ml induces a strong CL responsiveness in the two cell lines tested. Addition of similar quantities of murine IFN-α + β preparations or human recombinant IFN-γ did not induce any effect on the CL responsiveness of these cell lines. This agrees with other reports[25] indicating (1) that murine IFN-α + β are less potent than IFN-γ in induc-

[23] W. K. Roberts and A. Vasil, *J. Interferon Res.* **4,** 519 (1982).
[24] C. F. Nathan, H. W. Murray, M. E. Wiebe, and B. Y. Rubin, *J. Exp. Med.* **158,** 670 (1983).
[25] C. F. Nathan, T. J. Prendgast, M. E. Wiebe, E. R. Stanley, E. Platzer, H. G. Remold, K. Welte, B. Y. Rubin, and H. W. Murray, *J. Exp. Med.* **160,** 600 (1984).

TABLE II
INDUCTION OR ACTIVATION OF CL RESPONSIVENESS BY LYMPHOKINES

Cell line	Modulator[a]	CL response following stimulation with[b]		
		PMA	Zymosan	Micrococci
A. Lymphokine supernates				
2C11-12	1% murine LK-SN	7.1	8.4	11.7
	1% rabbit LK-SN	0.8	1	0.8
	1% human LK-SN	0.9	0.9	1
LA5-9	1% murine LK-SN	2.5	8.8	7.2
	1% rabbit LK-SN	0.8	1.0	1.3
	1% human LK-SN	0.9	0.8	1
B. Interferon preparations				
LA5-9	10 U murine IFN-γ	8	7.3	33
	10 U murine IFN-α + β	1.2	0.5	1
	10 U human IFN-γ	1	0.9	0.8
2C11-12	10 U murine IFN-γ	50	9	28
	10 U murine IFN-α + β	0.7	0.5	1.2
	10 U human IFN-γ	0.9	1.2	0.6
	1 U murine IFN-γ	20	4.2	4.2
	0.1 U murine IFN-γ	3.6	1.2	0.8

[a] LA5-9 or 2C11-12 cells were incubated for 24 or 48 hr, respectively, with lymphokine supernates or interferon preparations as indicated.
[b] The activation of luminol aided CL responsiveness is expressed as the stimulation index (CL with modulator divided by background CL).

ing tumoricidal activity or enhancing H_2O_2-releasing capacity and (2) a species specificity is observed for the expression of MAF activity in this bioassay. Using the C11-12 cell line as a cellular probe for CL modulation by recombinant IFN-γ we found that low IFN-γ levels (1–0.1 U/ml) could significantly increase the CL responsiveness of the C11-12 cell line (Table II,B). Taken together, these results indicate that IFN-γ is a potent activator of the oxidative metabolism of these macrophage cell lines as monitored by chemiluminogenic probing.

CL Analysis of the Interaction between IFN-γ and LPS on Macrophage Activation

Using the CL bioassay we have tested whether the induction of CL responsiveness reflects to a certain extent the functional activity of these macrophage cell lines such as their tumoricidal activity. Macrophages can be activated to kill tumor cells by a nonspecific, extracellular mechanism such as secretion of reactive oxygen intermediates[26] that may be important in host defense against neoplastic cells *in vivo*. It is thus of cardinal importance to optimize the conditions necessary for the full expression of tumoricidal activity. According to current views,[27] macrophage-mediated tumor cytolysis is a two-step process. First macrophages bind on the surface of tumor cells leading to inhibition of tumor cell proliferation. Following binding and cytostasis, macrophages, secrete cytolytic substances such as cytolytic proteases (CP), tumor necrosis factor (TNF), and active oxygen intermediates. The acquisition of binding capacity and cytostasis is inducible by one single external signal such as exposure to lymphokines. The manifestation of cytolytic activity, however, requires a second inducing signal such as LPS. As such, macrophages need two immunomodulators (LPS and MAF) to exert a full tumoricidal activity.[28] It was thus of interest to test, using the chemiluminometric assay (1) whether lymphokines and LPS would act synergistically on the induction or activation of CL responsiveness and (2) whether CL responsiveness reflects the tumoricidal state of a particular macrophage cell line following activation.

To test these possibilities, the 2C11-12 macrophage cell line was cultured in the presence of lymphokines (IFN-γ) and LPS and the CL responsiveness toward different triggering agents was assessed. The tumoricidal activity of the 2C11-12 macrophage cell line was tested by the ^{51}Cr release cytolytic test system using a ^{51}Cr-labeled tumor cell as target (the 3LL lung Lewis carcinoma). The results outlined in Fig. 3 indicate that IFN-γ and LPS act in a synergistic manner on the induction of CL responsiveness by the 2C11-12 cell line toward the different triggering agents (Fig. 3A, B, and C). Furthermore, this synergistic activity of LPS and lymphokines on CL responsiveness was reflected by the tumoricidal activity of this cell line (Fig. 3D). Thus, in confirmation of other reports,[28] optimal lysis or enhanced lysis was recorded when 2C11-12 macrophage cells were treated with lymphokines and LPS simultaneously. Hence the recorded effects of these immunomodulators on the tumoricidal activity of the macrophage cell lines are in accordance with those obtained by CL probing. In this context it should be stressed, however, that the activation of macrophage cell lines by endotoxins and lymphokines is temporarily variable, and that depending on the cell line used, different kinetics of activation were observed. Indeed, whereas the 2C11-12 cell lines reaches

[26] S. J. Weiss, A. F. Lobuglio, and H. B. Kessler, *Proc. Natl. Acad. Sci. U.S.A.* **77**, 584 (1980).
[27] D. O. Adams, J. G. Lewis, and W. J. Johnson, *Prog. Immunol., Int. Congr. Immunol., 5th, 1983* p. 1009 (1984).
[28] J. L. Pace and S. W. Russell, *J. Immunol.* **126**, 1863 (1983).

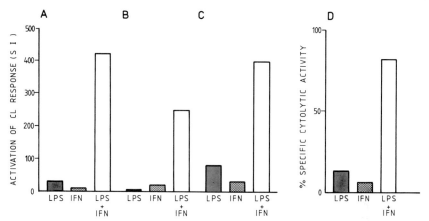

FIG. 3. Activation of CL responsiveness (A, B, C) and tumoricidal activity (D) of the
2C11-12 cell line following preincubation with LPS (1 ng), IFN-γ (1 U), or LPS and IFN-γ.
The CL response was assessed against PMA (A), zymosan (B), and micrococci (C).

a maximal level of activation following a 48 hr incubation with the two
immunomodulators, a completely different pattern of activation is observed by the LA5-9 cell line. As shown in Fig. 4 this cell line reaches a
maximal peak of activation 6–8 hr after addition of LPS and lymphokines.
LPS and lymphokines alone exert a lower CL inducing activity, which is
manifested after a longer incubation period. This hyperresponsiveness of

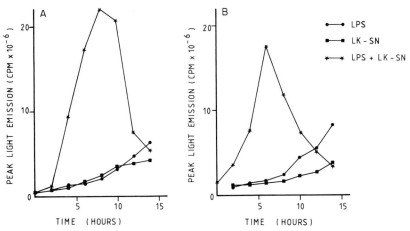

FIG. 4. CL responsiveness of the LA5-9 cell line following preincubation with LPS (1 ng),
LK-SN (5%), or LPS and LK-SN. The CL response was assessed against zymosan (A) or
micrococci (B) at the indicated time intervals.

LA5-9 to the synergistic activity by those two immunomodulators indicates that this cell line has been derived from a relatively mature macrophage.

Assessing Phagocytosis and Opsonization by CL Assay

The stimulation of enhanced oxidative metabolism and accompanying chemiluminescence, following the ingestion or phagocytosis of particulate substances, is dependent on both cellular and humoral factors. The humoral factors consist of a heterogeneous group of serum proteins collectively referred to as "opsonins." The main function of opsonins is to ensure the recognition and labeling of particulate substances in a specific manner prior to their attachment on and subsequent ingestion by phagocytic cells. This process of opsonization is accomplished via antigen-specific immunoglobulins which serve as the heat-stable or immune opsonins (IgG Fc-mediated opsonization) and/or the activities of the classical or alternative pathways of complement which are required for heat-labile opsonization (IgM C3-mediated or C3-mediated opsonization). Contact recognition of opsonified antigens via Fc and C3b receptors on the macrophage membrane results in phagocytosis, activation of O_2 redox metabolism, and finally CL emission. Measurements of CL emission allow as such the evaluation of the opsonic capacity of humoral serum components (i.e., antigen-specific antibodies and complement). Furthermore, CL assays can also be adopted to quantitate components that interfere with the triggering capacity of opsonized particles such as immune complexes, Fc or C3 fragments, natural or synthetic Fc fragments, and anti-Fc or C3 receptor antibodies.

Qualitative Assessment of Serum Opsonic Capacity

Allen has studied the utility of CL responses of phagocytozing polymorphonuclear leukocytes (PMNs) in quantitating serum opsonic capacity.[29,30] Similarly we have assessed Fc-mediated opsonization or C3-mediated opsonization through CL emission using activated macrophage cell lines. Opsonization by activation of the alternative pathway of complement was determined by utilizing for opsonization a substrate (zymosan) that activates the alternative pathway.[15,16] Antibody-dependent opsonization was determined by evaluating the triggering capacity of antibody-coated pathogens such as micrococci, plasmodia, and trypanosomes.[15,16] In particular we found this CL bioassay to be the method of choice in

[29] R. C. Allen, *Infect. Immun.* **15,** 828 (1977).
[30] R. C. Allen and M. N. Lieberman, *Infect. Immun.* **45,** 475 (1984).

evaluating the opsonizing capacity of polyclonal and monoclonal antibodies (Mab) directed against complex parasite antigens such as *P. chabaudi*. Indeed, as illustrated in Fig. 5, a clear-cut relation was observed between the optimal triggering capacity of *P. chabaudi* merozoites and the presence of opsonizing antibodies and complement. Of particular interest is the observed differential capacity of various anti-merozoite Mabs to induce CL emission. Furthermore, sera from patients infected with a human plasmodium, i.e., *Plasmodium falciparum,* were also potent opsonins for *P. chabaudi* merozoites. Hence these results demonstrate that CL probing is not only a sensitive diagnostic tool for the detection of opsonizing antibodies, it also allows discrimination between the biological activities of different antibodies, namely their opsonizing potential. This last aspect might be of tremendous value when screening Mabs for their *in vivo* protective activity. Indeed, the recognition between an Mab and a parasite antigen might not be necessarily related to its *in vivo* protective capacity if such recognition does not ensure an optimal stimulation of the catabolic activity by phagocytes.

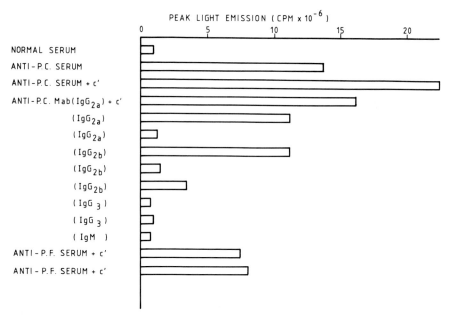

FIG. 5. CL responsiveness of the 2C11-12 cell line toward opsonized merozoites of *Plasmodium chabaudi.* Merozoites were opsonized by polyclonal anti-*P. chabaudi* (anti-P.C.) antibodies with or without complement (C′), monoclonal anti-*P. chabaudi* (anti-P.C. Mab) with complement, and polyclonal anti-*P. falciparum* (anti-P.F.) with complement. The IgG subclasses of the different anti-P.C. Mabs are indicated.

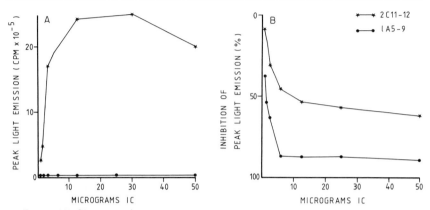

FIG. 6. (A) CL responsiveness of the macrophage cell lines toward immune complexes (IC). (B) Inhibition of the micrococci-induced CL responsiveness of macrophage cell lines by immune complexes (IC).

Detection of Immune Complexes by CL Assay

The detection of circulating immune complexes (IC) is of fundamental importance in gaining insight into the origin and evolution of many diseases. An alternative quantification method to currently used IC test systems such as the C1q assay can be provided by phagocyte CL tests. Indeed, according to a recent report, the binding of ICs on phagocytes through C3b and Fc receptors results in CL emission which appears to be an easy means to evaluate and quantitate ICs in sera.[31] Alternatively ICs could be determined in a CL bioassay by assessing their capacity to inhibit the CL induced by opsonized particles through competitive binding on the Fc or C3b receptor.

We have used the macrophage cell line CL assay to assess the capacity of ICs to induce CL emission or to inhibit the CL emission triggered by opsonized micrococci. When analyzing the intrinsic capacity of human ICs (tetanus toxoid/antitoxoid) to induce CL by activated 2C11-12 or LA5-9 cells, we found a dose-dependent induction of CL emission by the 2C11-12 cell line, but not by LA5-9 cells (Fig. 6A). Conversely, when testing the capacity of human ICs to inhibit CL emission triggered by opsonized micrococci, a dose-dependent inhibition of CL emission was observed (Fig. 6B). The level of inhibition was found to be more pronounced on the micrococci-induced CL emission by the LA5-9 cell line as

[31] J. Willems and M. Joniau, in "Analytical Applications of Bioluminescence and Chemiluminescence" (L. J. Kricka, P. E. Stanley, G. H. G. Thorpe, and T. P. Whitehead, eds.), p. 405. Academic Press, New York, 1984.

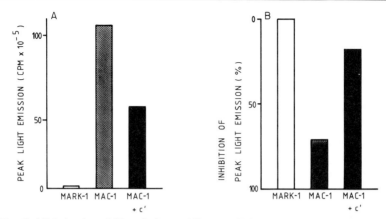

FIG. 7. (A) Induction of CL emission on 2C11-12 cells by control Mab antibodies (Mark-1), Mab M1/7.15.1 anti-macrophage antibodies (Mac-1), and anti-macrophage antibodies in the presence of complement (Mac-1 + C'). (B) Inhibition of the zymosan-induced CL emission of 2C11-12 cells by control Mab antibodies (Mark-1), Mab M1/7.15.1 anti-macrophage antibodies (Mac-1), and anti-macrophage antibodies in the presence of complement (Mac-1 + C').

compared to the 2C11-12 cell line. Hence on the basis of this protocol ICs can easily be quantitated through either inhibition of induced CL emission (LA5-9) or induction of CL emission (2C11-12).

Detection of Membrane Antigens by CL Assay

So far, in this chapter we have emphasized the role of the constant part of an antibody (i.e., Fc) to trigger CL emission following binding to the Fc receptor on the macrophage membrane. In terms of specificity, the binding of an antibody to an antigen located on the macrophage cell membrane might represent a similar event of membrane recognition phenomenon resulting in CL emission. The capacity of antibodies to induce CL emission through their antigen-specific Fab domains has been reported by Peterhans et al.[32] Accordingly, we have reported the CL-inducing capacity of polyclonal and monoclonal anti-macrophage antibodies on activated macrophage cell lines.[16] Hence, CL probing using activated cell lines might provide a sensitive screening system for either the identification of macrophage-specific Mabs and/or for the functional characterization of Mab-defined antigens that are directly or indirectly involved in the respiratory burst. These potential applications are exemplified in Fig. 7 whereby we have analyzed the capacity of the anti-macrophage Mab M1/

[32] E. Peterhans, H. Albrecht, and R. Wyler, J. Immunol. Methods 47, 295 (1981).

70.151 to induce CL by the 2C11-12 cell line or to modulate the CL emission triggered by opsonized zymosan. The results indicate that this anti-macrophage Mab is a potent inducer of CL emission (Fig. 7A) as well as a potent inhibitor of C3-receptor mediated CL (Fig. 7B). Such a finding might not be surprising since the M1/7.15.1 anti-macrophage Mab recognizes a differentiation antigen (Mac-1) expressed on different macrophages, that is closely related or identical to the macrophage C3 receptor.[33] The relation between the Mac-1 membrane antigen and the complement receptor is herein confirmed by CL probing assays, since complement abrogated partially the induction of CL emission (Fig. 7A) or the suppression of C3 receptor-mediated CL by the M1/7.15.1 Mab (Fig. 7B).

The Use of Chemiluminescence as an Index to Study the Effect of Pharmacological Agents

Since the chemiluminescent response of phagocytic cells is dependent on cellular metabolism and activation, the measurement of CL might represent a potentially useful index by which to assess the effect of pharmacological and toxicological agents on phagocytic cells.[4] We have previously reported that pharmacological activity can be assessed through CL emission by macrophage cell lines such as the CL-inducing activity of histamine[16] and the CL-inhibitory activity of local anesthetics such as lidocaine.[17] CL emission might thus provide a powerful tool in immunopharmacology for the following purposes: (1) analytical quantification of pharmacological mediators that either stimulate or modulate CL emission, and (2) screening of pharmacological drugs that manifest antagonistic or synergistic activities on the CL-inducing activity of inflammatory mediators. This last aspect is of particular importance in monitoring the effect of antiinflammatory drugs in phagocyte-mediated inflammatory reactions such as crystal-induced diseases.[34]

We have used CL by macrophage cell lines to investigate the interaction of monosodium urate crystals with phagocytic cells. Such crystals are found in the synovial fluid of patients with acute gout and the observation that the crystals are predominantly contained within PMNs has led to the suggestion that PMNs are of central importance in the induction of inflammation and tissue damage in these conditions.[34] Recently, monosodium urate crystals have been reported to stimulate efficiently the CL

[33] H. May-Kin and T. A. Springer, *J. Immunol.* **128**, 2281 (1982).
[34] D. J. McCarty and J. L. Hollander, *Ann. Intern. Med.* **54**, 452 (1961).

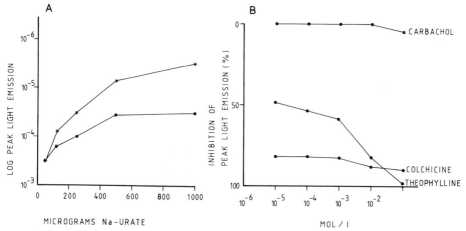

FIG. 8. (A) CL responsiveness of macrophage cell lines toward monosodium urate crystals (Na-urate). (B) Inhibition of the Na-urate-induced CL emission of the 2C11-12 cells by drugs.

emission by PMNs.[35] Using the macrophage cell lines 2C11-12 and LA5-9, we have obtained similar results, and according to the CL response curve to monosodium urate crystals in Fig. 8A, a dose-dependent increase in CL was seen in response to increasing concentrations of monosodium urate crystals. This crystal-mediated CL induction could be abrogated by agents that increase cellular levels of cyclic AMP (i.e., theophylline) or antimitotic drugs such as colchicine (Fig. 8B). Carbachol, an agent capable of reducing cellular levels of cyclic AMP, did not affect the crystal-induced CL response (Fig. 8B). Thus, in confirmation with other reports,[35] CL appears to be a useful technique for the study of the effect of antiinflammatory agents on crystal-induced inflammatory reactions.

Practical and Technical Considerations of the CL Bioassay

Practical Considerations

We have found the measurement of CL by macrophage cell lines to be a reliable and reproducible parameter to assess a variety of biological and chemical substances that either modulate or stimulate the CL response by

[35] P. N. Platt and I. Bird, *in* "Analytical Applications of Bioluminescence and Chemiluminescence" (L. J. Kricka, P. E. Stanley, G. H. G. Thorpe, and T. P. Whitehead, eds.), p. 319. Academic Press, New York, 1984.

macrophages. Obviously, this bioassay based on CL emission has widespread applicability to a variety of problems, both scientific and clinical in nature. It should be remembered that heterogeneous macrophage populations have also been used to assess phagocytosis-associated events by chemiluminogenic probing. Yet, monoclonal macrophage cell lines offer many advantages over heterogeneous macrophage cell populations on account of the following characteristics.

Uniformity. Conventional macrophage populations are notoriously difficult to standardize and no two cell populations, even when obtained from the same individual or animal, can be considered to be identical. Monoclonal macrophage cell lines are uniform, so that the way is open to obtain cellular reagents of definable reactivity, a matter of great importance for diagnostic work.

Availability. Macrophage cell lines can be maintained indefinitely in culture without loss of functional activity. Thus virtually unlimited quantities of homogeneous, functionally defined macrophage cells can be produced.

Specificity. Macrophage cell lines can be monitored for their selective reactivity toward well-defined triggering or modulating agents. An example of this kind is illustrated in Fig. 6A indicating the selective response of 2C11-12 cells toward immune complexes. The identification of macrophage cell lines that respond selectively to certain triggers and/or modulators will be useful to detect selectively certain agents present in complex biological mixtures such as culture supernates or serum. Such selective identification and quantification are impossible to perform with heterogeneous populations since different macrophage subpopulations might interact with different stimulators or modulators. Another aspect of specificity with broad scientific applications is the identification of macrophage cell line variants that are defective in oxidative metabolism and intracellular killing. We have recently isolated CL-negative variants from the LA5-9 cell line that arose spontaneously in culture probably as a result of random chromosome segregation. Such cell lines which manifest cellular defects in active oxygen metabolism, yet exert other macrophage-specific functions such as phagocytosis, will allow analysis of the susceptibility of pathogens and tumor cells to oxidative and nonoxidative killing. This approach is of obvious importance for the development of therapeutic strategies applying to well-defined pathogens and/or malignant cells.

Finally, besides the analytical screening applications, these different continuous macrophage-like cells may provide a useful model for the study of enzymes, membrane receptors, immune and nonimmune agents involved in the generation of a respiratory burst by phagocytic cells, as well as the role of immune and pharmacological agents in this fundamental defense mechanism.

Technical Considerations

One of the practical limitations encountered in the present work has been the availability of dedicated instrumentation. The only commercial instrument specifically designed for the measurement of phagocyte luminescence was developed by Berthold around 1980[36] and has been in use in our laboratory for more than 2 years. It is equipped with six complete detector units, each with its own photomultiplier and associated electronics, which unfortunately makes it a relative expensive instrument. Moreover, recording the luminescence of six samples in parallel continuously is not a prerequisite in the case of slow reactions and repeated measurements at regular time intervals will do as well. Such considerations prompted us to develop an instrument equipped with a single photomultiplier tube and a cell holder containing six sample tubes in a fixed position.[37]

Instruments equipped with sample changers can also be used for the measurement of phagocyte-associated luminescence. Scintillation counters, in the "off-coincidence" mode, have often been used for this purpose. However, the latter show serious drawbacks. (1) When counting the samples sequentially the time gaps between samples add up and tend to unduly increase the time elapsed before the same sample is counted again. (2) Liquid scintillation counters can only be used around room temperature and this temperature is difficult to control. (3) Injection in front of the detector is impossible.

The first automated luminometer to be described in the literature originated precisely from the need for an instrument better adapted to the measurement of phagocyte luminescence.[38] It consists of a metal carousel holding 12 reaction vials and kept at constant temperature by a heating element. Commercial instruments based on this principle appeared as recently as 1982: Picolite 6500 (Packard), 1251 Luminometer (LKB-Wallac), and Biocounter M3050 (Lumac/3M). Berthold's Auto-Biolumat LB

---

[36] F. Berthold, H. Kubisiak, M. Ernst, and H. Fischer, in "Bioluminescence and Chemiluminescence: Basic Chemistry and Analytical Applications" (M. A. DeLuca and W. D. McElroy, eds.), p. 699. Academic Press, New York, 1981.

[37] E. Schram, H. Roosens, and P. De Baetselier, in "Analytical Applications of Bioluminescence and Chemiluminescence" (L. J. Kricka, P. E. Stanley, G. H. G. Thorpe, and T. P. Whitehead, eds.), p. 289. Academic Press, New York, 1984.

[38] T. Rawlins and R. Peacock, Int. Lab. 48, 56 (1982).

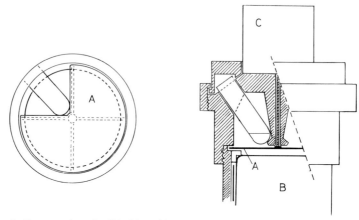

Fɪɢ. 9. Cross section of cell-holder with 4 tubes. (A) Rotating mask, (B) photomultiplier, (C) motor.

950 uses a conveyor belt instead of a carousel. These counters accommodate a number of samples that is actually much higher than can be used in the case of phagocyte luminescence: in the LKB luminometer the minimum time interval required to measure successive samples is 5.2 sec, which means that for 25 vials the measuring cycle amounts to more than 2 min.[39]

The instrument developed in our laboratory consists of a hollow copper block containing the cells at its periphery and fixed on top of a 2 in. photomultiplier. In a first set-up (Fig. 9) four tubes are introduced at an angle of 45° in such a way that their bottoms are in close proximity with the PM window. All tubes except one are masked by a cut disk rotating stepwise. The vials are of a commercially available standard type (12×47 mm) and are the same as those used for the macrophage cell preparations. Another design involves the use of a hexagonal block (Fig. 10) in which six vials can be easily fitted in a vertical position. With such a configuration the disk must be replaced by a rotating reflector. One of the advantages is that the whole surface of the PM window is seen in turn by each vial. Furthermore, the thermal isolation between the cell holder and the PM is more efficient and the thermostatization can easily be achieved by means of Peltier cells fixed on the outer surface. Temperature stability is achieved within a few hundredths of a degree between 16 and 40°. Light detection is achieved by means of a PM tube operated in the photon-counting mode. A 2-in. unselected EMI type 9635 B tube was chosen for the present purpose.

[39] B. R. Andersen and A. M. Brendzel, *J. Immunol. Methods* **19,** 279 (1978).

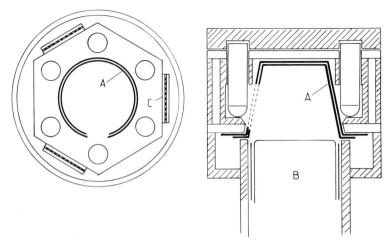

FIG. 10. Cross section of cell-holder with 6 tubes. (A) rotating cylinder, (B) photomultiplier, (C) Peltier cell.

The instrument may be operated in two ways. (1) "Single" mode: in this mode the same tube is counted repeatedly or its luminescence integrated over a given period of time; this mode is to be preferred for rapid kinetics. (2) "Group" mode: any number of tubes between 2 and 6 may be counted in repeated sequences. The counting time can be set between 1 and 99 sec but for the applications considered here, short periods will usually be preferred; since practically no time is lost between countings, a counting time of 3 sec will allow each tube to be counted three times a minute.

A printer is incorporated in the instrument, which delivers a listing of the successive results. By connecting the instrument to an external microcomputer, the results can be further processed in order to obtain for each sample the counts per minute in the course of time, the integrated value of the luminescence (by summation of the results), the slope (by subtraction of successive results), and a plot of the luminescence curve.

For the calibration of the individual positions of the cell holder as well as for checking the long-term stability of the instrument, use is made of a light source emitting single photons.[40] Such a light source is obtained by the incorporation of tritium or carbon-14 in a liquid scintillator quenched to such an extent that only single photons are left over.

Chemiluminometry involves kinetic assays, i.e., the registration over

[40] E. Schram, F. Demuylder, J. De Rycker, and H. Roosens, in "Liquid Scintillation, Science and Technology" (A. A. Noujaim, C. Ediss, and L. I. Wiebe, eds.), p. 243. Academic Press, New York, 1976.

certain periods of time of reaction rates which are likely to change with time. For obvious reasons a single channel instrument is not adequate for the measurement of numerous samples showing slow kinetics as in the case of phagocytes. Moreover, their reactivity changes in the course of time and parallel assays are therefore to be preferred for an easy comparison of the results. However, slow reactions do not need to be recorded continuously and discrete measurements at regular time intervals allow several reactions to be monitored simultaneously with a single detector. It further occurred to us that in the case of a limited number of samples, rather than moving the vials, it is more advantageous to use a rotating detector. In luminometry, instrumental configuration is fortunately very simple, involving only the sample vial and the detector, whereas in other photometric techniques three elements have to be carefully aligned: light source, photometric cuvettes, and detector.

Our instrument was built according to the above principles and offers the following advantages. (1) The compact shape of the cell holder allows an efficient control of the temperature from below room temperature up to 40°. (2) The time lost between successive countings is negligible. (3) The system is mechanically simple. (4) The vials are kept in a fixed position, avoiding unwanted shaking and possible temperature variations. (5) The heated part of the instrument can be placed above and be thermally isolated from the PM tube; heating of the photocathode by convection and conduction is hence limited. (6) Injection can occur in any position, before or while a sample is being counted. Besides its use for the measurement of phagocyte luminescence, the present instrument was found perfectly suited for all other chemiluminescent assays.

Acknowledgments

The authors are grateful to L. Remels, L. Mussche, L. Brijs, E. Vercauteren, M. Sieleghem, A. Chemtai, C. Hamers-Castermans, W. De Leersnijder, and K. Huyghen for their contribution of unpublished data and for providing materials for investigation. The luminometer described in this paper was built thanks to the skill and know-how of Henri Roosens whom the authors wish to thank for his efficient collaboration.

This work was partially supported by a grant of the A.S.L.K. Kankerfonds and a FKFO grant nr. 3.0096.83 (to E. Schram). P. De Baetselier is a fellow of the NFWO.

[39] Stable 1,2-Dioxetanes as Labels for Thermochemiluminescent Immunoassay

By JAN C. HUMMELEN, THEO M. LUIDER, and HANS WYNBERG

Introduction

Since 1976, when Schroeder *et al.* introduced a chemiluminescence immunoassay for biotin,[1] a variety of chemiluminescent techniques has been developed for immunochemical quantification of physiologically important substances. Yet, while fluorescence and enzyme immunoassays have taken a strong position as alternatives for radioimmunoassays, chemiluminescent assays (CIAs) have not grown as a routinely used, commercially available technique. In chemiluminescent techniques light is emitted in a multicomponent reaction. Consequently, reagent purity and mixing on the one hand, and sensitivity toward serum components on the other represent the most important problems in LIA techniques.

A new chemiluminescent system has now been developed in which all of the above-mentioned drawbacks are omitted because inherently chemiluminescent compounds, i.e., stable 1,2-dioxetanes, are used as the label. The emission of light from these compounds can be triggered by thermal activation. This chapter describes the principle of this thermochemiluminescence (TCL), the preparation and properties of TCL labels based on adamantylidene adamantane 1,2-dioxetane, the inclusion of 1,2-dioxetanes in γ-cyclodextrin, energy transfer TCL, the use of TCL in immunoassay techniques, and the detection of TCL.

Principle of Thermochemiluminescence

1,2-Dioxetanes (Fig. 1, **I**) decompose thermally into two carbonyl fragments. Because of the release of strain energy and because of a favorable reaction path, a fraction of the carbonyl fragments is formed in the first singlet or triplet electronically excited state (II-S_1 and II-T_1). Such electronically excited species can emit light either directly ("direct chemiluminescence," CL) or via energy transfer to a luminescent acceptor molecule A ("indirect chemiluminescence," ICL).

Only very few of the over 200 1,2-dioxetanes known today are stable

[1] H. R. Schroeder, P. O. Vogelhut, R. J. Carrico, R. C. Boguslaski, and R. T. Buckler, *Anal. Chem.* **48**, 1933 (1976).

METHODS IN ENZYMOLOGY, VOL. 133

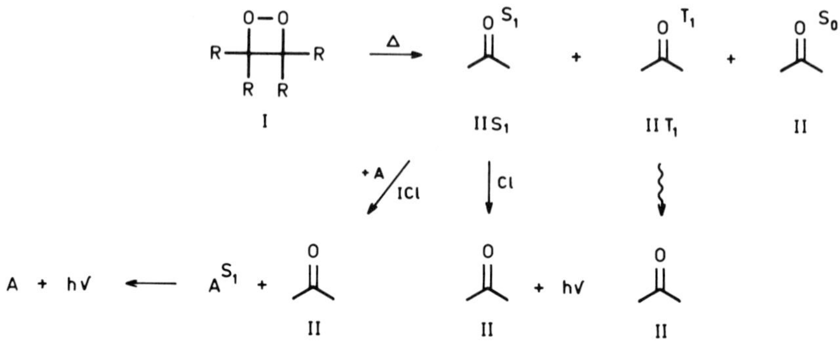

FIG. 1. Chemiluminescent thermal degradation of 1,2-dioxetanes.

(for days) at room temperature.[2] Adamantylidene adamantane 1,2-dioxetane (Fig. 2, **III**) is extremely stable (mp 174–176°)[3] and can be stored for years at room temperature or below without decomposition to adamantanone **IV** (a half-life of 1.2×10^4 years is calculated for **III** at 25°). The exponential relation between the half-life (τ) of **III** and the temperature is shown in Fig. 3. Since the decomposition is a first-order process, the kinetics of the chemiluminescent reaction are independent of the concentration of **III**; the shape of the chemiluminescence curve (I/I_{max} versus time) is a function of the temperature of the sample only. Autocatalytic degradation of **III** has never been observed. The efficiency of direct chemiluminescence (ϕ_{CL}) of **III** is 1×10^{-4}, i.e., 1 mol of **III** emits 6×10^{19} photons when heated.[4] This efficiency is a product of the efficiency of the decomposition reaction ($\phi_R = 1$), the efficiency of S_1-adamantanone formation ($\phi_S = 0.02$), and the efficiency of fluorescence of S_1-adamantanone ($\phi_F = 5.2 \times 10^{-3}$).[4,5] The T_1-adamantanone formed during the decomposition is quenched by collision with other molecules and cannot substantially contribute to the emission of light unless either a fluorescent triplet energy acceptor (e.g., 9,10-dibromoanthracene) is added or collision chances are minimized. The spectrum of direct CL from **III** (Fig. 4) shows a maximum at 425 nm and is identical to the fluorescence spectrum of **IV**.[4,6]

[2] W. Adam and G. Cilento, "Chemical and Biological Generation of Electronically Excited States." Academic Press, New York, 1982.
[3] J. H. Wieringa, J. Strating, H. Wynberg, and W. Adam, *Tetrahedron Lett.* p. 169 (1972).
[4] G. B. Schuster, N. J. Turro, H.-C. Steinmetzer, A. P. Schaap, G. Faler, W. Adam, and J. C. Liu, *J. Am. Chem. Soc.* **97**, 7110 (1975).
[5] A. M. Halpern and R. B. Walter, *Chem. Phys. Lett.* **25**, 393 (1974).
[6] C. A. Emeis and L. J. Oosterhoff, *J. Chem. Phys.* **54**, 4809 (1971).

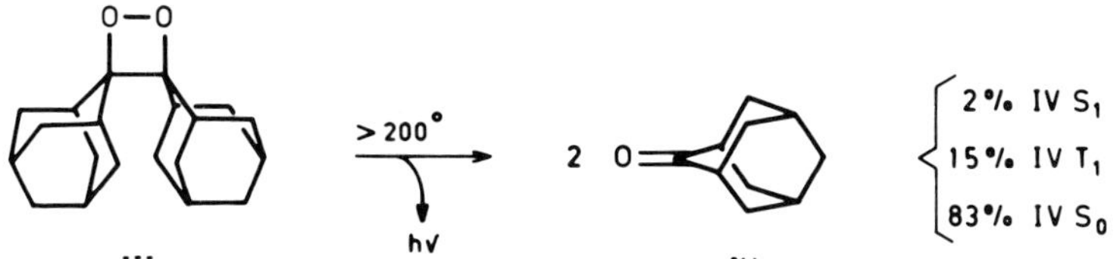

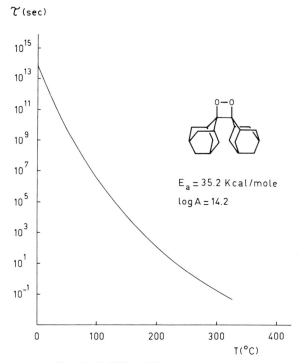

FIG. 2. Reaction of adamantylidene adamantane 1,2-dioxtane.

FIG. 3. Half-life of **III** vs temperature.

A detailed analysis of spectra and decay kinetics of TCL from **III** has been reported by Neidl and Stauff.[6a] Coincidentally, the emission spectrum nicely mimics the spectral response of bialkali photomultiplier tubes.

Both **III** and **IV** are colorless compounds [λ_{max}^{abs} (**III**) = 265 nm, ε = 21.5; and λ_{max}^{abs} (**IV**) = 280 nm, ε = 20].[4] As a consequence, ϕ_{CL} is independent of the concentration of **III**: no self-quenching or absorption occurs,

[6a] C. Neidl and J. Stauff, *Z. Naturforsch. B: Anorg. Chem., Org. Chem.* **33B**, 763 (1978).

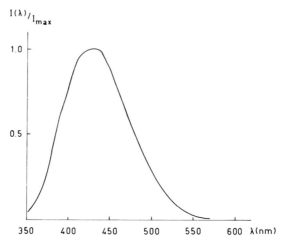

FIG. 4. TCL spectrum of **III** at 200° under an N_2 atmosphere.

even at very high local concentration. Therefore, macromolecular substrates can be heavily labeled with (colorless) derivatives of **III** without
relative loss of specific activity. The latter feature of **III** in particular is
rather unique for a label, and cannot be found among other chemiluminescent and fluorescent labels.

Synthesis of TCL Labels

A series of TCL labels based on adamantylidene adamantane 1,2-
dioxetane has been prepared. Some examples are shown in Fig. 5. Compounds **V** and **VI** react with strong nucleophiles (e.g., free thiol groups of
proteins), while the *N*-hydroxysuccinimide esters **VII** and **VIII** react with
amino group containing compounds. In this chapter attention will be focused on compound **VIII**.

The synthetic route for compound **VIII** is outlined in Fig. 6, starting
from bromoadamantylidene adamantane (**X**). Isomerically pure bromide
X is obtained upon reaction of *N*-bromosuccinimide with adamantylidene
adamantane **IX**, which in turn can be conveniently prepared on large scale
(>500 g) from commercially available adamantanone **IV** in a two-step
procedure.[7,8] The overall yield of the 7-step synthesis (from **IV** to **VIII**) is
50%.

In the last step of the synthetic route, sensitizer-generated singlet
oxygen attacks alkene **XIII** from both the top and bottom side to yield a

[7] E. W. Meijer and H. Wynberg, *J. Chem. Educ.* **59,** 1071 (1982).
[8] P. D. Bartlett and M. S. Ho, *J. Am. Chem. Soc.* **96,** 627 (1974).

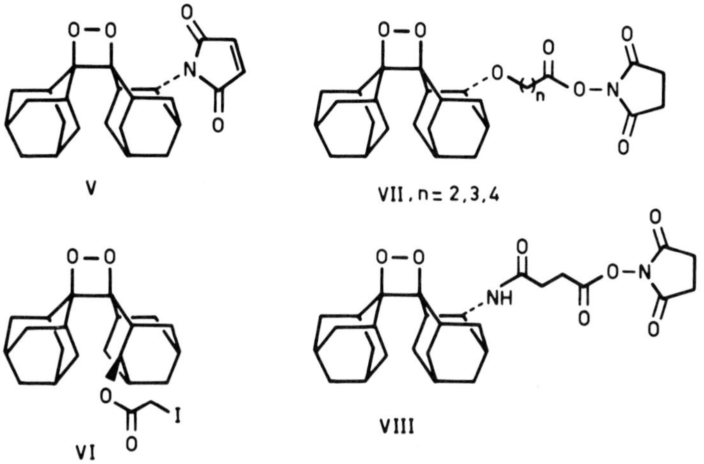

FIG. 5. TCL labels based on **III**.

(~1 : 1) mixture of isomeric 1,2-dioxetanes, *syn* and *anti* **VIII**. This mixture of isomers is used in labeling experiments. *N*-Hydroxysuccinimide ester **VIII** is a crystalline colorless compound which can be kept for months without decomposition if stored dry at or below room temperature.

4-eq-Bromoadamantylideneadamantane X

In a slightly modified previously published procedure,[9] 45 g (0.168 mol) **IX** and 55 g (2.1 eq) *N*-bromosuccinimide are dissolved in CH_2Cl_2 (600 ml) and acetic acid (15 ml) in a 1-liter round-bottom flask. The mixture is stirred magnetically and refluxed for 1 hr. The solvents are evaporated at a reduced pressure and the solid residue is dissolved with shaking in 3 liters of a 2 : 1 : 3 mixture of ether, pentane, and 0.5 N NaOH. After removal of the aqueous phase, the organic layer is shaken with 1 liter 0.5 N NaOH three times. After the addition of $MgSO_4$ (~5 g) and charcoal (~2 g), the suspension is filtered and concentrated to yield crude **X** (55 g) as a yellowish solid. Acetone (200 ml) is added and the suspension is stirred at 0° for 1 hr. The purified bromide is isolated by filtration and dried *in vacuo* (44.5 g, 76.5%); mp 128–130° (lit.[9] 130.5–131.5°).

4-eq-Aminoadamantylideneadamantane XI

Bromide **X** (7.0 g, 20 mmol) and dioxane (75 ml) are mixed in a stainless-steel cylinder (400 ml). Then liquid NH_3 (250 ml) is added slowly and

[9] E. W. Meijer, R. M. Kellogg, and H. Wynberg, *J. Org. Chem.* **47**, 2005 (1982).

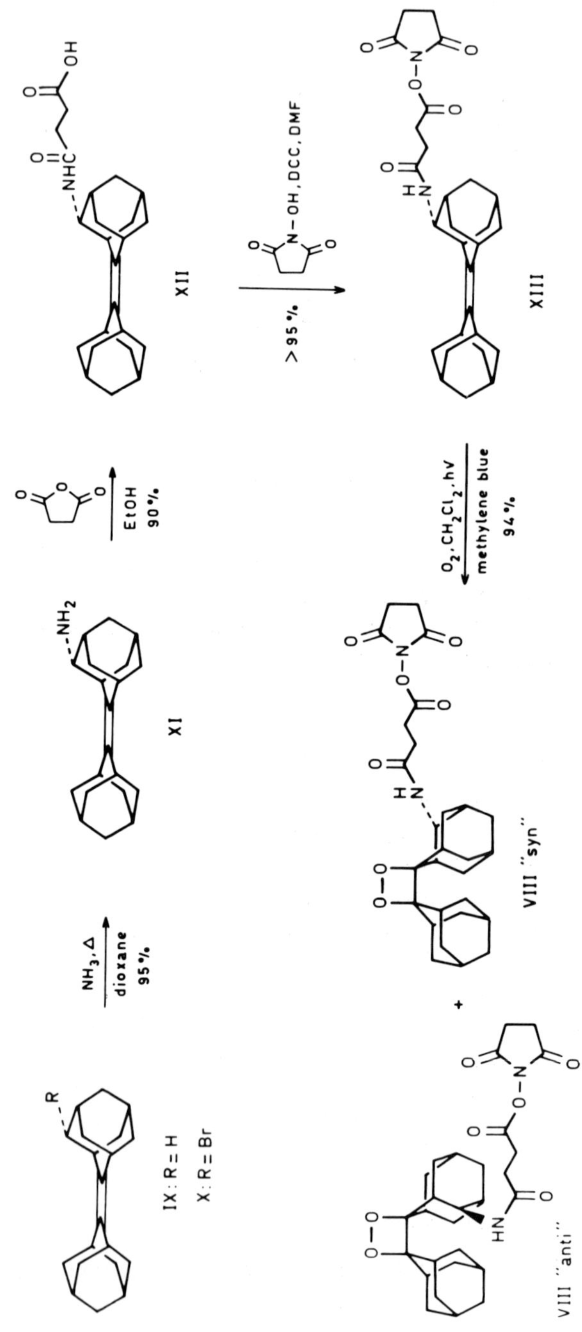

FIG. 6. Synthesis of **VIII**.

the cylinder is closed tightly. The mixture is shaken at 70° for 16 hr in a Carius oven. After cooling to $-50°$ (liquid N_2), opening of the cylinder, and evaporation of unreacted ammonia (this takes 2 hr), the residual solution is poured on 1 N NaOH (100 ml). Ether is added (250 ml) to extract the product and the ethereal solution is washed with H_2O (3 × 200 ml), dried over $MgSO_4$, filtered, and evaporated to produce 5.70 g (95%) spectroscopically pure **XI** as a white solid.

A small portion of this material was sublimed (0.001 mm Hg, 150°) to give white crystals; mp 126–160°. IR (KBr): 3350 (br), 2900 (s), 1600 (m), 1450 (s), 1090 (m) cm^{-1}. ^{1}H NMR (CDCl$_3$, TMS): δ3.15–2.50 (br, 5H); 2.45–0.80 (m, 24H). ^{13}C NMR (CDCl$_3$): δ134.4, 132.0, 56.0, 39.3, 38.7, 38.1, 37.0, 34.7, 32.0, 31.7, 30.5, 30.1, 28.2, 27.7. Mass: M$^+$ at m/e 283 (100%), 266 (20%), 135, 79. Exact mass: calculated for $C_{20}H_{29}N$, 283.230; found 283.228.

N-(Adamantylideneadamant-4-eq-yl)succinic Acid Monoamide **XII**

A solution of amine **XI** (9 g, 32 mmol) in absolute EtOH (75 ml) is added to a solution of succinic anhydride (3.2 g, 32 mmol; Merck) in absolute EtOH (150 ml). From the initially clear solution monoamide **XII** starts separating after a few minutes as an amorphous material. After stirring the mixture overnight, the product is filtered off, washed with a little absolute EtOH, and dried in a desiccator (12.2 g, 90%), mp 253–256°. Monoamide **XII** is insoluble in CCl$_4$, CHCl$_3$, CH$_2$Cl$_2$, ether, and alkanes and is soluble in DMSO and DMF.

IR (KBr): 3400 (m), 3300–2500 (s), 2900 (s), 1720 (s), 1640 (s), and 1550 (s) cm^{-1}. ^{1}H NMR (DMSO-d_6, TMS): δ7.5 (br, 1H); 3.6 (m, 1H); 2.7 (m, 4H), 2.3 (br, 4H), and 2.3–1.0 (m, 22H). ^{13}C NMR (DMSO-d_6): δ174.0, 170.5, 134.4, 131.5, 53.9, 37.9, 36.7, 35.2, 32.7, 31.7, 31.5, 31.4, 30.6, 30.4, 30.2, 29.4, 27.8, 27.0. Mass: M$^+$-peak at m/e 383 (100%) and 365, 283, 266. Exact mass: calculated for $C_{24}H_{33}NO_3$: 383.246; found: 383.244.

N-(Adamantylideneadamant-4-eq-yl)succinic Acid Monoamide
N-Hydroxysuccinimide Ester **XIII**

Monoamide **XII** (1.65 g, 4.3 mmol) is dissolved in warm dry DMF (50 ml). After the solution is cooled to room temperature, dicyclohexylcarbo-diimide (900 mg, 4.35 mmol) is added. After stirring the solution for 10 min, dry N-hydroxysuccinimide (660 mg, 5.75 mmol) is added and the reaction mixture is stirred under an atmosphere of N_2 for 16 hr. A little precipitated dicyclohexylurea is removed by filtration and the solvent is removed by high vacuum evaporation. The residue is washed with n-hexane (2 × 50 ml) and then dissolved in CH$_2$Cl$_2$ (50 ml). A little insoluble material is removed by filtration over a short column of seasand. After removal of the solvent at low pressure the N-hydroxysuccinimide ester remains as a white amorphous powder [mp 191–194° (EtOAc); 2.06 g, 100%].

IR (KBr): 3400 (m, NH), 2900 (s), 1805 (m), 1770 (m), 1720 (s), 1660 (m), 1535 (m), 1200 (m), 1060 (m) cm^{-1}. ^{1}H NMR (CDCl$_3$, TMS, 200 MHz): δ5.95 (t, J = 7 Hz, 1H), 3.9 (m, 1H), 3.02 (t, J = 7.5 Hz, 2H), 2.87 (m, 4H), 2.83 (s, 4H), 2.63 (t, J = 7.5 Hz, 2H), 2.2–1.0 (m, 22H). ^{13}C NMR (CDCl$_3$): δ168.8, 168.3, 136.0, 130.4, 54.3, 39.4, 39.0, 37.6, 37.1, 35.2, 33.8, 33.2, 32.2, 32.0, 31.6, 31.2, 30.9, 30.6, 28.3, 28.2, 27.3, 26.8, 25.4, 24.7. Mass: M$^+$ peak at m/e = 480, 365 (100%), 267, 266. Exact mass: calculated for C$_{28}$H$_{36}$N$_2$O$_5$: 480.262; found: 480.263.

Syn- and Anti-N-(Adamantylideneadamantane 1,2-Dioxetane-4-eq -yl)-succinic Acid Monoamide N-Hydroxysuccinimide Ester VIII

The active ester **XIII** (3.0 g, 6.38 mmol) is photooxygenated for 18 hr in a standard apparatus for photooxygenation of alkenes,[7] using CH$_2$Cl$_2$ (700 ml) as the solvent and methylene blue (5 mg) as the sensitizer. The solution is decolorized with charcoal, filtered, and concentrated with a flash evaporator. The yellowish residue is stirred with n-pentane (50 ml) and dried in a desiccator to yield the 1,2-dioxetane-active ester (3.0 g, 94%) **VIII** as a mixture of two isomers, spectroscopically pure, in the form of a slightly yellow solid. This material is used as thermochemiluminescent label without further purification (mp 96–110°).

IR (KBr): 3400 (br, NH), 2950 (s), 1810 (m), 1780 (m), 1730 (s), 1650 (m), 1540 (m), 1200 (m), 1080 (m) cm^{-1}. ^{1}H NMR (CDCl$_3$, TMS; 200 MHz): δ6.20 (d, J = 8 Hz, 0.5H), 6.00 (d, J = 8 Hz, 0.5H), 4.25 (m, 0.5H), 4.16 (m, 0.5H), 3.00 (t, J = 8 Hz, 2H), 2.83 (s, 4H), 2.65 (t, J = 8 Hz), 2.78–2.5 (m, 6H), 2.2–1.2 (m, 22H). ^{13}C NMR (CDCl$_3$): δ169.2, 169.1, 168.9, 168.2, 96.0, 95.7, 95.4, 95.3, 49.5, 48.6; signals between δ39.4 and 24.7. Mass: fragments at m/e = 362 and 150, 247, 219, 149, 100, 70. No M$^+$ peak. Exact mass fragment with m/e = 362: calculated for C$_{18}$H$_{22}$N$_2$O$_6$: 362.148; found: 362.150.

Application of TCL Labels: Characteristics of Labels and Labeled Compounds

Labeling of Proteins

Proteins that contain free amino groups can be labeled with ester **VIII** in a simple and fast one-step procedure. Because the solubility of **VIII** in aqueous buffers is low, a cosolvent is needed. 1,4-Dioxane is used routinely, but THF, EtOH 95%, DMF, and acetone suffice as well. When the concentration of cosolvent is kept below 5% (v/v), the immune reactivity of antibodies can be fully preserved. Bovine serum albumin (BSA) can be labeled with a high number of residues of **VIII** if the reaction is performed in dioxane/borate buffer (pH 8.5; 100 mM) 1 : 3.

Proteins, labeled with **VIII**, exhibit good solubility in aqueous buffers, e.g., BSA can be labeled with 25 residues **VIII** without significant loss of solubility.

General Procedure. The protein is dissolved in 100 mM borate buffer, pH 8.5 (1–20 mg/ml) at room temperature. Into this gently shaken solution is pipetted a fresh solution of **VIII** in dioxane [in such a concentration that a desired labeling ratio is obtained and the final concentration of dioxane is 5–10% (v/v) for antibodies]. The clear (or slightly turbid) solution is kept at 25° for 1 hr. Unreacted label is separated from the conjugate either by dialysis against borate buffer or by column chromatography (Sephadex LH-60/borate buffer). TCL activity of labeled proteins is preserved over a period of several months if stored at −20° in the dark.

Determination of Label Incorporation. Since TCL labels show neither UV absorption (at $\lambda > 280$ nm) nor fluorescence, the number of labels incorporated in proteins can best be determined by TCL measurement or by titration of residual amino groups in the conjugate. Amino group titration of TCL-labeled proteins can be performed using Habeeb's TNBS/SDS method.[10] An example is shown in Fig. 7. This titration indicates that when BSA reacts with 50 equivalents of **VIII**, 18 residues are incorporated, i.e., a labeling efficiency of 36% can be easily attained. Furthermore, the decrease of free amino groups upon conjugation is a good indication for the covalent nature of the label–protein linkage. When a mixture of hydrolyzed **VIII** (i.e., carboxylic acid) and a protein (e.g., BSA, human IgG) is dialyzed, the protein fraction does not exhibit TCL, indicating that strong aspecific (noncovalent) binding does not occur.

Specific Activity of TCL Labels and Labeled Compounds

The thermochemiluminescence spectra of **III**, **VIIa,b,c**, **VIII**, and TCL-labeled proteins are virtually identical. Additionally, the thermal stability of the 1,2-dioxane moiety is unaltered upon conjugation. Hence, the specific activity under standard conditions can be compared directly for these compounds.

Since the parent 1,2-dioxetane **III** and the free labels **VII** and **VIII** appear to be rather volatile compounds at ~240° (**VII** and **VIII** show

[10] A. F. S. A. Habeeb, *Anal. Biochem.* **14,** 328 (1966).

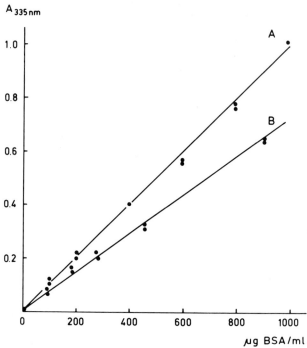

FIG. 7. Amino group titration of BSA and BSA–(**VIII**)$_n$ according to the method of Habeeb.[10] Thus the 100 μl of the protein solution were added 10% SDS (1 ml), 4% Na$_2$CO$_3$ (1 ml), and 0.1% (w/v) TNBS (1 ml) in H$_2$O, successively. After incubation at 37° for 2 hr, the reaction was stopped by addition of 1 ml 1 N HCl. UV absorption was measured at 335 nm. Since unlabeled BSA (A) has 61 free amino groups, it can be calculated that this BSA–(**VIII**)$_n$ (B) contains 18 amino group blocking residues of **VIII**.

thermochemical breakdown besides 1,2-dioxetane decomposition), TCL quantification of these compounds has to be carried out on strongly absorbing material, e.g., a small piece (~20 mm^2) of aluminum oxide thinlayer chromatography sheet (Neutral, Merck). The TCL of buffered solutions of labeled proteins can be measured by pipetting 1–10 μl of the solution on a disk of a thermoresistant polymer (Teflon, Kapton 500H), evaporation of solvent at 100°, and subsequent heating and detection at ~240°. Table I lists the specific activities of some TCL compounds, measured in the apparatus, described at the end of the chapter. This apparatus has a counting efficiency of 0.14%. It can be concluded that (1) the experimental values are in excellent agreement with the expected specific activity (1 mol **III** emits 6 × 10^{19} photons according to the reported values for ϕ_S and ϕ_F[4,5]); (2) substituted adamantylidene adamantane 1,2-dioxetanes show little variation in specific activity; (3) the specific activity is un-

TABLE I

SPECIFIC ACTIVITY OF SOME TCL COMPOUNDS[a]

Compound	Carrier	Measured specific activity (photon counts/mol)
III[b]	Al_2O_3	1.0×10^{17}
VIIC[b]	Al_2O_3	1.0×10^{17}
VIII[b]	Al_2O_3	5.0×10^{16}
BSA–(**VIII**)$_{18}$[c]	Al_2O_3	8.0×10^{17}
BSA–(**VIII**)$_{18}$[c]	Teflon[d]	8.0×10^{17}
BSA–(**VIII**)$_{18}$[c]	Kapton[e]	1.6×10^{17}

[a] Counting efficiency 0.14%; the sample is heated from 100 to ~240° within 15 sec and kept at that temperature for 45 sec under an atmosphere of nitrogen gas. During this period >90% of the total emission is detected.

[b] As a solution in dioxane.

[c] As a solution in 100 mM borate buffer, pH 8.5.

[d] Disk diameter 9 mm; thickness 0.5 mm; background 500 cpm.

[e] Disk diameter 9 mm; thickness 0.125 mm ("Kapton 500H," DuPont); background 170 cpm.

changed after conjugation to proteins; (4) labeling BSA with even 18 residues of **VIII** does not result in a relative decrease of specific activity.

Reproducibility and Linearity of TCL; the Use of γ-Cyclodextrin for the Prevention of Evaporation of Free Label

The factors that influence the reproducibility of TCL measurements are quite different from those ruling that of (room temperature) liquid phase bi- or trimolecular chemiluminescent reactions. So are the possible ways to eliminate disturbing factors and the ways to distinguish between true values and artifacts.

At the present state of the art of TCL chemistry and detection, the most important disturbing parameters in reproducibility and linearity are (in order of decreasing importance): (1) thermochemical degradation of the spacer moiety of the TCL labels, resulting in evaporation of the 1,2-dioxetane moiety before TCL occurs, (2) reproducibility of heat contact between the carrier material and the heating element in the TCL detection apparatus, (3) pipet precision (routinely 2–5 μl of a solution is pipetted on the carrier material), (4) sample position in the TCL detection apparatus, (5) inherent TCL detection apparatus reproducibility, and (6) counting statistics (only at lower concentrations).

The thermochemical degradation, and subsequent evaporation, is a parameter that operates predominantly at lower concentrations. It can be minimized in several ways: (1) by measuring a sample on a strongly absorbing material [e.g., aluminum oxide thin-layer chromatography material, Whatman glass filter (GS/A)], see Table II, entries 1 and 2; (2) by measuring samples as buffered solutions or as proteinaceous solutions: upon evaporation of the solvent this leaves a (very small) transparent but sticky residue on the carrier surface, thereby preventing evaporation; (3) by covering the sample with a nonvolatile transparent medium (e.g., silicon oil). Thus TCL-labeled proteins cannot be quantified accurately as solutions in pure water (below $\sim 10\ \mu g/ml$). The disappearance of material during the heating process can be followed by the technique of curve fitting, since the rate observed does not obey the kinetics established for adamantylidene adamantane 1,2-dioxetane decomposition. Both linearity and reproducibility are influenced by this parameter, as can be seen in Table II, entries 3–5 (see also Fig. 8A).

Evaporation of free label **VIII** is completely inhibited upon complexation with γ-cyclodextrin (γ-CD). A strong complex is formed by **VIII** and γ-CD, which can be chromatographed over a Sephadex LH-60 (40–120 μm) column. When dissociation of the complex upon dilution is prevented through dilution in 15 mM γ-CD/borate buffer, perfect linearity and reproducibility of TCL are observed (Table II, entry 8; Fig. 8B). Dissociation of the complex can also be prevented by incapsulation of the γ-CD complexes in liposomes. Solutions of liposomes (cholesterol/phosphatidyl-

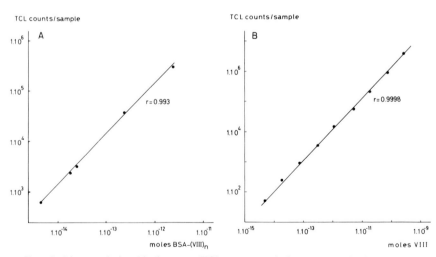

FIG. 8. Linear relationship between TCL output and the amount of BSA–(**VIII**)$_n$ in borate buffer (A), and the amount of free **VIII** in 15 mM γ-CD/borate buffer (B).

TABLE II
LINEARITY AND REPRODUCIBILITY OF TCL DETECTION

	TCL sample	Solvent	Carrier	Measured range	r^a	$CV (\%)^a$
1	**III**	Dioxane	Al$_2$O$_3$	60 pmol^b	—	8
2	**VIII**	Dioxane	Al$_2$O$_3$	40 pmol^b	—	6
3	BSA–(**VIII**)$_n$	0.1% aq. BSA	Kapton	250–2.5 fmol^c	0.996	5–10
4	BSA–(**VIII**)$_n$	B^d	Kapton	2 pmol–5 fmol^c	0.993	7–16
5	BSA–(**VIII**)$_n$	H$_2$O	Kapton	250–2.5 fmol^c	0.90	30–60
6	BSA–(**VIII**)$_n$	B	Teflon	70 pmol–70 fmol^c	0.998	7–28
7	BSA–(**VIIb**)$_n$	B	Teflon	70 pmol–70 fmol^c	0.996	6–50
8	**VIII**	15 mM γ-CD in B	Kapton	320 pmol–5 fmol^b	0.9998	1.4–3.1^e
9	**VIII**-γCD complex in liposomes	B	Kapton	60 pmol–60 fmol^b	0.996	0.9–2.6^e
10	**VIII**-γCD complex in liposomes	B	Teflon	400 pmol–40 fmol^b	0.998	4–8

^{a} r, correlation coefficient of linear regression; CV at highest to lowest concentration.
^{b} Amount of 1,2-dioxetane.
^{c} Amount of protein.
^{d} B, 0.1 M borate buffer, pH 8.5.
^{e} At lowest concentration CV corrected for counting statistics.

choline 1 : 1, sonificated to SUVs, centrifuged at 10,000 g and chromato-graphed over a Sephacryl S-300 column) that contain these complexes can be diluted without disturbance of the association equilibrium of **VIII** and γ-CD, since the local concentration inside the liposomes is kept constant. These TCL liposomes show excellent linearity and reproducibility when measured on Kapton as the carrier (Table II, entry 9). Potentially, these liposomes can be applied as macromolecular TCL labels, containing many γ-CD complexed 1,2-dioxetane molecules per liposome. A further advantage of liposome incapsulation of the γ-CD complexes is the crea-tion of a closed and constant surrounding for the 1,2-dioxetanes, protect-ing them from potential quenchers in analyte solutions.

The influence on the reproducibility by variation in heat contact/ transfer between the carrier material and the heating element of the TCL detector on the total reproducibility is seen in entries 4/6 and 9/10. Teflon disks (0.5 mm thickness) are stiff and slightly bent. In contrast, Kapton disks (0.125 mm thickness) are perfectly flat and thermostable up to >400°. As a result the thermochemiluminescence curve (TCL versus time) of TCL material, heated on Kapton disks is of a very reproducible shape (the same shape of I/I_{max} versus time at all concentrations!); there-fore the reproducibility is independent on the TCL integration period of time. When Teflon is used, however, disk to disk variation in heat contact results in pronounced variation in TCL/time curve shape and thus relatively long integration periods (~2 min) are needed to improve reproducibility of TCL quantification.

At present, pipet precision (CV ~0.5%), inherent apparatus reproducibility (CV ~0.1%), and counting statistics ($CV = 100\sqrt{N}/N\%$) determine the ultimate reproducibility of TCL detection.

Energy Transfer TCL

Improvement of Specific Activity through Isochromic Energy Transfer

The efficiency of direct TCL of **III** is only ~1% that of luminol and its analogs at their optimum. The difference in ϕ_{CL} is caused by the relatively low quantum efficiency of fluorescence $\phi_F = 5.2 \times 10^{-3}$) of adamantanone formed in the TCL reaction. At the theoretical maximum, the TCL intensity of **III** can be amplified by a factor 192 upon the addition of an efficient fluorescer (ϕ_F ~1), which acts as an acceptor of energy of singlet excited carbonyls (see also Fig. 1).

Since Belyakov and Vassil'ev[11] introduced this technique of excitation energy-transfer CL (ICL) as a way of visualization of poorly luminescent excited carbonyl products in hydrocarbon autoxydation and since Wilson and Schaap[12] showed that 9,10-diphenylanthracene (DPA) acts as a very efficient acceptor in 1,2-dioxetane decomposition, DPA has become a popular acceptor compound. As an acceptor of singlet energy in the decomposition of **III** (at ~200–250°!), DPA is especially very suitable for the following reasons: (1) it is a very efficient fluorescer ($\phi_F = 0.8–1$)[13,14]; (2) it is thermally stable; (3) it has a small negative temperature coefficient of ϕ_F; (4) it is an apolar nonbasic and nonacidic molecule, thus a strong influence on ϕ_F and λ_{em} (max) is not to be expected in the presence of proteins and other biological fluid components.[13] Such an influence is seen in the case of fluorescein and its analogs,[15] resulting in a pH optimum of ϕ_F and almost complete loss of fluorescence if bound to an antifluorescein antibody; (5) the absorption and emission spectra of DPA exhibit only a small overlap at 390–410 nm. Thus DPA shows minimal concentration quenching[16]: the concentration at which it has half its maximal ϕ_F:

[11] V. A. Belyakov and R. F. Vassil'ev, *Photochem. Photobiol.* **11,** 179 (1970).

[12] T. Wilson and A. P. Schaap, *J. Am. Chem. Soc.* **93,** 4126 (1971).

[13] A. Schmillen and R. Legler, *in* "Landolt-Börnstein Tables" (K.-H. Hellwege and A. M. Hellwege, eds.), Group II, Vol. 3, pp. 143 and 266. Springer-Verlag, Berlin and New York, 1967.

[14] P. S. Engel and B. M. Monroe, *Adv. Photochem.* **8,** 245 (1971).

[15] T. Förster, "Fluorescenz Organischer Verbindungen," p. 47. Vandenhoeck & Ruprecht, Göttingen, 1951.

[16] W. H. Melhuish, *J. Phys. Chem.* **65,** 229 (1961).

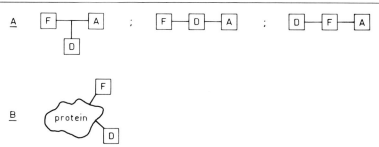

FIG. 9. Fluorescence amplified systems. F, Fluorescer (DPA); D, 1,2-dioxetane; A, reactive group.

C_h ~0.5 M [C_h (fluorescein) = 0.02 M, C_h (acridone) = 0.025 M[17]]; (6) DPA shows no excimer fluorescence.[18]

The end-point quantification step of a TCL immunoassay consists of heating a sample (for example as a precipitate or as a complex, bound to a carrier polymer or as a small volume of an aqueous solution). Sensitive detection requires a high signal–noise ratio rather than simply a strong signal. Upon addition of a constant amount of DPA to all samples before measurement, the background signal and the blank values would be increased significantly. Thus if DPA would be added as such it would not increase the sensitivity of the assays. Therefore DPA has to be linked either to the luminescent label to form a new label (A) or to the substance labeled with the luminescent label as well as a second label (B) (Fig. 9).

In this chapter only examples of system B will be discussed. In order to connect DPA to a protein, it first has to be transformed into a fluorescent label. The new fluorescent label 2-[O-(N-succinimidyl)carboxypropyl]-9,10-diphenylanthracene (**XIV**, SCP–DPA; Fig. 10) mimics all fluorescent properties of DPA to a great extent. SCP–DPA is prepared from the corresponding carboxylic acid by the standard procedure for such esters (N-hydroxysuccinimide, DCC, dioxane). The carboxylic acid, in turn, can be prepared from DPA in ~80% yield via a two-step procedure, described by Douris.[19] Both the emission and the absorption spectrum of SCP–DPA are shown in Fig. 11. Note that the the emission spectrum of SPC–DPA is almost identical to the emission spectrum of **III** and the corresponding TCL label **VIII** (Fig. 4). Thus the spectral distribution of TCL from a mixture of one of these 1,2-dioxetanes and SPC–DPA (or a derivative) is independent on the efficiency of energy transfer. The only visible effect of the addition of SPC–DPA is an amplification of the TCL signal.

[17] A. Schmillen and R. Legler, in "Landolt-Börnstein Tables" (K.-H. Hellwege and A. M. Hellwege, eds.), Group II, Vol. 3, p. 329. Springer, Verlag, Berlin and New York, 1967.
[18] J. B. Birks and L. G. Christophorou, *Proc. R. Soc. London, Ser. A* **277,** 571 (1964).
[19] R. G. Douris, *Ann. Chim. (Paris)* [13] **4,** 479 (1959).

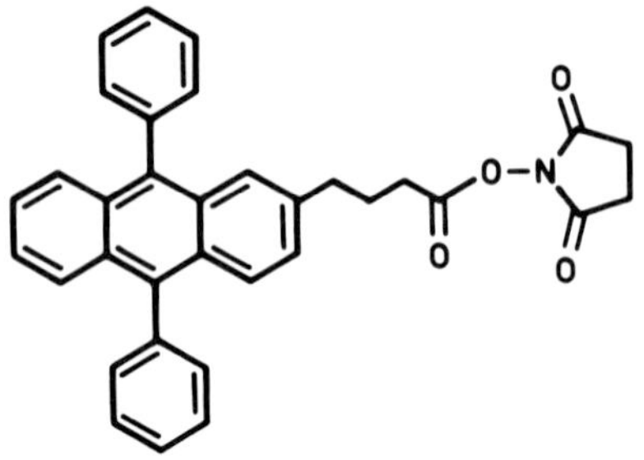

FIG. 10. Structure of SCP–DPA (**XIV**).

The critical distance R_0 for energy transfer from **III** to SPC–DPA (R_0 is the distance between donor and acceptor at which the efficiency of energy transfer is 0.5) is calculated to be 15.3 Å (in a medium with a refractive index of 1), according to the theory of Förster.[20,21] This value indicates that random labeling of smaller proteins with both **VIII** and SPC–DPA can result in a full energy transfer TCL protein, when only a few DPA residues are attached to the protein (in the case of BSA, approximately 6 DPA residues are needed). For efficient energy transfer TCL on IgGs, however, a minimum of approximately 20 residues of SPC–DPA would be needed. This is most likely to eliminate all immunological activity of the antibody.

In practice it appears that the maximal amplification of TCL from a labeled protein using SCP–DPA residues as energy acceptors is a factor 40. Thus BSA, labeled with 17 residues of **VIII**, shows a specific activity of 1×10^7 photon counts/μg in a TCL detection apparatus with a counting efficiency of 0.14%, when measured as 1 μl of a 1 mg/ml solution in 1% aqueous BSA, evaporated on a disk of Teflon; when this conjugate is labeled with even up to approximately 30 residues of SCP–DPA according to the general procedure for labeling proteins with TCL label **VIII**, the specific activity rises to a maximum of 4×10^8 counts/μg (i.e., ϕ_{CL} of this BSA dual conjugate = 4%). This very efficiently luminescent protein, which is readily soluble in aqueous buffers above pH 8.5, can be used as a label by covalently linking it to other proteins (IgGs) or small analytes, depending on the nature of the assay. A "fluorescence amplified thermo-chemiluminescence immunoassay," FATIMA, using this BSA dual conjugate as a label, is described in the next section.

[20] T. Förster, *Ann. Phys.* (*Leipzig*) [6] **2**, 55 (1947).
[21] L. Stryer, *Annu. Rev. Biochem.* **47**, 819 (1978).

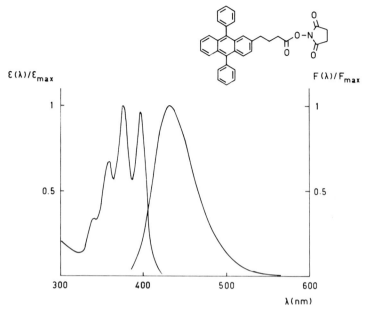

$\mathcal{E}(\lambda)/\mathcal{E}_{max}$ $F(\lambda)/F_{max}$

FIG. 11. Absorption and emission spectrum of SCP–DPA.

Red Shifted TCL through Energy Transfer

The many applications of energy transfer fluorescence and chemiluminescence in immunoassay have been based mainly on the color shift
of luminescence upon energy transfer. Energy transfer donor–acceptor
interactions between suitably labeled immunological counterparts can be
used as the basis for homogeneous assays.[22,23] A homogeneous TCL immunoassay, based on energy transfer, has not been developed yet. The
main obstacle is the temperature needed ($\sim240°$) for rapid TCL detection.
Hence, aqueous solutions cannot be measured neat.

It has been found, however, that very efficient energy transfer from **III**
and its derivatives to a number of acceptors (e.g., rubrene, perylene,
anthracene, 2-(carboxyethylcarbonyl)diphenylanthracene, 2-aminoan-
thracene, 9,10-dithienylanthracene, fluorescein, carboxyfluorescein, fluorescein isothiocyanate-labeled BSA, rhodamine B) takes place when a
sheet of thin-layer chromatography material (aluminum oxide, Merck) is
sprayed with a solution of **III** and an acceptor compound (which also may

[22] A. Patel, C. J. Davies, A. K. Campbell, and F. McCapra, *Anal. Biochem.* **129,** 162 (1983).
[23] A. K. Campbell and A. Patel, *Biochem. J.* **216,** 185 (1983).

be chromatographed) and heated to ~200–240°, subsequently. Spots containing fluorescer show bright luminescence of altered color.

Thermostable Solid Phase Materials for Immunoassay

Commonly used solid phase materials (e.g., polystyrene and polyvinylchloride) are not suitable for TCL immunoassay, since these materials degenerate at ~200–300°. Criteria for TCL solid phase materials are (1) thermostability up to ~300° under a nitrogen atmosphere, (2) low inherent thermoluminescence output (background), (3) no quenching effect on TCL emission at the surface, (4) a possibility for coating the surface with proteins, and (5) a convenient format (in the context of washing procedures).

Some materials that meet most or all of these criteria are microcrystalline cellulose, glass, silicon tube, Kapton, and Teflon. For the latter two polymers, that show very low thermoluminescence background and perfect thermostability below 250°, a convenient coating procedure has now been developed. This procedure yields antibody-coated polymer disks that show better immunoreactivity than coated polystyrene microtiter plates. From the two identical solid-phase sandwich immunoassays for hIgG, performed on polystyrene and Kapton 500H, shown in Fig. 12, it can be concluded that antibody-coated Kapton disks show both very little aspecific binding and a high immunoreactivity/cm². An identical assay, on Teflon disks, shows a precision and dynamic range identical to the ELISA on Kapton, however superimposed on a constant blank OD value of 0.2. Both ELISAs on Kapton and Teflon result in better precision, sensitivity, and dynamic range than ELISAs on polystyrene. Disks, coated with antibodies, show a TCL background of ~3 counts/sec (Kapton) and ~8 counts/sec (Teflon) at 240°. Antibodies show no thermoluminescence.

First Results of TCL Immunoassay: FATIMA

A solid-phase sandwich immunoassay for carcinoembryonic antigen, CEA, can be performed using the TCL technique described. The standard curve for a CEA–FATIMA in a clinically interesting range of concentrations is shown in Fig. 12.

Procedure. Kapton disks are coated with monoclonal αCEA (Roche) as described (Fig. 12). Polyclonal gαCEA is labeled with both **VIII** and SCP–DPA **XIV** by reaction with, respectively, 15 and 20 eq of both labels in borate buffer, pH 8.5/5% dioxane to yield an immunologically active, labeled antibody with a specific activity of 1.4×10^6 photon counts/μg (measured on Kapton, counting efficiency 0.14%). The αCEA-coated

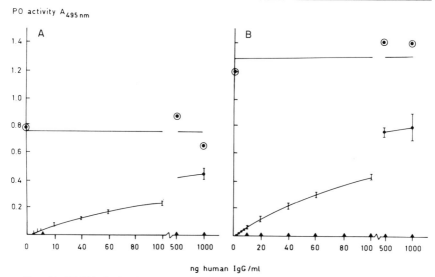

Fig. 12. ELISA for hIgG on a polystyrene microtiter plate (A) and Kapton 500 H (B). Coating procedures. (A) Physical coating with sαhIgG, 0.1 mg/ml in 0.1 M carbonate buffer, pH 9.6, for 1 week at 4°. (B) Disks (diameter = 0.9 cm; thickness 0.125 mm) of Kapton were refluxed with several changes of 96% EtOH and several changes of distilled water subsequently, and dried at ~300°. Thereafter these disks were incubated with sαhIgG, 0.1 mg/ml in Tris–HCl buffer, pH 7.6/0.02% NaN_3 at 4° for 2 days. Preincubation: both A and B with 4% BSA in 10 mM Tris–HCl (pH 8.0), 0.15 M NaCl, 0.05% Tween 20 (Buffer T) at 37° for 1 hr. Incubation: after washing with water, incubation with hIgG in buffer T for 1 hr at 37°. Second incubation: after washing with water, incubation with gαhIgG-peroxidase conjugate in buffer T for 1 hr at 37°. After 1 hr of color development with o-phenylenediamine (0.2 ml 2 × 10^{-3} M OPD–HCl in phosphate buffer, pH 5.0/0.0045% H_2O_2), the reaction was stopped with 0.4 ml 1 N H_2SO_4. In both assays all standards were measured in triplicate. Triangles, solid phase coated with BSA (blank); double circles, solid phase coated with hIgG (showing maximal capacity and reproducibility of coating).

disks are preincubated with BSA (4% BSA in 0.01 M Tris–HCl, pH 8.0, 0.3 M NaCl, 0.05% Tween 20) to prevent aspecific binding. The disks are incubated overnight at 37° with 0.2 ml of standard solutions of CEA (Roche) containing, respectively, 0.1, 0.25, 0.5, 2.5, 6.0, and 10 ng CEA/ml in PBS, pH 7.4/20% FCS. The disks are incubated with labeled antibody (0.2 ml of 5 μg antibody/ml PBS/20% FCS) for 3 hr at 37°. All washing procedures between and after the incubation are performed using distilled water. The disks are taken out of the assay wells and measured for TCL output directly without a previous drying procedure.

A similar assay for hIgG can be performed by using a second antibody coupled covalently to heavily labeled BSA [BSA(**VIII**)$_{16}$–(**XIV**)$_{~10}$], using N-ethyl-N'-(3-dimethylaminopropyl)carbodiimide–HCl (EDC) in a one-

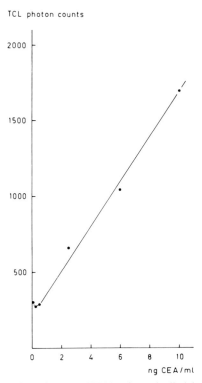

FIG. 13. Standard curve for a CEA–FATIMA using a dually labeled antibody conjugate. The number of photon counts shown is the total integration over a time interval of 14 sec, including the TCL peak intensity, corrected for a standard background by substraction of 145 counts.

step procedure.[24,25] This coupling procedure yields a high-molecular-weight conjugate according to SDS–gel electrophoresis on polyacrylamide. The procedure for the preparation of the protein–protein conjugate is as follows: to 2 ml of a mixture of 1.3 mg BSA(**VIII**)$_{16}$–(**XIV**)$_{\sim10}$ and 0.8 mg gαhIgG/ml 10 mM borate buffer, pH 8.5, is added 13 μl of a 100 mg/ml solution of EDC in distilled water. After stirring for 3 hr in the dark at room temperature, the reaction mixture is thoroughly dialyzed against PBS/0.02% NaN$_3$ at 4°. After dilution, this crude antibody conjugate solution is used in the assay without further purification.

The protocol for the standard curve for a hIgG–FATIMA shown in Fig. 14 is identical to the one described for the CEA–FATIMA, with the

[24] S. Bauminger and M. Wilchek, this series, Vol. 70, p. 151.
[25] T. L. Goodfriend, L. Levine, and G. D. Fasman, *Science* **144**, 1344 (1964).

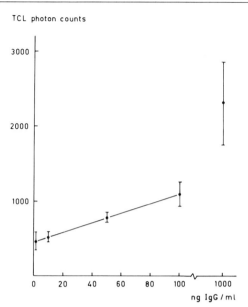

FIG. 14. Standard curve for hIgG–FATIMA using gαhIgG coupled covalently to BSA(**VIII**)$_{16}$–(**XIV**)$_{-10}$. The number of counts taken is the total integration of TCL output over a 32 sec time interval without background substraction.

exception that the period of both the first and the second incubation is 1 hr.

TCL Detection Apparatus

The stable 1,2-dioxetanes differ greatly from other chemiluminescent compounds with respect to the way in which the emission is stimulated: the thermochemiluminescent compounds have to be heated to elevated temperatures. For practical reasons as well as to obtain a high peak signal-to-noise ratio, a short detection period (a sharp luminescence peak) is advantageous. In order to decompose 99% of any amount of **III** (or a derivative of **III**) within 30 sec, heating of the sample to a temperature as high as ~250° is necessary (see Fig. 3). Thus the detection of TCL labels based on **III** cannot be carried out in aqueous solutions (at 1 atm). Therefore, all commercial luminometers for chemiluminescence and bioluminescence, designed for the detection of low level luminescence in (aqueous) solutions at ambient temperatures, are of no use for TCL detection.

This section describes a TCL detector, which is designed to meet the following criteria: (1) high sensitivity, (2) low (thermal) background,

(3) variable detection temperature (100–270°), (4) short heating period, and (5) high reproducibility and stability.

The technique of photon counting was chosen, since it offers the best possibilities for combination of high gain with the exclusion of electronic noise. Samples to be quantified consist of minute amounts of dry residues on small and thin disks ($\phi < 1$ cm, thickness 0.1–1 mm) of thermoresistant materials. Consequently, the sample compartment can be made extremely small in comparison with sample chambers of liquid phase CL detectors. Thermal insulation is the limiting factor in design of a TCL detector with optimal optical dimensions.

Four types of sources of background luminescence during TCL measurements can be distinguished: (1) CL from oxidative degradation of materials (in the sample) at ~250°, due to reaction with O_2, present in the air, (2) thermoluminescence, (3) blackbody radiation, and (4) X-ray radiation from the environment and from traces of radioactive isotopes. The oxidative degradation can be inhibited effectively by flushing the sample compartment with N_2 gas. The thermoluminescence of a series of carrier materials was investigated. Teflon, Kapton 500H, glass slides, and Whatman glass filter were found to be optimal with respect to thermostability and low thermoluminescence emission. The blackbody radiation of a perfect blackbody, in the spectral area of interest (~350–550 nm), was calculated to be only 0.7 photons sec^{-1} cm^{-2} at 250°. When the hot object is an aluminum heater or a disk of one of the above mentioned carrier materials, the expected blackbody radiation is even much lower, since these materials reflect (or transmit) light of this wavelength to a high extent. Any X-ray originated background is minimized by the use of a specially shielded photomultiplier housing and a sample compartment made of aluminum and brass. Figure 15 shows the block diagram of the TCL detector.

When the heater and sample are connected to the adaptor on the photomultiplier housing (Products for Research Te 1004/TS/110; cooling not used), the microswitch (Honeywell BZ-2RW822) gives the Apple II computer control over the multiplier power study (Öltronix A 2.5K-10HR, switched on negative output) via the protection unit. The signal from the photomutiplier tube (EMI 9893 QA/350, selected on low darkcount, i.e., 0.5 cps!) is preamplified (EG&G 1121A) and transported to the discriminator (EG&G 1121A)/photon counter (EG&G 1109) unit. The data enter the Apple II computer thereafter. These data are shown on the monitor after the measurement and printed subsequently. The computer switches four variables via an optocoupler during the measurement: (1) the oven controller can be set at 100° (on/off), (2) the power supply

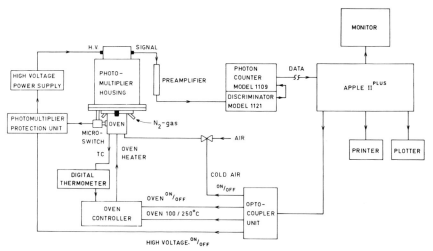

FIG. 15. Block diagram of TCL detection apparatus.

(on/off), (3) the oven controller at 240° [on = 240°, off = 100°; when (1) is "on"], and (4) an electric valve for cold air, which cools the oven at the end of a measurement. The temperature of the oven is shown continuously on a digital display (Analog Wakefield AN 2572).

The characteristic part of the TCL detector is shown in Fig. 16. This part is connected to the photomultiplier housing. A soldering iron element (50 W, 24 V; Weller) is used as the heating element. On top of this hollow element an aluminum "oven" is clamped. On this oven a piece of carrier material can be placed. A 1-mm rim on the oven prevents the sample from slipping away. A thermocouple (copper/constantan) is placed into a small hole in the bottom of the aluminum oven. A 3-mm-diameter hollow pipe enters the hollow soldering element to 0.5 cm distance from the aluminum oven. Through this pipe cold air is introduced after completion of each measurement in order to cool the oven to 100°. The soldering element is strengthened by a brass cylinder. On this brass cylinder a progressively protruding piece of brass switches the microswitch on upon connecting the heating device to the adaptor on the photomultiplier housing. The connection is a bayonet-type without a locking position: the heating device is pressed against the adaptor and subsequently rotated 60°. Because of the light-trap construction of the heater-adaptor combination, a perfectly dark sample compartment is obtained.

A horizontal flow of nitrogen gas (±0.4 liters/min) enters the sample compartment from two opposite positions. The nitrogen flows slowly around the oven through the light trap.

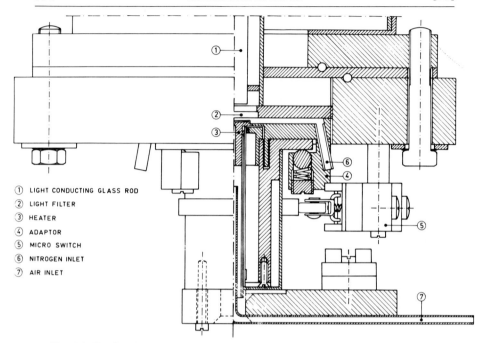

① LIGHT CONDUCTING GLASS ROD
② LIGHT FILTER
③ HEATER
④ ADAPTOR
⑤ MICRO SWITCH
⑥ NITROGEN INLET
⑦ AIR INLET

FIG. 16. Heating device and adaptor, filter, and light-conducting glass rod of TCL detector.

The heating device is coupled to the adaptor in such a way that the aluminum oven is positioned only 2 mm from the interference filter (Fairlight Edge filter, short wave pass; 50% transmittance at 500 nm) above the sample. Thus the sample compartment consists of the aluminum oven, brass "walls," and a glass topside.

A light-conducting glass rod (Scott LST-UV, 10 mm diameter, 90 mm length) transports the light, transmitted by the filter, to the photomultiplier cathode surface.

Apparatus Calibration

The apparatus was calibrated, using a disk (10 mm diameter and 1 mm thick) of scintillating plastic (Nuclear Enterprises Ltd. NE 134-^{14}C, activity: 0.0095 μCi). The plastic disk emits 1.7×10^5 photons/sec with a spectral distribution that mimics the thermochemiluminescence spectrum of adamantylidene adamantane 1,2-dioxetane **III** to a high extent. For this scintillator a counting efficiency of 0.14% was observed. Since the disks of Kapton and Teflon, used as solid phase supports, have the same diameter as the scintillator, we estimate the counting efficiency for labeled immune complexes on these polymers to be 0.14% as well.

TCL Measurement

A typical TCL plot for BSA, labeled with **VIII**, is shown in Fig. 17A. As a standard procedure for determination of TCL of solution, 1–5 μl of the solution is pipetted on a disk of Kapton 500H, which is placed on the heater (at 100°). Within a few seconds the solvent evaporates and thereafter the heater (+ sample) is connected to the adaptor on the PM housing. The sample is heated to ~250° and TCL is recorded over a ~0.5-min period.

The blank measurement (i.e., a clean disk of Kapton 500H) shows a very small increase of background signal with temperature: as a mean value a thermoluminescence noise of only 2.5 cps is observed for Kapton at 240° (Fig. 17B). Thus for a 32 sec measurement of TCL a total background of 57 ± 8 counts is observed. Kapton disks, coated with proteins (e.g., BSA, gαhIgG, αCEA) show a total blank value of 130 ± 15 counts for 32 sec. The latter value is the blank value for a solid phase sandwich immunoassay on Kapton 500H, if no aspecific binding of labeled second antibody occurs.

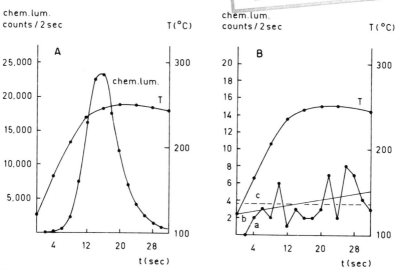

FIG. 17. (A) TCL intensity plotted versus time for BSA(**VIII**)$_n$, measured on Kapton 500H as 1 μl of a solution of the protein in phosphate buffer. (B) Typical blank measurements. (a) A single measurement; (b) mean value in time for clean Kapton disks; (c) mean value of background for 32 sec.

Conclusion

Stable 1,2-dioxetanes, derived from functionalized adamantylidene adamantanes, can be prepared and these compounds can be used as labels or tracers in analytical procedures since they are effective inherently chemiluminescent compounds.

These thermochemiluminescent labels exhibit some special features. (1) The light emission is the result of a unimolecular first order decomposition reaction. Thus, both the integration time for TCL detection is independent on the amount of label present in the sample, and the shape of the TCL curve (I/I_{max} versus time) is a function of the temperature of the sample only. Therefore, for a number of potentially disturbing factors, true and false TCL values can be distinguished by curve fitting. (2) Part of the background of a TCL measurement can be rejected because resolution in time is observed for background and 1,2-dioxetane TCL due to differences in activation energies for these processes. (3) Neither the 1,2-dioxetanes nor their degradation products absorp light in the spectral area of TCL emission. Hence, self-quenching does not occur and high local concentrations of these compounds can be used as is shown in the cases of heavily labeled proteins and liposomes without relative loss of specific activity. Although the TCL labels show a moderate specific activity ($\sim 6 \times 10^{19}$ photons emitted per mol), enhancement can be achieved through energy transfer to a variety of fluorescent compounds. The new fluorescent label SCP–DPA has been specially developed for enhancement of 1,2-dioxetane TCL, but it can also be used as fluorescent label or as a label that enhances peroxyoxalate ester luminescence.

Both the TCL labels and TCL-labeled compounds show good long-term stability at ambient temperature.

Thermochemical side reactions, leading to a deficiency in specific activity, take place under certain conditions. These reactions can be inhibited by complexation of the TCL labels with γ-cyclodextrin. The requirement of a sample temperature of $\sim 240°$ for rapid detection limits the sample format to that of a (dry) residue on a thermostable carrier. An advantageous aspect is that the size of the samples is minimal.

Linearity and reproducibility of TCL detection are dependent on a number of factors. Under optimal conditions, linearity is perfect over a range of five orders of magnitude and CV values of ~ 1–3% can be reached.

The FATIMAs for CEA and hIgG presented here are to be regarded as the first results of a new technique that required specific development of several components: a thermostable solid phase, a coating procedure for these specific solid-phase materials, and a TCL detection apparatus. All elements of the analytical procedures using 1,2-dioxetanes as tracers are subject to further improvement.

Acknowledgments

The authors would like to acknowledge the contributions of Dré Oudman, Jan Koek, Marten de Rapper, and Karin de Jager for the syntheses, the biochemistry, the TCL detection apparatus, and secretarial assistance, respectively.

[40] Synthesis of Chemiluminogenic Protein Conjugates with Dimethyl Adipimidate for Sensitive Immunoassays

By G. MESSERI, H. R. SCHROEDER, A. L. CALDINI, and C. ORLANDO

Introduction

Numerous substances of clinical interest have been determined with great sensitivity by radioimmunoassay for more than two decades. However, the inconvenience of licensing and safety, as well as stability problems with radiolabels have stimulated development of nonradioisotopic methods. As one successful alternative, chemiluminescence immunoassays have been developed for the determination of a number of haptens[1] and proteins[2] with sensitivities equivalent to those of radioimmunoassays. Furthermore, immunoassays for proteins monitored with chemiluminogenic labels are potentially more sensitive than those using a radiolabel. To avoid drastic loss of immunoreactivity, only one ^{125}I atom can be introduced per protein molecule.[3] Evidently, the reaction conditions necessary for radioiodination and radiolysis seem to account for the short life and the immunological features of the labeled protein. By contrast, several chemiluminogenic molecules may be coupled to a protein molecule with only moderate loss in immunoreactivity. This increased specific activity of the conjugate does not compromise conjugate stability. Thus, greater specific activity and stability of chemiluminogenic conjugates may

---

[1] M. Pazzagli, G. Messeri, A. L. Caldini, G. Moneti, G. Martinazzo, and M. Serio, *J. Steroid Biochem.* **19**, 407 (1983).

[2] F. Kohen, E. A. Bayer, M. Wilcheck, G. Barnard, J. B. Kim, W. P. Collins, I. Beheshti, A. Richardson, and F. McCapra, in "Analytical Applications of Bioluminescence and Chemiluminescence" (L. J. Kricka, P. E. Stanley, G. H. G. Thorpe, and T. P. Whitehead, eds.), p. 149. Academic Press, Orlando, 1984.

[3] H. A. Kemp and J. R. Woodhead, *Ligand Q.* **5**, 27 (1982).

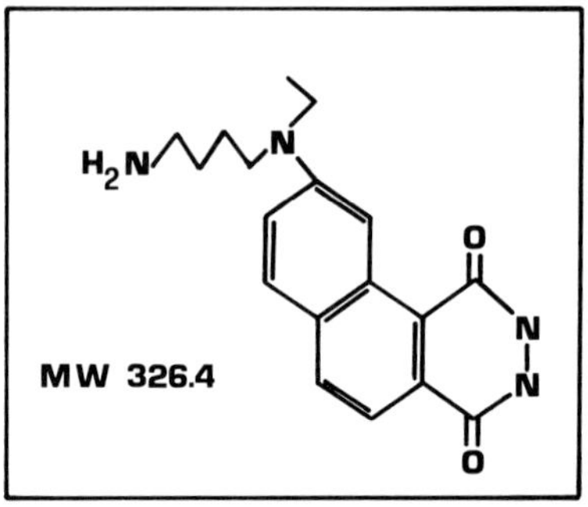

FIG. 1. Structure of ABENH.

permit immunoassays with lower detection limits than those with the respective radioiodinated protein.[4]

The aim of our study was to develop a better procedure for labeling proteins with our most efficient chemiluminogenic compound, 7-[(N-4-aminobutyl)-N-ethyl]aminonaphthalene-1,2-dicarboxylic acid hydrazide (ABENH) (Fig. 1).[5] We chose coupling via imidate chemistry because the positive charge of the protein is maintained, which helps preserve the integrity of the protein.[6] Furthermore, the amidine linkage should be easily cleavable at basic pH to liberate the label from chemiluminescence quenching environment of the protein.[7] Thus, the full efficiency of the label can be expected during chemiluminescence readout, which should improve the sensitivity. Transferrin, a medium molecular weight (73,000) protein, was selected as a model for labeling. The utility of ABENH–transferrin conjugates is demonstrated in a chemiluminescence immunoassay for transferrin in seminal fluid.

Materials and Methods

Reagents

Dimethyl adipimidate–2HCl, human transferrin, microperoxidase (EC 1.11.1.7; MP11, sodium salt from equine heart cytochrome c), bovine

[4] I. Weeks, I. Beheshti, F. McCapra, A. K. Campbell, and J. S. Woodhead, *Clin. Chem.* (*Winston-Salem, N.C.*) **29,** 1474 (1983).

[5] H. R. Schroeder, R. C. Boguslaski, R. J. Carrico, and R. T. Buckler, this series, Vol. 57, p. 424.

[6] M. J. Hunter and M. L. Ludwig, this series, Vol. 25, Part B, p. 585.

[7] H. R. Schroeder, C. M. Hines, D. D. Osborne, R. P. Moore, R. L. Hurtle, F. F. Wogoman, R. W. Rogers, and P. O. Vogelhut, *Clin. Chem.* (*Winston-Salem, N.C.*) **27,** 1378 (1981).

serum albumin, immunoglobulin G (HG 11, human Cohn fraction II), goat anti-rabbit IgG, and normal rabbit serum are obtained from Sigma Chemical Co. (St. Louis, MO). Hydrogen peroxide (300 ml/liter) is from Merck (Darmstadt, FRG). 7-[(N-4-Aminobutyl)-N-ethylamino]naphthalene-1,2-dicarboxylic acid hydrazide (ABENH), 6-[(N-4-aminobutyl)-N-ethyl]-amino-2,3-dihydrophthalazine-1,4-dione (ABEI), and thyroxine-ABEI are prepared as described previously.[5]

The Coomassie Blue reagent for the protein assay is provided by Bio-Rad Laboratories (Richmond, CA).

Rabbit anti-human transferrin serum is supplied by Dako Corporation (Santa Barbara, CA).

Disposable columns prepacked with Sephadex G-25 (1.5 × 5.2 cm) and Sephadex G-100 (1.5 × 25 cm) are from Pharmacia Fine Chemicals (Uppsala, Sweden).

The radioimmunoassay kit for transferrin is from Biodata (Rome, Italy).

The samples of seminal fluid were provided by the Andrology Unit of the University of Florence.

Reagent Solutions for the CL Reaction

The assay buffer is borate buffer, 0.1 mol/liter, pH 8.6, containing 9 g of NaCl, 80 mg of BSA, and 80 mg of human immunoglobulin G per liter.

A stock solution of microperoxidase at 400 mg/liter (0.2 mmol/liter) is prepared in Tris–HCl buffer (10 mmol/liter, pH 7.4) and stored at 4° for up to 4 months. It is diluted daily to 10 μmol/liter with assay buffer.

Hydrogen peroxide (1.5 ml/liter, 45 mmol/liter) solution is prepared daily by diluting the stock in doubly distilled water.

Light Measurements

Light-producing reactions are carried out in 12 × 47-mm polystyrene tubes (Lumacuvette P, Lumac Systems, Basel, Switzerland) containing the chemiluminogenic compound preincubated in 0.2 ml of 2 mol/liter NaOH at 60° for 30 min. One-tenth milliliter of 10 μmol/liter microperoxidase is added and the tubes are placed in the Biolumat LB 9500 luminometer (Berthold Instruments, Inc., Wildbad, FRG). Then the luminometer automatically injects 0.1 ml of 45 mmol/liter hydrogen peroxide to initiate chemiluminescence and displays the photon count, integrated for a 10 sec interval, in arbitrary units. Kinetics of chemiluminescence reactions are monitored on a storage oscilloscope (Model 5111; Tetronix, Beaverton, OR) connected to the luminometer.

Synthesis of the ABENH–Transferrin Conjugates

Activation of ABENH. ABENH (1.3 mg, 4 μmol) and dimethyl adipimidate (9.8 mg, 40 μmol) are dissolved in 1.8 ml of potassium carbonate buffer (1 mol/liter, pH 10, prewarmed to 37°) and incubated at 37° in a water bath for 5 min.

Coupling Reaction. Transferrin is dissolved at 12 g/liter, in 0.01 mol/liter phosphate, pH 7, and the solution is diluted serially to 6, 3, 1.5, 0.75, and 0.37 g/liter with the same buffer. Aliquots of the activated ABENH solution (0.2 ml) are mixed with the protein solutions (0.1 ml) and incubated for exactly 20 min at 37°. Under these conditions the respective molar ABENH/transferrin ratios are 23, 46, 92, 184, 368, and 736. Individual reaction mixtures are immediately applied to 1.5 × 5.2-cm columns packed with Sephadex G-25M and eluted with 0.1 mol/liter phosphate buffer, pH 7. The ABENH–transferrin conjugates are eluted in the fourth and fifth 0.5 ml fraction and are easily located by the yellow color and absorbance at 332 nm. The conjugates are purified further by gel filtration on a 1.5 × 25-cm column filled with Sephadex G-100 using the same phosphate buffer as the eluent.

Estimate of ABENH Incorporation into Transferrin

Spectral Ratio. The absorption spectra of ABENH–transferrin, an ABENH–transferrin mixture (molar ratio 3:1) and the conjugate solutions are recorded from 400 down to 200 nm on a DB 120 spectrophotometer (Beckman Instrument GMBH, Munich, FRG). Thus, the less intense absorption band of ABENH at 420 nm is not recorded. The molar extinction coefficients for ABENH at 420, 332, and 273 nm in 0.1 mol/liter NaOH are 6910, 11,700, and 29,860, respectively.

The concentration of ABENH in the conjugate is estimated from the absorbance at 332 nm and the extinction coefficient. Because the ABENH spectrum shows an absorption maximum at 273 nm, it strongly affects the protein absorbance peak at 280 nm. However, the contribution of the ABENH can be calculated as the product of the absorbance of ABENH at 332 nm and its 332/280 nm absorbance ratio (0.42) and subtracted.

Activity Ratio. The concentration of active ABENH in the protein conjugate is determined after the base cleavage step from the intensity of the light emission at several concentrations and using free ABENH as standard. The protein content of the pooled eluates is measured by the Coomassie Blue method[8] using transferrin as reference. Incorporation is

[8] M. M. Brandford, *Anal. Biochem.* **72**, 248 (1976).

defined as the molar ratio of ABENH and protein. A molecular weight of 73,000 is assumed for transferrin.

Immunoreactivity of ABENH–Transferrin Conjugates

Each conjugate is tested for its ability to compete with a standard preparation of native transferrin in displacing the radioiodine-labeled transferrin in a conventional radioimmunoassay system. The relative binding affinities are computed by comparing the conjugate doses necessary to produce 50% inhibition of binding of the radiolabeled transferrin.

Titration of Antibody With ABENH–Transferrin Conjugate

The ABENH–transferrin conjugate (40,000 counts, 0.1 ml) and serial diluted anti-transferrin serum (from 1:500 to 1:64,000, 0.1 ml) are combined in polystyrene tubes and incubated for 2 hr at room temperature. The antibody bound fraction is then precipitated by adding goat anti-rabbit IgG serum (titer 1:4000, 0.1 ml) and incubating an additional 2 hr. The reaction mixture is diluted with 1 ml of distilled water, the tubes are centrifuged at 1500 g for 20 min, and the supernatant is removed by aspiration. The pellet is finally resuspended with 0.2 ml of 2 mol/liter NaOH and incubated at 60° for 30 min. Luminescence measurements are then carried out as described earlier under "Light Measurements."

Chemiluminescence Immunoassay for Transferrin in Seminal Plasma

The samples are diluted (1/100) with doubly distilled water. Standards of human transferrin are prepared by diluting the stock solution (10 g/liter) to the appropriate concentration (1.56–2000 μg/liter) with assay buffer. A 0.1-ml aliquot of standard or diluted sample and 0.2 ml of ABENH–transferrin conjugate C (5 μg/liter) are mixed in polystyrene tubes and 0.1 ml of anti-transferrin (diluted to a titer of 1:8000) is added to start the competitive protein binding reaction. After 2 hr incubation at room temperature the assay is completed as described for the antibody titration.

Results

The ABENH–transferrin conjugates synthesized are isolated by gel filtration chromatography. The elution pattern shows all the luminogenic activity as a single peak, coincident with pure transferrin (Fig. 2). Furthermore, the absorbance spectra of these pure conjugates are identical

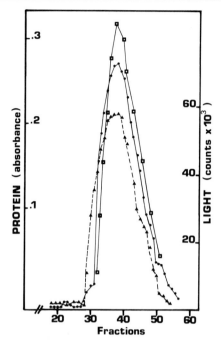

FIG. 2. Gel filtration chromatogram of transferrin and ABENH–transferrin conjugate. Elution of transferrin was monitored by the protein absorbance at 280 nm (squares). The conjugate elution was monitored by measuring both the protein content with the Coomassie Blue method (circles) and the light yielding ability (triangles).

with those of ABENH and transferrin mixed in the same proportions (Fig. 3).

Label incorporation into the protein is determined by both spectral and activity measurements and appears strictly related to the molar ratio of ABENH to transferrin in the reaction mixture (see the table). Neither the DMA/ABENH ratio (2:1, 5:1, and 10:1) nor the length of the activation reaction (from 1 to 20 min) has a marked effect on the efficiency of the subsequent coupling reaction (data not shown).

Chemiluminescence measurements permit detection of conjugates from 0.16 to as little as 0.013 fmol/tube, depending on the number of labels incorporated. The lower detection limit is defined as the amount of conjugate producing a light emission 3-fold the mean background.

Conjugates produced are stable in 0.1 mol/liter phosphate, pH 7, at 4° and can be used after at least 1 year storage without further purification.

The cross-reaction of the conjugates in the radioimmunoassay decreases with increasing label density (Fig. 4). Thus, the immunoreactivity

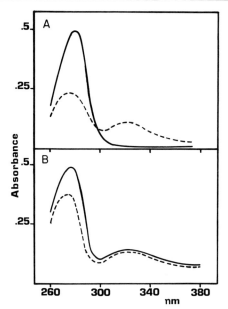

FIG. 3. Absorption spectra of transferrin (A, solid line), ABENH (A, dotted line), a physical mixture of ABENH and transferrin in 3:1 molar ratio (B, solid line) and an ABENH–transferrin conjugate (B, dotted line).

ABENH INCORPORATION INTO TRANSFERRIN AS A FUNCTION OF THE
ABENH/TRANSFERRIN RATIO IN THE COUPLING REACTION

		Incorporation[a]		
			Activity	
Conjugate	ABENH/ transferrin[a]	Spectral	pH 8.6[b]	pH 12.6[c]
A	23:1	1.8	0.3	1.5
B	46:1	2.5	0.6	2.3
C	92:1	3.9	0.8	3.0
D	186:1	5.5	1.7	5.0
E	373:1	9.6	3.1	8.8
F	746:1	20.0[d]	7.0	19.2

[a] Molar ratio.

[b] No hydrolysis step prior to chemiluminescence measurement in the borate buffer system at pH 8.6.[1]

[c] The conjugate is hydrolyzed and chemiluminescence is determined at pH 12.6 as described under "Light Measurements."

[d] The spectral incorporation of conjugate F could not be properly measured because of the low protein content.

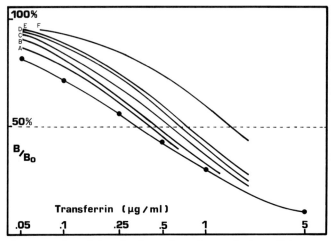

FIG. 4. Cross reaction of chemiluminogenic conjugates in the radioimmunoassay. The immunoreactivity of conjugates A–F (see the table) were tested as described earlier. The reference curve with human transferrin is shown with solid circles.

and lower detection limit are inversely related to label incorporated (Fig. 5). Conjugate C was chosen for development of the immunoassay because it has both a low detection limit by chemiluminescence (0.08 fmol/tube) and adequate immunoreactivity (68%).

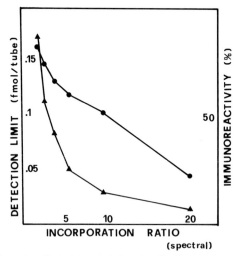

FIG. 5. Lower detection limit (triangles) by chemiluminescence and immunoreactivity (circles) of the conjugates by radioimmunoassay as a function of the incorporation ratio (spectral).

An antibody titration curve confirms the good chemiluminescence activity of conjugate C and demonstrates the specificity of the antibody through the competition of transferrin with conjugate C for anti-transferrin serum (Fig. 6). The sensitivity of the chemiluminescence immunoassay for transferrin was calculated from 10 different standard curves prepared in duplicate (Fig. 7). The amount of transferrin distinguishable is 2 ng/tube.

Transferrin concentrations in 212 seminal fluid samples have been determined with the chemiluminescence immunoassay and compared to those obtained by radioimmunoassay. As demonstrated by the correlation coefficient of 0.968, standard error of the estimate S_{xy} of 27.5 mg/liter (samples ranging from 7 to 693 mg/liter) and the equation for the line ($y = 1.00x + 1.65$) both methods agree well with each other.

The advantage of the hydrolysis step just prior to chemiluminescence readout was briefly investigated. In the absence of the hydrolysis step, ABENH in the conjugates already produces 20–35% of the chemiluminescence (activity) at pH 8.6 expected from the concentration determined by

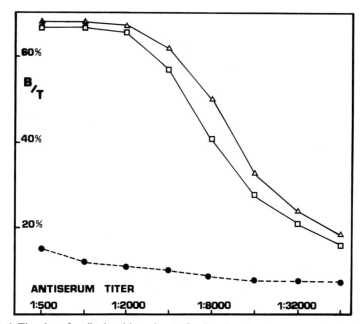

Fig. 6. Titration of antibody with conjugate C using chemiluminescense detection. Rabbit antiserum to transferrin at various dilutions indicated was incubated with 1 ng of conjugate C/tube in the absence (triangles) and presence of competing transferrin (25 ng/tube, squares; 20 μg/tube, circles) and assayed as described earlier.

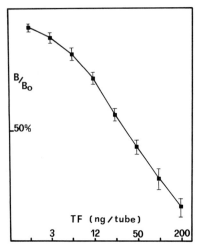

Fig. 7. Standard curve for transferrin by chemiluminescence immunoassay. As in Fig. 6 with 1 ng of conjugates C/tube, a dilution of antiserum of 1:8000 and varying concentrations of competing transferrin. The error bars indicate the standard deviation of the mean from 10 replicate curves (duplicate values for each point) performed on different days.

spectral measurement (see the table). Nearly all of the ABENH is liberated (active label) from all conjugates kept at pH 12.6 and 60° for 30 min, as determined with chemiluminescence measurements at pH 12.6 (see the table; used throughout this study) and at pH 8.6 (data not shown). In contrast, thyroxine–ABEI coupled by an amide bond releases only 23% of the ABEI in 1 hr in boiling 2 mol/liter NaOH. As controls ABEI lost no activity under these conditions and thyroxine–ABEI was only 2% active without the hydrolysis step. Therefore, as expected, labels coupled via amidine linkage are much more readily released than those with amide bonds.

Discussion

ABENH makes an ideal candidate for labeling proteins for use in sensitive immunoassays because it can be measured rapidly at very low concentration by chemiluminescence. However, unlike radioiodination procedures, the covalent attachment of a chemiluminogenic compound to a protein can strongly reduce its quantum efficiency,[7,9] while coupling to most low-molecular-weight ligands caused only moderate loss.[5] Quench-

[9] J. S. A. Simpson, A. K. Campbell, M. E. T. Ryall, and J. S. Woodhead, *Nature (London)* **279**, 646 (1979).

ing of the label is demonstrated by comparing the spectral incorporation with the active incorporation estimated from the chemiluminescence yield (without prehydrolysis). In ABENH–transferrin conjugates coupled with an amidine linkage, 20–35% of the label is active. By contrast, only 7% of seven isoluminol labels coupled to an antibody via active ester, one of the most successful methods to date, was active.[7] Also, only 1% of luminol remained active after diazotization and coupling to antibody.[9] Thus, the amidine linkage in ABENH–transferrin conjugates may preserve the light yielding capacity of the label better than other bonds. Probably the amidine coupler helps solubilize the hydrophobic ABENH labels. Also quenching may be less because ABENH has the chemiluminescence emission maximum at longer wavelength (515 nm) than luminol and isoluminol (about 420 nm).

In order to recover the inherent light yielding capacity of a chemiluminogenic conjugate, the label can first be released by chemical cleavage. Others have reported enhanced light production and improved sensitivity when the readout of the solid-phase immunoassay is performed after 30–60 min incubation at pH > 12 and elevated temperature.[1] Of course the dissociation of the antibody–antigen complex contributes to this, as postulated previously. However, most obviously the amide bond is partially hydrolyzed to release the chemiluminogenic label from the quenching environment when coupled to the protein. Indeed, we show that about one-fourth of thyroxine–ABEI conjugate molecules can be hydrolyzed under similar conditions. The ABENH–transferrin conjugates with an amidine linkage are fully cleaved under similar assay conditions employed in this study (see the table). However, shorter exposure at lower temperature and pH may be adequate to cleave the amidine bond,[10] which may be useful in a completely automated chemiluminescence immunoassay system. Other potential rapid cleavage methods include periodate cleavage of cis-diols and reduction of disulfide bonds with dithiothreitol.

The ABENH–transferrin conjugates prepared via imidate chemistry are useful for developing a sensitive immunoassay. The exceptionally low detection limit is mainly due to the quantum efficiency of ABENH, which is twice that of the acridium ester[4] and four to five times that of luminol and isoluminol derivatives.[5] Also, because the amidine linkage is fully cleaved in our assay the complete chemiluminescence efficiency is expressed. The immunoreactivity of transferrin in the conjugates is well preserved except when densely labeled. The immunoassay for transferrin developed here with a moderately labeled conjugate is specific, has excel-

[10] M. L. Ludwig and R. Byrne, J. Am. Chem. Soc. 84, 4160 (1962).

lent sensitivity, and gives good agreement with a radioimmunoassay reference procedure.

Unfortunately, the most efficient conjugates bearing more than 10 ABENH/transferrin have poor immunoreactivity and exhibit high non-specific binding, as has been observed for antibody labeled with isoluminol derivatives.[7] Possibly steric hindrance from the label as well as intra- and intermolecular crosslinking alters protein functionality in the immunoassay. According to a recent report, imidate activated label can be separated from a large excess of imidate before coupling to the protein.[11] This method permits use of high reagent concentrations to drive reactions and eliminates undesirable side reactions that compromise protein function. Additional investigation of the immunoassay conditions such as the solid phase, bound/free separation technique, and a detergent wash step may permit lowering the nonspecific binding with the highly labeled conjugates.

Conclusion

The coupling of ABENH to transferrin with imidate described here is a direct, rapid method for preparing conjugates that can be detected at low levels by chemiluminescence and serve to monitor ultrasensitive immunoassays. Conjugates are stable for at least 1 year. But the amidine linkage is easily cleaved to release the label from the protein to express its full light yielding potential during measurements.

The labeling of transferrin and the chemiluminescence immunoassay for transferrin developed should serve as a general model for future development of other chemiluminescence immunoassays for proteins. Improvements in the imidate coupling method and lowering nonspecific binding of conjugates in the immunoassay are feasible. Furthermore, similar chemiluminogenic conjugates of very high specific activity used with a noncompetitive format may permit even greater sensitivity. The sensitivity of an immunochemiluminometric assay, for example, is not limited by the antibody affinity, but directly related to the lower detection limit of labeled antibody.

[11] H. R. Schroeder, C. L. Dean, P. K. Johnson, D. L. Morris, and R. L. Hurtle, *Clin. Chem. (Winston-Salem, N.C.)* **31,** 1432 (1985).

[41] Chemiluminescent Probes for Singlet Oxygen in Biological Reactions

By AMBLER THOMPSON, HOWARD H. SELIGER, and GARY H. POSNER

Introduction

Since the first report of Howes and Steele[1] of chemiluminescence (CL) during lipid peroxidation by rat liver microsomal extracts and the observation by Allen *et al.*[2] of CL from phagocytosing polymorphonuclear leukocytes, the involvement of singlet molecular oxygen in these and many other biochemical processes has been hypothesized.

The unequivocal spectroscopic demonstration of singlet molecular oxygen has been shown both in the gas phase[3] following microwave excitation of oxygen and in the condensed phase, with oxygen excited via the photosensitization[4] of dyes and chemically via reactions such as the hypochlorite–hydrogen peroxide reaction.[5,6] The direct emission of singlet oxygen ($^1O_2$) has a low quantum efficiency in aqueous media ($<10^{-6}$) with a band in the near infrared (1268 nm) and the dimol emission of $^1O_2$ has major bands in the red at 634 and 703 nm. The use of these emissions to demonstrate $^1O_2$ production in biological reactions is problematic due to the low efficiency for singlet oxygen emission and the presence of other overlying red CL emissions from biological oxidations. Therefore most investigators employ chemical singlet oxygen traps and quenchers to corroborate the involvement of $^1O_2$ in biological reactions. For excellent reviews on the use of singlet oxygen traps see the reviews of Krinsky[7] and Foote.[8]

Foote initially showed that $^1O_2$ was highly reactive,[9,10] undergoing three basic types of reactions with organic molecules: allylic substitution

[1] R. M. Howes and R. H. Steele, *Res. Commun. Chem. Pathol. Pharmacol.* **2**, 619 (1971).
[2] R. C. Allen, R. L. Stjernholm, and R. H. Steele, *Biochem. Biophys. Res.* Commun. **47**, 679 (1972).
[3] S. J. Arnold, E. A. Ogryzlo, and H. Witzke, *J. Phys. Chem.* **40**, 1769 (1964).
[4] A. A. Krasnovskii, *Biofizika* **21**, 748 (1976).
[5] H. H. Seliger, *Anal. Biochem.* **1**, 60 (1960).
[6] A. U. Khan and M. Kasha, *J. Chem. Phys.* **39**, 2105 (1963).
[7] N. I. Krinsky, *in* "Singlet Oxygen" (H. H. Wasserman and R. W. Murray, eds)., p. 597. Academic Press, New York, 1979.
[8] C. S. Foote, *in* "Free Radicals in Biology" (W. A. Pryor, ed.), Vol. 2, p. 85. Academic Press, New York, 1976.
[9] C. S. Foote and S. Wexler, *J. Am. Chem. Soc.* **86**, 3879 (1964).
[10] C. S. Foote and S. Wexler, *J. Am. Chem. Soc.* **86**, 3880 (1964).

METHODS IN ENZYMOLOGY, VOL. 133

with olefins (ene reactions) forming allylic hydroperoxides, addition to dienes resulting in formation of endoperoxides ([2 + 4]cycloaddition), and reactions ([2 + 2]cycloaddition) forming dioxetanes (see Refs. 11 and 12 and the references therein). While $^1O_2$ existence and reactivity have been well characterized in simple chemical systems and the direct 1268 nm emission has been observed from some peroxidase reactions, the extent of $^1O_2$ involvement as the reactive intermediate in biochemical oxidation reactions remains to be conclusively demonstrated.

Assessment of the involvement of singlet oxygen in biological reactions is difficult and equivocal with currently used chemical $^1O_2$ traps. The steady-state concentration of $^1O_2$ is low in reactions of biological significance, due primarily to low production of $^1O_2$ and the microsecond lifetime of $^1O_2$ in aqueous environments. This necessitates the use of high concentrations of $^1O_2$ traps in order to observe significant loss of starting material or formation of reaction products by either standard chemical techniques such as absorption spectroscopy or chromatography. These high concentrations of $^1O_2$ chemical traps or quenchers (i.e., NaN_3) can affect other necessary biochemical components or reactions in the reaction mixture.[7,8] Thus the observation of inhibition or enhancement of a reaction could be due to these effects and not due to direct quenching or trapping of singlet oxygen. In addition, commonly used $^1O_2$ traps are easily oxidized by other species present in biological matrices and the loss of initial substrate is not always indicative of a specific reaction with enzymatically generated singlet oxygen. Due care must be taken in interpretation of these chemical trapping experiments.

The advantages of using chemiluminescent traps for $^1O_2$ are (1) that the species which is observed using chemiluminescent techniques is a specific reaction product, the dioxetane, whose formation is indicative of the presence of $^1O_2$ or $^1O_2$ equivalents such as a $Fe^{3+}-O-O^-$ complex, (2) that due to the high sensitivity for detection of CL, relatively low concentrations of the CL probe are necessary, thereby decreasing the likelihood of the artifactual inhibition of secondary biochemical reactions, and (3) since CL is usually measured as a function of time, the kinetics of the reaction giving rise to the chemiluminescent intermediate can be examined directly.

Requirements of Efficient Chemiluminescent Probes

Prior to using a chemiluminescent probe to study biological oxidation reactions, it is necessary to investigate the probe's chemical reactivity

[11] H. H. Wasserman and R. W. Murray, eds., "Singlet Oxygen." Academic Press, New York, 1979.
[12] A. Frimer, ed., "Singlet Oxygen," Vols I-IV. CRC Press, Boca Raton, Florida, 1985.

with $^1O_2$ and other reduced oxygen species and the dioxetane CL emission characteristics in well-defined chemical reactions. The observation of CL from a $^1O_2$ reaction with a probe is the result of several chemical and excited state molecular processes.[13] The overall CL quantum yield [Φ_{CL} (probe)] can be described by the following equation:

$$\Phi_{CL} \text{ (probe)} = \alpha\Phi_{EX}\Phi_{FL} \tag{1}$$

Where α is the chemical yield of dioxetane (**2**) production in the reaction of $^1O_2$ with the probe's (**1**) double bond. This chemical yield of the dioxetane is the product of the second-order reaction rate constant of the probe with $^1O_2$ and the efficiency of dioxetane formation relative to competing singlet oxygen pathways, such as endoperoxide and hydroperoxide formation. Φ_{EX} is the relative yield of the excited state carbonyl product molecule (*) produced by the decomposition of a dioxetane. Φ_{FL} is the chemiexcited fluorescence yield of the resultant excited state product.

There are two basic substituent effects to be considered in the production of a chemiluminescent dioxetane: (1) the effect of electron-donating groups increasing the second-order reaction rate constant for reaction of the double bond with $^1O_2$ and (2) the effect of easily oxidizable substituents such as pyrene which catalyze the decomposition of the dioxetane with high singlet excited state yields by donating an electron via an *intramolecular* electron-exchange luminescence mechanism.[14] This latter electron-donating substituent should also have a high fluorescent yield in order to maximize the overall CL yield.

The CL quantum yield for any probe species is experimentally determined by

$$\Phi_{CL} \text{ (probe)} = \text{photons emitted/probe molecules oxidized} \tag{2}$$

[13] W. Adam, *in* "Chemical and Biological Generation of Excited States" (W. Adam and G. Cilento, eds.), p. 115. Academic Press, New York, 1982.
[14] G. C. Schuster and K. A. Horn, *in* "Chemical and Biological Generation of Excited States" (W. Adam and G. Cilento, eds.), p. 229. Academic Press, New York, 1982.

It is very important that the photometer used for the quantitation of the CL photon emission from a reaction be calibrated[15] for absolute quantum efficiency over the wavelength region of the CL emission. Calibration using the luminol CL reaction[16] is appropriate for most CL reactions yielding excited state carbonyl products.

The efficiency of the dioxetane, once formed, to emit luminescence is defined from Eq. (1) as

$$\Phi_{CL} \text{ (dioxetane)} = \Phi_{EX}\Phi_{FL} \tag{3}$$

and from Eq. (2)

$$\Phi_{CL} \text{ (dioxetane)} = \text{photons emitted/dioxetanes produced} \tag{4}$$

A minimum estimate of the number of dioxetane molecules produced can be determined chromatographically as the amount of dioxetane decomposition (i.e., carbonyl) product that was formed from the reaction of the probe with $^1O_2$.

The photophysical fluorescence yield (Φ_{Fluor}) of the dioxetane decomposition product was measured in each solvent in which Φ_{CL} was determined and was assumed to be equal to the chemiexcited fluorescence yield Φ_{FL}. The factor Φ_{FL} for a product excited state is probably lower than Φ_{Fluor}. Photoexcited states are governed by photoselection rules and initially have the conformation of the ground state molecule in its ground state solvent cage, whereas chemiexcited states may not be governed by photoselection rules and their initial conformations relative to their ground state solvent cages may be quite different. Therefore the quenching of CL excited states may be quite different from the quenching of photoexcited states. Lee and Seliger measured Φ_{CL} (luminol) = 0.0125, whereas the photoexcited Φ_{Fluor} of the aminophthalic acid product was 0.3. The chemical yield of aminophthalic acid is quantitative. Whether this large difference is attributable to a difference between the chemiexcited and the photoexcited fluorescence yields or to an effect on Φ_{EX} has not been experimentally determined. The result is that the components of the product $\Phi_{EX}\Phi_{FL}$ cannot readily be evaluated from the substitution of Φ_{Fluor} measured by photoexcitation. Therefore the derived values of Φ_{EX} in Table II are minimum values.

Relative CL Yields of Substituted Vinylpyrenes and 7,8-Diol

Our initial interest in the production of chemiluminescent dioxetanes was to demonstrate that the CL observed during the liver microsomal

---

[15] H. H. Seliger, this series, Vol. 57, p. 560.
[16] J. Lee and H. H. Seliger, *Photochem. Photobiol.* **4**, 1015 (1965).

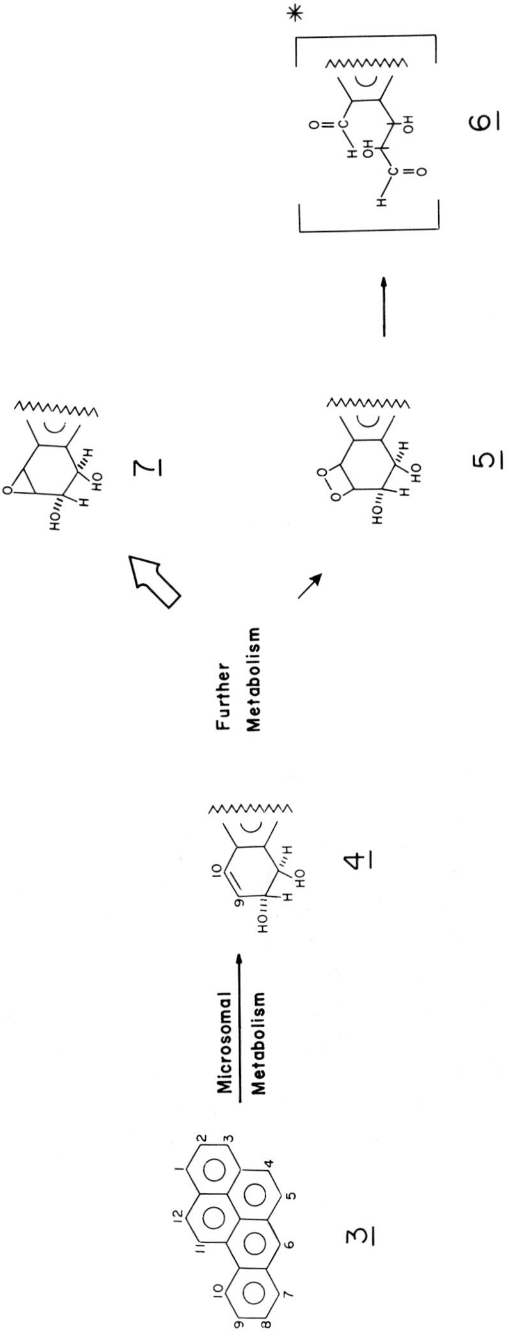
3
Microsomal Metabolism
4
Further Metabolism
7
5
6

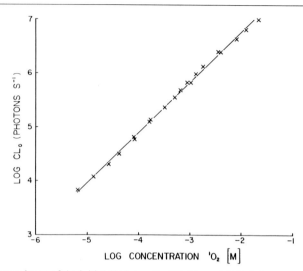

FIG. 1. Dependence of the initial CL intensity (CL_0) from 7,8-diol at room temperature on photochemically produced $^1O_2$. Solutions of methanol (2 ml) containing 20 nmol of rose bengal and 50 nmol of 7,8-diol were irradiated using either a 500 or 1000 W tungsten lamp. The amount of $^1O_2$ produced during the irradiation was varied by decreasing the lamp intensity using neutral density filters in the lamp beam and/or varying the time of irradiation (0.25 to 6.0 min).

metabolism of benzo[a]pyrene[17] (B[a]P) (**3**) was the result of the minor pathway production of a 9,10-dioxetane (**5**) during the remetabolism of a metabolite of B[a]P,[18] the *trans*-7,8-dihydroxy-7,8-dihydrobenzo[a]pyrene (7,8-diol) (**4**), to the carcinogenic diol epoxide (**7**).

In subsequent chemical experiments with $^1O_2$ it was shown that the singlet oxygen-initiated CL yield from 7,8-diol[19] was $10^2–10^5$ times higher than from B[a]P and all other B[a]P metabolites and in addition was 10^4 times more chemiluminescent than the 7,8-diol microsomal CL. The CL resulting from the reaction of singlet oxygen with 7,8-diol was first order for 5–6 half-lives of decay and the initial CL intensity (CL_0) was directly proportional (Fig. 1), for over four orders of magnitude, to the concentration of singlet oxygen produced by the photosensitization of rose bengal. This $^1O_2$-initiated CL of 7,8-diol was used to develop a sensitive assay[20]

[17] J. P. Hamman and H. H. Seliger, *Biochem. Biophys. Res. Commun.* **70**, 675 (1976).
[18] J. P. Hamman, H. H. Seliger, and G. H. Posner, *Proc. Natl. Acad. Sci. U.S.A.* **78**, 940 (1981).
[19] H. H. Seliger, A. Thompson, J. P. Hamman, and G. H. Posner, *Photochem. Photobiol.* **36**, 359 (1982).
[20] A. Thompson, H. H. Seliger, and G. H. Posner, *Anal. Biochem.* **130**, 498 (1983).

quantitating the amount of 7,8-diol produced in complex mixtures of B[a]P and its metabolites resulting from the microsomal metabolism of B[a]P, without the need for chromatographic separation. While the CL efficiency of the $^1O_2$-initiated CL of 7,8-diol was high, the chemical reactivity [α in Eq. (1)] of 7,8-diol with $^1O_2$ was low. Attempts to isolate the predicted fluorescent dialdehyde product (6) were unsuccessful. We therefore designed and synthesized a number of substituted vinylpyrene analogs (8) of 7,8-diol[21,22] in order to increase the reactivity toward $^1O_2$, to eliminate competing *ene* reactions and thereby to model the 7,8-diol CL. The product resulting in the singlet electronic excited state from the decomposition of a pyrenyl dioxetane (9) is pyrene-1-carboxaldehyde (10), a close structural homolog for the proposed dialdehyde product from the decomposition of 7,8-diol-9,10-dioxetane (5).

In a study of these vinylpyrene analogs (Table I), it was found that the *trans*-1-(2'-methoxyvinyl)pyrene analog (*t*-MVP) was 180 times more chemiluminescent[22] than 7,8-diol following a brief exposure to singlet oxygen. This higher *t*-MVP CL efficiency is presumably due to a higher reactivity with $^1O_2$ than the 7,8-diol. Methoxy groups are known to activate methoxyvinyl aromatic systems[23] toward [2 + 2]cycloaddition with $^1O_2$. Preliminary measurements of the second-order reaction rate constant of *t*-MVP with $^1O_2$ have indicated it to be on the order of $10^6 \ M^{-1}$ sec^{-1}. These measurements are in collaboration with Dr. John G. Parker of the JHU Applied Physics Laboratory and are currently in progress. Two products were isolated and identified[22] from the reaction of *t*-MVP

[21] G. H. Posner, J. R. Lever, K. Miura, C. Lisek, H. H. Seliger, and A. Thompson, *Biochem. Biophys. Res. Commun.* **123**, 869 (1984).

[22] A. Thompson, K. A. Canella, J. R. Lever, K. Miura, G. H. Posner, and H. H. Seliger, *J. Am. Chem. Soc.* **108**, 4498 (1986).

[23] D. S. Steichen and C. S. Foote, *Tetrahedron Lett.* p. 4363 (1979).

TABLE I
RELATIVE CL YIELD FROM A NUMBER OF
SUBSTITUTED VINYLPYRENES[a]

Substrate		τ	
R_1	R_2	(min)	Relative CL yield[b]
MeO	H	5.1	180
H	MeO	2.3	13
Chloro	H	2.3	2.7
t-Butyl	H	5.3	1.7
$S(CH_2)_3$	S	7.6	0.31
H	H	2.6	0.081

[a] Light yields are relative to that of 7,8-diol under the same experimental conditions. Reactions contained μM concentrations of the pyrenyl substrates and 10 nmol of rose bengal in 1.0 ml of methanol. Samples were irradiated for 2 min; 50-μl aliquots were transferred to a vial containing 1 ml of 0.1 M SDS and placed in a photometer. CL light yields were normalized to initial substrate concentrations. Mean lifetimes (τ) of the pyrenyldioxetanes are also shown. R_1 and R_2 positions are indicated by (8).
[b] Relative to CL yield of 7,8-diol under identical experimental conditions (5.5×10^{-5} photons/nmol).

with $^1O_2$: pyrene-1-carboxaldehyde (10% yield) and what appears to be the endoperoxide (87% yield).

The cis isomer of MVP was only 13 times more CL than 7,8-diol, even though it was 2–3 times more reactive with $^1O_2$ than the t-MVP. With the elimination of an *ene* reaction in these vinylpyrene analogs, the remaining singlet oxygen reaction pathway in competition with the formation of dioxetane is endoperoxide formation. This implies that endoperoxide formation is more highly favored in the reaction of $^1O_2$ with c-MVP than with the trans isomer.

In reactions of t-MVP with other reduced oxygen species (1) t-MVP did not react with excess H_2O_2 in aqueous micellar solutions,[21] (2) t-MVP did react with excess NaOCl (half-life of 2.5 hr), but no CL was observed,[22] (3) t-MVP produced no CL over the control when exposed to superoxide generated enzymatically using xanthine oxidase with hypoxanthine as the substrate,[21] and (4) t-MVP produced specific CL upon metabolism by induced rat liver microsomes with light yields slightly

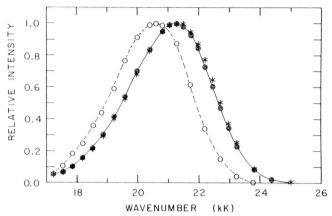

Fig. 2. $^1O_2$ initiated CL emission spectra from the *cis*- (asterisk) and *trans*- (solid circle) MVP and 7,8-diol (open circle) in 0.1 *M* SDS. For the CL emission spectra of *c*- and *t*-MVP, 100 nmol of each isomer was reacted separately with $^1O_2$ by injecting 50 μmol of NaOCl into 5 ml of 0.1 *M* SDS containing 300 μmol of H_2O_2. For the CL emission spectrum of 7,8-diol, 85.4 nmol of 7,8-diol was reacted with NaOCl/H_2O_2 in 3 ml of 0.1 *M* SDS. The fluorescence emission spectrum of 18.0 μM pyrene-1-carboxaldehyde in 0.1 *M* SDS is shown as the solid line; 366 nm excitation was supplied for the fluorescence spectrum of **(10)** by a Model 098 Perkin-Elmer prism monochrometer using a Leitz 200 W Hg light source isolated by 2 Corning 7-60 filters. Both CL and fluorescence emission spectra were determined on a 1.0 m F/3 Fastie-Ebert grating spectrometer.

higher than that observed during the metabolism of 7,8-diol[24] (this microsomal CL will be discussed later in this chapter). Therefore it appears that *t*-MVP is an extremely sensitive CL probe which is specific for singlet molecular oxygen and/or singlet oxygen electronic equivalents.[21,25]

CL Emission Spectra and CL Yield Solvent Dependences

In the initial work with 7,8-diol singlet oxygen-initiated CL a marked solvent dependence on the CL yield was noted. Aqueous solutions of 0.1 *M* sodium dodecyl sulfate (SDS) exhibited the highest 7,8-diol CL yields[22] and conversely the CL yields in nonpolar organic solvents were low.

The CL emission spectra of 7,8-diol, *c*-MVP, and *t*-MVP were measured[19,22] in 0.1 *M* SDS, this aqueous micellar solution exhibited the highest CL yields. Figure 2 shows the $^1O_2$-initiated CL emission spectra of

[24] A. Thompson, W. H. Biggley, G. H. Posner, J. R. Lever, and H. H. Seliger, *Biochim. Biophys. Acta* **882,** 210 (1986).
[25] N. Duran, *in* "Chemical and Biological Generation of Excited States" (W. Adam and G. Cilento, eds.), p. 345. Academic Press, New York, 1982.

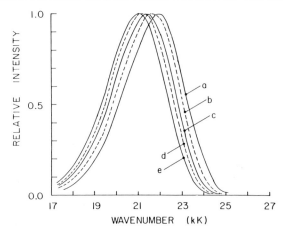

Fig. 3. Fluorescence emission spectra (relative quanta per wavenumber interval) of pyrene-1-carboxaldehyde in water: methanol mixtures (v/v): (a) methanol; (b) 20% H_2O; (c) 40% H_2O; (d) 60% H_2O; (e) 80% H_2O. Pyrene-1-carboxaldehyde concentrations were approximately $1\mu M$. Spectral measurements were made with a SLM 8000 photon counting spectrofluorometer with 347 nm excitation.

c-MVP, t-MVP, and 7,8-diol with the fluorescence emission spectrum of the dioxetane decomposition product, pyrene-1-carboxaldehyde (solid line) measured in 0.1 M SDS as well. The CL emission spectra of t-MVP and c-MVP agreed with the fluorescence of the product pyrene-1-carbox-aldehyde. The shapes of the MVP and 7,8-diol CL emission spectra were superimposable with a full width at half-maximum of 3.19 kK and 3.17 kK, respectively, implying similar aldehyde excited states. The CL emission maximum for 7,8-diol was red-shifted 0.67 kK from the emission maxima of the c- and t-MVP. This spectral shift amounts to an energy difference of 1.9 kcal/mol.

Pyrene-1-carboxaldehyde fluorescence emission spectra (Fig. 3) were strongly red shifted (from 450 to 475 nm) with identical full widths at half-maxima, when the water in water:methanol mixtures[22] was increased from zero to 80%. Over this same range in water concentration the photophysical fluorescence quantum yield (Φ_{Fluor}) increased linearly from 0.14 to 0.73. Spectral shifts and increased Φ_{Fluor} from pyrene-1-carboxaldehyde were also observed for the polar versus nonpolar solvents listed in Table II. These are attributed to hydrogen bond formation[26] between the pyrene-1-carboxaldehyde carbonyl group and water or the alcohol, lowering of

[26] K. Brederick, T. Forster, and H. G. Osterlin, in "Luminescence of Organic and Inorganic Materials" (H. P. Kallmann and G. M. Spruch, eds.), p. 161. Wiley, New York, 1962.

TABLE II
SOLVENT DEPENDENCE OF RELATIVE CL YIELDS[a]

Solvent[b]	Relative CL yield		t-MVP Φ_{CL} (dioxetane)[c]	Pyrene-1-carboxaldehyde Φ_{fluor}	Φ_{EX}
	7,8-Diol	t-MVP			
0.1 M SDS	1.0	1.0	$2.7 \pm 0.5 \times 10^{-2}$	0.44	0.061
Methanol	0.25	0.25	$6.8 \pm 1.3 \times 10^{-3}$	0.12	0.57
Ethanol	0.11	0.033	$8.9 \pm 1.5 \times 10^{-4}$	0.052	0.017
Acetonitrile	—	0.0056	$1.5 \pm 0.2 \times 10^{-4}$	0.044	0.0034
Chloroform	0.080	0.0044	$1.2 \pm 0.2 \times 10^{-4}$	0.021	0.0057
Chloro-benzene	0.075	—	—	—	—
Diethyl-ether	0.00011	0.00021	$5.7 \pm 0.5 \times 10^{-6}$	0.0036	0.0016
Dimethyl-sulfoxide	9×10^{-5}	2.6×10^{-7}	$7.2 \pm 1.2 \times 10^{-9}$	0.041	1.8×10^{-7}

[a] Solvent dependence of the relative CL yield from 7,8-diol and t-MVP and the Φ_{CL} (dioxetane) and Φ_{EX} for the CL from t-MVP dioxetanes and Φ_{Fluor} for pyrene-1-carboxaldehyde. In measurements of the relative CL from t-MVP and 7,8-diol the light yields in each solvent were normalized to the light yields in 0.1 M SDS. In measurements of Φ_{CL} (dioxetane) for t-MVP, 1 ml of methanol containing 10 nmol of rose bengal and 9.5 nmol of t-MVP was irradiated for 2 min and aliquots of the irradiated solution were added to the solvent of interest. Parallel samples were analyzed by HPLC to quantitate the amount of t-MVP reacted in the methanol with $^1O_2$ and the amount of (10) produced. Fluorescence quantum yields (Φ_{Fluor}) for pyrene-1-carboxaldehyde were measured relative to quinine sulfate in a SLM 8000 photon counting spectrofluorometer using the same solvent compositions as those used for Φ_{CL} (dioxetane). Φ_{EX} is derived from Eq. (3) utilizing values of Φ_{CL} (dioxetane) and Φ_{Fluor} obtained by direct measurements.
[b] Solvent concentration is 95%, 5% methanol from photosensitization reaction.
[c] SD ($n = 4$).

the energy of the $\pi \rightarrow \pi^*$ state relative to the $n \rightarrow \pi^*$ state. Therefore one of the factors responsible for the observed CL yield solvent dependency is the solvent dependence of ϕ_{FL} from Eqs. (1) and (3).

In Table II the singlet oxygen-initiated CL yield for 7,8-diol and t-MVP in a number of solvents was normalized to the yield in 0.1 M SDS; both compounds exhibited similar solvent dependencies. The amount of dioxetane [α in Eq. (1)] formed was held constant in these experiments by carrying out the photosensitizations in a single solvent, methanol, and adding an aliquot of the irradiated sample to the solvent of interest. The long lifetime of the 7,8-diol and t-MVP dioxetanes in methanol (minutes) permitted a precise extrapolation of measured CL to zero time (CL$_0$).

The advantage of studying t-MVP is that the emitting molecule, pyrene-1-carboxaldehyde, could be isolated and the photoexcited fluorescence properties of this compound could be studied in the same solvents as the CL. Since we could measure the amount of pyrene-1-carboxalde-hyde produced following an exposure to $^1O_2$, we could therefore determine Φ_{CL} (dioxetane) from Eq. (4). Assuming that the photophysical fluorescence quantum yield (Φ_{Fluor}) is identical (or proportional) to the chemiexcited fluorescence yield (Φ_{FL}), we solved for Φ_{EX} in Eq. 3, thereby estimating the solvent dependency of Φ_{EX} in Table II. These derived values for Φ_{EX} were also higher in polar hydrogen-bonding solvents, but as stated previously are minimum estimates of Φ_{EX}.

From Table II t-MVP Φ_{CL} (dioxetane) = 0.027 which is more than a factor of two higher than for luminol. However the yield of the luminol CL product aminophthalic acid is near unity,[27] whereas $^1O_2$ reactions with t-MVP result in the production of a dioxetane in approximately 10% yield. Therefore the overall Φ_{CL} (t-MVP) is ~0.003.

This high Φ_{CL} and the high specificity of this CL for $^1O_2$ make t-MVP an excellent probe for $^1O_2$. It should be possible to detect small amounts of cellular $^1O_2$ or singlet oxygen equivalents with trace amounts of t-MVP or other second generation (e.g., modified MVP) CL probes, where the presence of the probe may not interfere with normal physiological processes.

CL of MVP and 7,8-Diol in Liver Microsomal Extracts

One of the advantages of utilizing 7,8-diol and c- and t-MVP in the microsomal CL reactions is that they are CL singlet oxygen traps as well as substrates for the microsomal aryl hydrocarbon hydroxylase. Therefore the chemical quenching of $^1O_2$ as well as the microsomal CL are accomplished by the same molecule. Using the ratios of the observed relative CL yields (Table I) of these substrates in reactions with authentic singlet oxygen as criteria for the reaction with free singlet oxygen, we have a means of testing the extent of the involvement of singlet oxygen in biochemical reactions and perhaps distinguishing between reactions with $^1O_2$ and singlet oxygen equivalents. The use of these CL probes coupled with spectral measurements and inhibition experiments with catalase and superoxide dismutase should allow us to evaluate the involvement of free singlet oxygen and other reduced oxygen species during biological oxidations.

A substrate specific chemiluminescence was emitted during the metabolism of polycyclic aromatic hydrocarbons (PAH) by liver microsomes,[28] whose cytochrome P-450 aryl hydrocarbon hydroxylase had

[27] D. F. Roswell and E. H. White, this series, Vol. 57, p. 409.
[28] H. H. Seliger and J. P. Hamman, *J. Phys. Chem.* **80,** 2296 (1976).

been induced by specific inducers. Selecting B[a]P as a model PAH it was shown that this microsomal CL was due to the remetabolism of metabolites with 7,8-diol producing the highest microsomal CL yield of any metabolite including the parent molecule B[a]P.[18] The microsomal metabolism of the CL probes c- and t-MVP also produced significant CL.[24] The question is, "What is the extent of the involvement of free singlet molecular oxygen in the microsomal CL of 7,8-diol and other PAHs?"

The kinetics of the microsomal CL of t-MVP and 7,8-diol are shown in Fig. 4 and have two basic components: a rise time to a maximum, which is proportional to the rate of formation of the CL intermediate, and a decay proportional to the lifetime of the dioxetane intermediate. Neither the addition of catalase (857 units/ml) nor the addition of superoxide dismutase (112 units/ml) had any effect on the microsomal CL intensity or the CL yield from t-MVP and 7,8-diol.[24] Thus this CL is presumably not the result of the reaction with reduced oxygen species H_2O_2 and O_2^-. 7,8-Benzoflavone, a potent inhibitor of the aryl hydrocarbon hydroxylase, inhibited the microsomal CL from t-MVP with an apparent K_i of 1–2 μM, implying that this CL was the result of a specific enzymatic catalysis.[24]

The Aroclor 1254-induced microsomal CL yield from t-MVP (Fig. 5) was 3.4×10^{-8} photons/molecule of substrate, 10.6 times the CL yield of c-MVP (3.2×10^{-9}) and 1.3 times the CL yield of 7,8-diol (2.6×10^{-8}). The ratio of the singlet oxygen CL yields (Table I) between t-MVP and c-MVP was 14:1 and the ratio between t-MVP and 7,8-diol in singlet oxygen reactions was 180:1. Therefore this small difference in the ratios of

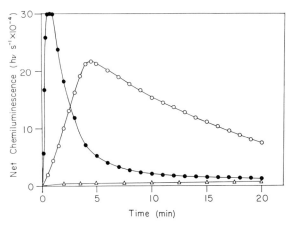

FIG. 4. Microsomal CL of 7,8-diol (●) (12.9 nmol) and t-MVP (○) (13.0 nmol) in absolute units of photon emission. Intrinsic CL of microsomes (△) without substrate subsequent to the addition of NADPH is also shown.

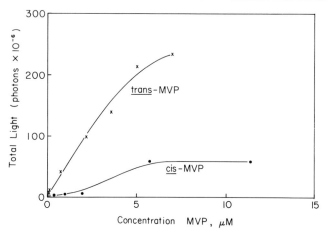

FIG. 5. Concentration dependence of total CL emission from (●) *cis*- and (×) *trans*-MVP in Aroclor-induced microsomes. Concentrations of *c*- and *t*-MVP were varied and the CL reaction was initiated at 37° by addition of NADPH and placed in the photometer.

the CL yields in the microsome reactions versus the large difference in the $^1O_2$ reactions implies that the CL observed in the microsomal reactions was not due to free $^1O_2$. The CL yields of *t*-MVP and 7,8-diol metabolized by noninduced microsomes were of the order of 10^{-10} photons/molecule of substrate, at least a 100-fold lower than the yields from induced microsomes.

When dioxetanes were formed by the rose bengal photosensitization of *t*-MVP and 7,8-diol and were added to solutions of either microsomes or phosphatidylcholine vesicles, the CL of these performed dioxetanes was quenched by a factor of greater than 10^5. Since the lifetime of the resulting CL decay was unchanged, this quenching was presumably not due to a competing dark *intermolecular* electron-exchange decomposition of the dioxetane[14] but rather to a quenching of Φ_{EX} and/or Φ_{FL}. This large quenching factor implies that the enzymatic pathway resulting in dioxetanes may be more significant than might be inferred from the low microsomal CL yields.

The microsomal CL emission spectrum (400–600 nm) of *t*-MVP (Fig. 6) was broad and unstructured, with a full width at half-maximum of 110 nm and an emission maximum of 500 nm.[24] Greater than 65% of the *t*-MVP CL emission was from wavelengths longer than 600 nm. Using various combinations of glass cut-on filters, the red emission extending beyond 600 nm appeared to be a broad continuous band. Applying the Krinsky[29] criterion for singlet oxygen dimol emission, we were unable to

[29] C. F. Deneke and N. I. Krinsky, *Photochem. Photobiol.* **25**, 299 (1977).

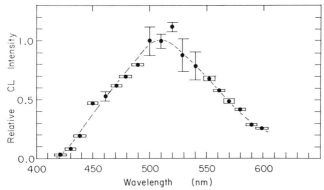

FIG. 6. CL emission spectrum resulting from the microsomal metabolism of t-MVP (5.2 μM). The position of a point represents the central transmission wavelength of the specific filter used and the mean of the relative intensity. When the number of replicate measurements, n, was equal to 2 the range of the measurements is indicated by the vertical dimensions of the box surrounding the point. When $n > 2$ the 95% confidence limits are indicated by error bars. In all cases, the bandwidths of the interference filters are shown by the widths of the boxes or bars. A smooth curve has been drawn through the data.

detect any statistically significant emission attributable to the 635 or 703 nm dimol bands of singlet oxygen. At the present, we have not examined the t-MVP microsomal CL for the $^1O_2$ infrared (1254 nm) emission. These experiments are now in progress.

Although CL of these probes was observed in the microsomal reaction, this CL was not from a reaction with free $^1O_2$. This evaluation is based on two criteria: the difference between the ratios of the microsomal CL yields and the ratios of CL yields with singlet oxygen, and the lack of spectral evidence for the dimol emission of singlet oxygen. The mere observation of CL upon addition of these CL singlet oxygen probes to a biological reaction is not a sufficient criterion to classify a reaction as a singlet oxygen reaction. Other corroborating experimental evidence is necessary as well.

It is probable that the reactive oxygen species formed by cytochrome P-450 in the production of a dioxetane during the oxidation of an electron-rich PAH is an $Fe^{3+}-O-O^-$ complex.[28,30] This perepoxide type reaction in which both oxygen atoms are transfered to the olefinic double bond would be equivalent to a singlet oxygen reaction but need not result in the same dioxetane:endoperoxide product ratios as those in the singlet oxygen reaction. We have observed significantly different product ratios in $-78°$ chemical reactions of t-MVP with a phosphite ozonide, a chemical singlet

[30] C. S. Foote, in "Biochemical and Clinical Aspects of Oxygen" (W. S. Caughey, ed.), p. 603. Academic Press, New York, 1979.

oxygen equivalent.[31] It is not necessary that free singlet oxygen be produced by cytochrome *P*-450.

Future Considerations

The CL probe *t*-MVP was specifically designed as an analog for the singlet oxygen-initiated CL of 7,8-diol. Due to *t*-MVP's structural similarity with 7,8-diol and the broad structural specificities of cytochrome *P*-450s, *t*-MVP was capable of being metabolized by the same enzymes which catalyzed the production of a chemiluminescent intermediate from 7,8-diol. This may not be the case in all biological reactions. One of the primary characteristics of enzymatic reactions is substrate specificity and therefore it may be necessary to tailor a CL probe specifically for application with a particular enzyme. There are two general scenarios for the production of free singlet oxygen in biological reactions. The first is where $^1O_2$ is produced via either photosensitization or enzymatic reactions and released free into solution to react randomly with cellular constituents. Since there is no substrate specificity, general CL probes should be sufficient to distinguish this case. The second is where $^1O_2$ or singlet oxygen equivalents are produced and reacted with a bound substrate within the active site of the enzyme, necessitating the use of a CL probe with molecular structure that can be recognized and bound by the enzyme prior to reaction. Fortunately most catabolic enzymes have broad substrate specificities.

We are currently involved in the design and synthesis of second generation CL probes for singlet oxygen: probes with increased reactivity toward $^1O_2$ and increased yield of dioxetane versus competing singlet oxygen reaction pathways and probes where the pyrene moiety has substituents which vary in electron-donating character and which therefore would affect the lifetime of the dioxetane and the emission characteristics, the CL yield, and the solvent dependency of the CL emission. Finally, we plan the preparation of water-soluble CL probes and probes attached to lipids to mimic the cellular microenvironments in which $^1O_2$ or $^1O_2$ equivalents are generated.

Acknowledgments

Research in the Departments of Biology and Chemistry at The Johns Hopkins University was supported by National Institute of Environmental Health Sciences Grant 1-POl-ES-02300. A.T. was supported by a N.I.E.H.S. postdoctoral fellowship. Special thanks are due to Dr. John R. Lever for synthesizing most of the compounds used in this research and to Mr. William H. Biggley for his assistance in the spectral measurements.

[31] A. P. Schaap and P. D. Bartlett, *J. Am. Chem. Soc.* **92,** 210 (1970).

Section III

Instrumentation

[42] Characteristics of Commercial Radiometers

By PHILIP E. STANLEY

In most studies involving bioluminescence or chemiluminescence the use of an instrument is generally required to measure the light output. Although one can construct such a unit to fit the need at hand, most users will find it more convenient and easier to purchase one. Since this chapter concerns commercially available instrumentation I will refer the reader elsewhere for reviews concerning custom-made instruments.[1,2]

In a previous publication[3] I discussed at some length the instrumentation available as of July 1982. Since that time a number of additional instruments have appeared and this chapter concerns instrumentation available as of September 1986. Recently a number of articles have been published concerning various aspects of instrumentation: commercial instruments,[3,4] video systems,[5] scintillation counters,[6] general review,[3,7] and multisample units and automation.[3,4,8–11]

The form of instrumentation discussed here is sometimes referred to as photometers or luminometers. According to Wampler[8] the correct term

[1] J. E. Wampler and J. C. Gilbert, *in* "Bioluminescence and Chemiluminescence: Instruments and Applications" (K. Van Dyke, ed.), Vol. I, p. 129. CRC Press, Boca Raton, Florida, 1985.

[2] J. M. Anderson, G. J. Faini, and J. E. Wampler, this series, Vol. 57, p. 529.

[3] P. E. Stanley, *in* "Clinical and Biochemical Luminescence" (L. J. Kricka and T. J. N. Carter, eds.), p. 219. Dekker, New York, 1982.

[4] K. Van Dyke, *in* "Bioluminescence and Chemiluminescence: Instruments and Applications" (K. Van Dyke, ed.), Vol. I, p. 83. CRC Press, Boca Raton, Florida, 1985.

[5] J. E. Wampler, *in* "Bioluminescence and Chemiluminescence: Instruments and Applications" (K. Van Dyke, ed.), Vol II, p. 123. CRC Press, Boca Raton, Florida, 1985.

[6] G. M. Pekoe, *in* "Bioluminescence and Chemiluminescence: Instruments and Applications" (K. Van Dyke, ed.), Vol. II, p. 147. CRC Press, Boca Raton, Florida, 1985.

[7] P. E. Stanley, J. Vossen, and S. E. Kolehmainen, *in* "Advances in Scintillation Counting" (S. A. McQuarrie, C. Ediss, and L. I. Wiebe, eds.), p. 271. Univ. of Alberta Press, Edmonton, Alberta, Canada, 1983.

[8] J. E. Wampler, *in* "Chemi- and Bioluminescence" (J. G. Burr, ed.), p. 1. Dekker, New York, 1985.

[9] F. Berthold, H. Kubisiak, M. Ernst, and H. Fischer, *in* "Bioluminescence and Chemiluminescence: Basic Chemistry and Analytical Applications" (M. A. DeLuca and W. D. McElroy, eds.), p. 699. Academic Press, New York, 1981.

[10] F. Berthold, *in* "Analytical Applications of Bioluminescence and Chemiluminescence" (L. J. Kricka, P. E. Stanley, G. H. G. Thorpe, and T. P. Whitehead, eds.), p. 457. Academic Press, New York, 1984.

[11] F. Berthold, *in* "Advances in Scintillation Counting" (S. A. McQuarrie, C. Ediss, and L. I. Wiebe, eds.), p. 230. Univ. of Alberta Press, Edmonton, Alberta, Canada, 1983.

is radiometer since photometers evaluate in photometric units and luminometers measure luminous flux. I will follow Wampler's nomenclature.

Radiometers are available in a wide range of sophistication and prices vary accordingly. Generally speaking modern instruments have approximately the same sensitivity, with the exception of the less sensitive units with photodiode detectors. Thus a radiometer with a photomultiplier as detector will be able to detect at least 0.1 pg ATP under ideal conditions and with modern firefly luciferase reagents. Photodiode based units will probably be 50 to 100 times less sensitive. In the detailed comments on each instrument which follow later, I have not mentioned the sensitivity claimed by the manufacturer. My recommendation to prospective purchasers is always to check the sensitivity themselves and with the reagents they intend to use.

Components of a Typical Radiometer

The configuration of a basic instrument comprises at minimum the following components: (1) detector and associated electronics, (2) detector chamber, and (3) injector assembly. A more advanced instrument may have some or all of the following: signal processing, multisample capacity, several injectors, automation, and data processing.

Detector

Spectral Range. In most models the radiometer accepts light from the entire spectrum of the emitting species and there is no provision to select a portion of the spectrum. However two new instruments[10,12] have been described that have filters which facilitate measuring light over a selected range of wavelengths. Such instruments find use in certain forms of very sensitive luminescence immunoassays and are fitted with two photomultipliers so that a ratio of two separate signals each derived for a separate wavelength range can be obained simultaneously.

One manufacturer (Tohuku) markets instruments fitted with filters so that a coarse spectrum of the emitted light can be obtained.

Photodiode and Photomultipliers. These are the two types of light detectors used. Photodiodes are currently far less sensitive than photomultipliers. For the present applications it is necessary to select carefully to obtain the very best in terms of not only sensitivity but also background signal. However photodiodes do have both a wide dynamic range and wavelength response and are quite rugged. As such they are therefore

---

[12] Literature from Turner Designs, 2247 Old Middlefield Way, Mountain View, California.

suited for instruments which can be portable and where the highest sensitivity is not required (e.g., ALL Monolight 100). See Wampler[8] for discussion and the table for a list of suppliers.

Photomultipliers are inherently more sensitive. Further they are available in a wide range of sizes, configurations, and sensitivities. However, they are far less rugged and generally considerably more expensive following the high degree of selection required.

Photomultipliers are characterized by having excellent amplification characteristics and this facilitates electronic handling of the signal. They have a photocathode which acts as a quantum detector and thus a portion of photons impinging upon it will give rise to a sizable signal. This may be had as current or as an electrical pulse for photon counting. Electronic amplification up to one hundred million is not uncommon.

Depending on the nature of the photocathode the photomultiplier will be sensitive in different parts of the spectrum. The so-called bialkali photocathodes have blue sensitivity around 400 nm and quantum efficiences around 25% at that wavelength. Red-sensitive (rubidium) photomultipliers have better sensitivity in the red range and are thus in some ways more suited to the luminescence reactions producing light above 500 nm. However, the latter have an inherently larger background and more often require cooling to produce satisfactory background characteristics. Many manufacturers of radiometers will allow the option of choosing your own response/supplier of photomultiplier. This can be a real advantage if for instance you wish only to make measurements in the red part of the spectrum. For details of photomultiplier suppliers see the table.

Most if not all commercially available radiometers are fitted with photomultipliers which have been carefully selected for high sensitivity and very low background. One or two inch photocathodes are used. They may

SOME MANUFACTURERS OF PHOTODIODES AND PHOTOMULTIPLIERS

Centronic Sales Ltd.	Hamamatsu Photonics K.K.
Centronic House	1126 Ichino-cho
King Henry's Drive	Hamamatsu City
New Addington,	Japan
Croydon CR9 0BG	
England	
Thorn EMI Electron Tubes Ltd.	RCA Electro Optics and Devices
Bury Street	Publication Services
Ruislip	P.O. Box 3200
Middlesex HA4 7TA	Somerville
England	New Jersey 08876

be fitted in special assemblies, e.g., mu metal shield to minimize the effect of magnetic fields, or they may be cooled, e.g., with Peltier elements, to reduce the inherent background signal.

For a discussion of the various types of photomultiplier and the various means that must be employed to keep them stable and operating under optimal conditions see the table and Wampler et al.[1,8,11,13,14]

Other Detectors. Developments at WIRA in Leeds, England have allowed them to recently introduce a new instrument to the market with photographic film as the detector.

Detector Chamber

A small light-tight chamber holds the cuvette (containing the reactants) which is inserted via a suitable shutter assembly. For most applications the chamber should be temperature controlled since most reaction rates are temperature dependent. In some instruments the temperature is controlled by circulated water, whereas in others it is controlled electrically with a Peltier element. Still others rely merely upon a heating element to raise the temperature to that selected and overheating, due for instance to siting the instrument in a very warm laboratory, cannot be controlled. Whatever the means of temperature control, it is usually necessary to allow a cuvette with its contents sufficient time to come to thermal equilibrium in the chamber before initiating the reaction. In addition if one is not injecting liquid at the same temperature as the cuvette contents then the injected volume should be small in comparison so as to avoid substantial temperature fluctuation and consequent variation in light output from the reaction. The inside walls of the chamber which surround the cuvette, all of which are in close proximity to the photomultiplier detector, are usually polished or coated with a reflective material. The purpose of this is to reflect photons emitted from the cuvette in 4π geometry toward the detector. The reflection coefficient of the chamber may be wavelength dependent. There have been few studies which permit optimization of this important sensitivity characteristic of the radiometer (see, however, Wampler[1] and Malcolm and Stanley[15-18]).

[13] G. T. Reynolds, *in* "Bioluminescence: Current Perspectives" (K. H. Nealson, ed.), p. 12. CEPCO Div., Burgess, Minneapolis, Minnesota, 1981.

[14] R. W. Engström, "Photomultiplier Handbook." RCA Solid State Division, Electro Optics and Devices, Lancaster, Pennsylvania, 1980.

[15] P. J. Malcolm and P. E. Stanley, *Int. J. Appl. Radiat. Isot.* **27**, 397 (1976).

[16] P. J. Malcolm and P. E. Stanley, *Int. J. Appl. Radiat. Isot.* **27**, 415 (1976).

[17] P. J. Malcolm and P. E. Stanley, *in* "Liquid Scintillation Counting" (M. A. Crook and P. Johnson, eds.), Vol. 4, p. 15. Heyden, London, 1977.

[18] P. E. Stanley and P. J. Malcolm, *in* "Liquid Scintillation Counting" (M. A. Crook and P. Johnson, eds.), Vol. 4, p. 44. Heyden, London, 1977.

Injector Assemblies

Injection of light-producing reagents into the sample cuvette must generally be done in the darkened detector chamber in front of the photomultiplier. For fast reactions ($t_{1/2} < 1$ sec) this is mandatory. Injection should be sufficiently vigorous to ensure mixing and this can be checked using colored materials. Bad mixing will usually manifest itself by giving results of poor reproducibility.

Manual injection can be by microsyringe through a light-tight septum. Results are generally not of the highest caliber since good reproducibility when depressing the plunger by hand is difficult to attain. The use of a Hamilton repeating syringe will improve reproducibility and several instrument manufacturers recommend these. The syringe can also be furnished with a microswitch so that signal acquisition can be synchronized with injection.

Care must be exercised in use of light-tight septa. Repeated perforation can lead to light leaks with resultant high background counts. Further, septa in poor condition can lead to contamination of samples by previously used reagents and by microbes. The latter is a problem when using the radiometer for rapid microbiology with ATP.

Automatic injection is available on many instruments. In some, the injector forms an integral part of the instrument, while in others it is available as an option. Some have fixed volumes while others are adjustable either mechanically or via software. Clearly it is important to be able to thoroughly cleanse, sterilize, and disassemble the injector and associated plumbing so as to avoid blockages, contamination and the like.

Details of Radiometers

Supplier. Amersham International plc, Amersham Place, Little Chalfont, Buckinghamshire HP7 9NA, England [tel: 02404-4444; telex: 838818 ACTIVA G].

Amersham International Amerlite Luminescence Plate Reader

An automatic instrument measuring light emitted from reactants in individual wells of a microtiter plate.

Photomultiplier detector and anode current measured by charge integration method. Long-term stability by reference to a pulsed LED and appropriate gain adjustment. Analog to digital conversion circuitry provides signal resolution greater than 19 bits thereby making it suited to chemiluminescence immunoassay work.

Microtiter plates are located with high precision (± 0.15 mm) on an X-Y table which can be moved to bring individual wells to the photomultiplier for measurement. Adjacent wells can be addressed within 200 msec. Microtiter plates are fitted with strips of white plastic wells into which reagents are placed prior to locating the plate in the instrument.

Z80 microprocessor is used to control all aspects of instrument operation and data reduction. There is a display and special keyboard. Software (in "C" language) is held in executable form on a 3-in. disc fitted in an in-built disc drive. Certain quality assurance information is written to disc but not data for individual samples. Hard copy of data and certain statistics are printed on an in-built printer.

Presently the software is configured for immunoassay work but some changes may be made by use of a "diagnostic disc" supplied by the manufacturer. An RS232 output port is provided for those who wish to connect the unit to a microcomputer.

Supplier. Analytical Luminescence Laboratory, 11760 Sorrento Valley Road, Suite E, San Diego, California 92121 [tel: (619) 455-9283].

Analytical Luminescence Laboratory Monolight 100

Manual operation.

A portable instrument with solid state silicon photodiode detector. Powered either by in-built rechargeable 6 V battery or mains. Six hours operation can be achieved from 16 hr charge cycle.

Pressing START button automatically zeros the instrument (background subtract).

Cuvette size 12 × 75 mm.

Injection is performed manually from a Hamilton repeating syringe.

Signal integration may be for 10 or 40 sec and there is an analog output for a chart recorder.

Digital display is provided (3 digit) and dynamic range is 4 decades (0.00 to 200) with autoranging.

Analytical Luminescence Laboratory Monolight 2001

Similar to Berthold Model LB9500.

Analytical Luminescence Laboratory Monolight 500/Micromate 501

Manual operation.

Photomultiplier detector with S4 response (others optional).

Cuvette size 12 × 45–55 mm.

Injection can be performed either manually (Hamilton repeating syringe) or automatically from a dual syringe unit (Micromate 501) capable of injecting 100 μl of each of two reagents. The START switch operates both Micromate 500 and 501 simultaneously.

Measuring modes are rate, peak height (maximum occurring during 10 or 30 sec integration starting 2 sec after injection), and integration (either 10 or 30 sec with 2 sec delay after injection.

There are three sensitivity ranges and the digital display has 4 digits. The dynamic range is 5.2 decades with blank subtract of 2 decades.

BCD output is optionally available.

Supplier. American Research Products Corporation, 10403 47th Avenue, Beltsville, Maryland 20705 [tel: (301) 595-5188].

American Research Products Corporation, Lumi-Tec

Manual operation.

Fitted with S4 response photomultiplier and operated in current mode. High voltage adjustable in 50 V steps from 700 to 900 V to vary sensitivity.

Cuvette size 12 × 47 mm (6 × 50 mm optional). Flow cell available. Ambient temperature operation.

Injection of reagent by repeating syringe (manual).

Signal output on digital display or chart recorder (optional).

Measuring modes peak height and integration.

Blank subtract available.

BCD option available.

Supplier. Laboratorium Prof. Dr. Berthold, D-7547 Wildbad 1, Calmbacher Strasse 22 (PO Box 160), Federal Republic of Germany [tel: (07081) 3981; telex: 724019].

Berthold Model Biolumat LB9500

A manual instrument.

Photomultiplier detector operating in photon counting mode.

Cuvette size 12 × 47 mm.

Counting chamber can be temperature stabilized at 25, 30, or 37°.

The injector is automatic and 100-μl aliquots of reagent can be dispensed with a SE of 0.5%.

Measuring modes are rate, peak height, and integral. Integration may be for 10, 30, or 60 sec.

There is a 6 digit display and the instrument's dynamic range is 50,000. Excessive light causes an overload light to be lit.

The analog output has three sensitivity settings.

Digital output may be connected to a desktop computer, e.g., Hewlett Packard HP97S, for external data processing.

Berthold Model Auto-Biolumat LB950

An automatic instrument available with various options. Operated by an 8-bit microprocessor. Designed to process large batches of samples in an automatic fashion with automatic addition of reagents and incubation.

Photomultiplier detector with photocathode cooled to 16° below ambient using a Peltier element to reduce photomultiplier background count. Operated in photon counting mode. Detector chamber with polished stainless-steel mirror to optimize light collection.

Cuvettes 12 × 47 mm are transported in the instrument using a flexible free-standing chain with a capacity for 200 cuvettes. Sections of chain can be added or removed for various operations, e.g., centrifugation.

Temperature control of the measuring chamber and the on-line incubation area is programmable in 0.1° steps between 20 and 43°.

Injection is automatic and up to three injectors may be fitted with up to four positions available: entrance to on-line incubation area, position immediately preceding measuring station and two ports at the measuring position. Volumes between 10 and 360 μl can be set independently for each injector.

Operation of the instrument is from the keyboard of an attached Apple IIe microcomputer with floppy disc drive, vdu, and graphics printer. All operations are interactive and in plain English. Multiuser operation is possible provided a new parameter set has been stored in advance. An empty sample position indicates new user set-up. Two empty positions indicate stop.

Extensive software is available for ATP work, phagocytosis, luminescence immunoassay, and enzyme determinations, etc. Certain results can be displayed in graphic format, e.g., phagocytosis luminescence.

A V24/RS232 interface is available to connect the unit to another computer.

Berthold Model LB9505 Six-Channel Luminescence Analyzer

An instrument to measure luminescence from six samples simultaneously. Particularly adapted for use for cell luminescence studies, e.g., phagocytosis.

This instrument has six measuring positions each with its own photomultiplier. These are operated in photon counting mode and are carefully matched for background and sensitivity. Each photocathode is cooled to 6° by a Peltier element to reduce background noise. For normalization purposes each of the six positions is calibrated once with a standard light source and the in-built microcomputer corrects results automatically. Up to 24 sensitivity constants may be stored, that is at four different wavelengths for each of six positions.

Cuvettes are 12 × 47 mm and are placed in detector chambers which are temperature controlled from the keyboard from 26 to 50° in steps of 0.1°. Reagents may be injected into the cuvettes through a light-tight septum using a syringe.

The microprocessor and keyboard allow a wide range of functions to be set including counting intervals (0.1 to 60 min), chamber temperature, normalization factor, output format, and start and stop configurations for each or all channels. Data and status are given on a 6-digit display.

Six individual analog outputs are available for connection to a chart recorder.

A V24/RS232 interface may be used to connect the instrument to a microcomputer, e.g., Apple II, for which a wide range of suitable software has been developed. Data may be processed and stored on floppy disc.

Supplier. Canberra Packard Inc., 2200 Warrenville Road, Downers Grove, Illinois 60515 [tel: (312) 969-6000; telex 270061].

Canberra Packard Inc. Picolite Model 6106 and 6112B

A six sample manual instrument with single photomultiplier detector. It is operated in the photon counting mode.

Cuvettes 6 × 50 mm (model 6106) or 12 × 55 mm (model 6112B) are held in a turret head assembly in one of six positions. The head is turned to bring the sample of choice to the measuring position. Temperature control is via a heating element or by circulated water. Heating element recommended for above ambient operation (max. 45°) and water for sub-ambient operation.

Injection is by syringe through a light-tight septum.

The instrument consists of two modules, the detector unit and the microprocessor unit. The latter has a system control panel with sensor keys which allows the user to set operating parameters such as delay, integration time, temperature, background subtraction, normalization constant, and selection of features such as automatic start (based on count) and averaging of repeat samples. There are seven preset measurement programs and the in-built printer gives program number, sample number, sequence number, delay and count time, counts, and result based on selected measurement program.

An analog output is available.

Canberra Packard Inc. Picolite Model 6500

An automatic instrument which will process and measure up to 48 samples held in a circular carousel.

Photomultiplier detector operated in the photon counting mode.

Temperature control of the sample compartment is from ambient to 37° and lower temperatures by use of a separate sample chamber cooling device. In addition all samples can be stirred including the one in front of the photomultiplier using a graduated speed control.

Injection of reagents can be from up to three optional injectors. Dispensed volumes can be selected from 50 to 2000 μl in 50 μl increments. Injections can be made at each of the two positions preceding the photomultiplier tube as well as into the sample being measured.

The Picolite 6500 has two digital displays and a sense touch control panel. A wide range of control parameters can be set including temperature, injection volumes, count delay, count interval, and injection delay. Sample positions to be counted are specified by the Group feature and data reduction is aided by programmable constants, background subtract, and Repeat functions.

The unit has a built-in printer.

In addition the instrument has nine built-in programs to cover the majority of most users requirements including ones for phagocytosis studies, internal standardization, and a Prime/Flush program to prepare injectors.

An RS232 interface is available for those wishing to connect the unit to an external computer. Packard's MP II-DAAS system with IBM PC is also available.

Supplier. Chrono-Log Corporation, 2 West Park Road, Havertown, Pennsylvania 19083 [tel: (215) 853-1130; telex: 831579].

Chrono-Log Series 400 and 500 Lumi-Aggregometers

Models 400, 450, and 460 allow monitoring of blood platelet aggregation in platelet-rich plasma samples by measuring both aggregation by optical density changes and ATP release with firefly luciferase.

Models 500, 550, and 560 allow monitoring of blood platelet aggregation by measuring both electrical impedance and optical density as well as ATP release with the firefly luciferase method. Instruments are suitable for whole blood or platelet-rich plasma studies.

Instruments are either one or dual channel.

Sample size 0.5 ml blood.

Stirring speed 1200 rpm or optionally (Series 500 only) 0-1200 in 11 settings.

Temperature of heater block, 37°.

Chart recorder dual channel required for analog output.

Supplier. CLEAR Ltd., Lewis Road, East Moors, Cardiff CF1 5EG, Wales, U.K. [tel: 0222-481877; telex: 497480 C LEAR G].

At the time of writing, the CLEAR radiometer is undergoing trials for rapid microbial testing in the clinical, food, and pharmaceutical industries. It is expected that the following features will be incorporated into this newly developed fully automatic instrument.

The detector is a photomultiplier which may be operated in photon counting or current mode. Temperature control is available for the photomultiplier, detector chamber, and sample station (incubation).

A number of injectors may be fitted which are software controlled. Injected volume may be varied from 5 μl–2.0 ml.

Sample capacity is 500 with a selection of disposable cuvettes to accommodate samples of size 2 μl–3 ml.

The instrument is fitted with a display unit and a simple keyboard for interactive use. Protocols are available for a number of routine applications and these are supplied on EPROMs.

A research utilities EPROM is available for those wanting to set the instrument parameters to their own specialized needs. An RS232 interface is fitted, thereby enabling the instrument to be controlled from an external microcomputer.

Supplier. LKB Wallac Oy, PO Box 10, 20101 Turku 10, Finland [tel: 921-67811; telex: 62333].

LKB Model 1250 Luminometer

Modular design.

Manual operation.

Fitted with side window photomultiplier and operated in current mode.

Gain can be adjusted according to need by a potentiometer. This allows signal size to be adjusted. Background of the photomultiplier may be backed off using a potentiometer.

Cuvette loaded into measuring head which is simply rotated for measurement.

Cuvette size 10–12.5 mm wide and up to 65 mm high.

Temperature control (optional) of sample chamber is by circulated water supplied from external source.

Cuvette contents may be mixed by device (optional) which rotates cuvette.

Injection may be by manual or automatic operation (optional); the latter requires a control unit.

Output can be to a chart recorder or to a separate module which gives a digital display and simple printout. Signal is smoothed by use of 1 or 10 sec time constant. Signal size may require adjustment by gain control applied to photomultiplier.

The sample head also contains a sector with a built in light standard which may be used to check instrument performance. It consists of a liquid scintillation sample spiked with radiocarbon.

LKB Luminometer Model 1251

An automatic instrument.

Available with various options.

Instrument has integral sample changer and a wide range of push buttons to set operating conditions.

Fitted with 2-in. end-window photomultiplier operated in current mode. Dynamic range 6–7 decades but dependent on time constant applied. Special circuitry is included to maintain gain stabilization.

Carousel will hold up to 25 special cuvettes which may be processed automatically or manually. Temperature control of sample incubator is within 0.1° over the range 20–40° for an ambient change of 10°. Cuvettes may be subjected to mixing (pulsed or continuous).

Signal output may be continuous, peak height, rate, or integral. In integration mode both delay and integration time may be set. A smoothing parameter (time constant) must be set appropriate for the range of signal sizes expected. Background subtraction may be set.

Digital display shows sample parameters. An optional printer and recorder may be used for hard copy output.

In the Auto mode the instrument will carry out a programmed list of operating commands and up to 64 assay protocols may be stored in an optional nonvolatile memory or EEPROM. These could be for special enzyme assays incorporating reagent addition and incubation steps.

In addition the Model 1251 can be connected to a microcomputer, vdu, recorder, and printer. A range of data reduction software is available.

Automatic reagent addition can be applied by use of up to three automatic dispensers (variable volume). Tubing to the measuring chamber can be readily accessed and cleaned. A heat exchanger preheats or cools reagent before injection.

Supplier. Lumac/3M bv, Postbus 31101, 6370 AC Schaesberg, The Netherlands [tel: 045-318335; telex: 56937].

Lumac/3M Model M2010 Biocounter

Manual operation.

Fitted with end-window photomultiplier operated in photon counting mode. Dynamic range 5 decades.

Cuvette size 12 × 47 mm.

Temperature control of detector chamber by heater only to 25, 30, or 37°.

Mixing of reagents dependent on injection speed. Subsequent *in situ* mixing not possible.

Injection of up to three reagents from three sealed dispensers. Volume 100 μl only. Injection actuated from front panel button or automatically if connected to optional external computer.

Overload indicator.

Output given on a digital display or printed if microcomputer and printer attached. Rate output may be given in analog form on a chart recorder.

Measuring modes are integral or rate. Integral times 10, 30, or 60 sec and rate updated about every 2.5 sec. Analog output three sensitivity settings.

Handshake operation is possible by connecting a suitable microcomputer. Interfaces available for several computers.

LUMAC Model M2010A Biocounter

Similar in general appearance and some specifications to the model M2010.

It differs in several respects.

There is no temperature control.

A wash and prime mode is selectable to automatically wash and prime each of the three injectors.

A programmed operation is selectable to process reagent injection and signal acquisition for a single sample.

An optional thermal printer may be attached to obtain hard copy.

Lumac/3M Autobiocounter M3000

Similar to Berthold LB950 but with the following different specification.

Injectors will dispense either 50- or 100-μl aliquots with volume accuracy of 2.5 μl.

Supplier. New Brunswick (UK) Ltd, 6 Colonial Way, Watford, Hertfordshire WD2 4PT, England [tel: (0923) 23293; telex: 894187 BRUNCO G].

New Brunswick (UK) Ltd. Lumitran II

A semiautomatic instrument.

Photomultiplier detector operated in photon counting mode.

The unit comprises two separate modules: (1) a detector assembly with sample turntable and a single reagent injector and (2) an electronics module which receives, processes, and displays the signals from the detector assembly.

A printer (optional) may be attached for data output.

The turntable, which can be manually indexed, is juxtaposed to the photomultiplier and can hold up to six cuvettes. The sample to be measured is elevated into the measuring position ready for injection of reagent. During sample transport the high voltage to the photomultiplier is switched off. Aliquots of reagent are dispensed from a Hamilton repeating syringe and a small electrical switch on it permits synchronization of signal acquisition.

Integration up to 999 sec and a delay may be set and it is possible to accumulate signal in 100 msec slices for up to 25 sec and thereafter to 999 sec in 200 msec. In this way a histogram may be built up. Peak height measurement is also possible.

Selection of working parameters is made through a digital keypad in response to interactive questions presented on a single line display.

Output to the printer can be automatically corrected for background value. A conversion factor may also be included.

An RS232 interface allows connection to an external microcomputer.

Supplier. Payton Associates, 120 Milner Avenue, Unit 9, Scarborough, Ontario, Canada M1S 3R2 [tel: (416) 298-9600; telex: 065-25312].

Payton Series 100 Lumi-Aggregation Module

This series of instruments (Models 1010, 1020, and 1015) allow monitoring of blood platelet aggregation by optical density change and simultaneous monitoring of ATP secretion using firefly luciferase.

Instruments can be single or dual channel (one or two patient) and ATP secretion is measured together with aggregation after addition of aggregating reagent.

Sample size 0.5 ml, minimum platelet concentration ~30,000 per ml.

Stirring speed adjustable 300–1200 rpm.

Temperature 32–42°, continuously adjustable.

Chart recorder (dual channel) required for analog output.

Supplier. Photon Technology International Inc., PO Box AA, Princeton, New Jersey 08542 [tel: (609) 921-0705; telex: 238818 RCA].

Photon Technology International Inc. Quantacount

This is a series of actinometers which permit determination of quantum yields for various reactions. It is said to be wavelength independent over the range 200 to 550 nm and is of modular construction. A silicon solar cell detector with rhodamine scintillator is used.

Supplier. SLM Instruments Inc., 810 West Anthony Drive, Urbana, Illinois 61801 [tel: (217) 384-7730; telex: 206079].

SLM Aminco, Chem Glow II

Manual instrument.
Photomultiplier operated in current mode. S4 response.
Cuvette size 6 × 50 mm, or 12 × 35 mm (dependent on sample chamber head).
Ambient temperature operation.
Injection of reagent by repeating syringe (manual) through septum.
Signal output on digital display or chart recorder (optional). Detector gain programmable in three decades. Background backoff available.
Light standard available.
Printer optional (Centronics).
Serial output RS 232C.

Supplier. Tohuku Electronic Industrial Co. Ltd., 36-4, 2 Chome Mukaiyama, Sendai, Japan 982 [tel: 0222-66-1611].

Tohuku Model OX-7 Chemiluminescence Analyzer

This is a single sample device for studying chemi- and bioluminescence under various physical conditions.
Photomultiplier (cooled) detector operated in photon counting mode. Counting periods of 1, 10, 30, and 60 sec may be selected.
The sample may be studied as a function of temperature which may be set from ambient to 200° using electronic temperature control.
The sample may be gassed from an external supply with a range of materials.
The spectrum of the luminescence may be estimated using a set of interference filters.
Digital display and printer are included.
Another model (OX-70) incorporates a low-pressure mercury lamp so that the sample can be irradiated.

An "Optical Box" is also available whereby large samples can be studied and complex manipulations made. An optical fiber carries the light signal to the chemiluminescence analyzer.

Supplier. Turner Designs, 2247 Old Middlefield Way, Mountain View, California 94043 [tel: (415) 965-9800; telex: 469852].

Turner Designs: Photometer Model 20

A manual instrument.

Photomultiplier unit operated in current mode.

Cuvettes range from 6 × 50 mm to 12 × 75 mm and 8 × 50 mm and scintillation vials and require appropriate holders. A continuous flow system is also available.

Temperature control of the detector chamber may be achieved from water circulated through the chamber from an external water bath.

Injection is by one of several methods: a single injection device with disposable tip, a 50-aliquot manually activated autostart Hamilton syringe, a motorized automatic injector, and an automatic pipet for use with a flow system.

Sensitivity of the instrument can be changed in precise multiples of 10 over a three decade range which with a 4 digit display gives a dynamic range of some six decades. Autoranging may be selected. There is also a continuously adjustable sensitivity control and automatic overload shutoff.

Output may be one of four selections: routine (final reading), full integral (total light/time), half integral (light in first half/time), or peak height. Integration time can be set as well as a smoothing factor.

Formatted data may be output to a printer. A time of day clock is inbuilt.

The instrument may be set in Repeat mode whereby a predetermined number of runs at predetermined times may be carried out. For flow work the Continuous mode may be selected.

A quenched tritium liquid scintillator light standard is available to set up the instrument.

An analog output is available for connection to a chart recorder and there is an RS232 interface for connecting an external computer.

Turner Designs: Photometer Model 25 Dual Wavelength Luminometer

This unit is designed to meet the needs of chemiluminescence energy transfer immunoassay wherein light at two different wavelength ranges must be measured independently but at the same time.

The instrument consists of two Model 20 radiometers with their optical assemblies placed at opposite sides of the sample cuvette. The cuvette holder also incorporates injection facilities. Light reaches each radiometer after passing through an appropriate light filter.

Supplier. WIRA Technology Group, WIRA House, West Park Ring Road, Leeds LS16 6QL, England [tel: (0532) 781381; telex: 551789 WIRALS G].

WIRA Technology Group Camera Luminometer

An instrument for detecting light emission and recording it on Polaroid Land film (ASA 20,000).

May be used to indicate light output from a two-dimensional device such as a microtiter plate or a Petri dish. Alternatively it may be used for microbiological testing of textiles using ATP luminescence.

A sample in its holder is placed into the device and the film exposed. Exposure can be monitored by means of the timer on the back of the device.

WIRA Technology Group are also developing instruments to estimate microbial load on surfaces using ATP firefly luciferase reactions and various types of optoelectronic hardware.

NOTE: The author has made every reasonable effort to ensure the data in this chapter are accurate.

Author Index

Numbers in parentheses are footnote reference numbers and indicate that an author's work is referred to although the name is not cited in the text.

Evans, J. F., 75, 78(15), 80, 82(15), 92, 94
Everett, L. J., 20
Ewetz, L., 331
Eyring, H., 452

F

Faini, G. J., 294, 587
Faler, G., 532
Fall, L., 28
Fam, C. F., 25, 27(7)
Farr, A. L., 58
Farr, R. S., 497
Farrelly, J., 436
Fasman, G. D., 550
Favaudon, V., 140
Fee, J. A., 455
Feinberg, H., 17
Feldman, M., 509
Fenical, W., 470
Ferguson, W. J., 64
Fertuck, J. C., 43
Festi, D., 218, 228
Fihn, S., 23
Fischer, H., 490, 527, 587
Fitzgerald, J. M., 149, 150(6)
Fletcher, R. D., 27
Flint, A., 307, 311(4)
Florini, J. R., 133
Fogel, M., 307, 310, 312, 315(14), 319(14), 324(14)
Fontijn, A., 438
Foote, C. S., 454, 470, 570(8), 575, 583
Ford, J., 217, 223(10), 230, 237(5), 246
Forrest, W. W., 17
Förster, T., 544, 546, 578
Francheschetti, F., 389, 393, 396(28), 428
Franco, C. M. M., 16
Franzen, J. S., 16, 18(14)
Frazer, K., 408, 410(6), 412(6), 420(6)
Freeman, T. M., 437
Freese, E., 18
Frei, R. W., 438, 439, 440(90)
Fricke, H., 355, 455
Fridovich, I., 453, 455, 473
Friedland, J., 100, 131
Frimer, A., 570
Fritsch, E. F., 5, 7, 8, 70, 71(5), 73(5), 74(5), 82(5), 92, 303

Fritsche, U., 435
Frölich, J. C., 402
Fujimoto, T. T., 470
Fujiwara, K., 437
Fukushima, H., 8
Fuller, F., 77
Fuwa, K., 437
Fynn, G. H., 22

G

Gabor, G., 332(14), 333
Gaddis, L. M., 58
Gadow, A., 355, 357, 358(6)
Galen, R. C., 23
Galitzer, S. J., 16, 18(12), 21(12)
Gallin, J. I., 496
Gandelman, M. S., 438
Gast, R., 118, 150, 151(9), 168(9), 170(9)
Gayer, B., 402
George, J., 351, 352(40), 412
Gerasimenko, M. I., 412
Gershman, L. C., 319
Ghini, S., 239, 240, 241(9), 244(7), 248(9)
Ghisla, S., 128, 135, 138, 140, 141, 142(7), 146(8), 149
Gibbons, J. E. C., 406, 407(4), 410(3, 4), 411(4), 412(4), 419(4), 420(4)
Gibson, Q. H., 100, 119, 128, 131
Gilad, S., 389, 390, 391(24)
Gilbert, J. C., 587, 590(1)
Gilbert, W., 293, 306
Gillespie, E., 332(16), 333, 335(16), 337(16), 338(16), 342(16), 345(16), 351, 352(37)
Girotti, S., 239, 240, 241(9), 244(7), 248(9)
Glasstone, S., 452
Gleu, K., 367, 465
Glorieux, F. H., 111
Goldberger, R. F., 300
Goldstein, L., 238, 240
Gooch, V. D., 307, 324, 325(27), 326(27)
Good, N. E., 64
Goodfriend, T. L., 550
Goodman, H. M., 291, 300, 304(6), 305(6), 306(6)
Gooljer, G., 439, 440(90)
Gordon, D. B., 384
Gordon, J. I., 300
Goto, T., 289, 290(12), 331

Subject Index

A

ABEI-H, 357–358
ABENH, 558
  absorption spectra, 561–563
  activation, 560
  immunoassays for labeling proteins for, 566
  incorporation into transferrin
    estimate of, 560–561
    as function of ABENH/transferrin ratio in coupling reaction, 563
  photographic assay, 414
  structure, 558
ABEN-hemisuccinamide, 357–358
ABENH-transferrin conjugates
  absorption spectra, 561–563
  activity ratio, 560–561
  with amidine linkage
    cleavage methods, 567
    light yield, 567
  chemiluminescence efficiency, 567–568
  elution pattern, 561–562
  immunoreactivity, 561, 568
  non-specific binding, 568
  spectral ratio, 560
  synthesis, 560
  titration of antibody with, 561
Absolute reaction rates, 452
Acetamide, bioluminescence test, 273
Acetate esterase, bacterial bioluminescence assay, 197
Acetate esters, bacterial bioluminescence assay, 194
Acetyl N-hydroxysuccinimide ester, preparation, 286
Acridine-9-carboxylic acid, 366
  preparation, 371–372
Acridine dyes, luminescence test, 276
Acridine orange, activity in bioluminescence test with SD-18, 272
Acridinium
  chemiluminescence, measurement, 369–370

excessive labeling with, loss of protein, 377–378
  reaction, rate of photo emission from, 370
Acridinium/carbinol base equilibrium, 368–369
Acridinium carbonyl nucleus, 367
Acridinium ester
  chemiluminescence, 366–370
    chemistry, 367–369
    mechanism, 367–368
  dedicated labels, synthesis, 371
  immunoassays using, methods, 379–386
  labeling, of purified monoclonal antibodies, 374–378
  labeling reaction, 374–375
  labels, 370–379
Acridinium-labeled antibodies, properties, 378–379
Acriflavin, activity in bioluminescence test with SD-18, 272
Actinomycin D, bioluminescence test, 273
Acyl-acyl-carrier protein, 183
Acyl-protein synthetase, assay, 174–175
Acyltransferase
  assay, 183–185
  bioluminescence-related, 183–188
    properties, 187–188
    purification, 185–187
    stimulation of the acyl-CoA cleavage activity, 188
  P. phosphoreum, purification, 185–186
  [³H]tetradecanoyl-CoA cleavage catalyzed by, 187–188
  V. harveyi, purification, 185–187
Adamantylidene adamantane 1,2-dioxetane, 531, 532
  half-life, vs. temperature, 533
  reaction of, 533
  thermochemiluminescence labels based on, 534–538
  thermochemiluminescence spectrum of, 532, 534

H

transcriptional regulation of, 93
use of gene fusions, 93–98
Lymphokines, 508

M

Macrophage-activating factors, chemiluminescence detection, 515–517
Macrophage activation
interaction between IFN-γ and LPS in
chemiluminescence analysis, 517–
520
regulation of, use of chemiluminescence as index of, 513–520
Macrophage cell lines
cellular chemiluminescence, 493
functions for immune responses, 507–
508
immortalized, 509
luminescent bioassays based on, 507–
530
availability, 526
chemiluminometric measurements,
511–513
instrumentation, 527–530
practical considerations, 525–526
principle, 510–511
procedure, 511–513
specificity, 526–527
technical considerations, 527–530
triggering and modulating agents,
513
uniformity, 526
restricted availability, 509
tumoricidal activity, 517–518
Macrophage hybridoma cells, culture, 511
Malaria-specific IgG, photographic immunoassay based on horseradish peroxidase conjugates, 416
Malate, assay for, 198–199
L-Malate, bioluminescent assay, using
coimmobilized enzymes, 202, 204–207
Malate dehydrogenase, 198–199
Malonylaldehyde, activity in bioluminescence test with P.f-13, 273
Manganese(II), photographic assay, 415
Marine bacteria
bioluminescence, 70
bioluminescence genes
analysis of mRNA, 78–82

cloned, 76–78, 81–83
cloning, 83–88
promoters, 77–80
screening for clones containing, 71–
75
transfer into E. coli, 70–71
genes, 70–83
luminescent system, light emission in
E. coli, 75–76
MAST chemiluminescent assay system,
420
MASTpette reaction vessel, 408–410
Meganyctiphanes norvegica, substance F,
323
Membrane antigens, detection of, by chemiluminescence assay, 523–524
6-Mercaptopurine, activity in bioluminescence test with P.f-13, 273
Mercury, photographic assay, 415
S-(Methanesulfonyl)-6-thiohexanoic acid,
synthesis, 254
Methotrexate, 265
activity in bioluminescence test with
P.f-13, 273
Methoxyvinyl pyrene
chemiluminescence, in liver microsomal
extracts, 580–584
chemiluminescence emission spectra
and chemiluminescence yield,
solvent dependences, 577–580
chemiluminescence yield, 575–577
trans-1-(2′-Methoxyvinyl)pyrene, chemiluminescence yield, 575–577
Methylmethane sulfonate, activity in bioluminescence test with SD-18, 272
N-Methyl-N-nitro-N-nitrosoguanidine,
effect on light development in P.
leiognathi SD-18, 268
N-Methyl-N-nitrosoguanidine, activity in
bioluminescence test with SD-18, 272
Micelle-producing surfactants, enhancement of light emission from chemiluminescent systems, 348–349
Microbial cells, ATP extraction, 20–21
in complex mixtures, 21
Microbial diaphorase, 198
Microbial growth, photographic assay,
414
Microcomputer-luminometer system, 421,
424–425, 433